AF361954

Advances in Multiple Sclerosis Research—Series I

Advances in Multiple Sclerosis Research—Series I

Editors

John Matsoukas
Vasso Apostolopoulos

MDPI • Basel • Beijing • Wuhan • Barcelona • Belgrade • Manchester • Tokyo • Cluj • Tianjin

Editors
John Matsoukas   Vasso Apostolopoulos
NewDrug, Patras Science Park   Victoria University
Greece   Australia

Editorial Office
MDPI
St. Alban-Anlage 66
4052 Basel, Switzerland

This is a reprint of articles from the Special Issue published online in the open access journal *Brain Sciences* (ISSN 2076-3425) (available at: https://www.mdpi.com/journal/brainsci/special_issues/Advance_MS_Research).

For citation purposes, cite each article independently as indicated on the article page online and as indicated below:

LastName, A.A.; LastName, B.B.; LastName, C.C. Article Title. *Journal Name* **Year**, *Volume Number*, Page Range.

ISBN 978-3-03943-947-8 (Hbk)
ISBN 978-3-03943-948-5 (PDF)

Cover image courtesy of Amanda Habib and Rhiannon Filippone, PhD students, Victoria University, Melbourne Australia (Supervisors, Professor Vasso Apostolopoulos andAssociate Professor Kulmira Nurgali). Immunohistochemical image of oligodendrocytes within theprefrontal cortex of the brain. Image showing oligodendrocytes stained for myelin basic protein. Inflammation of the brain results in down regulation of myelin which is associated with variousneurological disorders such as multiple sclerosis.

Contents

About the Editors

John Matsoukas has over 30 years of experience in research in the field of organic and peptide chemistry, nuclear magnetic resonance (NMR) and the chemistry of natural products. He has an extensive research background in NMR-based drug discovery, design and development. Professor Matsoukas has studied Chemistry at the University of Patras. He graduated from the University of New Brunswick in Canada with a MSc Degree in Chemistry. His dissertation was on the Total Synthesis of Natural Products and Nuclear Magnetic Resonance. He carried out his Ph.D. studies in Chemistry at the University of Patras, Greece in the peptide field. He joined the University of Calgary, Alberta, Canada, and the group of Professor Graham Moore, studying peptide hormones and peptide mimetics. He was the founder, director, and head of the successful Graduate Program "Medicinal Chemistry: Drug Discovery, Design and Development" of the University of Patras (1997–2013). Since 2000, he has been collaborating with Professor Vasso Apostolopoulos of Victoria University on multiple sclerosis peptides and the development of vaccines. He has published more than 500 articles in peer-reviewed journals, book chapters and conference proceedings. He has been granted many patents, awards and honors for his research and scientific activities.

Vasso Apostolopoulos's expertise is in the areas of immunology, crystallography, cellular biology, translational research, and the development of drugs and vaccines. Vasso has led/directed several research programs at various research centers and universities around the world. She is currently the Associate Provost, Research Partnership at Victoria University Australia. She has received more than 100 awards, published over 400 research papers, invented 18 patents and her current interests are treating chronic diseases with an immunological focus.

Editorial

Advances in Multiple Sclerosis Research–Series I

Vasso Apostolopoulos [1,*] **and John Matsoukas** [1,2,3]

1   Institute for Health and Sport, Victoria University, Melbourne 8001, Australia; imats1953@gmail.com
2   NewDrug, Patras Science Park, 26500 Patras, Greece
3   Department of Physiology and Pharmacology, Cumming School of Medicine, University of Calgary, Calgary, AB T2N 4N1, Canada
*   Correspondence: vasso.apostolopoulos@vu.edu.au; Tel.: +613-9919-2025

Received: 15 October 2020; Accepted: 25 October 2020; Published: 29 October 2020

Abstract: Designing immunotherapeutics, drugs, and anti-inflammatory reagents has been at the forefront of autoimmune research, in particular, multiple sclerosis, for over 20 years. Delivery methods that are used to modulate effective and long-lasting immune responses have been the major focus. This Special Issue, "Advances in Multiple Sclerosis Research—Series I", focused on delivery methods used for immunotherapeutic approaches, drug design, anti-inflammatories, identification of markers, methods for detection and monitoring MS and treatment modalities. The issue gained much attention with 20 publications, and, as a result, we launched Series II with the deadline for submission being 30 April 2021.

Keywords: multiple sclerosis; MS; vaccine; immunomodulation; carriers; MS drugs

---

1. Multiple Sclerosis

The World Health Organization estimates that globally, more than 2.5 million people are affected by multiple sclerosis (MS). With the global population growing to an unparalleled height of 7.0 billion in 2011 and recently reaching 7.8 billion (10 October 2020)—it is estimated to reach 8.5 billion by 2030 and 9.7 billion by 2050—the incidence and onset of MS in young adults is expected to rise exponentially, with an estimate of 2.3 million people living with MS globally. Clinical isolated syndrome is a type of MS which may or may not progress. As such, a person will experience a neurological episode lasting at least 24 h and resulting in damage to the central nervous system (CNS). There are three main subtypes of MS, (i) relapse/remitting MS (RRMS) accounting for 85% of MS cases, with 50% progressing to (ii) secondary progressive MS (SPMS), with (iii) 15% of those diagnosed at onset of primary progressive MS (PPMS) type. It is possible that RRMS patients can remain in that state for up to 30 years, whilst 8% develop a more aggressive disease, named highly active RRMS (HARRMS). In rare occasions, up to 5% are progressive relapsing MS type (PRMS), which is characterized by progressive worsening of the condition from the onset, similar to PPMS.

MS is characterized as a chronic demyelinating disorder of the CNS with inflammatory cells infiltrating around the nerve, leading to demyelination of the myelin sheath and immune attack to myelin basic protein (MBP), proteolipid protein (PLP) and myelin oligodendrocyte glycoprotein (MOG). Inflammatory cells which have been found to be involved in MS include macrophages, T helper type 1 (Th1) cells, Th17 cells, CD8+ T cells and B cells secreting auto-antibodies [1–3]. More recently, it has been shown that tetraspanin-32 is significantly downregulated in Th cells. Tetraspanin-32 controls the development of autoimmune responses, and in EAE models in mice, tetraspanin-32 is significantly expressed at lower levels on activated or encephalitogenic T cells compared to naïve Th cells. In the study by Basile and Cavalli et al., it was noted that tetraspanin-32 was downregulated in memory T cells and was further decreased upon ex vivo restimulation (Figure 1) [4]. Likewise, myelin-specific memory T cells and peripheral blood mononuclear cells (PBMC) from patients with MS also expressed lower levels of tetraspanin-32 compared to memory T cells from healthy subjects. In addition, MS patients with early relapses compared to those with a longer, stable disease expressed lower levels of tetraspanin-32 on their PBMC [4]. Hence, tetraspanin-32 is involved in immune responses underlying the pathophysiology of MS, and could be a viable diagnostic marker or therapeutic target against MS.

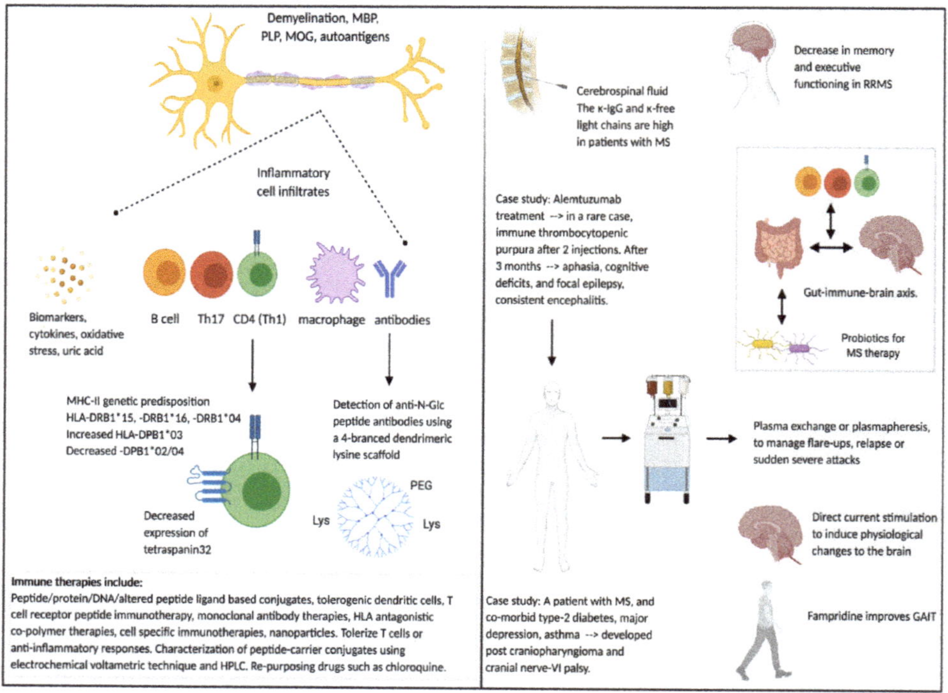

Figure 1. Summary of new advances in Multiple Sclerosis Research—Series I, papers in the Special Issue. Created with biorender.com.

A number of factors contribute to MS development, including genetic predisposition, especially those who are HLA-DR2 (HLA-DRB1*15, HLA-DRB1*16)- or HLA-DR4 (HLA-DRB1*04)-positive, environmental factors such as Epstein–Barr virus and human herpesvirus 6 exposure, and diet, such as low levels of vitamins B and D [1,5,6]. A number of health conditions are related to HLA phenotype, such as type-1 diabetes (HLA-DRB1*03 or HLA-DR3, HLA-DQB1*03 or HLA-DQ8), rheumatoid arthritis (HLA-DRB1*04), juvenile idiopathic arthritis (HLA-DRB1*08), celiac disease (HLA-DQ2, HLA-DQ8) and Graves' disease (HLA-DRB1*03, HLA-DQA1*0501). The paper by Maria Anagnostouli et al. studied the prevalence of HLA-DPB1 allele in MS patients from a Greek cohort and its association with HLA-DRB1 risk allele [7]. No significant differences were noted between early onset MS compared to adult onset MS for 23 distinct HLA-DPB1 and 12 HLA-DRB1 alleles. However, the frequency of HLA-DPB1*03 allele was significantly increased, and the frequency of HLA-DPB1*02 allele was significantly decreased, in AOMS patients compared to controls. Interestingly, the frequency of HLA-DPB1*04 allele was significantly decreased in both patients, with early onset and adult onset MS compared to controls, suggesting a protective role of this allele amongst Greek cohort patients (Figure 1) [7]. Koukoulitsa and colleagues present a nice review articulating the journey of the conformational complex between HLA-peptide with the T cell receptor of agonist peptides and their altered peptide ligands from MBP, MOG and PLP [8].

2. Detection and Monitoring of Patients with MS

Magnetic resonance imaging (MRI) has been the gold standard of diagnosing and monitoring disease by detecting brain lesions and the type of brain lesion which aids treatment decisions. In addition, other detection methods are used in combination with MRI, such as the Kurtzke Expanded Disability Status Scale (EDSS) which measures the body's function and how well it can move, as well as analysis of cerebrospinal fluid for free light chains and IgG. Together, these increase the accuracy of diagnosis of MS and are used to monitor disease progression. However, there are few simple assays available to follow up disease activity. As such, the detection of auto-antibodies from sera is a method to detect relevant biomarkers. The team by Nuti and Papini et al., developed a method to detect anti-N-glycosylated (N-Glc) peptide antibodies, using a four-branched dendrimeric lysine scaffold, linked to a polyethylene glycol-based spacer containing 19-amino acids. This efficient multivalent probe has specificity and high affinity for anti-N-Glc antibodies in patients with MS [9]. In addition, Gudowska-Sawczuk evaluated cerebrospinal fluid and sera from patients with either MS ($n = 34$) or other neurological disorders ($n = 42$) [10]. The concentrations of cerebrospinal fluid κ free light chains (κFLC) and λFLC, and sera κFLC, as well as κFLC, λFLC, and κIgG index, were significantly higher in patients with MS compared to those with other neurological disorders. The κIgG index showed the highest diagnostic power in the detection of MS with both κFLC index and κIgG indexes showing the highest diagnostic sensitivity. This study provides novel information about the diagnostic significance of four markers combined in the κIgG index [10] and shows that κFLC and κIgG combined in a novel algorithm may improve the detection and disease activity of MS (Figure 1).

Cognitive function refers to a range of high-level brain functions, such as the ability to learn and remember information, solve problems, focus, concentration, attention, and verbal fluency. Change in cognitive function is common in patients with advanced MS. However, Pitteri et al, showed that newly diagnosed RRMS patients ($n = 50$) performed worse than healthy controls ($n = 36$), in particular, in the domains of memory and executive functioning [11]. These data demonstrate that reduced cognitive functioning can be present early on during the course of disease, even in patients without evidence of cognitive impairment. As such, the cognitive impairment criteria for patients with MS should be re-evaluated and be monitored closely throughout the course of disease (Figure 1).

3. Treatments for MS

Treatments for MS include, interferon (IFN) beta-1a, IFN beta-1b (cytokines), fingolimod, ozanimod, siponimod (sphingosine-1-phosphate-receptor modulators), natalizumab (a monoclonal antibody against alpha4-integrin), dimethyl fumarate, glatiramer acetate, teriflunomide, cladribine, ocrelizumab (a humanized anti-CD20 monoclonal antibody) and, alemtuzumab (a humanized anti-CD52 monoclonal antibody) [1,2]. These drugs are focused on speeding recovery from relapse, slowing the progression of disease and managing MS symptoms, and in most cases, there are side effects and patients need to stop treatment due to non-tolerance of the treatment. In rare cases, more severe adverse events occur. In fact, Buscarinu et al., presented a case report of a 45 year old Italian woman with RRMS on alemtuzumab treatment who showed immune thrombocytopenic purpura after the second injection of alemtuzumab. Three months following treatment, the patient presented with transient aphasia, cognitive deficits, and focal epilepsy, consistent encephalitis [12]. Autoimmune complications following alemtuzumab treatment are generally rare, with only one previous case being reported. Furthermore, Sachinvala et al. reported a male patient with MS, and co-morbid type-2 diabetes, major depression, asthma, developed post craniopharyngioma and cranial nerve-VI palsy. Magnetic resonance imaging, Humphrey's visual filed and retinal nerve fiber thickening were used to determine changes to help the patient maintain productivity and mental state and mood (Figure 1) [13].

There is a need for the development of new treatment options which would stop progression and have little to no side effects. Immune therapies have come a long way in recent years, with a number of methods being tested in pre-clinical and clinical settings, such as peptide/protein/DNA based vaccines, tolerogenic dendritic cells, T cell receptor peptide immunotherapy, monoclonal antibody therapies (anti-integrin a-4, anti-leucine rich repeat and immunoglobin-like domain-containing protein 1 (LINGO-1), anti-CD52), HLA antagonistic co-polymer therapies, cell specific immunotherapies, peptide-carrier conjugates, all of which are extensively reviewed by Kammona and Kiparissides [14] and Metaxakis et al. [15] (Figure 1). An editorial entitled, the long road of immunotherapeutics against MS [16], highlighted 20 years of MS research of an international multi-disciplinary consortia including peptide chemistry, medicinal chemistry, protein synthesis, protein–peptide interactions, nuclear magnetic imaging, molecular modeling, molecular dynamics, molecular biology, immunology, cell biochemistry, animal research and clinical research. This multi-disciplinary consortia led to at least 10 immunotherapeutic peptide-carrier candidates to be tested in human clinical trials. In preclinical studies, these peptide-based immune modulating conjugates showed a safety profile whilst switching immune responses from pro-inflammatory to anti-inflammatory and protection against experimental autoimmune encephalomyelitis (EAE) in mouse models [3,17–24]. Characterization of peptide-carrier conjugates was demonstrated using electrochemical voltametric techniques and high-pressure liquid chromatography [25]. In addition, nanoparticles have been used to deliver MS antigens to the immune system to tolerize T cells or stimulate an anti-inflammatory responses, reviewed by Chountoulesi and Demetzos [26]. More recently, chloroquine, an anti-malarial drug, was shown to suppress EAE in mice by modulating dendritic cells, Th17 cells, astrocytes, oligodendrocytes and microglia. Microglia cells were also shown to secrete IL-10 and IL-12p70. These data provide evidence that drug repurposing of chloroquine may be useful to patients with MS (Figure 1) [27].

In the last ten years, the incidence of MS has increased considerably, with lifestyle and environmental factors being one of the main contributors. An informative review by Boziki and Grigoriadis et al., provide the current advances in the gut-microbiome-immune–brain axis in patients with MS with altered microbiome, and present the effects of MS treatments on gut microbiome (Figure 1) [28]. Thus, modification of gut microbiota by either dietary (such as, probiotics) or medicinal approaches is a promising approach for the management of MS. In fact, probiotics have been shown to have beneficial effects not only in the gut flora but also in modulating and maintaining a healthy immune system. Certain probiotics have been shown to have anti-inflammatory effects on immune cells (i.e., monocytes) and in disease settings, such as asthma and allergies [29,30]. The paper by Dargahi et al. showed that the probiotic *Streptococcus thermophilus* was able alter pro-inflammatory T cells responses against an agonist MBP$_{83-99}$ peptide to an anti-inflammatory profile (Figure 1) [31]. This study suggests that the consumption of *Streptococcus thermophilus* may be beneficial in the management and treatment of autoimmune diseases such as MS, and further research in this area is warranted.

In addition to intravenous or oral steroids that are used as the first line of therapy for MS relapse, therapeutic plasma exchange, or plasmapheresis, is another method used to treat patients with neuromyelitis optica spectrum disorders, autoimmune encephalitis and MS, especially those with sudden, severe attacks or relapse/flare-ups. It is used in MS patients to manage disease by exchanging their plasma with 'fresh' plasma to remove pro-inflammatory cytokines and other proteins involved in auto-immune attack. In a study published by Moser et al., in this Special Issue, they compare the indications, efficacy and safety of therapeutic plasma exchange treatment in MS, autoimmune encephalitis and other immune-mediated CNS disorders and noted consistent efficacy and safety [32]. Measuring biomarkers of inflammation and oxidative stress is important to understand the efficacy of treatments. As such, Moccia et al. studied 60 patients with RRMS who were treated with IFN beta-1a or Coenzyme Q10 and monitored patients for IL-1b, IL-2R, IL-3, IL-4, IL-5, IL-6, IL-7, IL-8, IL-13, RANTES, tumor necrosis factor and uric acid (Figure 1) [33]. These serum biomarkers could be used to determine the efficacy of treatments as well as their mechanisms of action.

It is believed that transcranial magnetic stimulation motors with direct current stimulation (tDCS) intensities induce physiological changes to the brain, although the mechanism of action, as well as its validity and efficacy, are not clear. In a pilot study by Workman and colleagues, they noted that there were no immediate changes in cerebral blood flow following direct current stimulation. Hence, further work is required to enable sufficient magnitudes of intracranial electrical fields to induce physiological changes in the brain to patients with MS (Figure 1) [34]. During disease progression, patients with MS develop walking limitations, and fampridine is usually recommended to improve gait. In the study by Ahdab et al., fampridine was evaluated for cortical excitability effects and whether changes could predict therapeutic responses in 20 patients with MS and gait impairment [35]. It was noted that fampridine increased the excitatory intracortical processes, as shown by paired-pulse transcranial magnetic stimulation, suggesting that this could be used to select patients with MS who would be likely to experience a favorable response to fampridine (Figure 1) [35].

4. Conclusions

The development of drugs, immunotherapeutics and vaccines against diseases is a long process often taking researchers a lifetime. In this Special Issue, "Advances in Multiple Sclerosis Research—Series I", a range of papers were published, including MS markers, treatments, detection, monitoring and the role of the microbiome in MS. Together, all this information advances our knowledge of MS research, with promising new leads being developed in the next few years and entering human clinical trials.

Author Contributions: Conceptualization, writing, review, editing: J.M., V.A. Both authors have read and agreed to the published version of the manuscript.

Funding: The writing of this editorial received no external funding.

Acknowledgments: V.A. would like to thank the Institute for Health and Sport, Victoria University for supporting her current efforts into MS research. J.M. would like to thank the General Secretariat for Research and Technology (GSRT) and Patras Science Park for supporting his MS research.

Conflicts of Interest: The authors declare no conflict of interest.

References

1. Dargahi, N.; Katsara, M.; Tselios, T.; Androutsou, M.E.; de Courten, M.; Matsoukas, J.; Apostolopoulos, V. Multiple Sclerosis: Immunopathology and Treatment Update. *Brain Sci.* **2017**, *7*, 78. [CrossRef]
2. Katsara, M.; Apostolopoulos, V. Editorial: Multiple Sclerosis: Pathogenesis and Therapeutics. *Med. Chem.* **2018**, *14*, 104–105. [CrossRef] [PubMed]
3. Katsara, M.; Matsoukas, J.; Deraos, G.; Apostolopoulos, V. Towards immunotherapeutic drugs and vaccines against multiple sclerosis. *Acta Biochim. Biophys. Sin.* **2008**, *40*, 636–642. [CrossRef] [PubMed]
4. Basile, M.S.; Mazzon, E.; Mangano, K.; Pennisi, M.; Petralia, M.C.; Lombardo, S.D.; Nicoletti, F.; Fagone, P.; Cavalli, E. Impaired Expression of Tetraspanin 32 (TSPAN32) in Memory T Cells of Patients with Multiple Sclerosis. *Brain Sci.* **2020**, *10*, 52. [CrossRef] [PubMed]
5. Nemazannikova, N.; Mikkelsen, K.; Stojanovska, L.; Blatch, G.L.; Apostolopoulos, V. Is there a Link between Vitamin B and Multiple Sclerosis? *Med. Chem.* **2018**, *14*, 170–180. [CrossRef] [PubMed]
6. Virtanen, J.O.; Jacobson, S. Viruses and multiple sclerosis. *CNS Neurol. Disord. Drug Targets* **2012**, *11*, 528–544. [CrossRef] [PubMed]
7. Anagnostouli, M.; Artemiadis, A.; Gontika, M.; Skarlis, C.; Markoglou, N.; Katsavos, S.; Kilindireas, K.; Doxiadis, I.; Stefanis, L. HLA-DPB1*03 as Risk Allele and HLA-DPB1*04 as Protective Allele for Both Early- and Adult-Onset Multiple Sclerosis in a Hellenic Cohort. *Brain Sci.* **2020**, *10*, 374. [CrossRef]
8. Koukoulitsa, C.; Chontzopoulou, E.; Kiriakidi, S.; Tzakos, A.G.; Mavromoustakos, T. A Journey to the Conformational Analysis of T-Cell Epitope Peptides Involved in Multiple Sclerosis. *Brain Sci.* **2020**, *10*, 356. [CrossRef]
9. Nuti, F.; Fernandez, F.R.; Sabatino, G.; Peroni, E.; Mulinacci, B.; Paolini, I.; Pisa, M.D.; Tiberi, C.; Lolli, F.; Petruzzo, M.; et al. A Multiple N-Glucosylated Peptide Epitope Efficiently Detecting Antibodies in Multiple Sclerosis. *Brain Sci.* **2020**, *10*, 453. [CrossRef]
10. Gudowska-Sawczuk, M.; Tarasiuk, J.; Kulakowska, A.; Kochanowicz, J.; Mroczko, B. Kappa Free Light Chains and IgG Combined in a Novel Algorithm for the Detection of Multiple Sclerosis. *Brain Sci.* **2020**, *10*, 324. [CrossRef]

11. Pitteri, M.; Ziccardi, S.; Dapor, C.; Guandalini, M.; Calabrese, M. Lost in Classification: Lower Cognitive Functioning in Apparently Cognitive Normal Newly Diagnosed RRMS Patients. *Brain Sci.* **2019**, *9*, 321. [CrossRef] [PubMed]

12. Buscarinu, M.C.; Fornasiero, A.; Pellicciari, G.; Renie, R.; Landi, A.C.; Bozzao, A.; Cappelletti, C.; Bernasconi, P.; Ristori, G.; Salvetti, M. Autoimmune Encephalitis and CSF Anti-GluR3 Antibodies in an MS Patient after Alemtuzumab Treatment. *Brain Sci.* **2019**, *9*, 299. [CrossRef]

13. Sachinvala, N.D.; Stergiou, A.; Haines, D.E.; Kocharian, A.; Lawton, A. Post-Craniopharyngioma and Cranial Nerve-VI Palsy Update on a MS Patient with Major Depression and Concurrent Neuroimmune Conditions. *Brain Sci.* **2019**, *9*, 281. [CrossRef] [PubMed]

14. Kammona, O.; Kiparissides, C. Recent Advances in Antigen-Specific Immunotherapies for the Treatment of Multiple Sclerosis. *Brain Sci.* **2020**, *10*, 333. [CrossRef]

15. Metaxakis, A.; Petratou, D.; Tavernarakis, N. Molecular Interventions towards Multiple Sclerosis Treatment. *Brain Sci.* **2020**, *10*, 299. [CrossRef] [PubMed]

16. Apostolopoulos, V.; Rostami, A.; Matsoukas, J. The Long Road of Immunotherapeutics against Multiple Sclerosis. *Brain Sci.* **2020**, *10*, 288. [CrossRef]

17. Katsara, M.; Deraos, G.; Tselios, T.; Matsoukas, J.; Apostolopoulos, V. Design of novel cyclic altered peptide ligands of myelin basic protein MBP83-99 that modulate immune responses in SJL/J mice. *J. Med. Chem.* **2008**, *51*, 3971–3978. [CrossRef]

18. Katsara, M.; Deraos, G.; Tselios, T.; Matsoukas, M.T.; Friligou, I.; Matsoukas, J.; Apostolopoulos, V. Design and synthesis of a cyclic double mutant peptide (cyclo(87-99)[A91,A96]MBP87-99) induces altered responses in mice after conjugation to mannan: Implications in the immunotherapy of multiple sclerosis. *J. Med. Chem.* **2009**, *52*, 214–218. [CrossRef]

19. Katsara, M.; Deraos, S.; Tselios, T.V.; Pietersz, G.; Matsoukas, J.; Apostolopoulos, V. Immune responses of linear and cyclic PLP139-151 mutant peptides in SJL/J mice: Peptides in their free state versus mannan conjugation. *Immunotherapy* **2014**, *6*, 709–724. [CrossRef]

20. Katsara, M.; Yuriev, E.; Ramsland, P.A.; Deraos, G.; Tselios, T.; Matsoukas, J.; Apostolopoulos, V. A double mutation of MBP(83-99) peptide induces IL-4 responses and antagonizes IFN-gamma responses. *J. Neuroimmunol.* **2008**, *200*, 77–89. [CrossRef]

21. Katsara, M.; Yuriev, E.; Ramsland, P.A.; Deraos, G.; Tselios, T.; Matsoukas, J.; Apostolopoulos, V. Mannosylation of mutated MBP83-99 peptides diverts immune responses from Th1 to Th2. *Mol. Immunol.* **2008**, *45*, 3661–3670. [CrossRef] [PubMed]

22. Katsara, M.; Yuriev, E.; Ramsland, P.A.; Tselios, T.; Deraos, G.; Lourbopoulos, A.; Grigoriadis, N.; Matsoukas, J.; Apostolopoulos, V. Altered peptide ligands of myelin basic protein (MBP87-99) conjugated to reduced mannan modulate immune responses in mice. *Immunology* **2009**, *128*, 521–533. [CrossRef] [PubMed]

23. Lourbopoulos, A.; Deraos, G.; Matsoukas, M.T.; Touloumi, O.; Giannakopoulou, A.; Kalbacher, H.; Grigoriadis, N.; Apostolopoulos, V.; Matsoukas, J. Cyclic MOG35-55 ameliorates clinical and neuropathological features of experimental autoimmune encephalomyelitis. *Bioorg. Med. Chem.* **2017**, *25*, 4163–4174. [CrossRef] [PubMed]

24. Lourbopoulos, A.; Matsoukas, M.T.; Katsara, M.; Deraos, G.; Giannakopoulou, A.; Lagoudaki, R.; Grigoriadis, N.; Matsoukas, J.; Apostolopoulos, V. Cyclization of PLP139-151 peptide reduces its encephalitogenic potential in experimental autoimmune encephalomyelitis. *Bioorg. Med. Chem.* **2018**, *26*, 2221–2228. [CrossRef]

25. Deskoulidis, E.; Petrouli, S.; Apostolopoulos, V.; Matsoukas, J.; Topoglidis, E. The Use of Electrochemical Voltammetric Techniques and High-Pressure Liquid Chromatography to Evaluate Conjugation Efficiency of Multiple Sclerosis Peptide-Carrier Conjugates. *Brain Sci.* **2020**, *10*, 577. [CrossRef]

26. Chountoulesi, M.; Demetzos, C. Promising Nanotechnology Approaches in Treatment of Autoimmune Diseases of Central Nervous System. *Brain Sci.* **2020**, *10*, 338. [CrossRef]

27. Thome, R.; Boehm, A.; Ishikawa, L.L.W.; Casella, G.; Munhoz, J.; Ciric, B.; Zhang, G.X.; Rostami, A. Comprehensive Analysis of the Immune and Stromal Compartments of the CNS in EAE Mice Reveal Pathways by Which Chloroquine Suppresses Neuroinflammation. *Brain Sci.* **2020**, *10*, 348. [CrossRef]

28. Boziki, M.K.; Kesidou, E.; Theotokis, P.; Mentis, A.A.; Karafoulidou, E.; Melnikov, M.; Sviridova, A.; Rogovski, V.; Boyko, A.; Grigoriadis, N. Microbiome in Multiple Sclerosis; Where Are We, What We Know and Do Not Know. *Brain Sci.* **2020**, *10*, 234. [CrossRef]

29. Dargahi, N.; Johnson, J.; Apostolopoulos, V. Streptococcus thermophilus alters the expression of genes associated with innate and adaptive immunity in human peripheral blood mononuclear cells. *PLoS ONE* **2020**, *15*, e0228531. [CrossRef]

30. Dargahi, N.; Johnson, J.; Donkor, O.; Vasiljevic, T.; Apostolopoulos, V. Immunomodulatory effects of probiotics: Can they be used to treat allergies and autoimmune diseases? *Maturitas* **2019**, *119*, 25–38. [CrossRef]

31. Dargahi, N.; Matsoukas, J.; Apostolopoulos, V. Streptococcus thermophilus ST285 Alters Pro-Inflammatory to Anti-Inflammatory Cytokine Secretion against Multiple Sclerosis Peptide in Mice. *Brain Sci.* **2020**, *10*, 126. [CrossRef] [PubMed]

32. Moser, T.; Harutyunyan, G.; Karamyan, A.; Otto, F.; Bacher, C.; Chroust, V.; Leitinger, M.; Novak, H.F.; Trinka, E.; Sellner, J. Therapeutic Plasma Exchange in Multiple Sclerosis and Autoimmune Encephalitis: A Comparative Study of Indication, Efficacy and Safety. *Brain Sci.* **2019**, *9*, 267. [CrossRef] [PubMed]

33. Moccia, M.; Capacchione, A.; Lanzillo, R.; Carbone, F.; Micillo, T.; Matarese, G.; Palladino, R.; Brescia Morra, V. Sample Size for Oxidative Stress and Inflammation When Treating Multiple Sclerosis with Interferon-beta1a and Coenzyme Q10. *Brain Sci.* **2019**, *9*, 259. [CrossRef]

34. Workman, C.D.; Ponto, L.L.B.; Kamholz, J.; Rudroff, T. No Immediate Effects of Transcranial Direct Current Stimulation at Various Intensities on Cerebral Blood Flow in People with Multiple Sclerosis. *Brain Sci.* **2020**, *10*, 82. [CrossRef]

35. Ahdab, R.; Shatila, M.M.; Shatila, A.R.; Khazen, G.; Freiha, J.; Salem, M.; Makhoul, K.; El Nawar, R.; El Nemr, S.; Ayache, S.S.; et al. Cortical Excitability Measures May Predict Clinical Response to Fampridine in Patients with Multiple Sclerosis and Gait Impairment. *Brain Sci.* **2019**, *9*, 357. [CrossRef] [PubMed]

Publisher's Note: MDPI stays neutral with regard to jurisdictional claims in published maps and institutional affiliations.

Article

Sample Size for Oxidative Stress and Inflammation When Treating Multiple Sclerosis with Interferon-β1a and Coenzyme Q10

Marcello Moccia [1,*], **Antonio Capacchione** [2], **Roberta Lanzillo** [1], **Fortunata Carbone** [3,4], **Teresa Micillo** [5], **Giuseppe Matarese** [4,6], **Raffaele Palladino** [7,8] and **Vincenzo Brescia Morra** [1]

[1]   Multiple Sclerosis Clinical Care and Research Centre, Department of Neuroscience, Reproductive Science and Odontostomatology, Federico II University, 80131 Naples, Italy; robertalanzillo@libero.it (R.L.); vincenzo.bresciamorra2@unina.it (V.B.M.)

[2]   Medical Affairs Department, Merck, 00176 Rome, Italy; antonio.capacchione@merckgroup.com

[3]   Neuroimmunology Unit, IRCCS Fondazione Santa Lucia, 00142 Rome, Italy; fortunata.carbone@alice.it

[4]   Laboratory of Immunology, Institute of Experimental Endocrinology and Oncology, National Research Council (IEOS-CNR), 80131 Naples, Italy; giuseppe.matarese@unina.it

[5]   Department of Biology, Federico II University, 80131 Naples, Italy; teresa.micillo2@unina.it

[6]   Treg Cell Lab, Department of Molecular Medicine and Medical Biotechnologies, Federico II University, 80131 Naples, Italy

[7]   Department of Primary Care and Public Health, Imperial College, London W68RP, UK; palladino.raffaele@gmail.com

[8]   Department of Public Health, Federico II University, 80131 Naples, Italy

*   Correspondence: moccia.marcello@gmail.com; Tel./Fax: +39-0817462670

Received: 23 August 2019; Accepted: 25 September 2019; Published: 27 September 2019

Abstract: Studying multiple sclerosis (MS) and its treatments requires the use of biomarkers for underlying pathological mechanisms. We aim to estimate the required sample size for detecting variations of biomarkers of inflammation and oxidative stress. This is a post-hoc analysis on 60 relapsing-remitting MS patients treated with Interferon-β1a and Coenzyme Q10 for 3 months in an open-label crossover design over 6 months. At baseline and at the 3 and 6-month visits, we measured markers of scavenging activity, oxidative damage, and inflammation in the peripheral blood (180 measurements). Variations of laboratory measures (treatment effect) were estimated using mixed-effect linear regression models (including age, gender, disease duration, baseline expanded disability status scale (EDSS), and the duration of Interferon-β1a treatment as covariates; creatinine was also included for uric acid analyses), and were used for sample size calculations. Hypothesizing a clinical trial aiming to detect a 70% effect in 3 months (power = 80% alpha-error = 5%), the sample size per treatment arm would be 1 for interleukin (IL)-3 and IL-5, 4 for IL-7 and IL-2R, 6 for IL-13, 14 for IL-6, 22 for IL-8, 23 for IL-4, 25 for activation-normal T cell expressed and secreted (RANTES), 26 for tumor necrosis factor (TNF)-α, 27 for IL-1β, and 29 for uric acid. Peripheral biomarkers of oxidative stress and inflammation could be used in proof-of-concept studies to quickly screen the mechanisms of action of MS treatments.

Keywords: multiple sclerosis; inflammation; oxidative; biomarker; sample size

---

1. Introduction

Monitoring multiple sclerosis (MS) and developing new disease modifying treatments (DMTs) requires the use of biomarkers for underlying pathological mechanisms [1,2]. Thus, it is crucial to define a set of biomarkers that can be easily measured (e.g., in accessible body fluids), are quickly responsive to change, and reflect MS clinical features accurately [2,3].

Experimental evidence supports the important role of inflammation and oxidative stress in the pathogenesis of MS [4]. In the initial relapsing-remitting (RR) phase, oxidative stress is strictly associated with inflammatory activity, whereas the progressive phase is characterized by chronic inflammation and neurodegeneration, further amplifying the oxidative damage [4,5]. In our recent study [6], supplementation with Coenzyme Q10, a natural anti-oxidant, along with Interferon-β1a 44 mcg treatment, was associated with an improved oxidative balance, with a shift toward an anti-inflammatory milieu and with related clinical benefits. However, in this study we used a large number of peripheral biomarkers of oxidative stress and inflammation, which was time-and resource-consuming, and ultimately resulted in a significant statistical challenge due to multiple comparisons [6]. Thus, future studies would benefit from a subset of biomarkers that are sensitive to change in a short time and on a small sample.

In the present post-hoc analysis of our previous longitudinal study, we aim to estimate the sample size needed in RR-MS for different peripheral biomarkers of oxidative stress and inflammation.

2. Materials and Methods

2.1. Study Design and Population

This is a post-hoc analysis on a prospective cohort that was fully described elsewhere [6]. Briefly, in 2016–2017, we included 60 RRMS patients on clinical stability and on treatment with subcutaneous high-dose Interferon-β1a (Rebif®, 44 mcg, Merck, Rome, Italy), either alone or with Coenzyme Q10 (Skatto®, 100 mg/ml, Chiesi Farmaceutici SpA, Parma, Italy) for 3 months, with a cross-over design. In particular, group 1 ($n = 30$) was treated with Interferon-β1a and Coenzyme Q10 from baseline to a 3-month visit, and then with Interferon-β1a alone until a 6-month visit; meanwhile, group 2 ($n = 30$) was treated with Interferon-β1a alone from baseline to a 3-month visit, and then with Interferon-β1a and Coenzyme Q10 until a 6-month visit. This design used within-subjects comparison of treatments, and therefore minimized confounding variables by removing any natural biological variation that may have occurred in the measurement of the outcome measures [6,7].

2.2. Laboratory Analyses

Blood samples were collected at baseline and after 3 and 6 months (60 patients with 3 laboratory measurements, with 180 measurements overall) in fasting conditions in lithium heparin tubes, immediately centrifuged, stored at $-80\ °C$, and then analyzed for:

(1) Markers of free radical scavenging activity: uric acid and bilirubin were measured by using the UA2 and the BILTS enzymatic methods (COBAS® c501 analyser, Roche Diagnostic, Mannheim, Germany);

(2) Markers of serum oxidative damage: 8-hydroxy-2-deoxyguanosine (8-OHdG, an end product of oxidative DNA damage) and protein carbonyls (an end product of oxidative protein damage) were measured by using the OxiSelect™ Oxidative DNA Damage ELISA kit, and the OxiSelect™ Protein Carbonyl ELISA Kit (both from Cell Biolabs, San Diego, CA, USA);

(3) Markers of inflammation: the Human Cytokine Magnetic 35-Plex Panel (Invitrogen by Thermo Fisher Scientific, Waltham, MA, USA) was used for the quantitative detection of epidermal growth factor (EGF), eotaxin, basic-fibroblast growth factor (FGF), granulocyte-colony stimulating factor (G-CSF), granulocyte-macrophage colony-stimulating factor (GM-CSF), hepatocyte growth factor (HGF), Interferon (IFN)-α, IFN-γ, interleukin (IL)-1α, IL-1β, IL-1RA, IL-2, IL-2R, IL-3, IL-4, IL-5, IL-6, IL-7, IL-8, IL-9, IL-10, IL-12, IL-13, IL-15, IL-17A, IL-17F, IL-22, IFN-γ-inducible protein (IP)-10, monocyte chemoattractant protein (MCP)-1, monokine induced by IFN-γ (MIG), macrophage inflammatory proteins (MIP)-1α, MIP-1β, regulated on activation-normal T cell expressed and secreted (RANTES), tumor necrosis factor (TNF)-α, and vascular endothelial growth factor (VEGF).

CellROX® Orange Reagent (Life Technologies, Carlsbad, CA, USA) was used for measuring intracellular reactive oxygen species (ROS) production in peripheral blood mononuclear cells (PBMCs) using a FACScanto II analyzer (Becton–Dickinson, San Diego, CA, USA) and Flow-Jo v10 software (Tree Star Inc., Ashland, OR, USA); intracellular ROS production (CellROX) was measured as percent positive cells (%) and mean fluorescence intensity (MFI).

2.3. Statistics

The sample size needed to detect a treatment effect on different markers of oxidative stress and inflammation was computed using the formula $n = \frac{2\left(Z_\alpha + Z_{1-\beta}\right)^2 \sigma^2}{\Delta^2}$, where n is the required sample size per treatment arm in 1:1 controlled trials, Z_α and $Z_{1-\beta}$ are constants (set at 5% alpha-error and 80% power, respectively), σ is the standard deviation, and Δ the estimated effect size [8,9]. The treatment effect was defined as the actual observed effect in our previous study (i.e., variation in each laboratory measure between treated and untreated groups), estimated using mixed-effect linear regression models (including age, gender, disease duration, baseline expanded disability status scale (EDSS), and duration of Interferon-β1a treatment prior to study inclusion as covariates; creatinine was also included for uric acid analyses) [6,8,9]. The crossover model included random effects for patient ID, and fixed-effects for time (baseline, 3 and 6 months), and for the visit after Coenzyme Q10 exposure, overall accounting for possible carry-over effects. Adjusted beta-coefficients of 3-month variations were obtained for each laboratory measure. We assumed that the observed variation, as estimated by the adjusted beta-coefficients, was the highest achievable treatment effect (100%) over 3 months. From there, with a conservative approach, we hypothesized a number of effect sizes—e.g., 30%, 50%, 70%, and 90%—that were smaller than the observed effect. Standard deviations were calculated from the variation of each laboratory measure after 3 months. Then, we hypothesized a clinical trial where two different biomarkers were included as primary outcome measures for sample size estimates (alpha-error was set at 2.5%). Finally, we considered that the study was designed to include one or two interim analyses in addition to the final analysis (alpha-error was set at 2.94% and 2.21%, respectively, according to the Pocock method) [10,11].

Stata 15.0 (StataCorp LLC, College Station, TX, USA) was used for data processing and analysis.

3. Results

Sixty RRMS patients were included in the present study (age: 41.5 ± 9.7 years; female: $n = 42$ (70%); disease duration: 11.0 ± 1.7 years; baseline EDSS: 2.5 (1.0–5.0)). Four patients presented with a clinical relapse (6.6%) during the study period.

Hypothesizing a clinical trial aiming to detect 70% effect in 3 months (power = 80% alpha-error = 5%), the sample size per treatment arm would be 1 for IL-3 and IL-5, 4 for IL-7 and IL-2R, 6 for IL-13, 14 for IL-6, 22 for IL-8, 23 for IL-4, 25 for RANTES, 26 for TNF-α, 27 for IL-1β, and 29 for uric acid (Figure 1, Table 1). Other investigated markers presented with a sample size per treatment arm larger than 30 (Table 1).

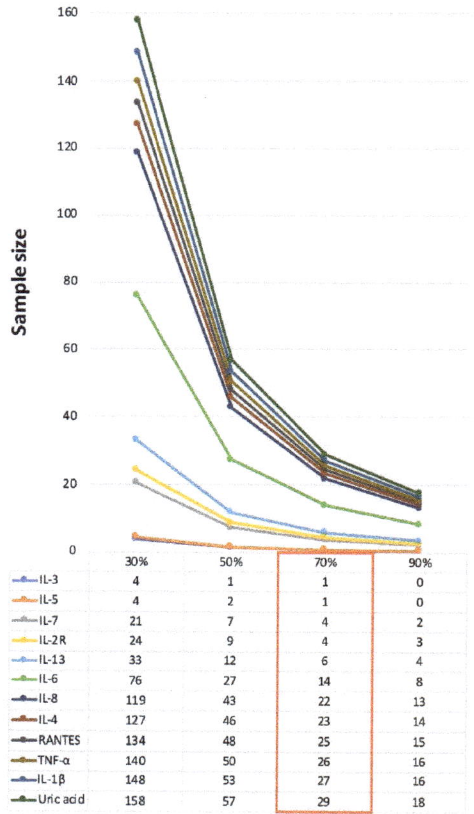

Figure 1. Profile plot for sample size estimates for a treatment arm. Figure shows sample sizes for laboratory markers of oxidative stress and inflammation (<30 patients for a treatment arm with a 70% treatment effect). Sample size per treatment arm is reported hypothesizing a 30%, 50%, 70%, and 90% treatment effect compared with the observed effect. Power was set at 80% and alpha-error at 5%. Abbreviations: interleukin (IL), regulated on activation-normal T cell expressed and secreted (RANTES), and tumor necrosis factor (TNF).

Hypothesizing the combination of two different biomarkers as primary outcome measures (alpha-error = 2.5%), sample size estimates per treatment arm remained substantially favorable (3 for IL-3 and IL-5, 7 for IL-7, 8 for IL-2R, 9 for IL-13, 19 for IL-6, 28 for IL-8, 30 for IL-4, 32 for RANTES, 33 for TNF-α, 35 for IL-1β, and 37 for uric acid) (Table 1).

Sample size estimates for a study with one or two interim analyses (Pocock method, setting alpha-error = 2.94% and 2.21% respectively), in addition to the final analysis, are presented in Table 1; this design would reduce study participants' exposure to an inferior or useless treatment.

Table 1. Sample size estimates for a treatment arm for 3-month variations of peripheral biomarkers of oxidative stress and inflammation.

	Baseline	Adj. Coeff. (3-Month Variation)	SD (3-Month Variation)	Sample Size (70% Treatment Effect)			
				One Primary Outcome	Two Primary Outcomes	Interim Analyses (Pocock Method)	
						One Interim	Two Interim
				5% alpha	2.5% alpha	2.94% alpha	2.21% alpha
Markers of scavenging activity							
Uric acid (mg/dL)	4.670 ± 0.566	0.123 *	0.117	29	37	15	11
Bilirubin (mg/dL)	1.466 ± 0.268	0.066	0.190	265	323	134	98
Markers of oxidative damage							
CellROX cells (%)	76.405 ± 9.348	−9.925 *	11.25	41	52	21	15
CellROX cells (MFI)	2605.320 ± 828.707	−523.308 *	1124.538	148	181	75	55
Protein carbonyls (nmol/mg)	2.976 ± 1.402	−0.266	1.393	878	1066	444	326
8-OHdG (ng/mL)	6.379 ± 1.140	−0.630 *	0.708	40	51	20	15
Markers of inflammation							
EGF (pg/mL)	6.597 ± 12.877	−3.637	8.513	175	214	89	65
Eotaxin (pg/mL)	116.432 ± 46.800	−18.669 *	31.968	94	116	47	35
Basic-FGF (pg/mL)	53.218 ± 282.165	−2.736	4.863	101	124	51	38
G-CSF (pg/mL)	80.445 ± 48.370	−4.692	61.503	5498	6667	2783	2041
GM-CSF (pg/mL)	5.791 ± 4.953	−1.751 *	2.524	66	82	34	25
HGF (pg/mL)	64.959 ± 77.650	−26.397 *	33.925	53	66	27	20
IFN-α (pg/mL)	80.869 ± 469.445	1.780	11.498	1335	1618	676	496
IFN-γ (pg/mL)	2.311 ± 1.952	−1.526 *	1.937	52	64	26	19
IL-1α (pg/mL)	4.427 ± 6.477	−2.460 *	2.526	34	43	17	13
IL-1β (pg/mL)	1.694 ± 7.274	−1.188	1.096	27	35	14	10
IL-1RA (pg/mL)	33.085 ± 39.824	−10.464	18.329	98	121	50	36
IL-2 (pg/mL)	20.979 ± 107.943	5.099	14.090	244	298	124	91

Table 1. *Cont.*

	Baseline	Adj. Coeff. (3-Month Variation)	SD (3-Month Variation)	Sample Size (70% Treatment Effect)			
				One Primary Outcome	Two Primary Outcomes	Interim Analyses (Pocock Method)	
						One Interim	Two Interim
				5% alpha	2.5% alpha	2.94% alpha	2.21% alpha
IL-2R (pg/mL)	105.950 ± 61.462	−29.971 *	11.182	4	8	2	2
IL-3 (pg/mL)	202.849 ± 1283.170	28.661	4.276	1	3	0	0
IL-4 (pg/mL)	4.421 ± 11.691	3.883 *	3.317	23	30	12	9
IL-5 (pg/mL)	8.177 ± 33.861	−12.890	2.069	1	3	0	0
IL-6 (pg/mL)	62.945 ± 363.131	5.559	3.671	14	19	7	5
IL-7 (pg/mL)	12.871 ± 40.625	−16.428	5.639	4	7	2	1
IL-8 (pg/mL)	12.095 ± 7.422	−11.418	9.425	22	28	11	8
IL-9 (pg/mL)	2.248 ± 4.814	−3.749 *	4.212	40	51	20	15
IL-10 (pg/mL)	1079.590 ± 6456.040	1615.546	2417.951	72	89	36	27
IL-12 (pg/mL)	58.932 ± 110.51	2.498	14.365	1058	1284	536	393
IL-13 (pg/mL)	1.714 ± 3.341	3.732 *	1.628	6	9	3	2
IL-15 (pg/mL)	117.149 ± 673.398	21.693	21.658	32	40	16	12
IL-17A (pg/mL)	1.460 ± 2.265	−0.453	0.941	138	169	70	51
IL-17F (pg/mL)	35.954 ± 86.735	−68.854 *	72.039	35	44	18	13
IL-22 (pg/mL)	250.425 ± 642.791	−8.406	40.134	729	886	369	271
IP-10 (pg/mL)	26.279 ± 16.844	5.699	30.460	914	1110	463	339
MCP-1 (pg/mL)	232.083 ± 79.633	39.540	96.247	190	232	96	70
MIG (pg/mL)	32.386 ± 13.580	−5.409	13.555	201	245	102	75
MIP-1α (pg/mL)	7.830 ± 11.718	−5.327 *	5.338	32	41	16	12
MIP-1β (pg/mL)	182.476 ± 1024.490	17.125	17.060	32	40	16	12
RANTES (pg/mL)	1739.970 ± 1475.350	−2331.281 *	2041.081	25	32	12	9

Table 1. *Cont.*

	Baseline	Adj. Coeff. (3-Month Variation)	SD (3-Month Variation)	Sample Size (70% Treatment Effect)			
				One Primary Outcome	Two Primary Outcomes	Interim Analyses (Pocock Method)	
						One Interim	Two Interim
				5% alpha	2.5% alpha	2.94% alpha	2.21% alpha
TNF-α (pg/mL)	2.725 ± 4.310	−1.795 *	1.608	26	33	13	10
VEGF (pg/mL)	0.619 ± 0.777	−0.398 *	0.519	54	68	28	20

Table shows absolute values of biomarkers of oxidative stress and inflammation at the baseline visit. Adjusted beta-coefficients (adj. coeff.) of 3-month variation for each laboratory measure were obtained with mixed-effect linear regression models (including age, gender, disease duration, baseline EDSS, and duration of Interferon-β1a treatment prior to study inclusion as covariates; creatinine was also included for uric acid analyses) (* indicates $p < 0.05$). Standard deviation (SD) was calculated from the variation of each laboratory measure after 3 months. Sample size per treatment arm is reported, hypothesizing a 70% treatment effect, compared with the observed effect, over 3 months (power was set at 80%, alpha-error was set at 5%). Then, we also performed calculations hypothesizing additional scenarios: (i) two different biomarkers were included as combined primary outcome measures for sample size estimates (alpha-error was set at 2.5%); (ii) the study was designed to include one or two interim analyses in addition to the final analysis in order to obtain early evidence of inferior or useless treatment (alpha-error was set to be 0.0294 and 0.0221, respectively, according to the Pocock method). Abbreviations: intracellular ROS production (CellROX), mean fluorescence intensity (MFI), 8-hydroxy-2-deoxyguanosine (8-OHdG), epidermal growth factor (EGF), eotaxin, basic-fibroblast growth factor (FGF), granulocyte-colony stimulating factor (G-CSF), granulocyte-macrophage colony-stimulating factor (GM-CSF), hepatocyte growth factor (HGF), interferon (IFN)- α, IFN-γ, interleukin (IL)-1α, IL-1β, IL-1RA, IL-2, IL-2R, IL-3, IL-4, IL-5, IL-6, IL-7, IL-8, IL-9, IL-10, IL-12, IL-13, IL-15, IL-17A, IL-17F, IL-22, IFN-γ-inducible protein (IP)-10, monocyte chemoattractant protein (MCP)-1, monokine induced by IFN-γ (MIG), macrophage inflammatory proteins (MIP)-1α, MIP-1β, regulated on activation-normal T cell expressed and secreted (RANTES), tumor necrosis factor (TNF)-α, and vascular endothelial growth factor (VEGF).

4. Discussion

Peripheral biomarkers of inflammation, scavenging activity, and oxidative damage gave realistically achievable sample size estimates, and could be used in exploratory clinical trials and observational studies to screen new or already existing medications with putative effects on inflammation and oxidative stress over a 3-month period. Not least, interim analyses could detect an inferior or useless treatment even earlier, with subsequent study termination or treatment switch within adaptive designs [12].

Current sample size calculations were rather conservative. In particular, in the Results (Section 3) and in Figure 1, we specifically focused on a 70% treatment effect, which was smaller than what we actually observed (100% treatment effect) [6,9]. However, greater treatment effects could be hypothesized with different medications and doses, leading to even smaller sample size estimates. Also, the inclusion of multiple markers as primary outcome measures would remain feasible for sample size calculations. Of note, present estimates are based on the combination of subcutaneous high-dose Interferon-β1a (Rebif®, 44 mcg, Merck, Rome, Italy) and Coenzyme Q10. For a subgroup of patients (50%), the Interferon-β1a treatment was also administered prior to study inclusion. Drug naïve patients were equally distributed between Coenzyme Q10 treatment groups, and we also included the duration of the Interferon-β1a treatment as a covariate in the statistical models, but, of course, we cannot exclude the possibility that previous treatment has affected the study outcomes. However, if we assume Interferon-β1a could have exerted its effects before inclusion in the study, we would have observed smaller Coenzyme Q10-related effects, resulting in subsequently more conservative sample size estimates. Interferon-β1a is an approved treatment for MS, with a well-established long-term efficacy and safety profile [13]. On the contrary, Coenzyme Q10 has proven effect on biomarkers of oxidative stress and inflammation and on MS symptoms [14–16], but its disease-modifying effect remains to be established. As such, future studies should evaluate the reproducibility of our findings on more recent medications (e.g., cladribine).

Most promising inflammatory biomarkers are strongly related to MS pathogenesis, and in particular, to acute (e.g., IL-1β, IL-3) and chronic inflammation (e.g., IL-2R, IL-6, IL-7, IL-8, TNF-α) within the central nervous system [17–21], to suppression of the activity of microglia toward brain repair (i.e., RANTES), and to neuroprotective modulation of pathologically-active macrophages and microglia (e.g., IL-4, IL-13) [17,22]. Markers of oxidative stress also resulted in rather small sample sizes, with particular regard to markers of serum scavenging activity (uric acid), and of oxidative damage in inflammatory cells and DNA (CellROX, %, and 8-OHdG). Biomarkers of oxidative stress and inflammation are not only related to MS pathogenesis [17,23], but are also clinically relevant to MS, being associated with MS risk and progression [6,17,21,24–26], and also being used as therapeutic targets [17]. For instance, IL-6, IL-8, and RANTES have been associated with the risk of clinical relapses [27], radiological activity (e.g., lesions, atrophy) [28], treatment switch, and disability progression after up to 6 years [20]. Interestingly, clinical associations might be particularly sound in patients in apparent clinical stability [29]. As such, longitudinal measurements of oxidative stress and inflammation can provide pathologically and clinically relevant information in MS observational studies and clinical trials.

Of note, for some inflammatory biomarkers (e.g., IL-3 and IL-5) sample size estimates were unexpectedly low and should be interpreted with caution. If we assume we are studying a compound with a specific molecular target (e.g., anti-TNF-α or anti-CD20 antibodies), then only a very small sample is necessary to detect biological effect [30,31]. On the contrary, for compounds with multimodal mechanisms of action, a larger sample would be needed or, at least, profiles of inflammatory pathology should be considered [26].

Limitations of this study include possible confounding factors. In our previous study, we excluded patients with possible confounding factors (e.g., contraceptive and immunosuppressive medication), we used within-patients comparison of treatments (minimizing confounding effects by removing any natural biological variation), and we accounted for a number of covariates in our statistical models [6], but factors influencing oxidative stress and inflammation are multiple and virtually impossible to exclude completely. For instance, four patients presented with a clinical relapse (6.6%) that we did not account for considering that patients were equally distributed in the Coenzyme-Q10-treated and untreated groups. Specificity of peripheral biomarkers to MS-related pathology remains to be further investigated, and based on current knowledge, these markers cannot replace conventional biomarkers of disability (e.g., neuroimaging) [32]. We included 180 measurements at three timepoints from 60 patients to estimate coefficients of variation for sample estimates. As such, included sample could have been larger, but was based on sample size calculations from our previous study, and not least, was in line with previous studies with similar goals [6,8,33]. Also, measurements over short intervals may be prone to increased measurement errors leading to a greater variability and larger sample, but apparently, this was not the case in our cohort. A control group (untreated or treated with a medication different from Interferon-β1a) was unfortunately not available, with difficulties in drawing formal conclusions on the observed effects.

5. Conclusions

In conclusion, peripheral biomarkers of oxidative stress and inflammation could be used in exploratory, proof-of-concept studies aiming to evaluate the activity profile of new or already existing medications. Medications with putative anti-oxidant and anti-inflammatory effects could be tested in a short time (3 months) and on small samples (<30 per treatment arm) by using a limited subset of biomarkers, before being moved toward larger and more expensive clinical trials.

Author Contributions: M.M., A.C. and V.B.M. conceived and designed the experiments; R.L., F.C., T.M., G.M. and R.P. performed the experiments; M.M., R.L., R.P. and V.B.M. analyzed the data; A.C., F.C., T.M. and G.M. contributed reagents/materials/analysis tools; M.M., A.C., R.L., F.C., T.M., G.M., R.P. and V.B.M. wrote the paper.

Funding: This research was partially supported by Merck S.p.A. (Italy), an affiliate of Merck KGaA, Darmstadt, Germany. The sponsor preliminarily approved the study design, and after independent data collection, analysis, and interpretation, approved the final version of the manuscript, which was subsequently sent to all co-authors for final approval.

Conflicts of Interest: Marcello Moccia has received research grants from MAGNIMS-ECTRIMS, United Kingdom and Northern Ireland MS Society, and Merck; and honoraria from Biogen, Sanofi-Genzyme, and Merck. Roberta Lanzillo has received honoraria from Almirall, Biogen, Novartis, Sanofi-Genzyme, Merck, and Teva. Fortunata Carbone is supported by the Ministero della Salute (grant no. GR-2016-02363725). Giuseppe Matarese is supported by the Fondazione Italiana Sclerosi Multipla (grant no. 2016/R/18), and Telethon (grant no. GGP17086), and has received research grants from Merck, Biogen, Novartis, and IBSA. Vincenzo Brescia Morra is supported by FISM (Fondazione Italiana Sclerosi Multipla) (cod. 2017/R/5) and financed or co-financed with the "5 per mille" public funding, and has received honoraria from Almirall, Bayer, Biogen, Novartis, Sanofi-Genzyme, Merck, Mylan, and Teva.

References

1. Tur, C.; Moccia, M.; Barkhof, F.; Chataway, J.; Sastre-Garriga, J.; Thompson, A.J.; Ciccarelli, O. Assessing treatment outcomes in multiple sclerosis trials and in the clinical setting. *Nat. Rev. Neurol.* **2018**, *14*, 75–93. [CrossRef]
2. Thompson, A.J.; Baranzini, S.E.; Geurts, J.; Hemmer, B.; Ciccarelli, O. Multiple sclerosis. *Lancet* **2018**, *391*, 1622–1636. [CrossRef]
3. Zaratin, P.; Comi, G.; Coetzee, T.; Ramsey, K.; Smith, K.; Thompson, A.; Panzara, M. Progressive MS Alliance Industry Forum: Maximizing Collective Impact to Enable Drug Development. *Trends Pharmacol. Sci.* **2016**, *37*, 808–810. [CrossRef]
4. Haider, L.; Zrzavy, T.; Hametner, S.; Höftberger, R.; Bagnato, F.; Grabner, G.; Trattnig, S.; Pfeifenbring, S.; Brück, W.; Lassmann, H. The topograpy of demyelination and neurodegeneration in the multiple sclerosis brain. *Brain* **2016**, *139*, 807–815. [CrossRef]
5. Friese, M.A.; Schattling, B.; Fugger, L. Mechanisms of neurodegeneration and axonal dysfunction in multiple sclerosis. *Nat. Rev. Neurol.* **2014**, *10*, 225–238. [CrossRef]

6. Moccia, M.; Capacchione, A.; Lanzillo, R.; Carbone, F.; Micillo, T.; Perna, F.; De Rosa, A.; Carotenuto, A.; Albero, R.; Matarese, G.; et al. Coenzyme Q10 supplementation reduces peripheral oxidative stress and inflammation in Interferon-Beta1a treated multiple sclerosis. *Ther. Adv. Neurol. Disord.* **2019**, *12*, 1–12. [CrossRef]

7. Sedgwick, P. What is a crossover trial? *BMJ* **2014**, *348*, 9–10. [CrossRef]

8. Altmann, D.R.; Jasperse, B.; Barkhof, F.; Beckmann, K.; Filippi, M.; Kappos, L.D.; Molyneux, P.; Polman, C.H.; Pozzilli, C.; Thompson, A.J.; et al. Sample sizes for brain atrophy outcomes in trials for secondary progressive multiple sclerosis. *Neurology* **2009**, *72*, 595–601. [CrossRef]

9. Moccia, M.; Prados, F.; Filippi, M.; Rocca, M.A.; Valsasina, P.; Brownlee, W.J.; Zecca, C.; Gallo, A.; Rovira, A.; Gass, A.; et al. Longitudinal spinal cord atrophy in multiple sclerosis using the generalised boundary shift integral. *Ann. Neurol.* **2019**. [CrossRef]

10. Li, G.; Taljaard, M.; Van den Heuvel, E.R.; Levine, M.A.; Cook, D.J.; Wells, G.A.; Devereaux, P.J.; Thabane, L. An introduction to multiplicity issues in clinical trials: The what, why, when and how. *Int. J. Epidemiol.* **2017**, *46*, 746–755. [CrossRef]

11. Pocock, S. Group sequential methods in the design and analysis of clinical trials. *Biometrika* **1977**, *64*, 191–199. [CrossRef]

12. Fox, R.; Chataway, J. Advancing Trial Design in Progressive Multiple Sclerosis. *Mult. Scler.* **2017**, *23*, 1573–1578. [CrossRef]

13. Moccia, M.; Palladino, R.; Carotenuto, A.; Saccà, F.; Russo, C.V.; Lanzillo, R.; Brescia Morra, V. A 8-year retrospective cohort study comparing Interferon-β formulations for relapsing-remitting multiple sclerosis. *Mult. Scler. Relat. Disord.* **2018**, *19*, 50–54. [CrossRef]

14. Sanoobar, M.; Eghtesadi, S.; Azimi, A.; Khalili, M.; Khodadadi, B.; Jazayeri, S.; Gohari, M.R.; Aryaeian, N. Coenzyme Q10 supplementation ameliorates inflammatory markers in patients with multiple sclerosis: A double blind, placebo, controlled randomized clinical trial. *Nutr. Neurosci.* **2015**, *18*, 169–176. [CrossRef]

15. Sanoobar, M.; Eghtesadi, S.; Azimi, A.; Khalili, M.; Jazayeri, S.; Gohari, M.R. Coenzyme Q10 supplementation reduces oxidative stress and increases antioxidant enzyme activity in patients with coronary artery disease. *Int. J. Neurosci.* **2013**, *123*, 776–782. [CrossRef]

16. Sanoobar, M.; Dehghan, P.; Khalil, M.; Azimi, A.; Seifar, F. Coenzyme Q10 as a treatment for fatigue and depression in multiple sclerosis patients: A double blind randomized clinical trial. *Nutr. Neurosci.* **2016**, *19*, 138–143. [CrossRef]

17. Göbel, K.; Ruck, T.; Meuth, S.G. Cytokine signaling in multiple sclerosis: Lost in translation. *Mult. Scler. J.* **2018**, *24*, 432–439. [CrossRef]

18. Lee, P.W.; Xin, M.K.; Pei, W.; Yang, Y.; Lovett-Racke, A.E. IL-3 Is a Marker of Encephalitogenic T Cells, but Not Essential for CNS Autoimmunity. *Front. Immunol.* **2018**, *9*, 1–7. [CrossRef]

19. Lin, C.-C.; Edelson, B.T. New Insights into the Role of IL-1β in Experimental Autoimmune Encephalomyelitis and Multiple Sclerosis. *J. Immunol.* **2017**, *198*, 4553–4560. [CrossRef]

20. Bassi, M.S.; Iezzi, E.; Landi, D.; Monteleone, F.; Gilio, L.; Simonelli, I.; Musella, A.; Mandolesi, G.; De Vito, F.; Furlan, R.; et al. Delayed treatment of MS is associated with high CSF levels of IL-6 and IL-8 and worse future disease course. *J. Neurol.* **2018**, *265*, 2540–2547. [CrossRef]

21. Tavakolpour, S. Interleukin 7 receptor polymorphisms and the risk of multiple sclerosis: A meta-analysis. *Mult. Scler. Relat. Disord.* **2016**, *8*, 66–73. [CrossRef]

22. Guglielmetti, C.; Le Blon, D.; Santermans, E.; Salas-Perdomo, A.; Daans, J.; De Vocht, N.; Shah, D.; Hoornaert, C.; Praet, J.; Peerlings, J.; et al. Interleukin-13 immune gene therapy prevents CNS inflammation and demyelination via alternative activation of microglia and macrophages. *Glia* **2016**, *64*, 2181–2200. [CrossRef]

23. Hu, W.T.; Howell, J.C.; Ozturk, T.; Gangishetti, U.; Kollhoff, A.L.; Hatcher-Martin, J.M.; Anderson, A.M.; Tyor, W.R. CSF Cytokines in Aging, Multiple Sclerosis, and Dementia. *Front. Immunol.* **2019**, *10*, 480. [CrossRef]

24. Moccia, M.; Lanzillo, R.; Palladino, R.; Russo, C.; Carotenuto, A.; Massarelli, M.; Vacca, G.; Vacchiano, V.; Nardone, A.; Triassi, M.; et al. Uric acid: A potential biomarker of multiple sclerosis and of its disability. *Clin. Chem. Lab. Med.* **2015**, *53*, 753–759. [CrossRef]

25. Moccia, M.; Lanzillo, R.; Costabile, T.; Russo, C.; Carotenuto, A.; Sasso, G.; Postiglione, E.; De Luca Picione, C.; Vastola, M.; Maniscalco, G.T.; et al. Uric acid in relapsing-remitting multiple sclerosis: A 2-year longitudinal study. *J. Neurol.* **2015**, *262*, 961–967. [CrossRef]

26. Magliozzi, R.; Howell, O.; Nicholas, R.; Cruciani, C.; Castellaro, M.; Romualdi, C.; Rossi, S.; Pittieri, M.; Benedetti, M.; Gajofatto, A.; et al. Inflammatory intrathecal profiles and cortical damage in multiple sclerosis. *Ann. Neurol.* **2018**, *83*, 739–755. [CrossRef]

27. Lanzillo, R.; Carbone, F.; Quarantelli, M.; Bruzzese, D.; Carotenuto, A.; De Rosa, V.; Colamatteo, A.; Micillo, T.; De Luca Picione, C.; Saccà, F.; et al. Immunometabolic profiling of patients with multiple sclerosis identifies new biomarkers to predict disease activity during treatment with interferon Interferon beta-1a. *Clin. Immunol.* **2017**, *183*, 249–253. [CrossRef]

28. Ziliotto, N.; Bernardi, F.; Jakimovski, D.; Baroni, M.; Bergsland, N.; Ramasamy, D.P.; Weinstock-Guttman, B.; Zamboni, P.; Marchetti, G.; Zivadinov, R.; et al. Increased CCL18 plasma levels are associated with neurodegenerative MRI outcomes in multiple sclerosis patients. *Mult. Scler. Relat. Disord.* **2018**, *25*, 37–42. [CrossRef]

29. Ghezzi, L.; Cantoni, C.; Cignarella, F.; Bollman, B.; Cross, A.H.; Salter, A.; Galimberti, D.; Cella, M.; Piccio, L. T cells producing GM-CSF and IL-13 are enriched in the cerebrospinal fluid of relapsing MS patients. *Mult. Scler.* **2019**, 1352458519852092. [CrossRef]

30. Gibellini, L.; De Biasi, S.; Bianchini, E.; Bartolomeo, R.; Fabiano, A.; Manfredini, M.; Ferrari, F.; Albertini, G.; Trenti, T.; Nasi, M.; et al. Anti-TNF-α drugs differently affect the TNFα-sTNFR system and monocyte subsets in patients with psoriasis. *PLoS ONE* **2016**, *11*, 1–16. [CrossRef]

31. Ellrichmann, G.; Bolz, J.; Peschke, M.; Duscha, A.; Hellwig, K.; Lee, D.H.; Linker, R.A.; Gold, R.; Haghikia, A. Peripheral CD19 + B-cell counts and infusion intervals as a surrogate for long-term B-cell depleting therapy in multiple sclerosis and neuromyelitis optica/neuromyelitis optica spectrum disorders. *J. Neurol.* **2019**, *266*, 57–67. [CrossRef] [PubMed]

32. Moccia, M.; de Stefano, N.; Barkhof, F. Imaging outcomes measures for progressive multiple sclerosis trials. *Mult. Scler.* **2017**, *23*, 1614–1626. [CrossRef] [PubMed]

33. Cawley, N.; Tur, C.; Prados, F.; Plantone, D.; Kearney, H.; Abdel-Aziz, K.; Ourselin, S.; Wheeler-Kingshott, C.A.M.G.; Miller, D.H.; Thompson, A.J.; et al. Spinal cord atrophy as a primary outcome measure in phase II trials of progressive multiple sclerosis. *Mult. Scler.* **2018**, *24*, 932–941. [CrossRef] [PubMed]

Article

Therapeutic Plasma Exchange in Multiple Sclerosis and Autoimmune Encephalitis: A Comparative Study of Indication, Efficacy, and Safety

Tobias Moser [1], Gayane Harutyunyan [1], Anush Karamyan [1], Ferdinand Otto [1], Carola Bacher [1], Vaclav Chroust [1], Markus Leitinger [1], Helmut F. Novak [1], Eugen Trinka [1] and Johann Sellner [1,2,3,*]

[1] Department of Neurology, Christian Doppler Medical Center, Paracelsus Medical University, 5020 Salzburg, Austria; t.moser@salk.at (T.M.); gayane.harutyunyan@stud.pmu.ac.at (G.H.); anush.karamyan@stud.pmu.ac.at (A.K.); f.otto@salk.at (F.O.); ca.bacher@salk.at (C.B.); info@neurologie-chroust.at (V.C.); ma.leitinger@salk.at (M.L.); h.novak@salk.at (H.F.N.); e.trinka@salk.at (E.T.)

[2] Department of Neurology, Klinikum rechts der Isar, Technische Universität München, 81675 München, Germany

[3] Department of Neurology, Landesklinikum Mistelbach-Gänserndorf, 2130 Mistelbach, Austria

* Correspondence: j.sellner@salk.at; Tel.: +43-5-7255-0

Received: 1 September 2019; Accepted: 8 October 2019; Published: 9 October 2019

Abstract: Therapeutic plasma exchange (TPE) is a well-established method of treatment for steroid-refractory relapses in multiple sclerosis (MS) and neuromyelitis optica spectrum disorders (NMOSD). Little is known about indications and clinical responses to TPE in autoimmune encephalitis and other immune-mediated disorders of the central nervous system (CNS). We performed a retrospective chart review of patients with immune-mediated disorders of the CNS undergoing TPE at our tertiary care center between 2003 and 2015. The response to TPE within a 3- to 6-month follow-up was scored with an established rating system. We identified 40 patients including 21 patients with multiple sclerosis (MS, 52.5%), 12 with autoimmune encephalitis (AE, 30%), and 7 with other immune-mediated CNS disorders (17.5%). Among patients with AE, eight patients had definite AE (Immunolobulin G for N-methyl-D-aspartate receptor $n = 4$, Leucine-rich, glioma inactivated 1 $n = 2$, Ma 2 $n = 1$, and Alpha-amino-3-hydroxy-5-methyl-4-isoxazolepropionic Acid $n = 1$). Intravenous immunoglobulins had been given prior to TPE in all but one patient with AE, and indications were dominated by acute psychosis and epileptic seizures. While TPE has a distinct place in the treatment sequence of different immune-mediated CNS disorders, we found consistent efficacy and safety. Further research should be directed toward alternative management strategies in non-responders.

Keywords: multiple sclerosis; autoimmune encephalitis; plasma exchange; autoimmunity; immunotherapeutics; clinical outcomes

---

1. Introduction

Apheresis therapies separate patients' plasma from the whole blood by using centrifugation devices or highly permeable filters. In this regard, therapeutic plasma exchange (TPE) aims to eliminate pathogenic antibodies and other proinflammatory mediators from the patient's circulation. This procedure is an established treatment for steroid-refractory relapses in multiple sclerosis (MS) [1]. Several studies corroborated its efficacy in about 66%–86% of patients undergoing TPE, after conventional high-dose glucocorticoid (GC) treatment had failed [2–4]. The rationale for treatment of acute and recurrent attacks in neuromyelitis optica spectrum disorders (NMOSD) is based upon evidence that humoral autoimmunity plays a key role in the pathogenesis. Of note, the interest to achieve rapid remission in NMOSD is driven by the high attack-related disability and -mortality [5]. Therefore, a more aggressive treatment concept based on immunosuppression, pulsed immunotherapy, or targeted disruption of the immunological cascade leading to neuroaxonal injury is maintained in NMOSD in order to preserve long-term neurological function [6].

The spectrum of immune-mediated disorders of the CNS widened over the recent decade. There is emerging evidence that GCs are less effective in B-cell-mediated diseases, including autoimmune encephalitis, and TPE is likely to be effective from a pathophysiological viewpoint in the treatment of antibody-mediated immune processes [7,8]. Autoimmune encephalitis (AE) is a clinical challenge, since presentation is unspecific, and therefore diagnostic consideration is often delayed. Moreover, some patients require treatment at the intensive care unit (ICU), and outcomes can be devastating [9]. Notably, due to the relatively recent discovery of anti-neuronal antibodies and the rarity of AE, treatment recommendations are based on retrospective reports and expert opinion. In a study of 30 patients with AE, 67% improved with TPE by at least 1 point in the modified ranking scale (mRS) [10]. There are, however, two other components of treatment of AE [11]. These include intravenous immunoglobulins (IVIG) and tumor removal. Of note, the ideal sequence for GC, IVIG, and TPE has not been established yet. In addition, there is emerging evidence for the efficacy of rituximab, a CD20 depleting antibody, for achieving long-term remission [12]. Moreover, TPE is not without risks and should only be carried out in conditions where there is good evidence of its effectiveness. Side-effects include disturbances of coagulation, vasovagal episodes, fluid overload or under-replacement, and allergic or anaphylactic reactions due to plasma infusion [13]. Immunoadsorption (IA) is a selective technique for the removal of autoantibodies and immune complexes with less adverse effects in contrast to TPE, which is a non-selective extracorporeal blood purification process with elimination of plasma and subsequent substitution. Recent studies have shown that IA is not only effective in GC-unresponsive MS relapses but also in exacerbations related to NMOSD [14,15].

Here, we hypothesized that TPE is effective in autoimmune encephalitis and therefore studied indication, efficacy, and safety in comparison with MS and other immune-mediated disorders of the CNS.

2. Materials and Methods

2.1. Study Design

We performed a retrospective chart review of all patients with immune-mediated disorders of the CNS who underwent TPE at the 9-bed neurological intensive care unit (NICU) of a tertiary university hospital (Christian Doppler Medical Center, Paracelsus Medical University, Salzburg, Austria). The study protocol was reviewed and approved by the local Ethics Committee (Ethikkommission für das Bundesland Salzburg; 415-EP/73/534-2015).

2.2. Study Population and Data Collection

We reviewed the electronic records for demographic data, neurological diagnosis, symptoms, complications, number of TPE cycles, and outcome and included patients according to the following inclusion criteria:

- acute immune-mediated disorder of the CNS
- TPE during the period of January 2003 to December 2015
- sufficient clinical documentation on underlying disease, indication, procedures, and complications

Multiple sclerosis was diagnosed according to the McDonald's criteria revised in 2010 [16]. For AE, we followed the diagnostic criteria set up by Graus and coworkers [4]. Briefly, diagnosis can be made when all three of the following criteria have been met:

- subacute onset (rapid progression of less than three months) of working memory deficits (short-term memory loss), altered mental status, or psychiatric symptoms

- at least one of the following: new focal CNS findings, seizures not explained by a previously known seizure disorder, CSF pleocytosis (white blood cell count of more than five cells per µL), MRI features suggestive of encephalitis

The cohort of patients with "other immune-mediated CNS disorders" comprised acute disseminated encephalomyelitis (ADEM), CNS lupus, optic neuritis not related to MS, and NMOSD. MS patients with progressive multifocal leukoencephalopathy (PML, $n = 3$) who received TPE for elimination of natalizumab were excluded. Three patients with AE were excluded for lack of sufficient follow-up (Figure 1).

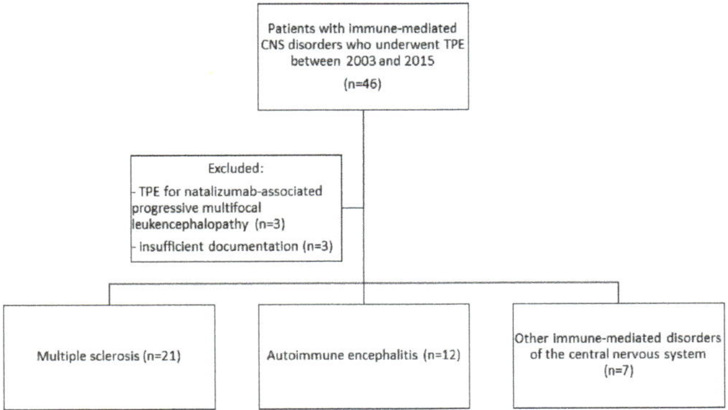

Figure 1. Flow chart for patient selection.

TPE was performed by experienced neurointensivists via a central venous access catheter. The clinical response to TPE was rated with a scoring system introduced by Magaña et al. [6,17,18], which proposes three different response categories; these were no, mild, and good improvement. Patients in the first group showed no recovery at all or even deterioration of symptoms. Mild recovery was defined as "improvement in neurological status without impacting function". Individuals with functionally relevant neurological recovery were considered to have a good improvement. We studied the outcome within six months from TPE. If patients were not seen regularly on follow-up at our center, we contacted individual patients or their caregivers to collect information on their recovery.

2.3. Statistical Analysis

All statistical analyses were conducted using IBM SPSS Version 21.0 (SPSS, Chicago, IL, USA). Descriptive statistics for clinical, demographic, and outcome data are provided. We report the median (interquartile range, IQR) for continuous variables and frequency (percent) for categorical variables. Intergroup comparisons were performed using Fisher's exact and Mann–Whitney U tests and one-way ANOVA where appropriate. All reported p-values were two-tailed and considered statistically significant at $p < 0.05$.

3. Results

3.1. Clinical and Demographic Characteristics

We identified a total of 40 patients with immune-mediated CNS disorders who underwent PLEX between 2003 and 2015. As shown in Figure 1, the most frequent condition was MS in 52.5% ($n = 21$), followed by AE in 30% ($n = 12$). The remaining disorders (summarized as "others" in the following) were CNS lupus ($n = 3$), optic neuritis (not otherwise specified, $n = 2$), ADEM ($n = 1$), and NMOSD ($n = 1$).

There was a general trend towards the increased utilization of TPE over time (Figure 2). Half ($n = 20$) of all patients received the treatment within the past four years, whereas in the period of 2003–2008, only eight patients were treated with TPE.

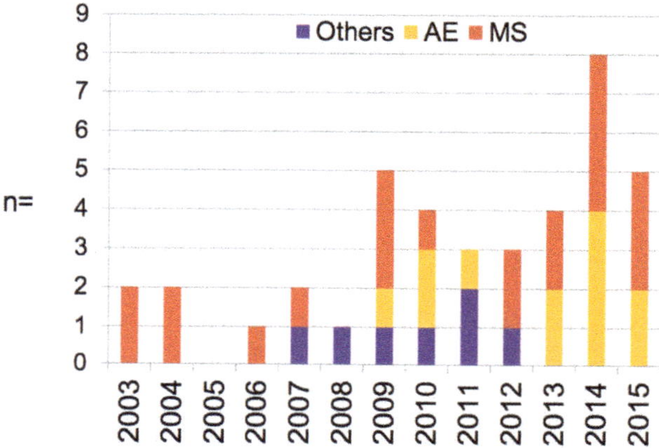

Figure 2. Time course of therapeutic plasma exchange (TPE) usage during the observation period.

The majority of patients were women ($n = 26$, 65%). In detail, women were more frequent in the MS group ($n = 15$, 71%), and the group of "other" immune-mediated CNS disorders comprised entirely women (100%). Of note, there were more male patients in the AE group ($n = 27$, 67%).

The mean age of patients with MS was 35.5 (standard deviation (SD) of 9.4). TPE was used in this group as second-line therapy except in one patient where steroids were contraindicated, as shown in Table 1. The most frequent indication was optic neuritis ($n = 9$), followed by pyramidal tract symptoms ($n = 6$). The mean number of TPE courses was 5.1 (range of 2–9). There were single patients with eight and nine TPE cycles, respectively.

Table 1. TPE in patients with multiple sclerosis.

No.	Age	Gender	Indication	GC Refractory	Courses of TPE	Clinical Response
1	42	M	GC contraindicated		5	good
2	37	F	optic neuritis	X	5	good
3	39	F	optic neuritis	X	5	good
4	44	F	optic neuritis	X	5	good
5	25	F	optic neuritis	X	5	mild
6	40	F	optic neuritis	X	5	mild
7	28	F	optic neuritis	X	5	mild
8	28	M	optic neuritis	X	5	mild
9	47	M	optic neuritis	X	5	no
10	33	F	optic neuritis	X	5	no
11	43	F	tetraparesis	X	2	good
12	23	M	tetraplegia	X	5	good
13	33	M	hemiparesis, dysarthria	X	5	good
14	17	M	hemiparesis	X	5	good
15	34	F	tetraparesis	X	5	mild
16	44	F	tetraplegia	X	8	no
17	50	M	tetrapresis, dysarthia, dysphagia	X	5	no
18	43	F	hemiparesis, ataxia	X	3	no
19	25	F	fulminant MRI, aphasia	X	5	good
20	29	F	fulminant MRI, natalizumab rebound	X	5	good
21	21	F	fulminant radiological findings	X	9	good

Legends: M, male; F, Female; MRI, magnetic resonance imaging; GC, glucocorticoid.

The mean age of patients with AE was 45.1 years (SD of 18.8), which was significantly higher than in MS ($p = 0.03$). The spectrum of clinical symptoms in patients with AE was broad, and the most frequent disturbances were psychiatric symptoms and epileptic seizures (Table 2). The average number of TPE cycles in patients with AE was 6.3 (SD of 2.7). Two patients had more than 10 cycles of TPE. In almost all patients, a prior treatment with intravenous immunoglobulins was performed. Some also received steroids prior to TPE. In the AE cohort ($n = 12$), eight (66%) patients had definite AE with positive antibodies (IgG for NMDA-R: $n = 4$, LGI1: $n = 2$, Ma 2: $n = 1$, and AMPA: $n = 1$). Six patients had inflammatory CSF and eight had pathologies on MRI. All except the patients with exclusively brainstem involvement developed neurocognitive signs, and seizures were common (60%). The four patients with anti-NMDAR-antibodies were women, two of whom had a histologically confirmed teratoma. All of the AE patients had additional treatments (surgery $n = 4$, steroids $n = 5$, IVIG $n = 11$, rituximab $n = 2$).

Table 2. TPE in patients with autoimmune encephalitis (AE).

n	Age	Gender	Details	Antibody	Memo	Epil	Psych	Move	Auto	Sleep	Pons	Detection of Lesion on MRI	CSF	Treatment	TPE Courses	Clinical Response
1	42	M	Paraneopl.AE	Ma1/Ma2	+	+	+			+		temporomesial	neg.	TPE/OP/Ritux	3	no
2	29	W	NMDAR-E	NMDAR		+	+	+				diffuse	IgG ↑	TPE/IVIG	10	good
3	26	W	NMDAR-E	NMDAR			+		+			none	33 cells	TPE/IVIG/OP	5	good
4	30	W	NMDAR-E	NMDAR		+	+					none	neg.	TPE/IVIG/OP	5	good
5	25	F	NMDAR-E	NMDAR		+	+	+				none	10 cells	TPE/IVIG/GC/OP	13	no
6	64	M	LE	LGI-1/VGKC	+	+	+					temporomesial	9 cells	TPE/IVIG/GC	5	mild
7	55	M	LE	none	+							temporomesial	21 cells	TPE/IVIG	5	mild
8	62	M	LE	AMPA-R1	+		+					temporomesial	16 cells	TPE/IVIG/Ritux	7	no
9	66	M	LE	LGI-1			+	+				none	neg.	TPE/ IVIG/ GC	4	good
10	24	F	probable LE	none		+	+			+		temperomesial	neg.	TPE/IVIG/GC	5	mild
11	71	M	Brainstem E	none							+	pons	neg.	TPE/IVIG	7	good
12	47	M	Brainstem E	none							+	pons	neg.	TPE/IVIG/GC	6	mild

Legends: LE: Limbic Encephalitis; NMDAR-E: NMDAR-Encephalitis; Brainstem E: Brainstem Encephalitis; memo: memory-deficit; epi: epileptic seizure; psych: psychiatric disorders; move: movement disorders; auto: autonomic dysfunction; sleep: sleep disturbance; pons: pontine signs; IVIG: intravenous immunoglobulins; GC: Glucocorticoids; OP: tumor surgery; Ritux: Rituximab.

The mean age of patients with other autoimmune CNS disorders was 50.7 years (SD 18.1). The mean number of TPE cycles was 5.7 (SD 2.1). Further details are presented in Table 3.

Table 3. TPE in patients with other immune-mediated disorders of the central nervous system (CNS).

No.	Age	Gender	Condition	TPE Courses (n)	Outcome
1	60	W	CNS-lupus	4	no
2	55	W	CNS-lupus	4	good
3	59	W	CNS-lupus	7	good
4	27	W	optic neuritis	5	good
5	47	W	optic neuritis	10	mild
6	78	W	NMOSD	5	no
7	29	W	ADEM	5	good

3.2. Time from Symptom Onset to Start of TPE

The time from relapse onset to the initiation of TPE was distinct among the three groups ($p = 0.03$). In detail, we calculated the median of 15 days (interquartile range (IQR) 10–27 days) for patients with MS, 77 days (24–203) for patients with AE, and 11 days (9–55) for patients with other immune-mediated disorders of the CNS.

3.3. Outcome

The overall rate of TPE responders was 75%. Further details are shown in Figure 3. A good or mild response was observed in 52.5% and 22.5% of the patients, respectively. The best outcome was observed within the "other CNS-ID" group, where 71% ($n = 5$) had a good recovery after PLEX, while only two patients showed no response (Table 3). The analysis of the MS and AE cohorts disclosed a functional improvement after plasma exchange treatment in 52% and 42%, respectively.

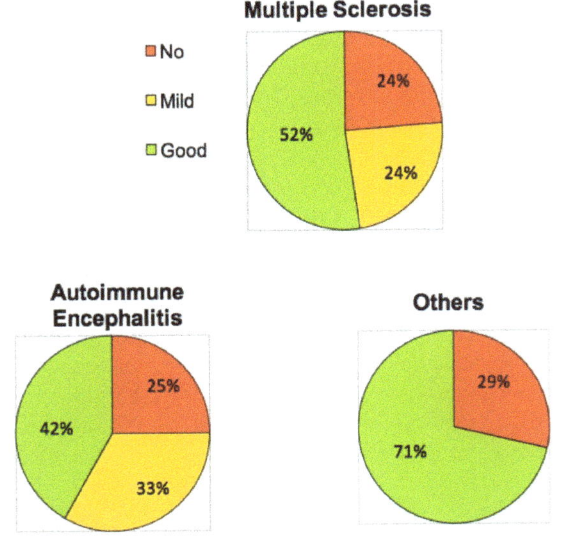

Figure 3. Clinical response to TPE across three subgroups of immune-mediated CNS disorders.

We also evaluated whether the period of symptom onset to the start of the TPE was distinct in patients with good response vs. mild/no response. In a pooled analysis of all three groups we did not find statistical differences (median of 16 days for the cohort with good response vs. 22 days for the cohort with mild or poor response).

3.4. Complications

In total, 219 TPEs were performed. Serious complications were found (1.8%): thrombosis of the subclavian vein, episode of bradycardia with transient loss of consciousness (one each), and two patients had impaired coagulation. No cases of death were reported.

4. Discussion

This study demonstrates that the majority of patients suffering from immune-mediated CNS disorders benefit from TPE, and the rate of non-responders is similar throughout the conditions. Moreover, the number of patients with AE treated with TPE is increasing over time, which reflects increased recognition of TPE as an effective treatment option. Of note, 50% of our patients with AE received TPE for intractable seizures and status epilepticus. Recent studies indicate that immune-mediated epilepsy such as in AE responds better to immunotherapies than to conventional epilepsy therapies [19]. Complications of TPE in our patient series was in the range of previously published studies in the real-life setting [3,20,21].

In our study, the rate of good recovery was higher in MS patients than in AE. This is could be related to the more complex pathogenesis and requirement of several lines of treatments for effective treatment of AE [10,22–24]. Thus, we cannot exclude the overlapping effects of prior treatment approaches. Among the patients with AE and no recovery after TPE, one suffered from a typical paraneoplastic syndrome with intracellular antibodies (MA1/2), which is to date believed to show no or limited response to TPE. In another patient with AE and poor response, the follow-up may have been too short. According to a study by Titulaer et al., almost half of the patients with NMDAR-encephalitis need a prolonged immune-suppressive treatment, and clinical improvement can be delayed [10]. Therefore, a delayed recovery cannot be excluded. Indeed, comparing two disorders with different pathogenesis is problematic. While AE may be monophasic, MS is mostly a relapsing-remitting disease. A B-cell-mediated pathogenesis is implicated in both AE and other CNS IDs, which backs the use of TPE, whereas this is the predominant process of acute inflammation in only a subset of MS patients [7,25]. Most importantly, time to TPE seems to be one of the most critical factors for observing an adequate treatment response [10,22,24]. Neurological involvement is relatively common in the majority of systemic autoimmune diseases and may lead to severe morbidity and mortality, if not treated promptly [26,27]. While our findings further support the use of TPE in neurological complications of systemic disease, our literature search revealed only a few case series for this indication [28,29]. Thus, prospective studies and establishment of registries are eagerly awaited. Open questions include the number of TPE cycles, definition of treatment goals and development of biomarkers. In addition, further studies should also elucidate the role of immunoadsorption as an alternative treatment option.

Moreover, the exact mechanism of action of TPE it is still incompletely understood. The removal of immunoglobulins seems to be a crucial factor. This process is likely to be followed by a shift from cerebral tissue towards systemic circulation [11]. TPE also modulates the immune system by changing the lymphocyte distribution, including changes in B and T cell numbers and activation, increased T suppressor function, and alteration in T-helper cell type 1/2 (Th1/Th2) ratio. In contrast, steroids alone often insufficiently resolve autoantibody-mediated pathologies [13]. A small retrospective review of 14 patients with NMDAR-encephalitis reports better outcomes in patients receiving TPE shortly after GCs than those with GC-treatment alone [12]. In our cohort, none of the 12 patients with AE were treated with TPE alone in the acute disease phase: 10 (83%) had additional IVIG and 6 (50%) had steroid treatment. Because adverse events are estimated to occur in about 6% of every single TPE procedure (the risk for the single individual is therefore multiplied by the number of exchanges), it was used as the ultima ratio treatment after failure of IVIG and/or GCs. Ehlers et al. studied a cohort of 37 GCS-unresponsive MS patients and reported a median time from symptom onset to begin of TPE of 44 days (range of 11–154) [30]. The median time in our MS cohort was 15 days, whereas 77 days had elapsed in patients with AE since the begin of clinical symptoms. This interval was even shorter for patient with other ID of the CNS (median 11 days). In the latter group were patients with known pathogenetic role of B-cells and humoral immunity including NMOSD and neurological manifestations of lupus. Moreover, due to this time lag, a delayed response to immunotherapies given prior to TPE is less likely. The long interval in AE is noteworthy and we assume that the approach for treatment escalation with TPE has changed over the recent years with knowledge about the key importance of an early TPE use in case of GCS-unresponsiveness and implementation in treatment guidelines [31]. In this regard, early initiation of TPE and lower patient age have been reported as predictors of a good response [22]. It also should be noted that none of our patients were treated with IA, which is an emerging treatment option in MS and CNS ID [2,15,32].

We report a low rate of adverse reactions. It should be noted that we exclusively used a central venous access which carries a significantly higher risk of complications than TPE via a peripheral line. In a recent study of complicated SLE and other autoimmune conditions (n = 66) using a central access device, the majority of complications were mild, with bleeding (25.8%) being the most common [28]. Electrolyte disturbances, hypotension, mild arrhythmia, and hypersensitivity were reported occasionally (6.1%, 9.1%, 3%, and 1.5%, respectively). Notably, 27.3% developed infections in the 14 days after TPE.

Improvement in function was achieved in 52.5% of all patients, which is in concert with TPE response rates found in literature ranging from 40% to 63%. Further subgroup analysis including analysis of predictors, however, could not be made due to the limited size of our cohort. Another limitation in this study is the retrospective character and the lack of a control-group.

5. Conclusions

Taken together we conclude that TPE shows similar response rates throughout different CNS IDs. Further studies should focus on the prediction of non-responders and development of alternative treatment strategies for these patients.

Author Contributions: Conceptualization, T.M. and J.S.; Data curation, G.H., A.K., F.O., C.B., M.L., V.C. and H.F.N.; Formal analysis, J.S.; Investigation, T.M. and J.S.; Methodology, T.M., G.H., Anush Karamyan and J.S.; Project administration, E.T. and J.S.; Resources, E.T. and J.S.; Writing—original draft, T.M. and J.S.; Writing—review & editing, G.H., A.K., F.O., C.B., M.L., V.C. and H.F.N.

Funding: This research received no external funding.

Conflicts of Interest: The authors declare no conflict of interest.

Abbreviations

ADEM	Acute disseminated encephalomyelitis
AE	Autoimmune encephalitis
CNS-ID	Central nervous system-inflammatory disorders
GC	Glucocorticosteroids
IVIG	Intravenous immunoglobulins
LE	Limbic encephalitis
MS	Multiple sclerosis
NMDAR	N-methyl-D-aspartate-receptor encephalitis
NMOSD	Neuromyelitis optica spectrum disorder
NICU	Neurological intensive care unit
ON	Optic neuritis
TPE	Therapeutic plasma exchange

References

1. Mauch, E.; Zwanzger, J.; Hettich, R.; Fassbender, C.; Klingel, R.; Heigl, F. Immunoadsorption for steroid-unresponsive multiple sclerosis-relapses: clinical data of 14 patients. *Nervenarzt* **2011**, *82*, 1590–1595. [CrossRef] [PubMed]

2. Trebst, C.; Bronzlik, P.; Kielstein, J.T.; Schmidt, B.M.; Stangel, M. Immunoadsorption Therapy for Steroid-Unresponsive Relapses in Patients with Multiple Sclerosis. *Blood Purif.* **2012**, *33*, 1–6. [CrossRef] [PubMed]

3. Correia, I.; Sousa, L. Plasma exchange in severe acute relapses of multiple sclerosis—Results from a Portuguese cohort. *Mult. Scler. Relat. Disord.* **2018**, *19*, 148–152. [CrossRef] [PubMed]

4. Graus, F.; Titulaer, M.J.; Balu, R.; Benseler, S.; Bien, C.G.; Cellucci, T.; Cortese, I.; Dale, R.C.; Gelfand, J.M.; Geschwind, M.; et al. A clinical approach to diagnosis of autoimmune encephalitis. *Lancet Neurol.* **2016**, *15*, 391–404. [CrossRef]

5. Ehrlich, S.; Fassbender, C.M.; Blaes, C.; Finke, C.; Gunther, A.; Harms, L.; Hoffmann, F.; Jahner, K.; Klingel, R.; Kraft, A.; et al. Therapeutic apheresis for autoimmune encephalitis: A nationwide data collection. *Nervenarzt* **2013**, *84*, 498–507. [CrossRef]

6. Magana, S.M.; Keegan, B.M.; Weinshenker, B.G.; Erickson, B.J.; Pittock, S.J.; Lennon, V.A.; Rodriguez, M.; Thomsen, K.; Weigand, S.; Mandrekar, J.; et al. Beneficial plasma exchange response in central nervous system inflammatory demyelination. *Arch. Neurol.* **2011**, *68*, 870–878. [CrossRef]

7. Lindner, M.; Klotz, L.; Wiendl, H. Mechanisms underlying lesion development and lesion distribution in CNS autoimmunity. *J. Neurochem.* **2018**, *146*, 122–132. [CrossRef]

8. Rommer, P.S.; Milo, R.; Han, M.H.; Satyanarayan, S.; Sellner, J.; Hauer, L.; Illes, Z.; Warnke, C.; Laurent, S.; Weber, M.S.; et al. Immunological Aspects of Approved MS Therapeutics. *Front. Immunol.* **2019**, *10*, 207–220. [CrossRef]

9. Harutyunyan, G.; Hauer, L.; Dünser, M.W.; Karamyan, A.; Moser, T.; Pikija, S.; Leitinger, M.; Novak, H.F.; Trinka, E.; Sellner, J. Autoimmune Encephalitis at the Neurological Intensive Care Unit: Etiologies, Reasons for Admission and Survival. *Neurocrit. Care.* **2017**, *27*, 82–89. [CrossRef]

10. Titulaer, M.J.; McCracken, L.; Gabilondo, I.; Armangue, T.; Glaser, C.; Iizuka, T.; Honig, L.S.; Benseler, S.M.; Kawachi, I.; Martinez-Hernandez, E.; et al. Treatment and prognostic factors for long-term outcome in patients with anti-NMDA receptor encephalitis: an observational cohort study. *Lancet Neurol.* **2013**, *12*, 157–165. [CrossRef]

11. Reeves, H.M.; Winters, J.L. The mechanisms of action of plasma exchange. *Br. J. Haematol.* **2014**, *164*, 342–351. [CrossRef] [PubMed]

12. DeSena, A.D.; Noland, D.K.; Matevosyan, K.; King, K.; Phillips, L.; Qureshi, S.S.; Greenberg, B.M.; Graves, D. Intravenous methylprednisolone versus therapeutic plasma exchange for treatment of anti-n-methyl-d-aspartate receptor antibody encephalitis: A retrospective review. *J. Clin. Apher.* **2015**, *30*, 212–216. [CrossRef] [PubMed]

13. Norda, R.; Berséus, O.; Stegmayr, B. Adverse events and problems in therapeutic hemapheresis. A report from the Swedish registry. *Transfus. Apher. Sci.* **2001**, *25*, 33–41. [CrossRef]

14. Kleiter, I.; Gahlen, A.; Borisow, N.; Fischer, K.; Wernecke, K.D.; Hellwig, K.; Florence Pache, K.R.; Pache, F.; Ruprecht, K.; Havla, J.; et al. Apheresis therapies for NMOSD attacks: A retrospective study of 207 therapeutic interventions. *Neurol. Neuroimmunol. Neuroinflamm.* **2018**, *5*, e504. [CrossRef] [PubMed]

15. Lipphardt, M.; Mühlhausen, J.; Kitze, B.; Heigl, F.; Mauch, E.; Helms, H.-J.; Müller, G.A.; Koziolek, M.J. Immunoadsorption or plasma exchange in steroid-refractory multiple sclerosis and neuromyelitis optica. *J. Clin. Apher.* **2019**, *34*, 381–391. [CrossRef]

16. Thompson, A.J.; Banwell, B.L.; Barkhof, F.; Carroll, W.M.; Coetzee, T.; Comi, G.; Correale, J.; Fazekas, F.; Filippi, M.; Freedman, M.S.; et al. Diagnosis of multiple sclerosis: 2017 revisions of the McDonald criteria. *Lancet Neurol.* **2018**, *7*, 162–173. [CrossRef]

17. Weinshenker, B.G.; O'Brien, P.C.; Petterson, T.M.; Noseworthy, J.H.; Lucchinetti, C.F.; Dodick, D.W.; Pineda, A.A.; Stevens, L.N.; Rodriguez, M. A randomized trial of plasma exchange in acute central nervous system inflammatory demyelinating disease. *Ann. Neurol.* **1999**, *46*, 878–886. [CrossRef]

18. Keegan, M.; Pineda, A.A.; McClelland, R.L.; Darby, C.H.; Rodriguez, M.; Weinshenker, B.G. Plasma exchange for severe attacks of CNS demyelination: predictors of response. *Neurology* **2002**, *58*, 143–146. [CrossRef]

19. De Bruijn, M.A.A.M.; van Sonderen, A.; van Coevorden-Hameete, M.H.; Bastiaansen, A.E.M.; Schreurs, M.W.J.; Rouhl, R.P.W.; van Donselaar, C.A.; Majoie, M.H.J.M.; Neuteboom, V.O.R.C.I.D.P.F.; Smitt, P.A.E.S.; et al. Evaluation of seizure treatment in anti-LGI1, anti-NMDAR, and anti-GABA(B)R encephalitis. *Neurology* **2019**, *92*, e2185–e2196. [CrossRef]

20. Jiao, Y.; Cui, L.; Zhang, W.; Zhang, Y.; Wang, W.; Zhang, L.; Tang, W.; Jiao, J. Plasma Exchange for Neuromyelitis Optica Spectrum Disorders in Chinese Patients and Factors Predictive of Short-term Outcome. *Clin. Ther.* **2018**, *40*, 603–612. [CrossRef]

21. Manguinao, M.; Krysko, K.M.; Maddike, S.; Rutatangwa, A.; Francisco, C.; Hart, J.; Chong, J.; Graves, J.S.; Waubant, E. A retrospective cohort study of plasma exchange in central nervous system demyelinating events in children. *Mult. Scler. Relat. Disord.* **2019**, *35*, 50–54. [CrossRef] [PubMed]

22. Llufriu, S.; Castillo, J.; Blanco, Y.; Ramió-Torrentà, L.; Rio, J.; Valles, M.; Lozano, M.; Castellà, M.D.; Calabia, J.; Horga, A.; et al. Plasma exchange for acute attacks of CNS demyelination: Predictors of improvement at 6 months. *Neurology* **2009**, *73*, 949–953. [CrossRef] [PubMed]

23. Zhang, M.; Li, W.; Zhou, S.; Zhou, Y.; Yang, H.; Yu, L.; Wang, J.; Wang, Y.; Zhang, L. Clinical Features, Treatment, and Outcomes Among Chinese Children with Anti-methyl-D-aspartate Receptor (Anti-NMDAR) Encephalitis. *Front. Neurol.* **2019**, *10*, 596. [CrossRef] [PubMed]

24. Harutyunyan, G.; Hauer, L.; Dünser, M.W.; Moser, T.; Pikija, S.; Leitinger, M.; Novak, H.F.; Trinka, E.; Sellner, J. Risk Factors for Intensive Care Unit Admission in Patients with Autoimmune Encephalitis. *Front. Immunol.* **2017**, *28*, 835. [CrossRef] [PubMed]

25. Huda, S.; Whittam, D.; Bhojak, M.; Chamberlain, J.; Noonan, C.; Jacob, A. Neuromyelitis optica spectrum disorders. *Clin. Med.* **2019**, *19*, 169–176. [CrossRef]

26. Kivity, S.; Agmon-Levin, N.; Zandman-Goddard, G.; Chapman, J.; Shoenfeld, Y. Neuropsychiatric lupus: a mosaic of clinical presentations. *BMC Med.* **2015**, *13*, 43. [CrossRef]

27. Chessa, E.; Piga, M.; Floris, A.; Mathieu, A.; Cauli, A. Demyelinating syndrome in SLE: review of different disease subtypes and report of a case series. *Reumatismo* **2017**, *69*, 175–183. [CrossRef]

28. Aguirre-Valencia, D.; Naranjo-Escobar, J.; Posso-Osorio, I.; Macía-Mejía, M.C.; Nieto-Aristizábal, I.; Barrera, T.; Obando, M.A.; Tobón, G.J. Therapeutic Plasma Exchange as Management of Complicated Systemic Lupus Erythematosus and Other Autoimmune Diseases. *Autoimmune Dis.* **2019**, *2019*. [CrossRef]

29. Pons-Estel, G.J.; Salerni, G.E.; Serrano, R.M.; Gómez-Puerta, J.A.; Plasín, M.A.; Aldasoro, E.; Lozano, M.; Cid, J.; Cervera, R.; Espinosa, G. Therapeutic plasma exchange for the management of refractory systemic autoimmune diseases: Report of 31 cases and review of the literature. *Autoimmun. Rev.* **2011**, *10*, 679–684. [CrossRef]

30. Ehler, J.; Blechinger, S.; Rommer, P.S.; Koball, S.; Mitzner, S.; Hartung, H.-P.; Leutmezer, F.; Sauer, M.; Zettl, U.K. Treatment of the First Acute Relapse Following Therapeutic Plasma Exchange in Formerly Glucocorticosteroid-Unresponsive Multiple Sclerosis Patients—A Multicenter Study to Evaluate Glucocorticosteroid Responsiveness. *Int. J. Mol. Sci.* **2017**, *18*, 1749. [CrossRef]

31. Cortese, I.; Cornblath, D.R. Therapeutic plasma exchange in neurology: 2012. *J. Clin. Apher.* **2013**, *28*, 16–19. [CrossRef] [PubMed]

32. Faissner, S.; Nikolayczik, J.; Chan, A.; Hellwig, K.; Gold, R.; Yoon, M.-S.; Haghikia, A. Plasmapheresis and immunoadsorption in patients with steroid refractory multiple sclerosis relapses. *J. Neurol.* **2016**, *263*, 1092–1098. [CrossRef] [PubMed]

Case Report

Post-Craniopharyngioma and Cranial Nerve-VI Palsy Update on a MS Patient with Major Depression and Concurrent Neuroimmune Conditions

Navzer D. Sachinvala [1,*,†], Angeline Stergiou [2], Duane E. Haines [3,4,5], Armen Kocharian [6] and Andrew Lawton [7]

[1] United States Department of Agriculture, Agricultural Research Service, USDA-ARS, New Orleans, Home 2261 Brighton Place, Harvey, LA 70058, USA
[2] Department of Medicine, Fairfield Medical Center, 401 North Ewing, Lancaster, OH 43130, USA; drangelinestergiou@gmail.com
[3] Department of Neurobiology and Anatomy, Wake Forest School of Medicine, 1 Medical Center Boulevard, Winston-Salem, NC 27157, USA; dhaines@wakehealth.edu
[4] Department of Neurology, Wake Forest School of Medicine, 1 Medical Center Boulevard, Winston-Salem, NC 27157, USA
[5] Department of Neurobiology and Anatomy, The University of Mississippi Medical Center, Jackson, MS 39216, USA
[6] Department of Radiology, Houston Methodist Hospital, 6565 Fannin Street, Houston, TX 77030, USA; akocharian@houstonmethodist.org
[7] Department of Ophthalmology, Ochsner Hospital, 1514 Jefferson Highway, New Orleans, 70121, USA; andrew.lawton@ochsner.org
* Correspondence: Sachinvala@aol.com
† This author is retired.

Received: 25 August 2019; Accepted: 11 October 2019; Published: 17 October 2019

Abstract: We report the case of a male multiple sclerosis (MS) patient with type 2 diabetes (T2D), asthma, major depression (MD or major depressive disorder, MDD), and other chronic conditions, after his recent difficulties with craniopharyngioma and cranial nerve-VI (CN6) palsy. In addition, we show magnetic resonance image and spectroscopy (MRI, MRS), Humphrey's Visual Field (HVF), and retinal nerve fiber layer thickness (RNFLT) findings to explain the changes in the patient's health, and discuss the methods that helped/help him sustain productivity and euthymia despite long-standing problems and new CNS changes.

Keywords: major depression; multiple sclerosis; bupropion; *S*-adenosylmethionine; vitamin D3; yoga; craniopharyngioma; fractionated stereotactic radiation treatments; sphenoid sinusitis; cranial nerve-VI palsy

---

1. Introduction

Earlier we discussed the case of a male multiple sclerosis (MS) patient (primary author, N.D.S.) who has many concurrent conditions including asthma, type 2 diabetes (T2D), Ehlers–Danlos syndrome (EDS), infections that take long to heal, and major depression (MD or MDD), and listed the methods he learned to be euthymic, by:

- Quelling pain and disability.
- Enduring medication side effects and noting what worked and what stopped working after a while.
- Subduing MD with bupropion, and supplements *S*-adenosylmethionine (SAMe) and vitamin-D3 (vit-D3).

- Maintaining routines for all medications, self-hypnosis, yoga, and physical exercises to stay fit and lucid.
- And academically studying symptoms and potential remedies for his ailments to engage in physician-assisted autoexperiments to discover solutions that provide him lasting relief [1].

Furthermore, the same methods that helped him attain and maintain euthymia with existing illnesses were successful in enabling him to be positive and productive despite new added medical diagnoses. For example, they worked:

- During his craniopharyngioma diagnosis and bitemporal vision loss.
- Through fractionated stereotactic radiation treatments (FSRT) to shrink his perisellar tumor and regain peripheral vision.
- And in his bout with cranial nerve-VI (CN6) palsy, diplopia, and their resolution.

In this report, we update our earlier case report with magnetic resonance imaging (MRI), magnetic resonance spectroscopy (MRS), Humphrey's Visual Field (HVF), and retinal nerve fiber layer thickness (RNFLT) data to explain the patient's status with regards to his MD, MS, tumor, vision, CN6 palsy, and overall health to suggest that even with new health difficulties the above methods he chanced upon continue to keep him positive and productive [1]. We write our reports to encourage readers (scientists, physicians, and patients) to discuss their own experiences in peer-reviewed forums to create insights that could empower the chronically ill to individualize their care with physician involvement and improve the quality of their lives.

2. Case Report

In Reference [1], we reported the case of a male MS patient (primary author, N.D.S.), now 63, who has battled MD since childhood. His early disablements, which continue to date, include: amblyopia (left eye, OS), dyslexia, color blindness, asthma, urticaria, allergies to multiple foods, drugs, and narcotic pain medications, and susceptibility to infections that take weeks to resolve with antibiotics [1]. At 19, he was diagnosed selective immunoglobulin-M (IgM) deficient [2], and by his mid-40 s, required multiple rhinoplasties for apnea, lens replacements for premature cataracts, and uvula removal for swelling and apnea. He belongs to a closed Indo-Iranian minority, Zoroastrians (aka Parsees or Parsis), wherein enforced consanguineous relations for over a century have produced some children with neuroimmune maladies, e.g., Ehlers–Danlos syndrome [3], multiple sclerosis [4,5], cancer [6], and Alzheimer's, cardiovascular, and Parkinson's diseases [6–8]. Notwithstanding health limitations, he grew-up in and enjoys a stable family, and cherished a productive career as an accomplished scientist [9]. At 49 (2005), he developed fulminant MS that left him depressed (Beck Depression Inventory, BDI scores 30–40) [10], mobility impaired, and unable to balance and coordinate his hands. Despite intense therapies, three years later, he accepted disability retirement, and to-date is unable to coordinate his hands to manipulate small objects, play his piano and guitar, and tolerate hot environments [1].

Soon after MS was diagnosed, the patient received guidance from his former teachers and colleagues who were with him in graduate school in Iran (1977–1979), and to whom this paper is dedicated, to:

- Academically study his immune-related illnesses through university courses, textbooks, and journals.
- Maintain regular depression inventories.
- Have consistent routines for yoga and self-hypnosis, sleep, and fitness exercises.
- Note daily changes in serum glucose levels so that with prescribed medications (see Box 2 in Reference [1]), diet, and above routines his glucose levels are maintained between 75 and 120 mg/dL [11–14].
- Quell pain and associated anger with yoga and self-hypnosis, and use NSAIDs when pain becomes disabling.

- Judiciously use prescribed medications and note what worked and what aggravated liver enzyme levels, skin rashes, or other side reactions.
- And study new literature methods that could improve health, mood, and physical and cognitive functioning to engage in physician-assisted autoexperiments that afford sustained relief.

By employing the above approaches between 2005 and 2015, the patient with much effort regained several lost functions, e.g., emerged from bed and wheelchair to walking briskly with crutches. Likewise, after years of experiments with antidepressants, he discovered that bupropion augmented with *S*-adenosylmethionine (SAMe) and vitamin-D3 (vit-D3), in addition to other medications (see reference [1], Box 2), and routines keep him pain-free, alert, and productive.

In late April 2017, he complained of heat intolerance, dizziness, falling, headaches, and blurry vision. Tests over the year revealed that he had craniopharyngioma, a benign perisellar tumor that was growing in place, pressed against his pituitary, anterior optic chiasm, surrounding tissues, and disrupted ophthalmic functions [15]. By late November, his conjoined papillary and adamantinous tumors had enlarged to ~2.3 and 0.7 mL, respectively, caused headaches and bitemporal hemianopsia (peripheral vision loss), and deteriorated R/S vision (from 20/25 and 20/40 in Sept-2016, to 20/60 and 20/100 in Nov-2017). However, his hypothalamic pituitary and adrenal, HPA, and axis hormone levels were unaffected [1,15–18].

From April 2017 to September 2018, the patient could not operate his automobile, and to-date reads text with the Kurzweil Reading Program using ≥14-point font projected on a 140 cm TV monitor, ~1 m from him [19].

Throughout the ordeal, he kept his BDI scores as low as possible (~mid 20 s) with daily: bupropion, SAMe, vit-D3, Yoga, self-hypnosis, exercises, regular schedule for other medications (see box 2 in Reference [1]), and visual imagery of himself scuba diving among coral and marine animals, studying, publishing, lecturing, and maintaining a positive outlook [20].

Given the patient's conditions, asthma, MS, T2D, EDS, and susceptibility to infections, almost all neurosurgeons he consulted advised to ablate his tumors with 30 rounds of fractionated stereotactic radiation treatments (FSRTs), instead of surgery [21]. Gradually his difficulties lessened.

Herein we explain the before and after statuses of his tumor, vision, recent bout with CN6 palsy, and diplopia (Figure 1, Figure 2 and Table 1), and conclude by describing his overall health, mood, and the challenges he must overcome. At present, challenges from old and new ailments are managed, by first becoming free of pain and disability (or lessening them), and then, engrossing self intellectually to problem-solving, so that new and long-standing maladies become puzzles to solve rather than factors that perpetuate helplessness.

3. Materials and Methods

3.1. Brain Imaging and Spectroscopy

Brain images and spectra of the tumor were obtained by coauthor AK at the Houston Methodist Hospital, on a General Electric (GE) Healthcare (Waukesha, WI, USA), Model: MR750, 3Tesla instrument with software level DV 25.1 R03. Images were recorded without and with intravenous gadolinium contrast to show new lesions, changes in existing MS lesions with time, as well as postradiation size changes in the suprasellar cyst using the pituitary diffusion imaging protocol. Single voxel MRS data on the tumor were collected using short and long times to echo (TE = 35 ms, and TE = 144 ms, respectively) [15–18].

3.2. Vision Changes

Vision changes were studied annually by coauthor AL at the Ochsner Hospital, Jefferson Avenue, New Orleans. RNFL measurements were performed using Spectralis® Tracking Laser, Model HRA/Spectralis with 870 nm diode laser, Heidelberg Engineering, Inc. (Franklin, MA, USA). A circular scan and a 3.4 mm circle were centered around the optic disc of each eye, and RNFLTs in microns were analyzed against data from age- and sex-matched Caucasians in the database (Table 1). Humphrey's Visual Field data were recorded using Ziess Humphrey HF analyzer Model 750i, Meditec Inc., Dublin, CA, USA. After patient data, time of test, and R- or S-eye to be tested were recorded, the Swedish interactive threshold algorithm, SITA, was set to perform fast automated kinetic perimetry (duration ~3 min). The stimulus was white light of intensity 30 dB that illuminated an area of ~2 mm^2 every 200 milliseconds on a chamber screen with background light intensity set at 31.5 apostilbs (ASB). The fixation monitor was gaze at blind spot, the target was central, and threshold was set at 24-2 (which is 24 degrees-temporal, 30 degrees-nasal, to record data at 54 points). Sensitivity at foveae for right and left eyes were measured at 38 and 37 decibels (dB), respectively. Right and left pupil diameters were 4.7 and 4.5 mm, respectively, and correction for both eyes was +4.75 diopter (D). Moreover, on most occasions, fixation losses for both eyes were 2/10, which suggested data were reliable. Thereafter, patient data were compared against data from age-matched Caucasian males in the HVF database (Figure 1).

4. Results and Discussion

4.1. Brain Imaging and Spectroscopy

The patient's MRIs before and after FSRT showed occipital lobe bending (OLB), a sign of chronic MD [1,22]. His conjoined tumor volume before FSRT was ~3 mL (Figure 1A). However, 15 months after FSRT, only the papillary tumor could be seen (V = 0.18 mL), and proton (^{1}H) spectra of the papilloma, before and after FSRT, showed the same alkyl resonances of cholesterol and fatty acids at δ ~1.5 ppm and of compounds with hydroxyl groups at δ ~4 ppm, likely water and inositol. In addition, proton spectra of adamantinous tumors are undefined. Since no other perisellar tumor shows these characteristics, his CTs, MRIs, MRSs, and endocrine workup were used to tentatively diagnose craniopharyngioma without surgical biopsy. This was supported by the patient's inability to efficiently control infections and asthma, and his difficulties with EDS, T2D, MS, and other immune problems [16–18].

4.2. Vision Changes

At maximum tumor volume (ca 3 mL, Figure 1A, December 2017), his R/S vision was 20/70 and 20/100, he had bitemporal hemianopsia (Figure 1B,C), and his left-temporal retinal nerve fiber layer thickness (RNFLT) was 44 microns (Table 1), and the statistical average for RNFLT for men his age was 72 microns. This suggested that in December 2017, the patient's temporal RNFLT was (1−44/72) ×100) ~39% below average for healthy Caucasian men of his age; however, RNFLTs in other sectors were within normal range at that time [23–25]. Statistical values of age- and sex-matched RNFLTs were from the Ziess Humphrey HF analyzer Model 750i, Meditec Inc., Dublin, CA database. At minimum tumor volume 2019 (ca 0.18 mL), his R/S vision had improved to 30/40 and 20/60, respectively; however, his RNFLT was 40 microns (Table 1), which was (1−40/72) ×100) ~ 44% below average thickness, and may have been due to inflammation of his sphenoid sinuses in April 2019 (Table 1).

Figure 1B–D show Humphrey's visual field (HVF) patterns for right (OD) and left (OS) eyes. The long red arrows point to blank regions (blind spots) in the temporal hemifields in Figure 1B,D. Statistical data in 1C from the HVF database enabled comparison of the patient's retinal sensitivity to light with male subjects of his age in the HVF database. Figure 1B showed that in 2017, the patient had bitemporal hemianopsia before FSRTs (because dark patterns appeared mostly in the R and L temporal sides of the grids). It is well known that when tumors grow just above the pituitary and press against the anterior optic chiasm, patients experience bitemporal vision loss. For an excellent tutorial and discussion on visual deficits caused by lesions in visual pathways, the reader is advised to consult Reference [25], pages 264–267. After FSRT, Figure 1D data showed that the patient had central scotomas (blue arrows, points along the central meridians of retinae where vision was impaired). HVF pattern deviation maps are used to compare a patient's retinal sensitivity to 30 decibel light with statistical information from healthy age- and sex-matched subjects in the HVF database. HVF tests evaluate retinal sensitivity at 54 points, in superior (S) and inferior (I) quadrants of temporal (T) and nasal (N) hemifields. Vision losses shown in HVF maps are graded in terms of <5%, <2%, <1%, and <0.5%. So, a four-dot rectangular pattern with a <5% designation in Figure 1C, when present in the patient's maps (Figure 1B,D), indicated that in that region of his retina, 95% of age-matched men had better vision than the patient. Likewise, the blackened rectangle with a <0.5% designation, when present in his maps Figure 1B,D, showed that in that retinal region, 99.5% of age-matched men had better vision than him. A patient's HVF pattern deviation maps ascribe her/his visual field defects in terms of anopsias and scotomas. Anopsias describe vision loss in quarter (quadrantanopia) or half fields (hemianopsias, Figure 1B), and scotomas describe vision loss at individual test points (Figure 1D, blue arrows). Since the highest collection of HVF patterns shown in Figure 1C was present in both temporal S and I quadrants in Figure 1B, his data showed that he had bitemporal hemianopsia before radiation treatments. Fifteen months after FSRT, as his tumor shrank, his visual acuity improved, and he showed central scotomas in both eyes (Figure 1D), which in time could resolve as inflammation lessens [15–18,26].

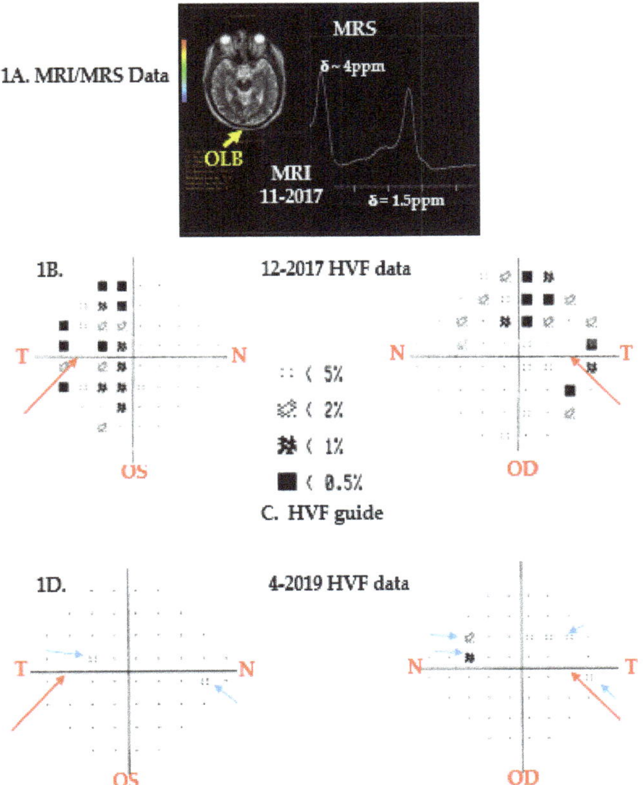

Figure 1. Patient's tumor and Humphrey's Visual Field Data.

The RNFL is a layer of retinal nerve fibers (axons) that eventually form the optic nerve, which enters the eye at the optic cup (blind spot) and proceeds to the brain. Table 1 compares the patient's RNFLTs in microns (μm) in S, I, N, and T regions (or sectors) of both retinas, with statistical data on healthy Caucasian men of his age in the RNFLT database. RNFLTs are measured using optical coherence tomography (OCT). Superior and inferior sectors of the eye are further divided into temporal superior and inferior (TS and TI), and nasal superior and inferior (NS and NI) sectors. The letter "G" in Table 1 indicates global or foveal RNFLT. Foveae are pits in retinal surfaces where visual acuity is highest due to the highest concentration of cone- and rod-shaped neuron cell bodies therein. Patient's RNFLTs that were within normal limits in comparison with statistical values from matched subjects are shown in black font. Those in blue bold font were borderline thicknesses, and those in red bold font indicated below normal retinal thicknesses (calculation shown above). Overall, his data showed that this patient could gradually lose central and temporal vision in his left eye, and needs to address the problem with a retina specialist [23–25,27].

Table 1. Patient's retinal nerve fiber layer thickness (RNFLT μm) data for December 2017 and April 2019 compared with statistical information from age- and sex-matched individuals in the database.

Right Eye (OD) Dec-2017						
Sector	**T**	**TS**	**TI**	**G**	**N**	**NS**	**NI**
Patient	85	176	136	101	72	93	91
Statistic	72	131	138	96	72	102	104

Overall Patient OD RNFLT: S = 135; I = 114; T 85; and N = 72

Left Eye (OS) Dec-2017						
Patient	**44**	124	126	90	82	116	99
Statistic	72	131	138	96	72	102	104

Overall Patient OS RNFLT: S = 120; I = 112; T = 44; N = 82

Right Eye (OD) Apr-2019						
Sector	**T**	**TS**	**TI**	**G**	**N**	**NS**	**NI**
Patient	55	127	128	87	70	94	94
Statistic	72	131	138	96	72	102	104

Overall Patient OD RNFLT: S = 111; I = 111; T = 55; and N = 70

Left Eye (OS) Apr-2019						
Patient	**40**	117	119	**80**	63	102	98
Statistic	72	131	138	96	72	102	104

Overall Patient OS RNFLT: S = 109; I = 108; T = 40; and N = 63

Numbers in bold red font showed that his left RNFLTs were 39% and 55% below average for years 2017 and 2019, respectively, in comparison with age-matched Caucasian men in the database. Value in bold blue font showed that his left foveal (G) RNFLT in April 2019 was 16% below average, but was within normal range in December 2017 (only ~6% below average). RNFLTs in black font were comparatively within normal limits.

Losses in RNFL thicknesses (Table 1) and visual fields (Figure 1) are known to occur with age in both women and men and are exacerbated in aging patients with chronic inflammatory comorbidities (obesity, diabetes, asthma, multiple sclerosis, and/or brain tumors). Furthermore, the likelihood of craniopharyngioma recurrence is also higher in obese diabetics. Therefore, it is imperative for this patient to curb inflammation, reduce weight, keep daily serum glucose values within normal ranges, and annually monitor his ophthalmic, metabolic, and CNS (brain and spine) changes to preclude recurrence of tumor and/or further retinal damage [15–18,23–25,27].

4.3. Bout with CN6 Palsy

In November 2018, the patient experienced sinus pain and intracranial pressure that did not abate with yoga, self-hypnosis, saline rinses, and naproxen. He experienced diplopia and orthostatic difficulties when fixating on vertical edges of walls to balance when standing-up, which to the patient felt like the two same vertical edges were side by side (binocular double vision). In addition, he had a slight convergent right eye squint, that is, his right eye was turning slightly inwards towards his nose. Furthermore, while driving, he saw phantom traffic lane lines obliquely cutting into his lane from the right. He stopped driving and managed diplopia with a right eye patch. Soon diplopia worsened, interfered with reading and daily functions; and by month's end (December), he was prescribed a 10-diopter base out Fresnel lens prism over his right eye (by coauthor, A.L.) after his misalignment was measured by alternate cover testing with hand-held prisms [28]. In addition, his neuro-ophthalmologist advised that the patient's diplopia might be due to right CN6 palsy, triggered by a sinus infection that might be causing his pain and intracranial pressure. Alternatively, given the patient's medical history, it could be due to his chronic diabetes and/or MS, and that such palsies are almost always temporary. At New Years, as nasal lavage thickened, appeared dark, and became more painful, the patient was prescribed amoxicillin to reduce infection. By late January, he was symptom free [22,26,28,29]. In March 2019, sinus pain, pressure, and diplopia returned, and at that time he was undergoing his postradiation therapy follow-up. His transaxial and coronal MRIs in addition to his tumor's status showed that his left sphenoid sinus was more inflamed than the right (Figure 2A and B, yellow triangles). Moreover, while MRIs showed no congestion and no legion(s) in the right pons CN6 region (Figure 2C, orange arrow), the effects of intracranial pressure on CN6 could not be ruled out. Since temporary diplopia is known to be due to ipsi- as well as contralateral CN6 palsies induced by sphenoid sinusitis and increased intracranial pressure, it was conjectured that his symptoms will subside as inflammation diminishes. Later that month, his pain and diplopia were resolved with saline rinses, naproxen, and yoga. CN3 (oculomotor nerve) palsy was ruled out because the patient's right eyelid was not droopy (no ptosis), the right pupil was not dilated, and the eye was not shifted temporally. CN4 (trochlear nerve) palsy was ruled out because diplopia did not worsen when looking down. In addition, his March MRIs showed that his tumor was greatly reduced, his MS was stable, and he had no new CNS lesions or infarctions [22–29].

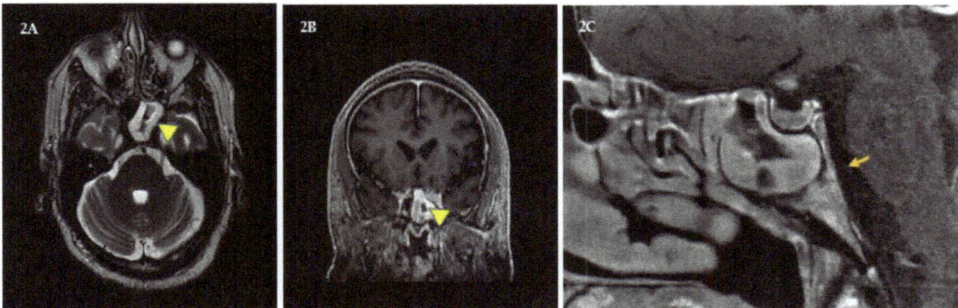

Figure 2. Sphenoid sinusitis. (**A** and **B**) shows transaxial and coronal views of inflammation in the patient's left sphenoid sinus (yellow triangles). (**C**) shows a sagittal view of his congested right sphenoid sinus and the plausible path of his right sixth cranial (abducens) nerve (orange arrow) along the pons. However, congestion in the right pons cranial nerve-VI (CN6) region was not seen, and increased intracranial pressure as cause of CN6 palsy was not ruled out.

5. Conclusions

This patient has struggled with multiple neuroimmune illnesses since childhood and as new difficulties emerge with age. Most of his maladies have a common theme, they are inflammatory, i.e., obesity, diabetes, asthma, MS, MDD, brain tumor, and the like. To control diabetes and obesity, he monitors his AM fasting glucose, which for example from January 1 to October 3, 2019, showed that his average (± standard error) serum glucose was 109.2 (± 1.3 mg/dL) and his percent glycated hemoglobin, A1C, was 5.9 (± 0.05%). Furthermore, his min–max glucose and % A1C values ranged between 71 and 188 mg/dL and 4.4% and 8.9%, respectively. These data were collected on 271 out of 275 calendar days. At present, with yoga, self-hypnosis, gym exercises, sleep, and medications (see Box 2, in Reference [1]), he manages his illnesses, weighs 231 lb (134.1 kg, BMI = 31.7), walks a mile in 20–22 min with crutches, keeps consistent routines, performs household tasks, operates his automobile, swims, scuba dives, participates in writing and editing scientific articles and reviews with former associates, and feels euthymic (BDI January 1–October 3, 2019, was between 6 and 17) [10–14]. His challenges are: reducing weight (to <200 lb, <91 kg, BMI <27), preventing tumor recurrence, not allowing further degradation of his retinae (by working with specialists), stopping exacerbation of asthma and MS symptoms, maintaining serum sugar levels between 75 and 120 mg/dL, and improving balance (to prevent falls) and hands coordination for small motor functions [23–25,27]. Important lessons he has learned to help him manage self are: to not give into overwhelm, discouragement, and anger when an existing disease is aggravated, or a new health problem arises. He understands that these will happen with age, given his background, and overall health. So, regardless the malady, he must act to quell pain, disability, and overwhelm with NSAIDs, stimulants (teas, ginko, coffee, etc.), dietary supplements, physically exhausting self in gym to sleep, and psychological methods. The patient has learned to carefully note how his prescribed medications work (i.e., absorption, distribution, metabolism, side effects, and elimination) and endures treatment hardships to heal. He enjoys working with individuals that help him discover new medications, supplements, and psychological methods to quickly emerge from illness, depression, and helplessness. He maintains routines for medications, self-hypnosis, yoga, physical exercise, mental imagery, and derives pleasure from enjoyable activities (like scuba diving). Furthermore, he academically studies symptoms and new treatments for extant and new problems, discusses them with physicians, and engages in guided autoexperiments that produce desired results.

While we have, at present, no control over our genetic inheritances, we can momentarily accept status quo, and then manage phenotype to live rewarding lives. Finally, we want to see similar case reports from readers (scientists, physicians, and patients) so much insight is created in the literature to help chronically ill patients personalize care for their health and wellbeing to live productive rewarding lives.

Author Contributions: Using his own clinical records, N.D.S. conceived, wrote, and edited the entire manuscript. A.S., D.E.H., A.K., and A.L. were invited as coauthors because of their expertise in psychiatry, neuroscience, radiology, and ophthalmology. In addition, A.K. and A.L. provided the radiology and ophthalmology data, respectively. All authors contributed toward analysis of the case and clinical data, drafting, and revising the paper, and all agree to be accountable for all aspects of this work.

Funding: This research received no external funding.

Acknowledgments: We thank: Bin Teh, Steve Fung, and David Baskin, Radiation Oncology, Radiology, and Neurosurgery Departments, Houston Methodist Hospital for managing NS's craniopharyngioma; Bridget Bagert and Diana Phan, Ochsner Hospital New Orleans, for managing NS's MS and overall health; Keki and Mazda Turel, Department of Neurosurgery, Bombay University for many helpful discussions on CNS changes with craniopharyngioma, diabetes, diplopia, MS, and MDD; and Naozumi Teramoto, Chiba Institute of Technology, for help with many references and edits that improved the quality of our manuscript.

Dedication: Dedicated to my teachers and friends then with me at the Institute of Biochemistry and Biophysics, University of Tehran, Iran (1977–1979).

Conflicts of Interest: The authors declare no conflict of interest.

References

1. Sachinvala, N.D.; Stergiou, A.; Haines, D.E. Remitting long-standing major depression in a multiple sclerosis patient with several concurrent conditions. *Neuropsychiatr. Dis. Treat.* **2018**, *14*, 2545–2550. [CrossRef] [PubMed]
2. Chovancova, Z.; Kralickova, P.; Pejchalova, A.; Bloomfield, M.; Nechvatalova, J.; Vlkova, M.; Litzman, J. Selective IgM Deficiency: Clinical and Laboratory Features of 17 Patients and a Review of the Literature. *J. Clin. Immunol.* **2017**, *37*, 559–574. [CrossRef] [PubMed]
3. Kose-Ozlece, H.; Ilik, F.; Huseyinoglu, N. Coexistence of Ehlers-Danlos syndrome and multiple sclerosis. *Iran. J. Neurol.* **2015**, *14*, 116–117. [PubMed]
4. Tripathy, V.; Reddy, B.M. Present status of understanding on the G6PD deficiency and natural selection. *J. Postgrad. Med.* **2007**, *53*, 193–202.

5. Naser Moghadasi, A. Multiple sclerosis in Parsis: A historical issue. *Iran. J. Public Health* **2014**, *43*, 387–388.
6. López, S.; Thomas, M.G.; van Dorp, L.; Ansari-Pour, N.; Stewart, S.; Jones, A.L.; Jelinek, E.; Chikhi, L.; Parfitt, T.; Bradman, N.; et al. The Genetic Legacy of Zoroastrianism in Iran and India: Insights into Population Structure, Gene Flow, and Selection. *Am. J. Hum. Genet.* **2017**, *101*, 353–368. [CrossRef]
7. Ahmadian-Attari, M.M.; Ahmadiani, A.; Kamalinejad, M.; Dargahi, L.; Shirzad, M.; Mosaddegh, M. Treatment of Alzheimer's disease in Iranian traditional medicine. *Iran. Red Crescent Med. J.* **2014**, *17*, e18052. [CrossRef]
8. Vazquez-Vidal, I.; Chittoor, G.; Laston, S.; Puppala, S.; Kayani, Z.; Mody, K.; Comuzzie, A.G.; Cole, S.A.; Voruganti, V.S. Assessment of cardiovascular disease risk factors in a genetically homogenous population of Parsi Zoroastrians in the United States: A pilot study. *Am. J. Hum. Biol.* **2016**, *28*, 440–443. [CrossRef]
9. Available online: https://patentscope.wipo.int/search/en/andthentypeSachinvala (accessed on 12 October 2019).
10. Benedict, R.H.; Fishman, I.; McClellan, M.M.; Bakshi, R.; Weinstock-Guttman, B. Validity of the Beck Depression Inventory-Fast Screen in multiple sclerosis. *Mult. Scler. J.* **2003**, *9*, 393–396. [CrossRef]
11. Cramer, H.; Lauche, R.; Langhorst, J.; Dobos, G. Yoga for depression: A systematic review and meta-analysis. *Depress. Anxiety* **2013**, *30*, 1068–1083. [CrossRef]
12. Otani, A. Eastern meditative techniques and hypnosis: A new synthesis. *Am. J. Clin. Hypn.* **2003**, *46*, 97–108. [CrossRef] [PubMed]
13. Tiers, M. *Integrative Hypnosis a Comprehensive Course in Change*; Melissa Tiers Publisher: New York, NY, USA, 2010.
14. Stetka, B. Available online: https://www.scientificamerican.com/article/do-vitamins-and-supplements-make-antidepressants-more-effective/ (accessed on 26 April 2016).
15. Osborn, A.G.; Salzman, K.L.; Jhaveri, M.D. Craniopharyngioma. In *Diagnostic Imaging: Brain*, 3rd ed.; Elsevier: Philadelphia, PA, USA, 2016; pp. 1048–1051.
16. Einstien, A.; Virani, R.A. Clinical Relevance of Single-Voxel (1) H MRS Metabolites in Discriminating Suprasellar Tumors. *J. Clin. Diagn. Res.* **2016**, *10*, TC01–TC04. [CrossRef] [PubMed]
17. Chernov, M.F.; Kawamata, T.; Amano, K.; Ono, Y.; Suzuki, T.; Nakamura, R.; Muragaki, Y.; Iseki, H.; Kubo, O.; Hori, T.; et al. Possible role of single-voxel (1) H-MRS in differential diagnosis of suprasellar tumors. *J. Neuro-Oncology* **2009**, *91*, 191–198. [CrossRef] [PubMed]
18. Saeger, W. New aspects of tumor pathology of the pituitary. *Pathologe* **2015**, *36*, 293–300. [CrossRef] [PubMed]
19. Schmitt, A.J.; McCallum, E.; Hawkins, R.O.; Stephenson, E.; Vicencio, K. The effects of two assistive technologies on reading comprehension accuracy and rate. *Assist. Technol.* **2019**, *31*, 220–230. [CrossRef] [PubMed]
20. de Voogd, E.L.; de Hullu, E.; Burnett-Heyes, S.; Blackwell, S.E.; Wiers, R.W.; Salemink, E. Imagine the bright side of life: A randomized controlled trial of two types of interpretation bias modification procedure targeting adolescent anxiety and depression. *PLoS ONE* **2017**, *12*, e0181147. [CrossRef]
21. Barber, S.M.; Teh, B.S.; Baskin, D.S. Fractionated Stereotactic Radiotherapy for Pituitary Adenomas: Single-Center Experience in 75 Consecutive Patients. *Neurosurgery* **2016**, *79*, 406–417. [CrossRef]
22. Koch, K.; Schultz, C.C. Clinical and pathogenetic implications of occipital bending in depression. *Brain* **2014**, *137*, 1576–1578. [CrossRef]
23. Takis, A.; Alonistiotis, D.; Ioannou, N.; Mitsopoulou, M.; Papaconstantinou, D. Follow-up of the retinal nerve fiber layer thickness of diabetic patients type 2, as a predisposing factor for glaucoma compared to normal subjects. *Clin. Ophthalmol.* **2017**, *11*, 1135–1141. [CrossRef]
24. Henricsson, M.; Heijl, A. Visual fields at different stages of diabetic retinopathy. *Acta Ophthalmol.* **1994**, *72*, 560–569. [CrossRef]
25. Haines, D.E. *Neuroanatomy in Clinical Context: An Atlas of Structures, Sections, Systems, and Syndromes*; 9ED; Walter Kluwer Health: Philadelphia, PA, USA, 2015.
26. Azarmina, M.; Azarmina, H. The six syndromes of the sixth cranial nerve. *J. Ophthalmic Vis. Res.* **2013**, *8*, 160–171. [PubMed]
27. Wijnen, M.; Olsson, D.S.; van den Heuvel-Eibrink, M.M.; Hammarstrand, C.; Janssen, J.A.M.J.L.; van der Lely, A.J.; Johannsson, G.; Neggers, S.J.C.M.M. Excess morbidity and mortality in patients with craniopharyngioma: A hospital-based retrospective cohort study. *Eur. J. Endocrinol.* **2018**, *178*, 93–102. [CrossRef] [PubMed]
28. Weblink Cover Tests. Available online: https://www.aao.org/bcscsnippetdetail.aspx?id=5051fd44-4106-4b1a-bf19-00566baa0b07 (accessed on 12 October 2019).

Brain Sci. **2019**, 9, 281

29. Siu, J.; Sharma, S.; Sowerby, L. Unilateral isolated sphenoid sinusitis with contralateral abducens nerve palsy—A rare complication treated in a low-resource setting. *J. Otolaryngol. Head Neck Surg.* **2015**, *1*, 44–49. [CrossRef] [PubMed]

Case Report

Autoimmune Encephalitis and CSF Anti-GluR3 Antibodies in an MS Patient after Alemtuzumab Treatment

Maria Chiara Buscarinu [1], Arianna Fornasiero [1], Giulia Pellicciari [2], Roberta Reniè [2], Anna Chiara Landi [2], Alessandro Bozzao [1], Cristina Cappelletti [3], Pia Bernasconi [3], Giovanni Ristori [1,*] and Marco Salvetti [1,4,*]

1   Department of Neurosciences, Mental Health and Sensory Organs, Centre for Experimental Neurological Therapies (CENTERS), Faculty of Medicine and Psychology, Sapienza University, 00189 Rome, Italy; mchiara.buscarinu@gmail.com (M.C.B.); ari.fornasiero@gmail.com (A.F.); alessandro.bozzao@uniroma1.it (A.B.)
2   Faculty of Medicine and Psychology, Sapienza University, 00189 Rome, Italy; giuliapellicciari11@gmail.com (G.P.); roberta.renie@gmail.com (R.R.); landiannachiara@gmail.com (A.C.L.)
3   Neurology IV–Neuroimmunology and Neuromuscular Diseases Unit, Fondazione IRCCS Istituto Neurologico Carlo Besta, 20133 Milan, Italy; cristina.cappelletti@istituto-besta.it (C.C.); pia.bernasconi@istituto-besta.it (P.B.)
4   IRCCS Istituto Neurologico Mediterraneo (INM) Neuromed, 86077 Pozzilli, Italy
*   Correspondence: giovanni.ristori@uniroma1.it (G.R.); marco.salvetti@uniroma1.it (M.S.); Tel.: +39-063-377-6044 (G.R.); +39-063-377-659-94 (M.S.); Fax: +39-063-377-5900 (G.R.); +39-063-377-59-00 (M.S.)

Received: 10 September 2019; Accepted: 28 October 2019; Published: 30 October 2019

Abstract: A 45-year-old Italian woman, affected by relapsing–remitting multiple sclerosis (RR-MS) starting from 2011, started treatment with alemtuzumab in July 2016. Nine months after the second infusion, she had an immune thrombocytopenic purpura (ITP) with complete recovery after steroid treatment. Three months after the ITP, the patient presented with transient aphasia, cognitive deficits, and focal epilepsy. Serial brain magnetic resonance imaging showed a pattern compatible with encephalitis. Autoantibodies to glutamate receptor 3 peptide A and B were detected in cerebrospinal fluid and serum, in the absence of any other diagnostic cues. After three courses of intravenous immunoglobulin (0.4 mg/kg/day for 5 days, 1 month apart), followed by boosters (0.4 mg/kg/day) every 4–6 weeks, her neurological status improved and is currently comparable with that preceding the encephalitis. Autoimmune complications of the central nervous system during alemtuzumab therapy are relatively rare: only one previous case of autoimmune encephalitis following alemtuzumab treatment has been reported to date.

Keywords: multiple sclerosis; autoimmune diseases; immune thrombocytopenic purpura; autoimmune encephalitis; alemtuzumab; antibodies against GluR3 peptide

---

1. Introduction

Alemtuzumab, a humanized monoclonal antibody indicated for the treatment of patients with relapsing–remitting multiple sclerosis (RR-MS), increases the risk of autoimmune adverse events, including thyroid disorder, renal disease, and immune thrombocytopenic purpura (ITP) [1]. Recently, new complications after alemtuzumab treatment have been described, like stroke, myocardial infarction, diffuse alveolar hemorrhage, and hemophagocytic lymphohistiocytosis (as reported in the European Medicine Agency note EMEA/H/A-20/1483/C/3718/0028).

We here report a case of presumed autoimmune encephalitis (AE) after the second course of alemtuzumab. AE is one of the most common causes of non-infectious encephalitis, with a variety of clinical manifestations, including behavioral and psychiatric symptoms, autonomic disturbances, movement disorders, and seizures. First-line immune therapies in AE consist of corticosteroids (intravenous and oral), sometimes coupled with intravenous immunoglobulin (IVIG) and/or plasma exchange (PE). Second-line treatments, including rituximab, cyclophosphamide, azathioprine, and mycophenolate mofetil, are administered when the first-line therapies fail to produce adequate benefits, when the disease is severe or relapsing, or, even in case of response to first-line treatments, with the goal of decreasing the risk of relapse in AE [2].

2. Case Report

We report the case of a 45-year-old Italian woman affected by RR-MS from 2011, when she had a diplopia and underwent a magnetic resonance imaging (MRI) showing multiple contrast-enhancing lesions in her brain and spinal cord white matter. After a spontaneous recovery, she later had another clinical attack and was treated with high intravenous steroids. Having fulfilled the criteria of definite diseases, a spinal tap was not performed and a disease-modifying therapy was started. After the failure of two first-line therapies (glatiramer acetate and dimethylfumarate) with clinical reactivations and new lesions identified after a new MRI, she started alemtuzumab in July 2016. Other second-line treatments, including natalizumab and fingolimod, were contraindicated for the presence of anti-JC virus antibodies at high titer (stratify index 3.20) and bradycardia. The alemtuzumab schedule (12 mg once daily (QD) for 5 days, followed by 12 mg QD for 3 days after one year) was approved for MS treatment.

In June 2018, 9 months after the second alemtuzumab infusion cycle, she reported a longer and more abundant menstrual period, bleeding from the gums, and scattered red spots on the skin. She was then referred to the emergency department: her platelet level was 1000/µL (normal range: 150,000–450,000/µL), with positive direct and indirect Coombs tests, and a normal bone marrow biopsy. A diagnosis of ITP was made and steroid treatment (methyl-prednisolone 40 mg daily for 7 days, followed by tapering) was promptly started with improvement: her platelet count became normal and the symptoms regressed in approximately 30 days.

In September 2018, 3 months after ITP, the patient presented with progressive aphasia and underwent a brain MRI that showed a pattern compatible with encephalitis (Figure 1a). She was hospitalized and her neurological examinations showed a change in neurological status with anomic aphasia and motor apraxia. Cerebrospinal fluid (CSF) was clear, with a slight increase in glucose (73 mg/dl) and protein (61 mg/dl) and normal cell numbers (4 cells/mmc; normal range: 0–5 cell/mmc). Immunoelectrophocusing showed an IgG index of 1.45 (0.00–0.65) and the presence of 17 oligoclonal bands. The PCR for herpes viruses (HSV (herpes simplex virus), CMV (cytomegalovirus), VZV (varicella-zoster virus), EBV (Epstein-Barr virus), HHV6 (human herpesvirus 6)) and the JC virus (JCV) was negative. Autoimmune screening (anti-gliadin IgG e IgA, anti-transglutaminase, anti-cardiolipin, antibodies to double-stranded DNA, extractable nuclear antigens, and anti-neutrophil cytoplasmic antibodies) was negative. Serology for common and neurotropic infectious agents (Toxoplasma, B. Burgdorferi, HIV), levels of oncotumor markers (CEA (carcino-embryonic antigen), AFP (alpha fetoprotein), CA (cancer antigen) 125, CA 15-3, CA 19-9, and Cyfra (cytokeratin 19 fragment antigen) 21-1 NSE (neuron-specific enolase)), antibodies against onconeural antigens (anti-amphiphysin, anti-MA2, anti-Yo, anti-Ri, anti-Hu, anti-GAD65, anti-titin, anti-recoverin, anti-Sox1, and anti-Zic4), and a total-body computerized tomography were all normal.

Three days after hospitalization, aphasia recovered completely; however, a new brain MRI showed increased edema in the left fronto-temporal subcortical white matter, without diffusion restriction or contrast enhancement (Figure 1b). Five days after hospitalization, a focal epilepsy started with clonic movements in her left upper limb, associated with the worsening of working memory and mood changes. An electroencephalogram showed non-specific electrical alterations in the bilateral temporo-occipital lobes. The patient started therapy with oral levetiracetam (1500 mg daily), with a stop in seizures and improvement of neurological status. We chose not to start steroid or other immunomodulatory therapies, planning strict clinical and neuroradiological follow-up instead.

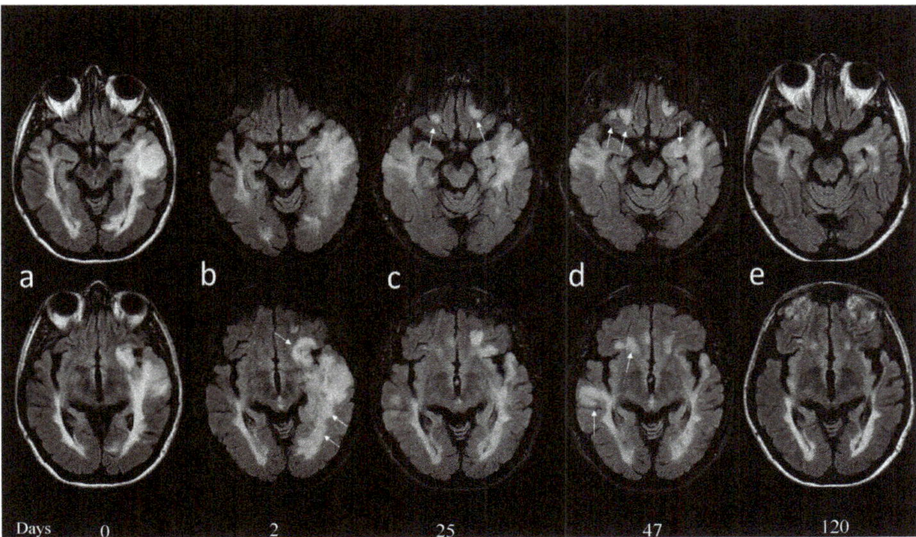

Figure 1. At admission (**a**) showed a large T2 signal alteration involving the left temporal lobe and expanding the superior temporal gyrus. (**b**) Two days later, a new brain magnetic resonance imaging (MRI) showed increased edema also involving the frontal subcortical and periventricular white matter (**b**, arrows); compression of the ventricular system was increased; and no diffusion restriction or contrast enhancement was demonstrated. (**c**) Twenty-five days after admission a new MRI showed reduction of the previously described T2 temporal lobe signal alteration with reduced compression of the temporal horn of the ventricle. Five new lesions were demonstrated and two of these were located in the fronto-orbital regions bilaterally (**c**, arrows). (**d**) Forty-seven days after admission, a new MRI documented a worsening of the T2 signal alterations in the fronto-orbital and in the temporal region on the right, with mild contrast enhancement in the right hippocampus and cingulum cortex (not shown). (**e**) Four months after the beginning of symptomatology all the signal alterations were markedly reduced and no enhancement was evident.

After twenty days, the patient presented with vomiting and mental confusion. A new MRI showed the reduction of a T2 hyperintense lesion previously described, but the appearance of five similar lesions (Figure 1c). Neurological examination showed a worsening of cognitive (especially executive) function and mood status. A second CSF examination was performed that showed 1 cell/mmc, 29 proteins mg/dl, glucose 54 mg/dl, and also included the search for a panel of autoantibodies known to be associated with AE (not investigated in the previous CSF examination). Autoantibodies to glutamate receptor 3 (GluR3) peptide A and B were detected both in the CSF (0.143 and 0.140, respectively, at CSF dilution 1:2) and in the serum (1.074 and 1.155, respectively, at serum dilution 1:200) by an enzyme-linked immunosorbent assay, as described [3,4]. No other findings emerged from the second CSF examination and we therefore started the first course of intravenous immunoglobulin (0.4 mg/kg/day for 5 days).

After a strict follow-up of about 3 weeks, during which the patient showed a partial recovery, a new MRI documented a worsening condition with an extension of signal alterations in the right frontal-orbital and temporal-basal region, with evident contrast enhancements in the hippocampus and cingulum cortex (Figure 1d). The patient underwent another two courses of intravenous immunoglobulin (0.4 mg/kg/day for 5 days), one month apart, with progressive clinical-MRI improvement (Figure 1e).

Her neurological status is currently comparable with that preceding the encephalitis. Given the response to IVIG treatment, we decided to continue that treatment (0.4 g/kg/day, every 4–6 weeks), while we stopped any treatment for MS. The patient is currently on monthly follow-up visits, possibly planning B cell-depleting treatments.

3. Discussion

Autoimmune complications of the central nervous system (CNS) during alemtuzumab therapy are relatively rare: one case of AE was reported to date [5]. The AE case occurred seven months after the second course of alemtuzumab, presenting with a polymorphic epilepsia partialis continua/status epilepticus in a patient with previous autoimmune hypothyroidism and ITP. No autoantibodies were reported in this case.

The clinical-MRI pattern of our patient is comparable with that recently reported. However, the peculiarity of our case was the positivity for anti-GluR3 autoantibodies. This is one of the antibodies directed against ionotropic glutamate receptors. They are present in 25%–30% of patients with different types of epilepsy, underpinning forms of 'autoimmune epilepsy' with frequent cognitive, psychiatric, and behavioral impairments [6]. The finding of anti-GluR3 autoantibodies was also associated with Rasmussen encephalitis, where they cause complement-mediated neuronal damage, irrespective of an excitotoxic effect [7]. A recent case of intractable myoclonus associated with anti-GluR3 antibodies was reported after allogeneic bone marrow transplantation [8].

Overall, our case and that described by Giarola et al. [5] strongly suggest a relationship between AE and previous therapy with alemtuzumab. Especially in patients with other well-known alemtuzumab-associated autoimmune complications (such ITP), monitoring of clinical events that may encompass the AE spectrum is advisable.

Author Contributions: M.C.B., M.S., and G.R. contributed equally to the writing and critical reading of the manuscript; A.B. provided interpretation of MRI imaging; C.C. and P.B. provided interpretation of immunological data; A.F., G.P., R.R., and A.C.L. provided critical reading of the manuscript. All authors read and approved the final manuscript.

Funding: This research received no specific grant from any funding agency in the public, commercial, or not-for-profit sectors.

Conflicts of Interest: The authors declare no conflict in interests. The authors report no disclosures relevant to the manuscript.

Consent: Written informed consent was obtained from the patient for publication of this case report and any accompanying images. A copy of the written consent is available upon request.

References

1. Coles, A.J.; Cohen, J.A.; Fox, E.J.; Giovannoni, G.; Hartung, H.P.; Havrdova, E.; Schippling, S.; Selmaj, K.W.; Traboulsee, A.; Compston, D.A.S.; et al. Alemtuzumab CARE-MS II 5-year follow-up: Efficacy and safety findings. *Neurology* **2017**, *89*, 1117–1126. [CrossRef] [PubMed]

2. Nosadini, M.; Mohammad, S.S.; Ramanathan, S.; Brilot, F.; Dale, R.C. Immune therapy in autoimmune encephalitis: A systematic review. *Expert Rev. Neurother.* **2015**, *15*, 1391–1419. [CrossRef] [PubMed]

3. Mantegazza, R.; Bernasconi, P.; Baggi, F.; Spreafico, R.; Ragona, F.; Antozzi, C.; Bernardi, G.; Granata, T.; Italian Rasmussen's Encephalitis Study Group. Antibodies against GluR3 peptides are nots pecific for Rasmussen's encephalitis but are also present in epilepsy patients with severe, early onset disease and intractable seizures. *J. Neuroimmunol.* **2002**, *131*, 179–185. [CrossRef]

4. Borroni, B.; Stanic, J.; Verpelli, C.; Mellone, M.; Bonomi, E.; Alberici, A.; Bernasconi, P.; Culotta, L.; Zianni, E.; Archetti, S.; et al. Anti-AMPA GluA3 antibodies in frontotemporal dementia: A new molecular target. *Sci. Rep.* **2017**, *7*, 6723. [CrossRef] [PubMed]

5. Giarola, B.; Massey, J.; Barnett, Y.; Rodrigues, M.; Sutton, I. Autoimmune encephalitis following alemtuzumab treatment of multiple sclerosis. *Mult. Scler. Relat. Disord.* **2019**, *28*, 31–33. [CrossRef] [PubMed]

6. Levite, M. Glutamate receptor antibodies in neurological diseases: Anti-AMPA-GluR3 antibodies, Anti-NMDA-NR1 antibodies, Anti-NMDA-NR2A/B antibodies, Anti-mGluR1 antibodies or Anti-mGluR5

antibodies are present in subpopulations of patients with either: Epilepsy, Encephalitis, Cerebellar Ataxia, Systemic Lupus Erythematosus (SLE) and Neuropsychiatric SLE, Sjogren'ssyndrome, Schizophrenia, Mania or Stroke. These autoimmune anti-glutamate receptor antibodies can bind neurons in few brain regions, activate glutamate receptors, decrease glutamate receptor's expression, impair glutamate-induced signaling and function, activate Blood Brain Barrier endothelial cells, kill neurons, damage the brain, induce behavioral/psychiatric/cognitive abnormalities and Ataxia in animal models, and can be removed or silenced in some patients by immunotherapy. *J. Neural. Transm.* **2014**, *121*, 1029–1075. [PubMed]

7. Frassoni, C.; Spreafico, R.; Franceschetti, S.; Aurisano, N.; Bernasconi, P.; Garbelli, R.; Antozzi, C.; Taverna, S.; Granata, T.; Mantegazza, R. Labeling of rat neurons by anti-GluR3 IgG from patients with Rasmussen encephalitis. *Neurology* **2001**, *57*, 324–327. [CrossRef] [PubMed]

8. Solaro, C.; Mantegazza, R.; Bacigalupo, A.; Uccelli, A. Intractable myoclonus associated with anti-GluR3 antibodies after allogeneic bone marrow transplantation. *Haematologica* **2006**, *91*, ECR62. [PubMed]

Article

Lost in Classification: Lower Cognitive Functioning in Apparently Cognitive Normal Newly Diagnosed RRMS Patients

Marco Pitteri *, Stefano Ziccardi, Caterina Dapor, Maddalena Guandalini and Massimiliano Calabrese *

Neurology section, Department of Neurosciences, Biomedicine and Movement Sciences, University of Verona, 37134 Verona, Italy; stefano.ziccardi@univr.it (S.Z.); caterina.dapor@univr.it (C.D.); maddalena.guandalini@univr.it (M.G.)
* Correspondence: marco.pitteri@univr.it (M.P.); massimiliano.calabrese@univr.it (M.C.); Tel.: +39-045-8124678 (M.P.)

Received: 11 October 2019; Accepted: 11 November 2019; Published: 13 November 2019

Abstract: Cognitive functioning in multiple sclerosis (MS) patients is usually related to the classic, dichotomic classification of impaired vs. unimpaired cognition. However, this approach is far from mirroring the real efficiency of cognitive functioning. Applying a different approach in which cognitive functioning is considered as a continuous variable, we aimed at showing that even newly diagnosed relapsing–remitting MS (RRMS) patients might suffer from reduced cognitive functioning with respect to a matched group of neurologically healthy controls (HCs), even if they were classified as having no cognitive impairment (CI). Fifty newly diagnosed RRMS patients and 36 HCs were tested with an extensive battery of neuropsychological tests. By using Z-scores applied to the whole group of RRMS and HCs together, a measure of cognitive functioning (Z-score index) was calculated. Among the 50 RRMS patients tested, 36 were classified as cognitively normal (CN). Even though classified as CN, RRMS patients performed worse than HCs at a global level ($p = 0.004$) and, more specifically, in the domains of memory ($p = 0.005$) and executive functioning ($p = 0.006$). These results highlight that reduced cognitive functioning can be present early in the disease course, even in patients without an evident CI. The current classification criteria of CI in MS should be considered with caution.

Keywords: multiple sclerosis; cognitive impairment; diagnosis; neuropsychological assessment

---

1. Introduction

Multiple Sclerosis (MS) is one of the most common inflammatory neurodegenerative disorders of the human central nervous system (CNS), characterized histologically by multifocal areas of inflammation, demyelination, and neurodegeneration [1] within the white matter (WM) [2] as well as within cortical and deep gray matter (GM) [3].

In addition to physical disability, cognitive impairment (CI) is common in MS patients, with frequencies ranging from 43% to 70% [4] depending on the studied population, the tests used, and the applied cut-off scores [5,6]. CI can occur early in the disease course [7] and has been strongly associated with both focal and diffuse GM damage [8,9] and WM lesion measures [10,11]. The mainly affected cognitive domains are verbal learning and memory, attention, information processing speed, and executive functions [12]. CI can alter MS patients' behavior and quality of life [13,14], leading to social and personal difficulties, despite minimal physical disability [15]. Longitudinal studies have shown that CI detected at the time of diagnosis can predict the conversion from clinically isolated syndrome to definite MS [16], the progression of physical disability [17], the transition to the secondary progressive (SP) phase [18], and the worsening of physical disability and GM atrophy in the long term [19]. These studies suggest that assessing cognitive functioning since the early phases of the disease is of paramount importance [20,21].

Despite the different batteries of neuropsychological tests used to assess cognitive functioning, the classification of CI is undoubtedly affected by the chosen cut-off applied [6,22] and by the number of neuropsychological tests used. Usually, MS patients are classified as having either "normal cognition" or "impaired cognition" in a perspective of dichotomous classification (unimpaired vs. impaired). This approach, however, is far from being meaningful considering the real life [5], in which measures of functional aspects, such as cognitive functioning, resemble continuous variables, as also underlined in other neurological populations (e.g., see [23,24]).

Dichotomizing continuous variables, such as cognitive functions, carries the risk of losing information that might increase the number of false positive results, as well as of underestimating the extent of variation in patients' performance [25], rendering difficult the diagnosis and the subsequent clinical decisions. For this reason, it would be more appropriate to use different psychometric methods, switching from a "cognitive impairment-based" to a "cognitive functioning-based" approach, considering cognitive functioning as a continuum variable as it is in real life, ranging from a minimum to a maximum level of performance. This is of particular interest given that cognitive decline may develop as a result of gradual progression, related to neurodegeneration and brain atrophy, or of acute disease activity, for which decline in cognitive performance can be often followed by incomplete recovery, thus contributing to the burden of CI in the long term [11].

In order to investigate the usefulness of this approach, with the present study we aimed at investigating the cognitive performance of a group of newly diagnosed MS patients with relapsing–remitting (RR) course as compared to a group of healthy controls (HCs). We expected that also the newly diagnosed MS patients, even if classified as being "cognitively normal" when referring to the classic, dichotomous approach, would rather show reduced cognitive functioning with respect to HCs.

2. Materials and Methods

2.1. Participants

Fifty consecutive newly diagnosed RRMS patients (37 females, mean ± SD age = 38.2 ± 11.6 years; mean ± SD education = 14.2 ± 2.7 years; mean ± SD disease duration from onset = 3.5 ± 5.2 years; median [range] effects of disability (EDSS) = 1.5 (0–4)) were tested with an extensive battery of neuropsychological tests near the time of MS diagnosis (average: 6 months). At the time of neuropsychological testing, 31 RRMS patients were still untreated, whereas 14 were treated with dymethilfumarate, 1 with fingolimod, 1 with natalizumab, 1 with interferon beta1-a, 1 with peg-interferon beta1-a, and 1 with azathioprine. Inclusion criteria for RRMS patients comprised diagnosis of RRMS [26], no relapse or steroid treatment in the 30 days before neuropsychological assessment, no concomitant neurological or other pathological health conditions, no substance abuse or other MS concomitant medication (as benzodiazepines or antidepressant drugs), and no visual impairment.

A group of 36 HCs, matched with RRMS patients for age, education, and gender, was recruited and tested with the same battery of neuropsychological tests used to assess RRMS patients. Inclusion criteria for HCs comprised no cognitive deficits measured with the Montreal Cognitive Assessment (MoCA) test [27], a test of global cognitive functioning; no neurologic, psychiatric, or other concomitant pathologies; normal or corrected to normal vision; no substance abuse or other prior or concomitant medications.

All participants were recruited at the MS Center of the Verona University Hospital (Verona, Italy). The study was approved by the local Ethics Committee, and written informed consent was collected from all participants. Demographic and clinical characteristics of RRMS and HCs are listed in Table 1.

Table 1. Demographic and clinical characteristics of relapsing–remitting multiple sclerosis (RRMS) patients and healthy controls (HCs).

	CI (*n* = 14)	CN (*n* = 36)	HCs (*n* = 36)	*p*
Gender (M/F)	3/11	10/26	13/23	0.547
Age (years)	39.3 ± 14.0	37.8 ± 10.8	33.6 ± 10.4	0.170
Education (years)	13.8 ± 4.0	14.4 ± 2.0	15.1 ± 2.6	0.229
EDSS [1]	2.0 (0–4)	1.0 (0–3)	/	/
Disease duration (years)	4.4 ± 8.2	3.1 ± 3.6	/	/
Time between diagnosis and neuropsychological assessment (months)	6 (±3)	6 (±2)	/	/

[1] Means ± SDs were provided for continuous variables. Median (range) was provided for effects of disability (EDSS). EDSS = Expanded Disability Status Scale; CI = cognitive impairment; CN = cognitive normal.

2.2. Neuropsychological Assessment

RRMS patients and HCs were tested with an extensive battery of neuropsychological tests, which included the Brief Repeatable Battery (BRB) of neuropsychological tests [28]; the Stroop Test, ST [29]; the Phonological, Semantic, and Alternate Verbal Fluency test, (VF [30]); and the Modified Five Point Test (MFPT; [31]). The BRB is composed of tests of verbal learning and delayed memory recall (Selective Reminding Test, SRT); visuospatial learning and delayed memory recall (10/36 Spatial Recall Test, SPART); visual information processing speed and attention (Symbol Digit Modalities Test, SDMT); auditory information processing speed, attention, and calculation (Paced Auditory Serial Addition Task, PASAT); and semantic verbal fluency (Word List Generation, WLG). The ST is a test of attention and of automatic response inhibition; the VF test is a test of verbal fluency (phonemic, semantic, and alternate); the MFPT is a test of figurative fluency and use of strategies.

Depression, anxiety, and stress were evaluated with the 21-item Depression Anxiety Stress Scale (DASS-21; [32]) and subjective fatigue with the Fatigue Severity Scale (FSS; [33]). According to the most used method [6], RRMS patients were classified as "cognitive normal" (CN) if they scored below the cut-off (5° percentile; z-score = −1.65) on zero, one, or two neuropsychological tests administered; otherwise, if RRMS patients obtained a score below the cut-off on three or more neuropsychological tests, they were classified as having CI.

For each neuropsychological test and for each RRMS patient and HC, we calculated the Z-score index (for details see [34]), in which we did not use the normative data of the Italian validation of each test but, rather, the mean and standard deviation (SD) of scores of the RRMS patients and the HCs together. Considering the mean and SD of both groups together, in which MS patients and HCs compose the same population, allows to normalize the dependent variable (Z-score index) in a unique gaussian distribution with overlapped curves, mimicking a more real-life condition. Following this procedure, we calculated: (1) a global cognitive functioning index (Z-global) considering the average of the Z-scores of each neuropsychological test; and (2) three domain-specific Z-score indexes: memory (Z-MEM), attention/information processing speed (Z-ATT/IPS), and executive functions (Z-EF). For the detailed classification of each cognitive domain, see Table 2.

Table 2. Neuropsychological tests considered for each Z-score domain index.

Z-MEM	Z-ATT/IPS	Z-EF
SRT-LTS	SDMT	ST (average EIT and EIE)
SRT-CLTR	PASAT-3	Phonemic VF
SRT-D	PASAT-2	Alternate VF
SPART-I		MFPT-UDs
SPART-D		MFPT-CSs

Z-MEM = Z-score–Memory; Z-ATT/IPS = Z-score–Attention/Information Processing Speed; Z-EF = Z-score–Executive Functions; SRT-LTS = Selective Reminding Test-Long-Term Storage; SRT-CLTR = Selective Reminding Test-Consistent Long-Term Retrieval; SRT-D = Selective Reminding Test-Delayed; SPART-I = Spatial Recall Test-Immediate; SPART-D = Spatial Recall Test-Delayed; SDMT = Symbol Digit Modalities Test; PASAT = Paced Auditory Serial Addition Test; ST-EIT = Stroop Test-Effect Interference Time; ST-EIE = Stroop Test-Effect Interference Error; VF = Verbal Fluency; MFPT-UDs = Modified Five Point Test-Unique Designs; MFPT-CSs = Modified Five Point Test-Cumulative Strategies.

2.3. Statistical Analyses

ANOVA models with Tukey post-hoc analysis and chi-square test were applied to compare demographic, clinical, and Z-index scores among CI, CN, and HCs. Effects of EDSS, disease duration, emotional state (DASS-21), and fatigue (FSS) on the global cognitive functioning index (Z-score global index) and on the three cognitive domains (Z-MEM, Z-ATT/IPS, Z-EF) were controlled for RRMS patients' by using a stepwise multiple regression analysis.

3. Results

Among the 50 RRMS patients tested, 14 were classified as having CI, and 36 as being CN. The majority (12/14: 86%) of the CI patients were impaired in the domains of ATT/IPS (64%) and EF (71%).

Group comparison results between CI ($n = 14$), CN ($n = 36$), and HCs ($n = 36$) showed no significant difference between the three groups in terms of age ($p = 0.170$), education ($p = 0.229$), and gender ($p = 0.547$).

The Z-score global index was significantly different among the three groups ($p < 0.001$). Post-hoc comparisons showed a significant difference between CI and HCs ($p < 0.001$), between CI and CN ($p < 0.001$), and also between CN and HCs ($p = 0.004$) (Figure 1).

Significant difference was found among the three groups also for Z-MEM ($p < 0.001$), Z-ATT/IPS ($p < 0.001$), and Z-EFs ($p < 0.001$). Post-hoc analysis showed a significant difference between CI and HCs for Z- MEM ($p < 0.001$), Z-ATT/IPS ($p < 0.001$), and Z-EF ($p < 0.001$); between CI and CN for Z-MEM ($p = 0.009$), Z-ATT/IPS ($p < 0.001$), and Z-EF ($p < 0.001$); and between CN and HCs for Z-MEM ($p = 0.005$) and Z-EF ($p = 0.006$). No significant difference was found between CN and HCs for Z-ATT/IPS ($p = 0.087$), as shown in Figure 1.

Considering CI and CN patients together, the results of the multiple regression analysis (final model $R^2 = 0.254$, $p = 0.170$) showed no significant effects of age ($\beta = -0.255$, $p = 0.155$), education ($\beta = 0.196$, $p = 0.240$), gender ($\beta = 0.224$, $p = 0.185$), disability ($\beta = -0.125$, $p = 0.437$), disease duration ($\beta = 0.085$, $p = 0.591$), emotional state ($\beta = -0.197$, $p = 0.288$), and fatigue ($\beta = -0.145$, $p = 0.477$) on the Z-global index. No significant effect of these variables was also found on Z-MEM ($R^2 = 0.289$, $p = 0.099$), Z-ATT/IPS ($R^2 = 0.208$, $p = 0.311$), and Z-EF ($R^2 = 0.252$, $p = 0.173$).

Considering each single neuropsychological test, we found a significant difference among the three groups (CI, CN, and HCs) in all the neuropsychological tests (all $p < 0.05$), except for the WLG test ($p = 0.180$). Post-hoc analysis showed a significant difference between CI and HCs in all neuropsychological tests (all $p < 0.05$). Moreover, we found a significant difference between CI and CN for the SRT-CLTR ($p = 0.013$), SRT-D ($p = 0.026$), SDMT ($p = 0.040$), PASAT-3 ($p = 0.001$), PASAT-2 ($p = 0.043$), ST-Effect Interference Time (EIT) ($p = 0.049$), ST-Effect Interference Error (EIE) ($p < 0.001$), Phonemic Verbal Fluency ($p = 0.023$), Semantic Verbal Fluency ($p = 0.042$), MFPT-Unique Designs (UDs)

($p = 0.001$), and MFPT-Error Index ($p = 0.020$), as shown in Table 3. Finally, comparing CN and HCs, we found significant difference for the SRT-LTS ($p = 0.012$), SRT–CLTR ($p = 0.016$), SRT-D ($p = 0.007$), SDMT ($p = 0.014$), Phonemic Verbal Fluency ($p = 0.034$), MFPT-UDs ($p = 0.003$), and MFPT-Cumulative Strategies (CSs) ($p = 0.019$), as shown in Table 3.

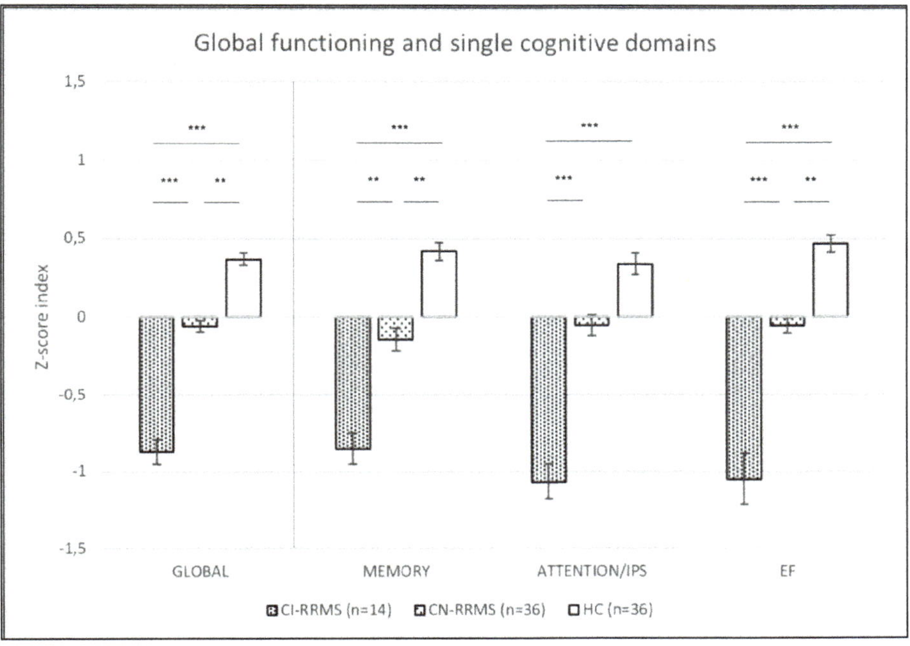

Figure 1. Functioning at a global level and on single cognitive domains of patients with cognitive impairment (CI), cognitive normal (CN) patients and healthy controls (HCs). $** = p < 0.01$, $*** = p < 0.001$.

Table 3. Neuropsychological performance of RRMS patients and HCs and results of the comparison between the Z-score indexes of each subtest. Means ± SDs are provided.

| NP Battery/NP Test | Subtest | CI ($n = 14$) | | CN ($n = 36$) | | HCs ($n = 36$) | | p CI vs. CN | p CN vs. HCs |
		Raw Scores	Z-score Index	Raw Scores	Z-score Index	Raw Scores	Z-score Index		
BRB (Brief Repeatable Battery)	SRT-LTS	38.5 ± 15.7	−0.8 ± 1.2	47.3 ± 12.8	−0.2 ± 0.9	55.7 ± 10.2	0.5 ± 0.8	0.076	0.012 *
	SRT-CLTR	26.2 ± 13.1	−1.0 ± 0.8	39.9 ± 15.2	−0.1 ± 0.9	49.7 ± 14.3	0.5 ± 0.9	0.013 *	0.016 *
	SRT-D	6.8 ± 2.5	−0.9 ± 1	8.7 ± 2.5	−0.2 ± 1	10.3 ± 1.7	0.5 ± 0.7	0.026 *	0.007 *
	SPART	20.3 ± 4.4	−0.7 ± 1	22.7 ± 4.4	−0.1 ± 1	25 ± 4.1	0.4 ± 0.9	0.211	0.054
	SPART-D	6.9 ± 1.7	−0.6 ± 0.9	7.9 ± 2	−0.1 ± 1	8.8 ± 1.8	0.3 ± 0.9	0.246	0.145
	SDMT	45.1 ± 11.3	−0.8 ± 0.9	53.9 ± 9.7	−0.1 ± 0.8	61.7 ± 12.9	0.5 ± 1	0.04 *	0.014 *
	PASAT-3	31.2 ± 11.5	−1.1 ± 1	43.6 ± 10.6	0.009 ± 0.9	47.8 ± 9.6	0.4 ± 0.8	0.001*	0.198
	PASAT-2	26.1 ± 10.8	−0.9 ± 1.1	35.3 ± 9.9	0.009 ± 0.1	37.1 ± 8.7	0.2 ± 0.9	0.043 *	0.689
	WLG	23.8 ± 7	−0.5 ± 1.1	27.2 ± 6.6	0.06 ± 1	27.5 ± 6	0.1 ± 0.9	0.231	0.974
Stroop Test (ST)	ST-EIT	17.3 ± 7.8	−0.7 ± 1.4	13.4 ± 5.4	−0.04 ± 1	11.3 ± 3.5	0.3 ± 0.6	0.049 *	0.202
	ST-EIE	1.6 ± 2.3	−1.1 ± 2.1	0.2 ± 0.5	0.2 ± 0.4	0.1 ± 0.4	0.3 ± 0.3	0.000 *	0.882
Verbal Fluency Test (VF)	Phonemic	34.1 ± 11.9	−0.8 ± 0.9	44.7 ± 12.7	−0.01 ± 0.9	53.9 ± 10.6	0.7 ± 0.8	0.023 *	0.034 *
	Semantic	47.5 ± 11.5	−0.7 ± 1	56.2 ± 11.3	−0.005 ± 0.9	63.5 ± 9.1	0.6 ± 0.8	0.042 *	0.071
	Alternate	37.8 ± 12.3	−0.6 ± 1.1	43.2 ± 10.8	−0.09 ± 0.9	50.5 ± 9.9	0.5 ± 0.9	0.350	0.073
	Shifting Index	0.9 ± 0.3	0.06 ± 1.5	0.9 ± 0.1	−0.06 ± 0.9	0.9 ± 0.1	−0.02 ± 0.9	0.944	0.993
Modified Five Point Test (MFPT)	MFPT-UDs	23.2 ± 12.7	−1.1 ± 1.2	34.3 ± 7.7	−0.07 ± 0.7	41.9 ± 7.5	0.6 ± 0.7	0.001 *	0.003 *
	MFPT-CSs	8.7 ± 11.9	−0.7 ± 1	14.9 ± 10.6	−0.1 ± 0.9	22.8 ± 11.1	0.5 ± 0.9	0.193	0.019 *
	MFPT-Error Index	15.3 ± 16.8	−0.7 ± 1.7	6.7 ± 6.9	0.2 ± 0.7	7.1 ± 8.4	0.1 ± 0.8	0.02 *	0.985

* = significant result.

4. Discussion

With the present study, we aimed at investigating the cognitive performance of a group of newly diagnosed RRMS patients as compared to a matched group of HCs by using a cognitive "functioning-based" approach instead of the classic "impairment-based" approach, in order to obtain a better real-life picture of RRMS patients' effective cognitive functioning.

Considering a functioning-based approach (i.e., Z-score index), the results of the present study showed that newly diagnosed RRMS patients can differ significantly from a group of HCs both on a global level and with reference to the cognitive domains of attention/processing speed, memory, and executive functioning. However, the most interesting finding is related to the fact that this significant difference between RRMS patients and HCs persists even after isolating those patients classified as CN, considering the classic categorization criterion [6]. Specifically, the group of CN patients showed a significant decrease in cognitive performance as compared to HCs at the global level as well as in the domains of memory and executive functions. The grading scores assigned on the basis of this cognitive "functioning-based" approach, as opposed to the classic "impairment-based" approach, highlight that also newly diagnosed CN RRMS patients can show worse cognitive performance as compared to HCs since the early stages of the disease, independently of the effect of other clinical and demographical variables like age, education, physical disability, disease duration, fatigue, or emotional state. The classic cognitive "impairment-based" approach is undoubtedly affected by different cut-offs threshold and by the different number of neuropsychological tests used, which can render the diagnosis of CI uncertain. Given that cognitive decline can occur as a result of gradual progression related to neurodegeneration or of more transient changes related to inflammatory (i.e., relapses) disease activity, by using a functioning-based approach (i.e., Z-score index) we expected that also newly diagnosed RRMS patients would perform worse with respect to HCs. In fact, it has been found that brain alterations due to GM and WM lesions and inflammatory phenomena can be observed since the time of diagnosis and are related to differences in the inflammatory profile [35,36]. Considering previous studies that showed that early neurodegeneration phenomena affect mainly the frontal and the temporal lobes since the early stage of the disease [8], it is remarkable that a significant difference between newly diagnosed CN patients and HCs was found specifically in the domains of memory and executive functions, that are mainly related to the activity of frontal and temporal brain areas, respectively. We would like to strongly highlight the alterations in executive functioning, since this domain is often neglected and not included in the most used batteries of neuropsychological tests in MS (i.e., the BRB and the Brief International Cognitive Assessment for MS, BICAMS).

The Z-score index, in which cognitive performance is considered as a continuum, seemed to effectively reflect the accumulation of cognitive alterations even in those RRMS patients that would be classified as "cognitively normal". As recently highlighted [37], if we accept that cognitive deficits in MS patients, or cognitive decline from baseline, reflect mainly cerebral dysfunctions related to MS disease, after excluding other confounding factors such as physical disability, fatigue, and emotional state, then cognitive functioning merits clinical attention as would any other indication of disease activity.

With this perspective, the classic impairment-based approach, usually limited by outdated and less representative normative data, can be overcome, optimizing the identification of slight alterations in cognitive performance already evident in newly diagnosed RRMS patients classified as being "cognitively normal" according to the traditional classification method. As underlined in previous studies, the early detection and monitoring of cognitive dysfunction may be crucial to identify MS patients with a probable worse prognosis and more severe disease progression [18,19], enabling early pharmacological and non-pharmacological interventions aimed at preventing further cognitive decline and disability in the long term [38]. According to this, a complete neuropsychological assessment in terms of level of performance, not just prone to classification criteria, seems of paramount importance not only in patients that show evident cognitive impairment [20], but also in apparently "cognitively normal" patients, as highlighted in the present study.

As recently underlined by Weber et al. [39], neuropsychological tests have shown a significant predictive value also regarding everyday-life activity and can be used in the clinical setting as one of several measures to help the clinician understand the impact of MS disease on the patients and their families. This view of considering patients' "cognitive performance" instead of patients' "cognitive impairment" might be an invaluable window on the real-life performance of MS patients since the time of diagnosis, given that early cognitive alterations can be considered as a signal of increased risk of disease progression [20].

We are aware that this study has some limitations. First, considering the variability of the MS population, further studies should include a larger number of both MS patients and matched HCs to substantiate the results of the present study. Second, the study is limited by the lack of a longitudinal neuropsychological assessment; this functioning-based approach should be tested more extensively with follow-up measures. Third, this study focused only on patients with RR course; future studies should extend this approach by investigating different MS populations. However, this is a proof-of-concept study, with which we aimed at highlighting the limitation of using the dichotomic approach derived from the classic neuropsychological assessment, frequently used for MS patients.

5. Conclusions

The results of the present study suggest that cognitive dysfunction in RRMS is a phenomenon that can be detected also in newly diagnosed patients. Extensive cognitive assessment since the early phase of the disease would be then of critical importance. This would support an accurate judgement of decline in cognitive functioning and would be clinically meaningful to determine a baseline cognitive profile to be monitored in the follow-up. We suggest approaching with extreme caution the traditional classification method of cognitive impairment: this classification criterion might fail in measuring the actual cognitive performance and should be interpreted with caution. In this regard, preferring an approach based on the evaluation of cognitive functioning as a continuous variable should be therefore recommended, also considering computerized devices [40,41].

Author Contributions: Conceptualization, M.P.; methodology, M.P. and S.Z.; formal analysis, S.Z.; investigation, M.P., S.Z., M.G., C.D., and M.C.; data curation, M.P., S.Z., M.G., C.D., and M.C.; writing—original draft preparation, S.Z.; writing—review and editing, M.P., S.Z., M.G., C.D., and M.C.; supervision, M.P. and M.C.; project administration, M.P.

Funding: This research received no external funding.

References

1. Kutzelnigg, A.; Lassmann, H. *Pathology of Multiple Sclerosis and Related Inflammatory Demyelinating Diseases*, 1st ed.; Elsevier: Amsterdam, The Netherlands, 2014; Volume 122, ISBN 9780444520012.

2. Noseworthy, J.; Lucchinetti, C.; Rodriguez, M.; Weinshenker, B. Multiple Sclerosis. *N. Engl. J. Med.* **2000**, *343*, 938–952. [CrossRef] [PubMed]

3. Calabrese, M.; Magliozzi, R.; Ciccarelli, O.; Geurts, J.J.G.; Reynolds, R.; Martin, R. Exploring the origins of grey matter damage in multiple sclerosis. *Nat. Rev. Neurosci.* **2015**, *16*, 147–158. [CrossRef] [PubMed]

4. Chiaravalloti, N.D.; DeLuca, J. Cognitive impairment in multiple sclerosis. *Lancet Neurol.* **2008**, *7*, 1139–1151. [CrossRef]

5. Amato, M.P.; Morra, V.B.; Falautano, M.; Ghezzi, A.; Goretti, B.; Patti, F.; Riccardi, A.; Mattioli, F.; Amato, M.P. Cognitive assessment in multiple sclerosis—An Italian consensus. *Neurol. Sci.* **2018**, *39*, 1317–1324. [CrossRef]

6. Fischer, M.; Kunkel, A.; Bublak, P.; Faiss, J.H.; Hoffmann, F.; Sailer, M.; Schwab, M.; Zettl, U.K.; Köhler, W. How reliable is the classification of cognitive impairment across different criteria in early and late stages of multiple sclerosis? *J. Neurol. Sci.* **2014**, *343*, 91–99. [CrossRef]

7. Amato, M.P.; Portaccio, E.; Goretti, B.; Zipoli, V.; Hakiki, B.; Giannini, M.; Pastò, L.; Razzolini, L.; Pia, M.; Portaccio, A.E. Cognitive impairment in early stages of multiple sclerosis. *Neurol. Sci.* **2010**, *31*, 211–214. [CrossRef]

8. Calabrese, M.; Agosta, F.; Rinaldi, F.; Mattisi, I.; Grossi, P.; Favaretto, A.; Atzori, M.; Bernardi, V.; Barachino, L.; Rinaldi, L.; et al. Cortical lesions and atrophy associated with cognitive impairment in relapsing-remitting multiple sclerosis. *Arch. Neurol.* **2009**, *66*, 1144–1150. [CrossRef]

9. Eijlers, A.J.C.; van Geest, Q.; Dekker, I.; Steenwijk, M.D.; Meijer, K.A.; Hulst, H.E.; Barkhof, F.; Uitdehaag, B.M.J.; Schoonheim, M.M.; Geurts, J.J.G. Predicting cognitive decline in multiple sclerosis: A 5-year follow-up study. *Brain* **2018**, *141*, 2605–2618. [CrossRef]

10. Preziosa, P.; Rocca, M.A.; Pagani, E.; Stromillo, M.L.; Enzinger, C.; Gallo, A.; Hulst, H.E.; Atzori, M.; Pareto, D.; Riccitelli, G.C.; et al. Structural MRI correlates of cognitive impairment in patients with multiple sclerosis. *Hum. Brain Mapp.* **2016**, *37*, 1627–1644. [CrossRef]

11. Rocca, M.A.; Amato, M.P.; De Stefano, N.; Enzinger, C.; Geurts, J.J.; Penner, I.-K.; Rovira, A.; Sumowski, J.F.; Valsasina, P.; Filippi, M. Clinical and imaging assessment of cognitive dysfunction in multiple sclerosis. *Lancet Neurol.* **2015**, *14*, 302–317. [CrossRef]

12. Benedict, R.H.B.; Cookfair, D.; Gavett, R.; Gunther, M.; Munschauer, F.; Garg, N.; Weinstock-Guttman, B. Validity of the minimal assessment of cognitive function in multiple sclerosis (MACFIMS). *J. Int. Neuropsychol. Soc.* **2006**, *12*, 549–558. [CrossRef]

13. Amato, M.P.; Zipoli, V.; Portaccio, E. Multiple sclerosis-related cognitive changes: A review of cross-sectional and longitudinal studies. *J. Neurol. Sci.* **2006**, *245*, 41–46. [CrossRef] [PubMed]

14. Rao, S.M. Cognitive function in patients with multiple sclerosis: Impairment and treatment. *Int. J. MS Care* **2004**, *6*, 9–22. [CrossRef]

15. Benedict, R.H.B.; Zivadinov, R. Risk factors for and management of cognitive dysfunction in multiple sclerosis. *Nat. Rev. Neurol.* **2011**, *7*, 332–342. [CrossRef] [PubMed]

16. Zipoli, V.; Goretti, B.; Hakiki, B.; Siracusa, G.; Sorbi, S.; Portaccio, E.; Amato, M.P. Cognitive impairment predicts conversion to multiple sclerosis in clinically isolated syndromes. *Mult. Scler.* **2010**, *16*, 62–67. [CrossRef] [PubMed]

17. Deloire, M.; Ruet, A.; Hamel, D.; Bonnet, M.; Brochet, B. Early cognitive impairment in multiple sclerosis predicts disability outcome several years later. *Mult. Scler.* **2010**, *16*, 581–587. [CrossRef]

18. Moccia, M.; Lanzillo, R.; Palladino, R.; Chang, K.C.-M.M.C.-M.; Costabile, T.; Russo, C.; De Rosa, A.; Carotenuto, A.; Sacca, F.; Maniscalco, G.T.; et al. Cognitive impairment at diagnosis predicts 10-year multiple sclerosis progression. *Mult. Scler.* **2016**, *22*, 659–667. [CrossRef]

19. Pitteri, M.; Romualdi, C.; Magliozzi, R.; Monaco, S.; Calabrese, M. Cognitive impairment predicts disability progression and cortical thinning in MS: An 8-year study. *Mult. Scler.* **2017**, *23*, 848–854. [CrossRef]

20. Kalb, R.; Beier, M.; Benedict, R.H.; Charvet, L.; Costello, K.; Feinstein, A.; Gingold, J.; Goverover, Y.; Halper, J.; Harris, C.; et al. Recommendations for cognitive screening and management in multiple sclerosis care. *Mult. Scler. J.* **2018**, *24*, 1665–1680. [CrossRef]

21. Sumowski, J.F.; Benedict, R.; Enzinger, C.; Filippi, M.; Geurts, J.J.; Hamalainen, P.; Hulst, H.; Inglese, M.; Leavitt, V.M.; Rocca, M.A.; et al. Cognition in multiple sclerosis. *Neurology* **2018**, *90*, 278–288. [CrossRef]

22. Binder, L.M.; Iverson, G.L.; Brooks, B.L. To err is human: "abnormal" Neuropsychological scores and variability are common in healthy adults. *Arch. Clin. Neuropsychol.* **2009**, *24*, 31–46. [CrossRef] [PubMed]

23. Azouvi, P. The ecological assessment of unilateral neglect. *Ann. Phys. Rehabil. Med.* **2017**, *60*, 186–190. [CrossRef] [PubMed]

24. Pitteri, M.; Chen, P.; Passarini, L.; Albanese, S.; Meneghello, F.; Barrett, A.M. Conventional and functional assessment of spatial neglect: Clinical practice suggestions. *Neuropsychology* **2018**, *32*, 835. [CrossRef] [PubMed]

25. Altman, D.G.; Royston, P. The cost of dichotomising continuous variables. *BMJ* **2006**, *332*, 1080. [CrossRef] [PubMed]

26. Polman, C.H.; Reingold, S.C.; Banwell, B.; Clanet, M.; Cohen, J.A.; Filippi, M.; Fujihara, K.; Havrdova, E.; Hutchinson, M.; Kappos, L.; et al. Diagnostic criteria for multiple sclerosis: 2010 revisions to the McDonald criteria. *Ann. Neurol.* **2011**, *69*, 292–302. [CrossRef] [PubMed]

27. Santangelo, G.; Siciliano, M.; Pedone, R.; Vitale, C.; Falco, F.; Bisogno, R.; Siano, P.; Barone, P.; Grossi, D.; Santangelo, F.; et al. Normative data for the Montreal Cognitive Assessment in an Italian population sample. *Neurol. Sci.* **2015**, *36*, 585–591. [CrossRef]

28. Amato, M.P.; Portaccio, E.; Goretti, B.; Zipoli, V.; Ricchiuti, L.; De Caro, M.F.; Patti, F.; Vecchio, R.; Sorbi, S.; Trojano, M.; et al. The Rao's Brief Repeatable Battery and Stroop test: Normative values with age, education and gender corrections in an Italian population. *Mult. Scler.* **2006**, *12*, 787–793. [CrossRef]

29. Caffarra, P.; Vezzadini, G.; Dieci, F.; Zonato, F.; Venneri, A. Una versione abbreviata del test di Stroop: Dati normativi nella popolazione italiana. *Riv. Neurol.* **2002**, *12*, 111–115.

30. Costa, A.; Bagoj, E.; Monaco, M.; Zabberoni, S.; De Rosa, S.; Papantonio, A.M.; Mundi, C.; Caltagirone, C.; Carlesimo, G.A. Standardization and normative data obtained in the Italian population for a new verbal fluency instrument, the phonemic/semantic alternate fluency test. *Neurol. Sci.* **2014**, *35*, 365–372. [CrossRef]

31. Cattelani, R.; Dal Sasso, F.; Corsini, D.; Posteraro, L. The Modified Five-Point Test: Normative data for a sample of Italian healthy adults aged 16–60. *Neurol. Sci.* **2011**, *32*, 595–601. [CrossRef]

32. Bottesi, G.; Ghisi, M.; Altoè, G.; Conforti, E.; Melli, G.; Sica, C. The Italian version of the Depression Anxiety Stress Scales-21: Factor structure and psychometric properties on community and clinical samples. *Compr. Psychiatry* **2015**, *60*, 170–181. [CrossRef] [PubMed]

33. Krupp, L.B.; Larocca, N.G.; Muir Nash, J.; Steinberg, A.D. The fatigue severity scale: Application to patients with multiple sclerosis and systemic lupus erythematosus. *Arch. Neurol.* **1989**, *46*, 1121–1123. [CrossRef] [PubMed]

34. Strauss, E.; Sherman, E.M.S.; Spreen, O. *A Compendium of Neuropsychological Tests: Administration, Norms and Commentary*, 3rd ed.; Oxford University Press: New York, NY, USA, 2006.

35. Frischer, J.M.; Bramow, S.; Dal-bianco, A.; Lucchinetti, C.F.; Rauschka, H.; Schmidbauer, M.; Laursen, H.; Sorensen, P.S.; Lassmann, H. Neurodegeneration in multiple sclerosis brains. *Brain* **2009**, *132*, 1175–1189. [CrossRef]

36. Magliozzi, R.; Howell, O.W.; Nicholas, R.; Cruciani, C.; Castellaro, M.; Romualdi, C.; Rossi, S.; Pitteri, M.; Benedetti, M.D.; Gajofatto, A.; et al. Inflammatory intrathecal profiles and cortical damage in multiple sclerosis. *Ann. Neurol.* **2018**, *83*, 739–755. [CrossRef]

37. Weinstock-Guttman, B.; Eckert, S.; Benedict, R.H.B. A decline in cognitive function should lead to a change in disease-modifying therapy—Yes. *Mult. Scler. J.* **2018**, *24*, 1681–1682. [CrossRef]

38. Forn, C.; Rocca, M.A.; Valsasina, P.; Boscá, I.; Casanova, B.; Sanjuan, A.; Ávila, C.; Filippi, M. Functional magnetic resonance imaging correlates of cognitive performance in patients with a clinically isolated syndrome suggestive of multiple sclerosis at presentation: An activation and connectivity study. *Mult. Scler. J.* **2012**, *18*, 153–163. [CrossRef]

39. Weber, E.; Goverover, Y.; DeLuca, J. Beyond cognitive dysfunction: Relevance of ecological validity of neuropsychological tests in multiple sclerosis. *Mult. Scler. J.* **2019**, *25*, 1412–1419. [CrossRef]

40. Golan, D.; Wilken, J.; Doniger, G.M.; Fratto, T.; Kane, R.; Srinivasan, J.; Zarif, M.; Bumstead, B.; Buhse, M.; Fafard, L.; et al. Validity of a Multi-Domain Computerized Cognitive Assessment battery for Patients with Multiple Sclerosis. *Mult. Scler. Relat. Disord.* **2019**, *30*, 154–162. [CrossRef]

41. De Meijer, L.; Merlo, D.; Skibina, O.; Grobbee, E.J.; Gale, J.; Haartsen, J.; Maruff, P.; Darby, D.; Butzkueven, H.; Van der Walt, A. Monitoring cognitive change in multiple sclerosis using a computerized cognitive battery. *Mult. Scler. J. Exp. Transl. Clin.* **2018**, *4*, 205521731881551. [CrossRef]

Article

Cortical Excitability Measures May Predict Clinical Response to Fampridine in Patients with Multiple Sclerosis and Gait Impairment

Rechdi Ahdab [1,2,3], Madiha M. Shatila [1,3], Abed Rahman Shatila [4], George Khazen [1,5], Joumana Freiha [1,3], Maher Salem [1,3], Karim Makhoul [1,3], Rody El Nawar [1,3], Shaza El Nemr [1], Samar S. Ayache [6,7] and Naji Riachi [1,3,*]

[1] Lebanese American University Gilbert and Rose Mary Chagoury School of Medicine,
Byblos P.O. Box 36, Lebanon; chadahdab@gmail.com (R.A.); mado.shatila@gmail.com (M.M.S.);
gkhazen@lau.edu.lb (G.K.); joumanafreiha@gmail.com (J.F.); Maher.Salem@lau.edu (M.S.);
karim.makhoul@lau.edu (K.M.); roudynawar@hotmail.com (R.E.N.); shaza.el-nemr@hotmail.com (S.E.N.)
[2] Hamidy Medical Center, Tripoli 1300, Lebanon
[3] Neurology Division, Lebanese American University Medical Center Rizk Hospital,
Beirut P.O Box 11-3288, Lebanon
[4] Makassed General hospital, Beirut P.O. Box 11-6301, Lebanon; ashatila1@hotmail.com
[5] Computer Science and Mathematics Department, Lebanese American University,
Byblos P.O. Box 36 , Lebanon
[6] Service de Physiologie-Explorations Fonctionnelles, Hôpital Henri Mondor, Assistance Publique—Hôpitaux
de Paris, 51 avenue de Lattre de Tassigny, 94010 Créteil, France; samarayache@gmail.com
[7] EA 4391, Excitabilité Nerveuse et Thérapeutique, Université Paris-Est-Créteil, 94010 Créteil, France
[*] Correspondence: naji.riachi@lau.edu.lb

Received: 18 October 2019; Accepted: 3 December 2019; Published: 5 December 2019

Abstract: *Background:* Most multiple sclerosis (MS) patients will develop walking limitations during the disease. Sustained-release oral fampridine is the only approved drug that will improve gait in a subset of MS patients. *Objectives:* (1) Evaluate fampridine cortical excitability effect in MS patients with gait disability. (2) Investigate whether cortical excitability changes can predict the therapeutic response to fampridine. *Method:* This prospective observational study enrolled 20 adult patients with MS and gait impairment planned to receive fampridine 10 mg twice daily for two consecutive weeks. Exclusion criteria included: Recent relapse (<3 months), modification of disease modifying drugs (<6 months), or Expanded Disability Status Scale (EDSS) score >7. Neurological examination, timed 25-foot walk test (T25wt), EDSS, and cortical excitability studies were performed upon inclusion and 14 days after initiation of fampridine. *Results:* After treatment, the mean improvement of T25wt (ΔT25wt) was 4.9 s. Significant enhancement of intra-cortical facilitation was observed (139% versus 241%, $p = 0.01$) following treatment. A positive correlation was found between baseline resting motor threshold (rMT) and both EDSS ($r = 0.57$; $p < 0.01$) and ΔT25wt ($r = 0.57$, $p = 0.01$). rMT above 52% of the maximal stimulator output was found to be a good predictor of a favorable response to fampridine (accuracy: 75%). *Discussion:* Fampridine was found to have a significant modulatory effect on the cerebral cortex, demonstrated by an increase in excitatory intracortical processes as unveiled by paired-pulse transcranial magnetic stimulation. rMT could be useful in selecting patients likely to experience a favorable response to fampridine.

Keywords: short intracortical inhibition; intracortical facilitation; fampridine; multiple sclerosis; walking disability

1. Introduction

Most patients with multiple sclerosis (MS) will develop walking limitations at some point in the course of the disease [1]. The impact of such limitations on daily activities and quality of life is substantial [2]. Currently, sustained-release oral fampridine is the only approved drug for the symptomatic treatment of walking disability due to MS. Fampridine (4-aminopyridine) is a voltage-dependent potassium channel-blocker that restores action potential conduction in poorly myelinated central nerve fibers, with a positive impact on synaptic transmission and neuronal excitability [3,4]. Unfortunately, only a subset of patients with MS and walking disabilities are responders to fampridine [5]. Limited data suggest that more advanced walking disability [6,7] and prolonged central conduction times [6,8] are predictive of a favorable response to treatment. The predictive value of more advanced measurements of cortical dysfunction have not been studied.

Changes in cortical excitability are frequent in MS, especially in more advanced stages of the disease [9–12]. Such changes are influenced by several factors, including the clinical form, stage of the disease, and degree of disability. It has been suggested that some excitability parameters correlate with the level of disability. This concerns more particularly the short-interval intracortical inhibition (SICI), intracortical facilitation (ICF), and resting motor threshold (rMT) [9–12].

This study was designed to evaluate the effect of slow-release fampridine on cortical excitability in patients with MS and gait disabilities. In addition, we set out to investigate whether cortical excitability changes could serve as predictors of therapeutic response to fampridine.

2. Materials and Methods

2.1. Study Design

This prospective observational study included adult patients with MS and gait impairment. All patients planned to receive fampridine in the period between April 2016 and August 2017 were asked if they were willing to participate in the study. If so, the protocol was thoroughly explained and written informed consent was obtained. The study was approved by our institutional review board.

2.2. Subjects

Eligible patients were adults (aged 18 and above) with a definite diagnosis of MS according to the 2010 McDonald criteria [13]. All patients had gait disturbance and were planned to receive a 2-week fampridine trial to assess the clinical impact of the drug on walking speed. This is an essential step needed to secure approval for the treatment from third party payers in our country. Patients were required to have recordable motor evoked potentials (MEPs) in at least one hand. Those with a recent relapse (<3 months), a recent modification of disease modifying drugs (prior 6 months), or an EDSS above 7 were excluded. Patients with contraindication to fampridine (seizures, pregnancy, breastfeeding, and renal impairment) or transcranial magnetic stimulation (TMS) (seizure, pacemaker, or ferromagnetic material in the head area) were also excluded. Eligible patients received oral fampridine 10 mg twice daily for two consecutive weeks.

2.3. Clinical and Neurophysiological Assessments

Assessments were performed at baseline (T0) and 14 days after the commencement of fampridine (T1). The physicians performing the clinical assessment were blinded to the cortical excitability results and vice versa.

Clinical assessment was performed by N.R., S.E.N. and M.M.S.. At baseline (T0), a detailed medical history was obtained with particular emphasis on key MS related elements: Onset and type of MS, number and dates of clinical relapses, prior disease modifying drugs, comorbidities, and current medications. At the follow-up visit (T1), the interview was focused on drug side effects and subjective improvement. Particular attention was paid to the occurrence of any new symptom or any worsening of a pre-existing deficit that could be indicative of a new relapse. A thorough neurological examination and an assessment of disability status using the Kurtzke Expanded Disability Status Scale (EDSS) were done at baseline (T0) [14]. Cortical excitability assessment and a timed 25-foot walk test (T25wt) [15] were done during both visits (T0 and T1).

Decrease in T25wt was calculated as follows: ΔT25wt = T25wt at T0 − T25wt at T1. Then, the percentage of improvement was calculated as follows: % of improvement = $\frac{\Delta T25wt}{T25wt\,at\,T0} \times 100$.

Patients who showed a percentage of improvement greater than 20% were considered responders. This threshold was based on the finding that a 20% improvement in walking time is considered clinically meaningful for patients with MS [16].

Cortical excitability studies were performed by R.A. and included the resting motor threshold (rMT), short-interval cortical inhibition (SICI), intracortical facilitation (ICF), and cortical silent period (CSP). MEPs were recorded from the first dorsal interosseous muscle (FDI) using pre-gelled disposable surface electrodes in a tendon-belly montage (ref 019-400400, Natus, Pleasanton, CA, USA). Recording was done on the side affected least by MS or on the left side if both sides were normal or equally affected.

MEPs were filtered (20 Hz to 2 KHz), amplified, and stored for off-line analyses (Nicolet EDX, Natus, Pleasanton, CA, USA). Patients were seated in a comfortable chair with the arms at rest. Baseline muscle activity was continuously monitored to ascertain complete muscle relaxation (except for CSP measurement). A tight-fitting cap was placed on the patient's head to help mark the coil position. Magnetic stimulation was delivered using an eight-shaped coil with an inner diameter of 70 mm (M200^2 D70 double coil 3190-00, Magstim Co, Carmarthenshire, U.K.) placed tangentially to the scalp (handle pointing backwards) and connected to a Magstim Bistim^2 stimulator (Magstim Co, Carmarthenshire, U.K.).

The motor hot spot was determined by scanning the scalp for the coil position associated with the largest MEPs. This position was marked on the cap and used for all subsequent excitability measurements. It was determined in a fully relaxed FDI muscle and defined as the lowest stimulus intensity required to produce MEPs larger than 50 μV (peak-to-peak) in 5 out of 10 consecutive trials [17].

A paired stimulus paradigm was used to test SICI using interstimulus intervals (ISIs) of 2 and 4 ms (SICI2 and SICI4 respectively) [18]. ICF was tested using ISIs of 10 and 15 (ICF10 and ICF15, respectively) [18]. The conditioning stimulus was delivered at 80% of rMT and the test stimulus was delivered at 120% of rMT. Eight trials were performed for each condition. The average of eight trials of unconditioned TMS pulses delivered at 120% of rMT was used as control. The amplitudes of conditioned MEPs were expressed as the percentage of the mean unconditioned MEP amplitude. The mean degree of inhibition and facilitation from all tested ISIs were retained for analysis.

CSP was defined as the duration of electromyogram (EMG) activity interruption following a single TMS pulse delivered at 140% of the rMT. The degree of FDI activation was controlled by visual feedback. Five trials were performed and averaged. The minimal CSP duration was measured from the end of the MEP until the first reoccurrence of EMG activity on highly magnified traces [19].

This study was sponsored by an investigator-initiated trial grant from Biogen, who reviewed the protocol. Biogen did not participate in patient recruitment or study implementation, had no access to the data, and did not participate in the statistical analysis.

2.4. Statistical Analysis

The analysis was conducted using the R statistical program 3.5.3 (R Foundation for statistical computing, Vienna, Austria). Descriptive statistics are presented as mean ± standard deviation (SD) and median (interquartile range) for continuous variables.

The Wilcoxon rank sum paired test was used to assess changes in cortical excitability parameters (comparison between measurements at T0 and T1), and Pearson correlation coefficient was used to analyze the relationship between ΔT25wt and baseline cortical excitability measures (rMT, MEPs amplitude, SICI, ICF, CSP) and between ΔT25wt and EDSS scores. The relationship between EDSS and baseline excitability measures was also examined. A *p*-value <0.05 was considered significant.

A classification model using the C5.0 decision tree algorithm (Quinlan R (1993). C4.5: Programs for Machine Learning. Morgan Kaufmann Publishers [20] was used to classify the samples as fampridine responders and non-responders based on their rMT value. As stated previously, patients were considered responders if they had at least a 20% reduction in their T25wt at the follow-up visit (T1), and non-responders otherwise. A simple model was built using the formula "responder status ~ rMT" while keeping the default values for all the remaining parameters in the model. The C5.0 decision tree algorithm relies on the maximum information gain to identify the rMT cutoff value that best splits the samples as responders and non-responders to estimate a threshold value that discriminates between fampridine responders and non-responders. The model was built using the C5.0 from the C50 package in R.

3. Results

3.1. Clinical and Sociodemographic Data

A total of 20 patients (11 male) completed the study (Table 1). They had a mean age of 49.75 ± 11.36 (median interquartile range (IQR): 50.00 (18.00); overall range: 25–65) years. Thirteen had relapsing remitting, 6 had secondary progressive, and one had primary progressive MS. Mean disease duration was 13.75 ± 8.17 years (median (IQR): 12.50 [10]; overall range: 3–39). Their mean EDSS score was 4.70 ± 1.31 (median (IQR): 4.00 (1.88); overall range 2.5–7). Nineteen patients were on disease modifying treatment: 3 on interferon beta 1a, 12 on fingolimod, 3 on natalizumab, and 1 on dimethyl fumarate.

Table 1. Baseline characteristics of subjects.

Subjects	Age	Sex	Disease Duration	EDSS	MS Type	Present Medication
1	53	F	7 years	5	RRMS	Fingolimod
2	46	M	11 years	4	SPMS	Fingolimod
3	25	M	5 years	4	RRMS	Fingolimod
4	49	F	10 years	7	RRMS	Natalizumab
5	62	M	18 years	5	RRMS	Fingolimod
6	62	F	22 years	5.5	SPMS	Interferon β 1a
7	42	F	17 years	4	SPMS	Natalizumab
8	47	F	8 years	5.5	RRMS	Fingolimod
9	58	M	19 years	4	SPMS	Fingolimod
10	58	F	12 years	7	SPMS	Fingolimod
11	62	M	11 years	4	PPMS	None
12	51	F	13 years	6	SPMS	Dimethyl Fumarate
13	65	M	18 years	6	RRMS	Fingolimod
14	50	F	18 years	6.5	RRMS	Fingolimod
15	35	M	3 years	3.5	RRMS	Fingolimod
16	32	M	3 years	3	RRMS	Natalizumab
17	50	M	18 years	4	RRMS	Interferon β 1a
18	64	M	39 years	3.5	RRMS	Interferon β 1a
19	36	M	15 years	4	RRMS	Fingolimod
20	48	F	8 years	2.5	RRMS	Fingolimod

EDSS: Expanded Disability Status Scale; F: female; M: male; MS, Multiple sclerosis; PPMS: Primary progressive multiple sclerosis; RRMS: Relapsing remitting multiple sclerosis; SPMS: Secondary progressive multiple sclerosis.

3.2. Clinical Response to Fampridine

Significant improvement in T25wt was found following fampridine (21.5 ± 5.1 versus 16.64 ± 1.40 , $p < 0.001$). Mean ΔT25wt was 4.86 ± (range 0.04–17.62) seconds (s). Eleven patients (55%) were responders to fampridine, as defined previously.

The treatment was well tolerated, and no serious adverse effects were reported. Thirteen patients (65%) reported minor side effects, mostly nausea, abdominal pain, lumbar/cervical pain, and dizziness. There were no instances of patient withdrawal owing to treatment side effects.

3.3. Effects of Fampridine on Cortical Excitability Measures

As depicted in Table 2, the follow-up visit (T1) showed a statistically significant enhancement of ICF15 (139.17 ± 110.98% versus 241.69 ± 131.46%, $p = 0.01$). ICF10 also tended to increase, but this effect did not reach statistical significance (161.58 ± 154.12% versus 242.94 ± 125.79%, $p = 0.06$) (Table 2). No significant changes were observed for the rMT (60.95 ± 12.47% versus 62.45 ± 15.20%, $p = 0.81$), SICI2 (53.08 ± 68.24% versus 73.62 ± 69.95%, $p = 0.11$), SICI4 (88.18 ± 107.10% versus 121.51 ± 88.48%, $p = 0.15$), or CSP (132.15 ± 55.67 ms versus 138.60 ± 51.90 ms, $p = 0.92$).

Table 2. Observed measures of excitability parameters and timed 25-foot walk test (T25wt) at baseline in seconds (s) and after fampridine treatment.

Parameter	Baseline Mean ± SD Median (IQR)	After Fampridine Mean ± SD Median (IQR)	Wilcoxon Paired-Test p-Value
rMT	60.95 ± 12.47 57.50 (19.75)	62.45 ± 15.20 58.50 (30.50)	0.81
SICI2	53.08 ± 68.24 31.95 (53.22)	73.62 ± 69.95 50.00 (70.94)	0.11
SICI4	88.18 ± 107.10 47.05 (52.77)	121.51 ± 88.48 107.90 (150.17)	0.15
ICF10	161.58 ± 154.12 134.61 (172.06)	242.94 ± 125.79 217.67 (210.46)	0.06
ICF15	139.17 ± 110.98 114.75 (139.63)	241.69 ± 131.46 235.00 (182.49)	**0.01**
CSP	132.15 ± 55.67 119.00 (54.25)	138.60 ± 51.90 147.50 (71.25)	0.92
T25wt	21.5 ± 5.1	16.64 ± 1.40	**<0.001**

The short intracortical inhibition at interstimulus intervals (ISIs) of 2 and 4 ms (SICI2, SICI4), intracortical facilitation at ISIs of 10 and 15 ms (ICF10, ICF15) are expressed as the percentage of the unconditioned motor evoked potential (MEP) amplitude. The cortical silent period (CSP) is expressed in ms. The resting motor threshold (rMT) is expressed as percentage of the maximal stimulator output. p-values in bold reflect significant difference ($p < 0.05$). IQR, interquartile range.

3.4. Relationship Between EDSS and Baseline Excitability Measures

At baseline (T0), a positive correlation was found between EDSS and rMT ($r = 0.57$; $p < 0.01$) (Table 3). Conversely, no statistically significant correlation was found between EDSS and any of the other cortical excitability measures.

Table 3. Correlation between the Expanded Disability Status Scale (EDSS) and baseline measures of resting motor threshold (rMT), short-interval intracortical inhibition at interstimulus intervals (ISIs) of 2 and 4 ms (SICI2 and SICI4, respectively), intracortical facilitation at ISIs of 10 and 15 (ICF10 and ICF15, respectively), cortical silent period (CSP), and timed 25-foot walking test (T25wt).

	Variables	Correlation	p-Value
	rMT	0.57	<0.01
	SICI2	−0.24	0.30
	SICI4	−0.27	0.24
EDSS	ICF10	−0.18	0.42
	ICF15	0.007	0.97
	CSP	−0.08	0.71
	ΔT25wt	0.75	<0.01

3.5. Relationships Between Improvement of T25wt (ΔT25wt) and Each of EDSS and Baseline Cortical Excitability Measures

A statistically significant positive correlation was found between baseline rMT and improvement of walking speed at the follow-up visit ($r = 0.57$, $p = 0.01$). No correlation was found between ΔT25wt and other baseline excitability parameters.

A positive correlation was also found between ΔT25wt and EDSS. ($r = 0.75$; $p < 0.01$) (Table 3).

3.6. Predictors of a Favorable Response to Fampridine

The predictive value of baseline rMT in terms of response to fampridine was studied. The threshold to predict good response to fampridine was found to be 52 (with 14 patients with rMT above 52 were found to be good responders to the treatment). Mean ± SD of rMT in responder and non-responder groups were 64.10 ± 11.99 and 57.11 ± 12.62, respectively. The classification of responders and non-responders had an accuracy of 75% (specificity: 83.3% (5/6) and sensitivity 71.4% (10/14)).

4. Discussion

Three interesting findings emerged from our study. First, higher rMT at baseline was predictive of a favorable response to fampridine, translating clinically into improved gait speed. Conversely, the other excitability parameters were not found to have any predictive value. From a clinical standpoint, patients with higher EDSS showed enhanced drug effects compared to those with lower EDSS. At a more mechanistic level, fampridine was found to induce a significant increase in intracortical excitatory mechanisms as revealed by paired-pulse TMS, with no significant effect on the other excitability parameters.

The positive impact of fampridine on gait speed is now well established, but only 40% of patients are expected to experience a clinically meaningful improvement [15,21]. Proper selection of patients most likely to experience clinical benefits from fampridine has been the subject of rare dedicated studies [6–8]. It has been suggested that patients with more disability at baseline have the best outcome. This seems logical, since advanced ambulatory impairment in MS is associated with a higher amount of axonal demyelination within the neural locomotor networks, providing more targets for fampridine to reinforce gait function [6,8]. Filli and collaborates tested the predictive value of a set of demographic and clinical criteria including T25wt, walking endurance 6-min walk test (6MWT), and the 12-item multiple sclerosis walking scale (MSWS-12) [7]. Their findings suggest that walking function at baseline (6MWT and T25wt) accurately predicts the responder status. EDSS only weakly correlated with the outcome. In this study, reduced walking endurance as measured by the 6MWT was the best predictor of a good outcome with an accuracy of 80% [7]. In comparison, we found that EDSS correlated with improvement in T25wt; however, we did not test walking endurance, since we included patients with advanced gait impairment who were unable to complete the 6MWT.

At the physiological level, high rMT was found to be correlated with high EDSS score, and seemed to be a good predictor of outcome. rMT is a global measurement of excitability and membrane properties of cortical pyramidal cells [22,23]. It can be regarded as the electrophysiological signature of clinical impairment in MS patients. Higher rMT was found to correlate with clinical relapses [24], disease progression [25], secondary progressive disease [10,26], and the presence of fatigue [11]. Another neurophysiological measure of potential predictive value is prolonged motor central conduction [6,8]. The latter parameter was not evaluated in our study.

Studies of cortical excitability, a surrogate measure of ion channel and synaptic functions, have repeatedly shown significant excitability changes in patients with MS [10–12]. In the earlier stages of relapsing remitting MS, this state of altered excitability can be improved by immunomodulatory therapy [27]. As the disease progresses and the degenerative process takes place, excitability changes could become permanent [10,26]. At this stage, drugs having a more direct effect on cortical excitability are more likely to reverse these changes than those with immunomodulatory effects. Our study clearly demonstrates a cortico-modulatory effect of fampridine, a rather predictable finding given its putative mechanism of action. Cortical excitability measures are dependent on networks of interneurons (both excitatory and inhibitory) and their synaptic interaction with each other. SICI reflects the recruitment of inhibitory pathways with GABAergic mediation (GABA-A receptors) and ICF reflects the recruitment of excitatory pathways with glutamatergic mediation. Healthy excitatory and inhibitory systems are essential for proper functioning of brain circuits. Fampridine was found to increase ICF in our study but had no impact on the other excitability parameters. Such findings would indicate an increase in facilitatory intracortical mechanisms aiming to improve the clinical outcomes, as reflected by an increase in walking speed [28,29].

Our study has many shortcomings, including an open label design, small sample size, and short follow-up period. Another shortcoming is that cortical excitability changes were correlated to changes in walking speed, but not other cortical functions. Since cortical excitability studies are classically measured in the hand, correlations with hand function (i.e., nine-hole pegboard test) would have been more appropriate, especially as fampridine has been shown to have a positive impact on hand function [30]. Alternatively, we could have chosen a lower extremity muscle (rather than the FDI) to measure lower extremity cortical excitability. This choice of the FDI was based on the many challenges associated with measuring cortical excitability in lower extremities, such as the deeper location of the lower extremity motor areas (making them significantly more difficult to activate), the difficulty of targeting individual lower extremity muscles, and the paucity of studies investigating the reliability and reproducibility of lower extremity excitability studies. Yet another limitation is the short follow-up period, which was limited to 2 weeks. It has been demonstrated that walking speed can continue to improve up to 4 weeks after fampridine initiation [15,21]. As such, a longer period of assessment might have been required to more reliably define the responder status. These shortcomings were inevitable, given the observational nature of the study. Indeed, the inclusion criteria and follow-up period were dictated by our national guidelines for selecting candidates for fampridine therapy. As such, inclusion was restricted to patients with gait impairment, most of which had little or no baseline abnormalities in the hands, and the follow-up period was limited to 2 weeks. Our study could also be criticized for including 10 patients on fingolimod, which was found to reduce ICF [4]. Since no changes in disease modifying and symptomatic treatments were allowed during the period extending from 6 months prior to inclusion to the end of the follow-up, this drug would have mostly affected the baseline recordings rather than the changes observed after fampridine administration. Patients receiving high-dose intravenous steroids, on the other hand, were excluded from the study, given its significant impact on cortical excitability (reduction of SICI and enhancement ICF) [9]. Lastly, it is important to mention the limitations of EDSS as a disability measure in MS. The latter has well documented weaknesses in reliability and sensitivity to change [31].

5. Conclusions

In conclusion, our work suggests that fampridine has a significant modulatory effect on the cerebral cortex, demonstrated by an increase in excitatory intracortical mechanisms as unraveled by paired-pulse TMS paradigms. rMT could be useful in selecting patients more likely to experience a favorable response to fampridine. Larger studies with more reliable outcomes (i.e., hand function) are needed to better define the cortical excitability parameters that best discriminate between potential responders and non-responders.

Author Contributions: Conceptualization, R.A., S.S.A. and N.R.; Formal analysis, G.K.; Investigation, R.A., M.M.S., A.R.S., J.F., M.S., K.M., R.E.N., S.E.N. and N.R.; Methodology, R.A., S.S.A. and N.R.; Writing—original draft, R.A., M.M.S., A.R.S., G.K., J.F., M.S., K.M., R.E.N., S.E.N., S.S.A. and N.R.; Writing—review & editing, R.A. and N.R.

Funding: This research was funded by Biogen grant number LBN-FMP-14-10739.

Acknowledgments: This study was supported by an investigator-initiated grant from Biogen.

Conflicts of Interest: SSA declares having received travel grants or compensation from Genzyme, Biogen, Novartis and Roche. The remaining authors declare no conflict of interest.

References

1. Bethoux, F.; Bennett, S. Evaluating walking in patients with multiple sclerosis: Which assessment tools are useful in clinical practice? *Int. J. MS Care* **2011**, *13*, 4–14. [CrossRef]
2. Heesen, C.; Bohm, J.; Reich, C.; Kasper, J.; Goebel, M.; Gold, S.M. Patient perception of bodily functions in multiple sclerosis: Gait and visual function are the most valuable. *Mult. Scler.* **2008**, *14*, 988–991. [CrossRef]
3. Kaji, R.; Sumner, A.J. Effects of 4-aminopyridine in experimental CNS demyelination. *Neurology* **1988**, *38*, 1884–1887. [CrossRef]
4. Mainero, C.; Inghilleri, M.; Pantano, P.; Conte, A.; Lenzi, D.; Frasca, V.; Bozzao, L.; Pozzilli, C. Enhanced brain motor activity in patients with MS after a single dose of 3,4-diaminopyridine. *Neurology* **2004**, *62*, 2044–2050. [CrossRef]
5. Pikoulas, T.E.; Fuller, M.A. Dalfampridine: A medication to improve walking in patients with multiple sclerosis. *Ann. Pharmacother.* **2012**, *46*, 1010–1015. [CrossRef] [PubMed]
6. Brambilla, L.; Rossi Sebastiano, D.; Aquino, D.; Torri Clerici, V.; Brenna, G.; Moscatelli, M.; Frangiamore, R.; Giovannetti, A.M.; Antozzi, C.; Mantegazza, R.; et al. Early effect of dalfampridine in patients with MS: A multi-instrumental approach to better investigate responsiveness. *J. Neurol. Sci.* **2016**, *368*, 402–407. [CrossRef] [PubMed]
7. Filli, L.; Werner, J.; Beyer, G.; Reuter, K.; Petersen, J.A.; Weller, M.; Zorner, B.; Linnebank, M. Predicting responsiveness to fampridine in gait-impaired patients with multiple sclerosis. *Eur. J. Neurol.* **2019**, *26*(2), 281–289. [CrossRef] [PubMed]
8. Zeller, D.; Reiners, K.; Brauninger, S.; Buttmann, M. Central motor conduction time may predict response to fampridine in patients with multiple sclerosis. *J. Neurol. Neurosurg. Psychiatry* **2014**, *85*, 707–709. [CrossRef] [PubMed]
9. Ayache, S.S.; Creange, A.; Farhat, W.H.; Zouari, H.G.; Mylius, V.; Ahdab, R.; Abdellaoui, M.; Lefaucheur, J.P. Relapses in multiple sclerosis: Effects of high-dose steroids on cortical excitability. *Eur. J. Neurol.* **2014**, *21*, 630–636. [CrossRef]
10. Conte, A.; Lenzi, D.; Frasca, V.; Gilio, F.; Giacomelli, E.; Gabriele, M.; Bettolo, C.M.; Iacovelli, E.; Pantano, P.; Pozzilli, C.; et al. Intracortical excitability in patients with relapsing-remitting and secondary progressive multiple sclerosis. *J. Neurol.* **2009**, *256*, 933–938. [CrossRef]
11. Liepert, J.; Mingers, D.; Heesen, C.; Baumer, T.; Weiller, C. Motor cortex excitability and fatigue in multiple sclerosis: A transcranial magnetic stimulation study. *Mult. Scler.* **2005**, *11*, 316–321. [CrossRef] [PubMed]
12. Rossi, S.; Furlan, R.; De Chiara, V.; Motta, C.; Studer, V.; Mori, F.; Musella, A.; Bergami, A.; Muzio, L.; Bernardi, G.; et al. Interleukin-1beta causes synaptic hyperexcitability in multiple sclerosis. *Ann. Neurol.* **2012**, *71*, 76–83. [CrossRef] [PubMed]
13. Polman, C.H.; Reingold, S.C.; Banwell, B.; Clanet, M.; Cohen, J.A.; Filippi, M.; Fujihara, K.; Havrdova, E.; Hutchinson, M.; Kappos, L.; et al. Diagnostic criteria for multiple sclerosis: 2010 revisions to the McDonald criteria. *Ann. Neurol.* **2011**, *69*, 292–302. [CrossRef] [PubMed]
14. Kurtzke, J.F. Rating neurologic impairment in multiple sclerosis: An expanded disability status scale (EDSS). *Neurology* **1983**, *33*, 1444–1452. [CrossRef] [PubMed]
15. Goodman, A.D.; Brown, T.R.; Krupp, L.B.; Schapiro, R.T.; Schwid, S.R.; Cohen, R.; Marinucci, L.N.; Blight, A.R. Sustained-release oral fampridine in multiple sclerosis: A randomised, double-blind, controlled trial. *Lancet* **2009**, *373*, 732–738. [CrossRef]
16. Hobart, J.; Blight, A.R.; Goodman, A.; Lynn, F.; Putzki, N. Timed 25-foot walk: Direct evidence that improving 20% or greater is clinically meaningful in MS. *Neurology* **2013**, *80*, 1509–1517. [CrossRef]
17. Rossini, P.M.; Barker, A.T.; Berardelli, A.; Caramia, M.D.; Caruso, G.; Cracco, R.Q.; Dimitrijevic, M.R.; Hallett, M.; Katayama, Y.; Lucking, C.H.; et al. Non-invasive electrical and magnetic stimulation of the brain, spinal cord and roots: Basic principles and procedures for routine clinical application. Report of an IFCN committee. *Electroencephalogr. Clin. Neurophysiol.* **1994**, *91*, 79–92. [CrossRef]
18. Kujirai, T.; Caramia, M.D.; Rothwell, J.C.; Day, B.L.; Thompson, P.D.; Ferbert, A.; Wroe, S.; Asselman, P.; Marsden, C.D. Corticocortical inhibition in human motor cortex. *J. Physiol.* **1993**, *471*, 501–519. [CrossRef]
19. Lefaucheur, J.P.; Drouot, X.; Menard-Lefaucheur, I.; Keravel, Y.; Nguyen, J.P. Motor cortex rTMS restores defective intracortical inhibition in chronic neuropathic pain. *Neurology* **2006**, *67*, 1568–1574. [CrossRef]

20. Quinlan, J.R. *C4.5: programs for machine learning*; Morgan Kaufmann Publishers Inc.: San Francisco, CA, USA, 1993.
21. Goodman, A.D.; Brown, T.R.; Edwards, K.R.; Krupp, L.B.; Schapiro, R.T.; Cohen, R.; Marinucci, L.N.; Blight, A.R. A phase 3 trial of extended release oral dalfampridine in multiple sclerosis. *Ann. Neurol.* **2010**, *68*, 494–502. [CrossRef]
22. Chen, R.; Samii, A.; Canos, M.; Wassermann, E.M.; Hallett, M. Effects of phenytoin on cortical excitability in humans. *Neurology* **1997**, *49*, 881–883. [CrossRef] [PubMed]
23. Ziemann, U.; Lonnecker, S.; Steinhoff, B.J.; Paulus, W. Effects of antiepileptic drugs on motor cortex excitability in humans: A transcranial magnetic stimulation study. *Ann. Neurol.* **1996**, *40*, 367–378. [CrossRef] [PubMed]
24. Caramia, M.D.; Palmieri, M.G.; Desiato, M.T.; Boffa, L.; Galizia, P.; Rossini, P.M.; Centonze, D.; Bernardi, G. Brain excitability changes in the relapsing and remitting phases of multiple sclerosis: A study with transcranial magnetic stimulation. *Clin. Neurophysiol. Off. J. Int. Fed. Clin. Neurophysiol.* **2004**, *115*, 956–965. [CrossRef] [PubMed]
25. Ayache, S.S.; Créange, A.; Farhat, W.H.; Zouari, H.G.; Lesage, C.; Palm, U.; Abdellaoui, M.; Lefaucheur, J.P. Cortical excitability changes over time in progressive multiple sclerosis. *Funct. Neurol.* **2015**, *30*, 257–263. [CrossRef]
26. Vucic, S.; Burke, T.; Lenton, K.; Ramanathan, S.; Gomes, L.; Yannikas, C.; Kiernan, M.C. Cortical dysfunction underlies disability in multiple sclerosis. *Mult. Scler.* **2012**, *18*, 425–432. [CrossRef]
27. Mori, F.; Kusayanagi, H.; Buttari, F.; Centini, B.; Monteleone, F.; Nicoletti, C.G.; Bernardi, G.; Di Cantogno, E.V.; Marciani, M.G.; Centonze, D. Early treatment with high-dose interferon beta-1a reverses cognitive and cortical plasticity deficits in multiple sclerosis. *Funct. Neurol.* **2012**, *27*, 163–168.
28. Ayache, S.S.; Chalah, M.A. Cortical excitability changes: A mirror to the natural history of multiple sclerosis? *Neurophysiolie Clin.* **2017**, *47*, 221–223. [CrossRef]
29. Savin, Z.; Lejbkowicz, I.; Glass-Marmor, L.; Lavi, I.; Rosenblum, S.; Miller, A. Effect of Fampridine-PR (prolonged released 4-aminopyridine) on the manual functions of patients with Multiple Sclerosis. *J. Neurol. Sci.* **2016**, *360*, 102–109. [CrossRef]
30. Rizzo, V.; Quartarone, A.; Bagnato, S.; Battaglia, F.; Majorana, G.; Girlanda, P. Modification of cortical excitability induced by gabapentin: A study by transcranial magnetic stimulation. *Neurol. Sci.* **2001**, *22*, 229–232. [CrossRef]
31. Meyer-Moock, S.; Feng, Y.S.; Maeurer, M.; Dippel, F.W.; Kohlmann, T. Systematic literature review and validity evaluation of the Expanded Disability Status Scale (EDSS) and the Multiple Sclerosis Functional Composite (MSFC) in patients with multiple sclerosis. *BMC Neurol.* **2014**, *14*, 58. [CrossRef]

Article

Impaired Expression of Tetraspanin 32 (TSPAN32) in Memory T Cells of Patients with Multiple Sclerosis

Maria Sofia Basile [1], Emanuela Mazzon [2], Katia Mangano [1], Manuela Pennisi [1], Maria Cristina Petralia [2], Salvo Danilo Lombardo [1], Ferdinando Nicoletti [1], Paolo Fagone [1,*] and Eugenio Cavalli [2]

[1] Department of Biomedical and Biotechnological Sciences, University of Catania, Via S. Sofia 89, 95123 Catania, Italy; sofiabasile@hotmail.it (M.S.B.); kmangano@unict.it (K.M.); manuela.pennisi@unict.it (M.P.); salvo.lombardo.sdl@gmail.com (S.D.L.); ferdinic@unict.it (F.N.)
[2] IRCCS Centro Neurolesi "Bonino-Pulejo", Via Provinciale Palermo, Contrada Casazza, 98124 Messina, Italy; emanuela.mazzon@irccsme.it (E.M.); m.cristinapetralia@gmail.com (M.C.P.); eugenio.cavalli@irccsme.it (E.C.)
* Correspondence: paolofagone@yahoo.it

Received: 29 November 2019; Accepted: 14 January 2020; Published: 17 January 2020

Abstract: Tetraspanins are a conserved family of proteins involved in a number of biological processes. We have previously shown that Tetraspanin-32 (TSPAN32) is significantly downregulated upon activation of T helper cells via anti-CD3/CD28 stimulation. On the other hand, TSPAN32 is marginally modulated in activated Treg cells. A role for TSPAN32 in controlling the development of autoimmune responses is consistent with our observation that encephalitogenic T cells from myelin oligodendrocyte glycoprotein (MOG)-induced experimental autoimmune encephalomyelitis (EAE) mice exhibit significantly lower levels of TSPAN32 as compared to naïve T cells. In the present study, by making use of ex vivo and in silico analysis, we aimed to better characterize the pathophysiological and diagnostic/prognostic role of TSPAN32 in T cell immunity and in multiple sclerosis (MS). We first show that TSPAN32 is significantly downregulated in memory T cells as compared to naïve T cells, and that it is further diminished upon ex vivo restimulation. Accordingly, following antigenic stimulation, myelin-specific memory T cells from MS patients showed significantly lower expression of TSPAN32 as compared to memory T cells from healthy donors (HD). The expression levels of TSPAN32 was significantly downregulated in peripheral blood mononuclear cells (PBMCs) from drug-naïve MS patients as compared to HD, irrespective of the disease state. Finally, when comparing patients undergoing early relapses in comparison to patients with longer stable disease, moderate but significantly lower levels of TSPAN32 expression were observed in PBMCs from the former group. Our data suggest a role for TSPAN32 in the immune responses underlying the pathophysiology of MS and represent a proof-of-concept for additional studies aiming at dissecting the eventual contribution of TSPAN32 in other autoimmune diseases and its possible use of TSPAN32 as a diagnostic factor and therapeutic target.

Keywords: TSPAN32; tetraspanins; multiple sclerosis; cellular immunity; memory T cells

---

1. Introduction

Tetraspanins are a conserved family of proteins involved in several biological processes, such as the regulation of cellular adhesion, motility, cancer metastasis, signal transduction, and activation [1,2]. Tetraspanins comprise four transmembrane (TM) domains. TM domains 1 and 2 flank a small extracellular loop (SEL), while TM3 and TM4 flank a large extracellular loop (LEL). TM domains are typically involved in the interaction with non-tetraspanin molecules. The juxtamembrane cysteine residues in the cytoplasmic domains contribute to the formation of tetraspanin-enriched microdomains

(TEMs), while the cytoplasmic regions provide links to cytoskeletal and signaling molecules [3]. Several immune-related proteins take part in TEMs, including pattern recognition receptors, co-stimulatory molecules, Major Histocompatibility Complex molecules and T cell receptor-associated proteins (reviewed in [2]). The tetraspanins Cluster of Differentiation 82 (CD82), CD9, CD63, CD81, and CD53 exert a co-stimulatory role in T cells [4,5], whereas cells deficient for CD37, CD151, and CD81 have been shown to be hyperproliferative following stimulation [6–8]. Tarrant and colleagues [9] have shown that T cells from Tssc6 Tetraspanin-32 (TSPAN32)-deficient mice have increased responses upon stimulation, and have proposed that TSPAN32 may negatively regulate peripheral T-lymphocyte activation. Along the same lines, we have previously shown that TSPAN32 expression is significantly reduced upon cell activation, although in Treg cells, TSPAN32 levels undergo minor changes. Moreover, significantly lower levels of TSPAN32 were found in encephalitogenic T cells from myelin oligodendrocyte glycoprotein (MOG)-Induced experimental autoimmune encephalomyelitis (EAE) mice. Finally, ex vivo-activated circulating CD4 T cells from MS patients showed lower levels of TSPAN32 as compared to cells from healthy donors [10].

Multiple sclerosis (MS) is the most frequent immuno-inflammatory disorder of the central nervous system, characterized by immune cell infiltration, microglia activation and progressive demyelination, with consequent neurological deficits. It is well-known that increased conversion from naïve to memory cells can be observed in MS [11] and that most of the myelin-reactive T cells are present in the memory T cell subset [12]. It has been also shown that memory T cells are activated independently of CD28 co-stimulation [13,14]. In the present paper, we aimed to better characterize the pathophysiological role of TSPAN32 in cellular immunity and in MS. To this aim, by making use of ex vivo and in silico analysis, we have evaluated the expression levels of TSPAN32 in memory T cells from healthy donors and MS patients, both in inactive state and upon activation. Next, we determined the diagnostic and prognostic value of TSPAN32 in the peripheral blood mononuclear cells (PBMCs) of MS patients. Our analysis demonstrates that TSPAN32 is significantly downregulated in memory T cells as compared to naïve T cells, and that it is further diminished upon ex vivo restimulation. In addition, following antigenic stimulation, myelin-specific memory T cells from MS patients exhibited significantly lower expression of TSPAN32 as compared to memory T cells from healthy donors (HD). Further, the expression levels of TSPAN32 was significantly downregulated in PBMCs from drug-naïve MS patients as compared to HD, irrespective of the disease state. Finally, we observed a moderate but significantly reduced expression of TSPAN32 in PBMCs from MS patients undergoing early relapses in comparison to those from patients with a longer course of stable disease.

2. Materials and Methods

2.1. Ex Vivo Study

2.1.1. Cell Isolation and Real-Time PCR

Mononuclear cells were obtained from the peripheral blood of healthy donors (HD) ($n = 7$) by step-gradient centrifugation, using the Ficoll–Hypaque medium (Sigma Aldrich, Milano, Italy), as per manufacturer's instructions. CD4 + CD45RA + CD45RO – CD25 + CD127^{low} cells (naive Treg cells), CD4 + CD45RA – CD45RO + CD25 + CD127^{low} cells (memory Treg cells), CD4 + CD45RA – CD45RO + CD25 – CD127 + cells (memory Teff cells), and CD4 + CD45RA + CD45RO – CD25 – CD127 + cells (naive Teff cells) were enriched by magnetic beads sorting, obtaining a cell purity of at least 95%. In another set of experiments, memory Teff cells from 3 healthy donors were activated by plate-bound anti-CD3 (10 μg/mL) and anti-CD28 (5 μg/mL) for 12 h.

2.1.2. Real-Time PCR

Total RNA was extracted and gene expression levels were determined by real-time PCR. 2 μg of total RNA were reverse-transcribed with a High-Capacity cDNA Reverse Transcription Kit (Applied

Biosystems, Monza, Italy) in a 20 μL reaction volume, and real-time PCR was performed using the SYBR Green PCR Master Mix (Applied Biosystems, Monza, Italy), 200 nM forward and 200 nM reverse primers, and 20 μg cDNA. Relative gene expression levels were obtained using the formula: $2^{-\Delta\Delta Ct}$, where $\Delta\Delta Ct = (Ct_{\text{target gene}} - Ct_{\text{beta-actin}})$ stimulated cells – $(Ct_{\text{target gene}} - Ct_{\text{beta-actin}})$ control cells.

2.2. In Silico Analysis

The Gene Expression Omnibus (GEO; https://www.ncbi.nlm.nih.gov/gds) browser was interrogated using the MeSH (Medical Subject Headings) term "Multiple Sclerosis". Datasets were manually excluded if the studies were not performed on human subjects, if the patients enrolled were under immunosuppressive/immunomodulatory treatment, and if the cell types analyzed were not immune cells. For the evaluation of the expression levels of TSPAN32 in encephalitogenic memory T cells, and the evaluation of the diagnostic role of TSPAN32, datasets carried out only on one cohort of subjects (i.e., MS patients and healthy donors) were excluded. For the afore-mentioned reasons, the analysis was then carried out on the GSE66763 and the GSE138064, respectively. For the determination of the prognostic properties of TSPAN32 in predicting MS relapses, the GSE15245 was selected as it is the only dataset including prospective data on disease evolution. A flowchart of the in silico study design is provided as Figure 1. The characteristics of the datasets used are described in the following sections.

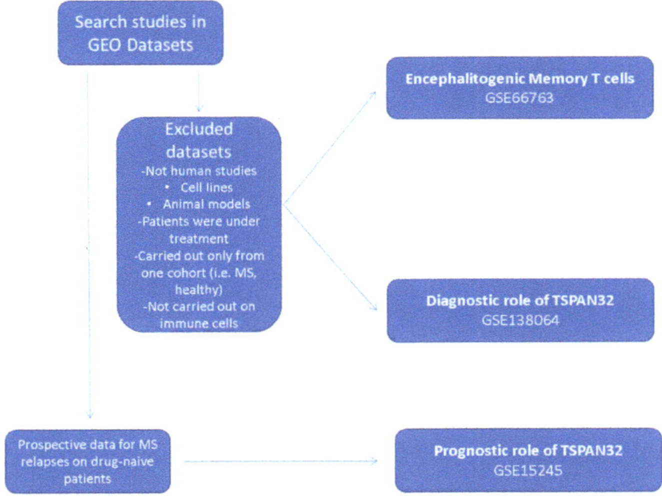

Figure 1. Flowchart of the in silico study.

2.2.1. TSPAN32 in Memory T Cells from MS Patients

The GSE66763 dataset was used to investigate the expression levels of TSPAN32 in circulating memory T cells from MS patients [15]. The dataset included whole-genome RNA sequencing data of C-C Motif Chemokine Receptor 6 (CCR6)$^+$ memory (CD45RA – CD45RO + CD25 – CCR6+) CD4+ T from 3 Human Leukocyte Antigen – DR isotype (HLA-DR)4+ healthy subjects and 5 HLA-DR4+ MS patients. Cells were amplified by PhytoHaemAgglutinin (PHA) and Interleukin (IL)-2 and stimulated by irradiated autologous monocytes and DR4 myelin peptides Myelin Oligodendrocyte Glycoprotein ((MOG)$_{97-109}$ and ProteoLipid Protein (PLP)$_{180-199}$). Patients were immunotherapeutic naïve or had not received treatment for at least 12 months. Cell proliferation was determined and the highest proliferated wells were chosen for DR4 tetramers staining (MOG$_{97-109}$-tetramers and PLP$_{180-199}$-tetramers). Then, myelin tetramer+ and tetramer− cells were sorted and lysed for the extraction of RNA and subsequent

RNA sequencing. Gene expression is shown as $\log_2$ Fragments Per Kilobase of transcript per Million mapped (FPKM) values.

2.2.2. TSPAN32 in PBMCs from MS Patients

In order to investigate the expression levels of TSPAN32 in PBMCs from MS patients in both stable and active disease, as compared to healthy donors, we interrogated the GSE138064 dataset [16]. The dataset included transcriptomic data from therapy-naïve Relapse-Remitting (RR) MS patients (10 with stable MS, age 45.2 ± 2.6, 8/2 female/male, and 9 during relapse, age 46.3 ± 3.5, 8/2 female/male). Eight healthy controls were included, age 42.3 ± 4.8, 5/3 female/male.

2.2.3. Predictive Analysis of TSPAN32 in MS

In order to evaluate the relationship between expression levels of TSPAN32 and the time to relapse in MS patients, we interrogated the GSE15245 dataset that included whole-genome transcriptomic profiles of PMBCs from 51 drug-naïve MS patients [17]. The patient's age was 38.5 ± 1.4, with a mean Expanded Disability Status Scale (EDSS) score of 2.4 ± 0.2. The Affymetrix Human Genome U133A 2.0 Array was used for the generation of the dataset and raw data were preprocessed using the robust multi-array average (RMA) algorithm. Sample population was sorted based on the expression levels of TSPAN32 and log-rank test was applied to evaluate differences in the percentage of patients developing acute relapses in a 1500-day time frame.

2.3. Statistical Analysis

Data are shown as mean ± SD and statistical analysis was performed using either a Student's *t*-test or one-way ANOVA followed by Fisher's Least Significant Difference test. Correlation analysis was performed using the non-parametric Spearman's test. Hierarchical clustering was used to determine the relative distance of samples using Pearson's correlation as similarity comparison. The self organizing map (SOM) algorithm was used for the unsupervised identification of clusters of commonly modulated genes [18]. Distance metric for SOM was Pearson's correlation, with random genes initialization, Gaussian neighborhood, and 2000 iterations. The linear model for microarray (LIMMA) algorithm was used to evaluate statistical significance for differences in RNA sequencing data [19]. As the experimental design and the information provided are different for the three whole-genome transcriptomic datasets here analyzed, and in consideration that no additional datasets with overlapping experimental layouts are currently available in publicly available databases, a meta-analysis cannot be performed. Gene ontology and gene term enrichment analysis was conducted using the web-based utility, Metascape [20]. GraphPad Prism 8 and MeV (version 4.9) software programs were used for the statistical analysis and the generation of the graphs.

3. Results

3.1. TSPAN32 in Memory T Cells

When analyzing the expression levels of TSPAN32 in memory CD4+ T cells from healthy donors, we observed significantly lower levels of TSPAN32 in memory T effector cells as compared to naïve T cells ($p < 0.01$) (Figure 2A). On the other hand, no modulation was observed in memory Treg cells (Figure 2A). We also wanted to determine whether a modulation of TSPAN32 could be found upon restimulation. As shown in Figure 2B, restimulation of memory T cells is associated to a significant down regulation in TSPAN32 levels ($p < 0.001$) (Figure 2B). Similar data have been obtained from the analysis of the GSE22886 dataset (Table S1).

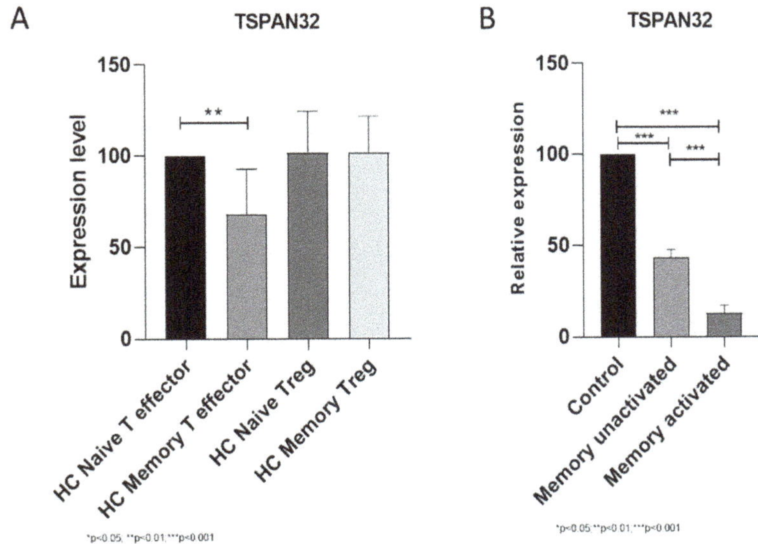

Figure 2. (**A**) Basal expression levels of Tetraspanin-32 (TSPAN32) in naïve T effector, Treg, memory T effector, and memory Treg cells from healthy donors; (**B**) modulation of TSPAN32 expression in memory T cells upon reactivation. ** $p < 0.01$, *** $p < 0.001$.

Next, we wanted to determine the expression of TSPAN32 in memory CD4+ T cells from HLA-DR4+ MS patients, following amplification by PHA and IL-2 and stimulation by irradiated autologous monocytes and DR4 myelin peptides. As shown in Figure 3A, significant lower levels of TSPAN32 were observed in tetramer+ memory T cells from MS patients as compared to tetramer- memory T cells from HD ($p < 0.05$). Similarly, comparable levels of TSPAN32 were observed in tetramer+ memory T cells from HD (Figure 3A). SOM analysis identified 599 genes that clustered together with TSPAN32 (Cluster 5) (Figure 3B). Gene ontology revealed that the most significant enriched terms were "Small GTPase-mediated signal transduction", "Meiosis", "DNA repair", "BARD1 pathway" and "Membrane lipid biosynthetic process" (Figure 3B–D). Interestingly, significantly lower TSPAN32 levels were also observed in tetramer- memory T cells from MS patients (Figure 3A). As LIMMA analysis revealed significant transcriptomic differences between tetramer- MS memory T cells and tetramer- HD memory T cells, with enrichment of several immune-related biological processes (Figure S1A,B), and HCL analysis clustered together tetramer- and tetramer+ memory T cells from MS patients (Figure S1C), the reduced TSPAN32 levels may be associated to a reduced activation threshold of memory T cells from MS patients, and could explain the underlying autoimmune process.

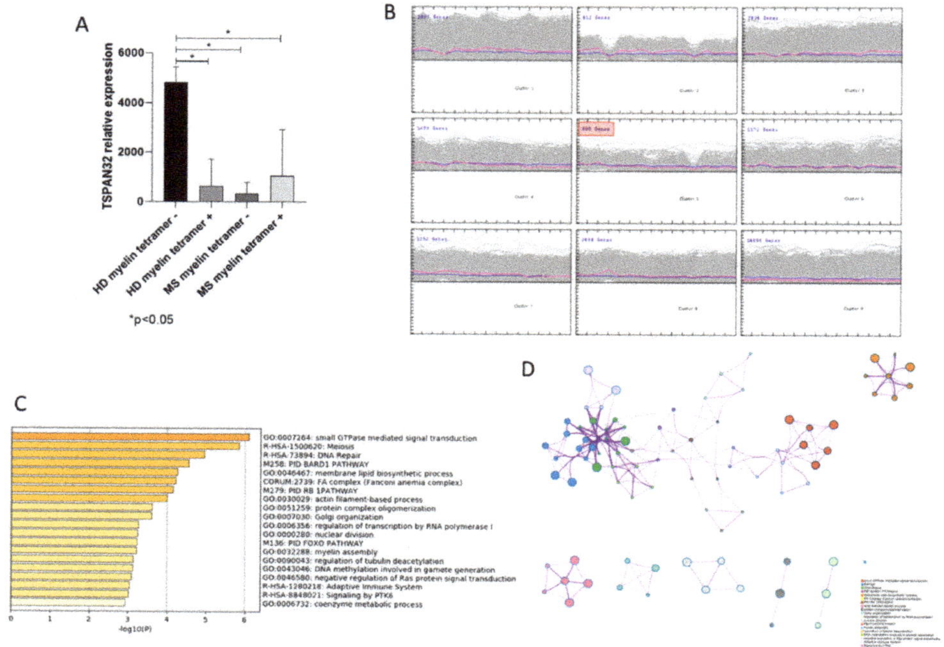

Figure 3. (**A**) TSPAN32 expression in memory T cells from healthy donors and multiple sclerosis (MS) patients; * $p < 0.05$; (**B**) clusters of genes obtained from the self organizing map (SOM) analysis; (**C**) most enriched biological processes by genes commonly regulated with TSPAN32, as obtained from SOM analysis; (**D**) network showing the interconnection among the most enriched biological processes by genes commonly regulated with TSPAN32, obtained from SOM analysis.

3.2. TSPAN32 Expression in PBMCs from MS Patients

In order to evaluate whether a modulation in TSPAN32 levels could be observed in peripheral immune cells from MS patients, we interrogated the GSE138064 dataset. As shown in Figure 4A, a significant reduction in TSPAN32 expression was observed in PBMCs from MS patients in both stable and relapsing disease ($p < 0.001$) (Figure 4A). Receiver operating characteristic (ROC) analysis confirmed the diagnostic ability of TSPAN32 to discriminate MS from HD, entailing a $p < 0.001$ (Figure 4B,C). No significant differences were instead observed when comparing TSPAN32 levels in PBMCs from patients in stable disease as compared to PBMCs from patients in exacerbation (Figure 4A). Accordingly, ROC curve area was 0.6889, entailing a $p = 0.1651$ (Figure 4D). This is in accordance with data from the GSE19224 dataset, that show an adjusted p value > 0.99 and a $\log_2$(fold) change of 0.276 for TSPAN32 expression levels when comparing PBMCs from MS patients in stable versus relapsing disease (https://www.ncbi.nlm.nih.gov/geo/geo2r/?acc=GSE19224).

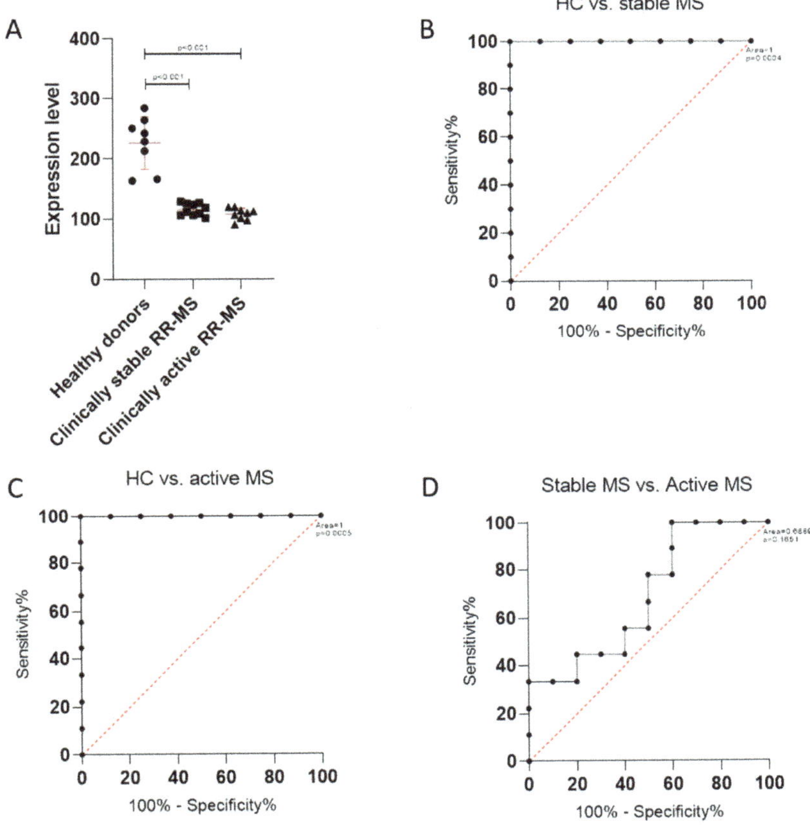

Figure 4. (**A**) TSPAN32 in peripheral blood mononuclear cells (PBMCs) from healthy donors and RRMS patients in stable and relapsing disease; (**B**) receiver operating characteristic (ROC) curve for TSPAN32 in healthy controls (HC) and multiple sclerosis patients in stable disease; (**C**) receiver operating characteristic (ROC) curve for TSPAN32 in healthy controls (HC) and multiple sclerosis patients in active disease; (**D**) receiver operating characteristic (ROC) curve for TSPAN32 in multiple sclerosis patients in stable and active disease.

Finally, we evaluated whether the different transcriptional levels of TSPAN32 in PBMCs from MS patients could promote disease exacerbation or protect MS patients from acute relapses. Non-parametric correlation between TSPAN32 and the time-to-relapse revealed a trend of direct correlation, which did not reach the statistical significance ($p = 0.0856$) (Figure 5A). ROC curve area was 0.6036, entailing a $p = 0.3695$ (Figure 5B). However, Log-rank analysis performed on patients divided into two groups based on the expression level of TSPAN32 in PBMCs (referred as High and Low TSPAN32) showed that a trend of protection from acute relapses was observed in patients expressing higher TSPAN32 levels (Figure 5C). In addition, significantly lower levels of TSPAN32 were found in PBMCs from MS patients developing exacerbation of the disease before 300 days as compared with patients who underwent relapses later than 1500 days (Figure 5D).

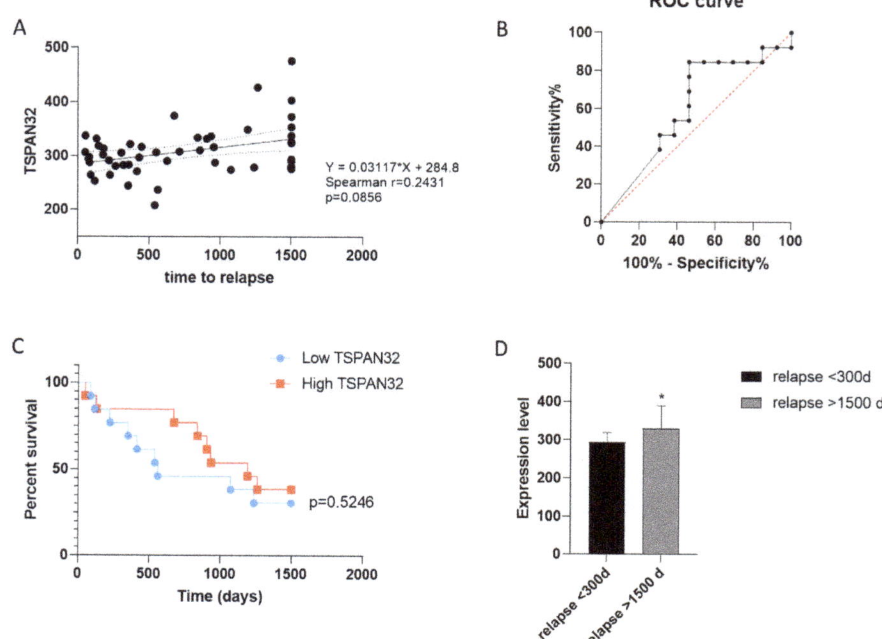

Figure 5. (**A**) Correlation between TSPAN32 expression in PBMCs from drug-naïve MS patients and the time-to-relapse; Spearman correlation value (r) and p value (p) are indicated. (**B**) receiver operating characteristic (ROC) curve for the evaluation of the prognostic value of TSPAN32 in predicting relapses in MS patients; (**C**) log-rank analysis for time-to-relapse in patients expressing low and high levels of TSPAN32 in PBMCs, respectively; (**D**) expression levels of TSPAN32 in MS patients undergoing early exacerbation of the disease (<300 days) or with longer stable disease (relapse > 1500 days).

4. Discussion

Diverse members of the TSPAN family have been shown to be involved in the regulation of both the innate and adaptive immune responses. For instance, CD81 is involved in the formation of the immune synapse, providing a link between the Antigen Presenting Cells and the T cells [21,22], while CD37 and CD151 promote antigen presentation and regulate the costimulatory signaling pathways [23].

In a TSPAN32 (Tssc6)-deficient mouse model, despite normal hemopoiesis, T cell proliferation and responses are significantly augmented [9]. It has also been observed that the activity of T cells from mice double knockout for CD37 and TSPAN32 are upregulated, and that the dendritic cell stimulation capacity is increased as compared to single knockout, suggesting a cooperative role for these two tetraspanins in controlling T cell-mediated immunity [24]. These findings suggest that TSPAN32 might contribute to shape cellular immunity. In our previous work, we have described that T cells express a baseline level of TSPAN32, favoring the maintenance of an inactive state, which is decreased following CD3-mediated signaling [10].

By means of in silico and ex vivo analyses, in this study we wanted to gain further insights into the role of TSPAN in the biology and physiology of memory T cells and evaluated whether its expression was altered in memory T cells from MS patients as compared to HD. We also studied the possible diagnostic and prognostic value of TSPAN32 expression in PBMC of MS patients, on the course of the disease. The use of whole-genome expression databases has been largely exploited [25–28] for the characterization of pathogenic pathways and to identify therapeutic targets for a variety of disorders, including immunoinflammatory and autoimmune diseases [29–36], cancer [37–39], and has allowed dismantling pathogenetic pathways [40–42], along with the identification of novel tailored therapeutic targets [43–46].

MS is an autoimmune/immunoinflammatory disorder sustained by activated, myelin-specific T cells that migrate into the central nervous system (CNS), promoting inflammation. The characterization of the phenotype of myelin-specific immune cells is, therefore, crucial for the elucidation of MS pathogenesis [47–49].

In the present study, we have first analyzed the expression levels of TSPAN32 in circulating memory T cells from HD and we show that significantly lower levels of TSPAN32 can be observed in memory T effector cells but not in memory Treg cells. This is in line with our previous observation that only a marginal downregulation of TSPAN32 occurs in Treg cells, upon activation [10]. Interestingly, following the in vitro reactivation of memory T effector cells, TSPAN32 expression levels further decreased. The differential pattern of expression and modulation of TSPAN32 in Treg cells has yet to be deciphered.

Next, we analyzed the expression of TSPAN32 in autoreactive T cells from MS patients. As shown by Cao et al., myelin-reactive T cells from MS patients are prevalently from the memory CCR6+ T cell population, and are characterized by the secretion of larger amounts of proinflammatory cytokines as compared to T cells from HD. As expected, the expression levels of TSPAN32 in myelin-reactive MS tetramer-positive T cells resulted significantly lower than those in tetramer-negative memory T cells from HD. On the other hand, a significant lower expression of TSPAN32 was found in MS tetramer-negative memory T cells, comparable to that of MS tetramer-positive T cells. Although this may be counterintuitive, the observation that the transcriptomic features of the MS tetramer-negative memory T cells are more closely related to those of the MS tetramer-positive memory T cells than those of HD tetramer-negative memory T cells suggests that a lower activation threshold characterizes memory T cells from MS patients. Notably, a similar trend of reduced TSPAN32 levels was also observed in myelin-reactive HD tetramer-positive T cells. It is already known that MS patients and healthy subjects share a similar number of circulating myelin-reactive T cells. The lower levels of TSPAN32 in these cells suggest that regardless of the activation state of the T cells, the engagement of the TCR with the cognate ligand is sufficient to modulate the expression of TSPAN32. This is in accordance with our previous data, showing that the CD3-mediated signaling is sufficient to downregulate TSPAN32 gene expression.

Finally, we found diminished TSPAN32 levels in PBMCs from MS patients, both in stable and active disease, as compared to HD. No differences were however found between patients in stable versus relapsing disease. In addition, only a moderate reduction in the time-to-relapse was observed in patients expressing higher levels of TSPAN32 when MS patients were divided into two groups on the basis of their level of TSPAN32 expression in PBMC (referred as High and Low TSPAN32).

However, it was possible to observe that those with high expression had a moderate but significant protection from acute relapses. In agreement with this observation, significantly lower levels of TSPAN32 were found in PBMCs from MS patients developing exacerbation of the disease before 300 days, as compared with patients who underwent relapses later than 1500 days. Overall, our data suggest that the defective expression of TSPAN32 may characterize different T cell subsets of MS patients, including memory T cells, and that this may contribute to trigger anti-myelin immune responses. Along with our previous publication [10], this new transcriptomic analysis strengthens our hypothesis that defective TSPAN32 expression may represent an additional important immunopathogenetic abnormality that may play a role in the pathogenesis of at least some cases of MS.

It should also be pointed out that, although all of the in silico data have been generated from third-party reanalysis of whole-genome transcriptomic datasets previously validated by the respective original authors, the number of biological replicates in each of these datasets is relatively low. Therefore, although statistical significance has been achieved in most cases of our analyses, the data warrant to be confirmed from a larger population of MS patients. Along this line of research, it will be interesting to study if and how the current disease modifying therapies influence the course of the disease by modulating TSPAN32 expression. In a similar manner, the expression profile of TSPAN32 in secondary progressive and primary progressive MS seems of interest. Additional studies may also be warranted to dismantle whether defective expression of TSPAN32 is also observed in other T cell-mediated autoimmune diseases.

Moreover, further effort is required to understand the molecular pathways involved in the regulation of the immune responses exerted by TSPAN32. Up to now, no drugs targeting tetraspanins have received approval for the use in the clinical setting, but many strategies have been explored, including the use of monoclonal antibodies, recombinant soluble large extracellular loops or RNA interference (RNAi) (reviewed in [50]). Therefore, it is reasonable that several chances for tailored-specific intervention will be available in the future. Additionally, a deeper understanding of the mechanisms that control TSPAN32 expression could be pursued for their possible efficacy in patients suffering from MS.

5. Conclusions

Our data suggest a role for TSPAN32 in the immune responses underlying the pathophysiology of MS and represent a proof-of-concept for additional studies aiming at dissecting the eventual contribution of TSPAN32 in other autoimmune diseases and its possible use of TSPAN32 as a diagnostic factor and therapeutic target.

Supplementary Materials: The following are available online at http://www.mdpi.com/2076-3425/10/1/52/s1, Figure S1: Comparative transcriptomic profile of MS tetramer- memory T cells; Table S1: Significant genes in the GSE22886 dataset.

Author Contributions: Conceptualization, M.S.B., F.N., P.F., and E.C.; methodology, P.F.; formal analysis, K.M., M.P., M.C.P., and E.C.; investigation, M.S.B., K.M., S.D.L., M.C.P.; resources, E.M., and F.N.; writing—original draft preparation, M.S.B., M.P., S.D.L, P.F., E.C.; writing—review and editing, E.M., K.M., and F.N.; visualization, M.S.B., M.C.P., and P.F.; supervision, P.F. and F.N.; project administration, F.N.; funding acquisition, E.M. All authors have read and agreed to the published version of the manuscript.

Funding: This study was supported by current research funds 2019 of IRCCS "Centro Neurolesi Bonino-Pulejo", Messina, Italy.

Conflicts of Interest: The authors declare no conflicts of interest.

References

1. Termini, C.M.; Gillette, J.M. Tetraspanins Function as Regulators of Cellular Signaling. *Front. Cell Dev. Biol.* **2017**, *5*, 34. [CrossRef] [PubMed]

2. Charrin, S.; Jouannet, S.; Boucheix, C.; Rubinstein, E. Tetraspanins at a glance. *J. Cell Sci.* **2014**, *127*, 3641–3648. [CrossRef] [PubMed]

3. Levy, S.; Shoham, T. The tetraspanin web modulates immune-signalling complexes. *Nat. Rev. Immunol.* **2005**, *5*, 136–148. [CrossRef] [PubMed]

4. Lebel-Binay, S.; Lagaudrière, C.; Fradelizi, D.; Conjeaud, H. CD82, member of the tetra-span-transmembrane protein family, is a costimulatory protein for T cell activation. *J. Immunol.* **1995**, *155*, 101–110.

5. Tai, X.G.; Yashiro, Y.; Abe, R.; Toyooka, K.; Wood, C.R.; Morris, J.; Long, A.; Ono, S.; Kobayashi, M.; Hamaoka, T.; et al. A role for CD9 molecules in T cell activation. *J. Exp. Med.* **1996**, *184*, 753–758. [CrossRef]

6. van Spriel, A.B.; Puls, K.L.; Sofi, M.; Pouniotis, D.; Hochrein, H.; Orinska, Z.; Knobeloch, K.-P.; Plebanski, M.; Wright, M.D. A regulatory role for CD37 in T cell proliferation. *J. Immunol.* **2004**, *172*, 2953–2961. [CrossRef]

7. Miyazaki, T.; Müller, U.; Campbell, K.S. Normal development but differentially altered proliferative responses of lymphocytes in mice lacking CD81. *EMBO J.* **1997**, *16*, 4217–4225. [CrossRef]

8. Wright, M.D.; Geary, S.M.; Fitter, S.; Moseley, G.W.; Lau, L.-M.; Sheng, K.-C.; Apostolopoulos, V.; Stanley, E.G.; Jackson, D.E.; Ashman, L.K. Characterization of mice lacking the tetraspanin superfamily member CD151. *Mol. Cell. Biol.* **2004**, *24*, 5978–5988. [CrossRef]

9. Tarrant, J.M.; Groom, J.; Metcalf, D.; Li, R.; Borobokas, B.; Wright, M.D.; Tarlinton, D.; Robb, L. The absence of Tssc6, a member of the tetraspanin superfamily, does not affect lymphoid development but enhances in vitro T-cell proliferative responses. *Mol. Cell. Biol.* **2002**, *22*, 5006–5018. [CrossRef]

10. Lombardo, S.D.; Mazzon, E.; Basile, M.S.; Campo, G.; Corsico, F.; Presti, M.; Bramanti, P.; Mangano, K.; Petralia, M.C.; Nicoletti, F.; et al. Modulation of Tetraspanin 32 (TSPAN32) Expression in T Cell-Mediated Immune Responses and in Multiple Sclerosis. *Int. J. Mol. Sci.* **2019**, *20*, 4323. [CrossRef]

11. Zaffaroni, M.; Rossini, S.; Ghezzi, A.; Parma, R.; Cazzullo, C.L. Decrease of CD4+ CD45+ T-cells in chronic-progressive multiple sclerosis. *J. Neurol.* **1990**, *237*, 1–4. [CrossRef] [PubMed]

12. Burns, J.; Bartholomew, B.; Lobo, S. Isolation of myelin basic protein-specific T cells predominantly from the memory T-cell compartment in multiple sclerosis. *Ann. Neurol.* **1999**, *45*, 33–39. [CrossRef]

13. Perrin, P.J.; Lovett-Racke, A.; Phillips, S.M.; Racke, M.K. Differential requirements of naïve and memory T cells for CD28 costimulation in autoimmune pathogenesis. *Histol. Histopathol.* **1999**, *14*, 1269–1276. [PubMed]

14. Lovett-Racke, A.E.; Trotter, J.L.; Lauber, J.; Perrin, P.J.; June, C.H.; Racke, M.K. Decreased dependence of myelin basic protein-reactive T cells on CD28-mediated costimulation in multiple sclerosis patients. A marker of activated/memory T cells. *J. Clin. Invest.* **1998**, *101*, 725–730. [CrossRef] [PubMed]

15. Cao, Y.; Goods, B.A.; Raddassi, K.; Nepom, G.T.; Kwok, W.W.; Love, J.C.; Hafler, D.A. Functional inflammatory profiles distinguish myelin-reactive T cells from patients with multiple sclerosis. *Sci. Transl. Med.* **2015**, *7*, 287ra74. [CrossRef]

16. Feng, X.; Bao, R.; Li, L.; Deisenhammer, F.; Arnason, B.G.W.; Reder, A.T. Interferon-β corrects massive gene dysregulation in multiple sclerosis: Short-term and long-term effects on immune regulation and neuroprotection. *EBioMedicine* **2019**, *49*, 269–283. [CrossRef]

17. Gurevich, M.; Tuller, T.; Rubinstein, U.; Or-Bach, R.; Achiron, A. Prediction of acute multiple sclerosis relapses by transcription levels of peripheral blood cells. *BMC Med. Genom.* **2009**, *2*, 46. [CrossRef]

18. Jansen, C.; Ramirez, R.N.; El-Ali, N.C.; Gomez-Cabrero, D.; Tegner, J.; Merkenschlager, M.; Conesa, A.; Mortazavi, A. Building gene regulatory networks from scATAC-seq and scRNA-seq using Linked Self Organizing Maps. *PLoS Comput. Biol.* **2019**, *15*, e1006555. [CrossRef]

19. Ritchie, M.E.; Phipson, B.; Wu, D.; Hu, Y.; Law, C.W.; Shi, W.; Smyth, G.K. limma powers differential expression analyses for RNA-sequencing and microarray studies. *Nucleic Acids Res.* **2015**, *43*, e47. [CrossRef]

20. Zhou, Y.; Zhou, B.; Pache, L.; Chang, M.; Khodabakhshi, A.H.; Tanaseichuk, O.; Benner, C.; Chanda, S.K. Metascape provides a biologist-oriented resource for the analysis of systems-level datasets. *Nat. Commun.* **2019**, *10*, 1523. [CrossRef]

21. Unternaehrer, J.J.; Chow, A.; Pypaert, M.; Inaba, K.; Mellman, I. The tetraspanin CD9 mediates lateral association of MHC class II molecules on the dendritic cell surface. *Proc. Natl. Acad. Sci. USA* **2007**, *104*, 234–239. [CrossRef] [PubMed]

22. Mittelbrunn, M.; Yanez-Mo, M.; Sancho, D.; Ursa, A.; Sanchez-Madrid, F. Cutting Edge: Dynamic Redistribution of Tetraspanin CD81 at the Central Zone of the Immune Synapse in Both T Lymphocytes and APC. *J. Immunol.* **2002**, *169*, 6691–6695. [CrossRef]

23. Sheng, K.-C.; van Spriel, A.B.; Gartlan, K.H.; Sofi, M.; Apostolopoulos, V.; Ashman, L.; Wright, M.D. Tetraspanins CD37 and CD151 differentially regulate Ag presentation and T-cell co-stimulation by DC. *Eur. J. Immunol.* **2009**, *39*, 50–55. [CrossRef] [PubMed]

24. Gartlan, K.H.; Belz, G.T.; Tarrant, J.M.; Minigo, G.; Katsara, M.; Sheng, K.-C.; Sofi, M.; van Spriel, A.B.; Apostolopoulos, V.; Plebanski, M.; et al. A Complementary Role for the Tetraspanins CD37 and Tssc6 in Cellular Immunity. *J. Immunol.* **2010**, *185*, 3158–3166. [CrossRef] [PubMed]

25. Gustafsson, M.; Edström, M.; Gawel, D.; Nestor, C.E.; Wang, H.; Zhang, H.; Barrenäs, F.; Tojo, J.; Kockum, I.; Olsson, T.; et al. Integrated genomic and prospective clinical studies show the importance of modular pleiotropy for disease susceptibility, diagnosis and treatment. *Genome Med.* **2014**, *6*, 17. [CrossRef] [PubMed]

26. Falzone, L.; Scola, L.; Zanghì, A.; Biondi, A.; Di Cataldo, A.; Libra, M.; Candido, S. Integrated analysis of colorectal cancer microRNA datasets: Identification of microRNAs associated with tumor development. *Aging* **2018**, *10*, 1000–1014. [CrossRef] [PubMed]

27. Candido, S.; Lupo, G.; Pennisi, M.; Basile, M.; Anfuso, C.; Petralia, M.; Gattuso, G.; Vivarelli, S.; Spandidos, D.; Libra, M.; et al. The analysis of miRNA expression profiling datasets reveals inverse microRNA patterns in glioblastoma and Alzheimer's disease. *Oncol. Rep.* **2019**, *42*, 911–922. [CrossRef]

28. Lombardo, S.D.; Mazzon, E.; Mangano, K.; Basile, M.S.; Cavalli, E.; Mammana, S.; Fagone, P.; Nicoletti, F.; Petralia, M.C. Transcriptomic Analysis Reveals Involvement of the Macrophage Migration Inhibitory Factor Gene Network in Duchenne Muscular Dystrophy. *Genes* **2019**, *10*, 939. [CrossRef]

29. Lombardo, S.D.; Mazzon, E.; Basile, M.S.; Cavalli, E.; Bramanti, P.; Nania, R.; Fagone, P.; Nicoletti, F.; Petralia, M.C. Upregulation of IL-1 Receptor Antagonist in a Mouse Model of Migraine. *Brain Sci.* **2019**, *9*, 172. [CrossRef]

30. Petralia, M.C.; Mazzon, E.; Fagone, P.; Falzone, L.; Bramanti, P.; Nicoletti, F.; Basile, M.S. Retrospective follow-up analysis of the transcriptomic patterns of cytokines, cytokine receptors and chemokines at preconception and during pregnancy, in women with post-partum depression. *Exp. Ther. Med.* **2019**, *18*, 2055–2062. [CrossRef]

31. Fagone, P.; Mazzon, E.; Cavalli, E.; Bramanti, A.; Petralia, M.C.; Mangano, K.; Al-Abed, Y.; Bramati, P.; Nicoletti, F. Contribution of the macrophage migration inhibitory factor superfamily of cytokines in the pathogenesis of preclinical and human multiple sclerosis: In silico and in vivo evidences. *J. Neuroimmunol.* **2018**, *322*, 46–56. [CrossRef]

32. Nicoletti, F.; Mazzon, E.; Fagone, P.; Mangano, K.; Mammana, S.; Cavalli, E.; Basile, M.S.; Bramanti, P.; Scalabrino, G.; Lange, A.; et al. Prevention of clinical and histological signs of MOG-induced experimental allergic encephalomyelitis by prolonged treatment with recombinant human EGF. *J. Neuroimmunol.* **2019**, *332*, 224–232. [CrossRef]

33. Cavalli, E.; Mazzon, E.; Basile, M.S.; Mangano, K.; Di Marco, R.; Bramanti, P.; Nicoletti, F.; Fagone, P.; Petralia, M.C. Upregulated Expression of Macrophage Migration Inhibitory Factor, Its Analogue D-Dopachrome Tautomerase, and the CD44 Receptor in Peripheral CD4 T Cells from Clinically Isolated Syndrome Patients with Rapid Conversion to Clinical Defined Multiple Sclerosis. *Medicina* **2019**, *55*, 667. [CrossRef]

34. Fagone, P.; Mazzon, E.; Mammana, S.; Di Marco, R.; Spinasanta, F.; Basile, M.S.; Petralia, M.C.; Bramanti, P.; Nicoletti, F.; Mangano, K. Identification of CD4+ T cell biomarkers for predicting the response of patients with relapsing-remitting multiple sclerosis to natalizumab treatment. *Mol. Med. Rep.* **2019**, *20*, 678–684. [CrossRef]

35. Cavalli, E.; Mazzon, E.; Basile, M.S.; Mammana, S.; Pennisi, M.; Fagone, P.; Kalfin, R.; Martinovic, V.; Ivanovic, J.; Andabaka, M.; et al. In Silico and In Vivo Analysis of IL37 in Multiple Sclerosis Reveals Its Probable Homeostatic Role on the Clinical Activity, Disability, and Treatment with Fingolimod. *Molecules* **2019**, *25*, 20. [CrossRef]

36. Petralia, M.C.; Mazzon, E.; Basile, M.S.; Cutuli, M.; Di Marco, R.; Scandurra, F.; Saraceno, A.; Fagone, P.; Nicoletti, F.; Mangano, K. Effects of Treatment with the Hypomethylating Agent 5-aza-2′-deoxycytidine in Murine Type II Collagen-Induced Arthritis. *Pharmaceuticals* **2019**, *12*, E174. [CrossRef]

37. Cavalli, E.; Mazzon, E.; Mammana, S.; Basile, M.S.; Lombardo, S.D.; Mangano, K.; Bramanti, P.; Nicoletti, F.; Fagone, P.; Petralia, M.C. Overexpression of Macrophage Migration Inhibitory Factor and Its Homologue D-Dopachrome Tautomerase as Negative Prognostic Factor in Neuroblastoma. *Brain Sci.* **2019**, *9*, 284. [CrossRef]

38. Petralia, M.C.; Mazzon, E.; Fagone, P.; Russo, A.; Longo, A.; Avitabile, T.; Nicoletti, F.; Reibaldi, M.; Basile, M.S. Characterization of the Pathophysiological Role of CD47 in Uveal Melanoma. *Molecules* **2019**, *24*, 2450. [CrossRef]

39. Lombardo, S.D.; Presti, M.; Mangano, K.; Petralia, M.C.; Basile, M.S.; Libra, M.; Candido, S.; Fagone, P.; Mazzon, E.; Nicoletti, F.; et al. Prediction of PD-L1 Expression in Neuroblastoma via Computational Modeling. *Brain Sci.* **2019**, *9*, 221. [CrossRef]

40. Mangano, K.; Mazzon, E.; Basile, M.S.; Di Marco, R.; Bramanti, P.; Mammana, S.; Petralia, M.C.; Fagone, P.; Nicoletti, F. Pathogenic role for macrophage migration inhibitory factor in glioblastoma and its targeting with specific inhibitors as novel tailored therapeutic approach. *Oncotarget* **2018**, *9*, 17951–17970. [CrossRef]

41. Mammana, S.; Fagone, P.; Cavalli, E.; Basile, M.; Petralia, M.; Nicoletti, F.; Bramanti, P.; Mazzon, E. The Role of Macrophages in Neuroinflammatory and Neurodegenerative Pathways of Alzheimer's Disease, Amyotrophic Lateral Sclerosis, and Multiple Sclerosis: Pathogenetic Cellular Effectors and Potential Therapeutic Targets. *Int. J. Mol. Sci.* **2018**, *19*, 831. [CrossRef]

42. Mammana, S.; Bramanti, P.; Mazzon, E.; Cavalli, E.; Basile, M.S.; Fagone, P.; Petralia, M.C.; McCubrey, J.A.; Nicoletti, F.; Mangano, K. Preclinical evaluation of the PI3K/Akt/mTOR pathway in animal models of multiple sclerosis. *Oncotarget* **2018**, *9*, 8263–8277. [CrossRef]

43. Fagone, P.; Mangano, K.; Quattrocchi, C.; Cavalli, E.; Mammana, S.; Lombardo, G.A.G.; Pennisi, V.; Zocca, M.-B.; He, M.; Al-Abed, Y.; et al. Effects of NO-Hybridization on the Immunomodulatory Properties of the HIV Protease Inhibitors Lopinavir and Ritonavir. *Basic Clin. Pharmacol. Toxicol.* **2015**, *117*, 306–315. [CrossRef]

44. Paskas, S.; Mazzon, E.; Basile, M.S.; Cavalli, E.; Al-Abed, Y.; He, M.; Rakocevic, S.; Nicoletti, F.; Mijatovic, S.; Maksimovic-Ivanic, D. Lopinavir-NO, a nitric oxide-releasing HIV protease inhibitor, suppresses the growth of melanoma cells in vitro and in vivo. *Invest. New Drugs* **2019**, *37*, 1014–1028. [CrossRef]

45. Basile, M.; Mazzon, E.; Krajnovic, T.; Draca, D.; Cavalli, E.; Al-Abed, Y.; Bramanti, P.; Nicoletti, F.; Mijatovic, S.; Maksimovic-Ivanic, D. Anticancer and Differentiation Properties of the Nitric Oxide Derivative of Lopinavir in Human Glioblastoma Cells. *Molecules* **2018**, *23*, 2463. [CrossRef]

46. Lazarević, M.; Mazzon, E.; Momčilović, M.; Basile, M.S.; Colletti, G.; Petralia, M.C.; Bramanti, P.; Nicoletti, F.; Miljković, Đ. The H_2S Donor GYY4137 Stimulates Reactive Oxygen Species Generation in BV2 Cells While Suppressing the Secretion of TNF and Nitric Oxide. *Molecules* **2018**, *23*, 2966. [CrossRef]

47. McFarland, H.F.; Martin, R. Multiple sclerosis: A complicated picture of autoimmunity. *Nat. Immunol.* **2007**, *8*, 913–919. [CrossRef]

48. Goverman, J. Autoimmune T cell responses in the central nervous system. *Nat. Rev. Immunol.* **2009**, *9*, 393–407. [CrossRef]

49. Nylander, A.; Hafler, D.A. Multiple sclerosis. *J. Clin. Invest.* **2012**, *122*, 1180–1188. [CrossRef]

50. Hemler, M.E. Targeting of tetraspanin proteins–potential benefits and strategies. *Nat. Rev. Drug Discov.* **2008**, *7*, 747–758. [CrossRef]

Communication

No Immediate Effects of Transcranial Direct Current Stimulation at Various Intensities on Cerebral Blood Flow in People with Multiple Sclerosis

Craig D. Workman [1], Laura L. Boles Ponto [2], John Kamholz [3] and Thorsten Rudroff [1,3,*]

[1] Department of Health and Human Physiology, University of Iowa, Iowa City, IA 52242, USA; craig-workman@uiowa.edu
[2] Department of Radiology, University of Iowa Hospitals and Clinics, Iowa City, IA 52242, USA; laura-ponto@uiowa.edu
[3] Department of Neurology, University of Iowa Hospitals and Clinics, Iowa City, IA 52242, USA; john-kamholz@uiowa.edu
* Correspondence: thorsten-rudroff@uiowa.edu; Tel.: +1-319-467-0363

Received: 10 January 2020; Accepted: 2 February 2020; Published: 4 February 2020

Abstract: Animal and transcranial magnetic stimulation motors have evoked potential studies suggesting that the currently used transcranial direct current stimulation (tDCS) intensities produce measurable physiological changes. However, the validity, mechanisms, and general efficacy of this stimulation modality are currently being scrutinized. The purpose of this pilot study was to investigate the effects of dorsolateral prefrontal cortex tDCS on cerebral blood flow. A sample of three people with multiple sclerosis underwent two blocks of five randomly assigned tDCS intensities (1, 2, 3, 4 mA, and sham; 5 min each) and [^{15}O]water positron emission tomography imaging. The relative regional (i.e., areas under the electrodes) and global cerebral blood flow were calculated. The results revealed no notable differences in regional or global cerebral blood flow from the different tDCS intensities. Thus, 5 min of tDCS at 1, 2, 3, and 4 mA did not result in immediate changes in cerebral blood flow. To achieve sufficient magnitudes of intracranial electrical fields without direct peripheral side effects, novel methods may be required.

Keywords: tDCS; neuroimaging; positron emission tomography; cerebral blood flow; multiple sclerosis

---

1. Introduction

Multiple sclerosis (MS) is a chronic central nervous system disease that affects approximately 2.3 million people worldwide [1]. Because some MS symptoms (e.g., neuropathic pain) are treatment-resistant [2], practical and inexpensive adjunctive therapies, like transcranial direct current stimulation (tDCS), are of high interest. Despite promising findings in tDCS studies in people with multiple sclerosis (PwMS) [3–5] (see [6] for a review), the validity and utility of tDCS is under scrutiny. For example, a critical review [7] did not support the idea that tDCS has a reliable neurophysiological effect beyond motor evoked potential (MEP) amplitude modulation. Though MEP amplitude appears to be sensitive to tDCS modulation, other reliable transcranial magnetic stimulation (TMS) measures that rely on similar neural mechanisms (e.g., short interval intracortical inhibition (SICI), intracortical facilitation (ICF), and cortical silent period (cSP)) have all shown no tDCS effect. Questions concerning the mechanistic foundations and general efficacy of this stimulation modality are on the rise.

Animal studies have suggested that the intensities that are typically employed in human research are sufficient to produce measurable physiological changes [8]. Unfortunately, findings from animal models, especially in vitro approaches, may have poor applicability to human studies. Thus, a vital concern is that tDCS may not induce a sufficient current in the cortex to have a measurable effect on neural function—at least not for the commonly used stimulation intensities ($\leq$2 mA). Therefore, it is necessary to combine tDCS with human neuroimaging to complement animal studies and to further clarify whether tDCS can affect neural function.

Cerebral activity in PwMS has been investigated with [^{15}O]water [9]. A close coupling of perfusion and metabolism was assumed, as this reflects the oxidative phosphorylation of glucose as the predominant energy source. Consequently, cerebral blood flow (CBF) is often considered an indirect measure of neuronal function and integrity [10]. This is supported by the significant association of glucose metabolism with regional CBF (rCBF) across different brain regions and global CBF (gCBF) across varying states of consciousness.

2. Materials and Methods

We conducted a pilot investigation on the effects of tDCS on CBF with semi-quantitative [^{15}O]water Positron Emission Tomography (PET), the gold standard for CBF measurements [11–15], in a sample of three people with relapsing–remitting MS (Table 1). Scans were completed in two blocks, one at baseline and five immediately after randomized tDCS stimulation intensities (1, 2, 3, 4 mA, and sham; six scans per block). The second block (intensity re-randomized) helped verify the reliability of the first. Therefore, each subject experienced 12 total scans in one session (Figure 1). The subjects performed a simple counting task for the duration of each scan (100 sec). This study was approved by the University of Iowa's Institutional Review Board (IRB-01: IRB#201905826; clinicaltrials.gov NCT04033133). All subjects provided written informed consent before participating.

Table 1. Subject demographic information (n = 3). Data are mean $\pm$ SD.

Sex (M/F)	2/1
Age (years)	45.3 $\pm$ 19.0
Height (cm)	171.9 $\pm$ 18.7
Weight (kg)	83.5 $\pm$ 19.9
Time since diagnosis (years)	8.0 $\pm$ 5.3
Patient-Determined Disease Steps [1]	2.3 $\pm$ 2.1

[1] Provides an indication of disease severity. A score of 2–4 out of 8 indicates moderate disability.

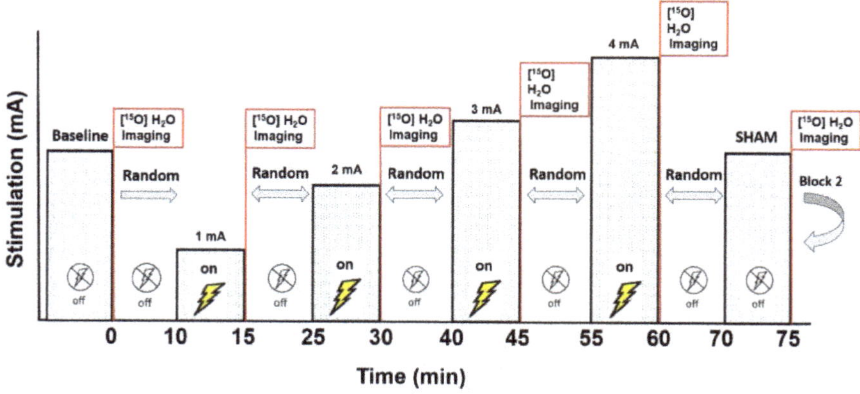

Figure 1. The stimulation and scan protocol. After the baseline scan, the five stimulation intensities (1, 2, 3, 4 mA, and sham) were delivered for 5 min each in a random order (six total scans per block). Ten minutes after the final scan of Block 1, Block 2 began with a new baseline scan and a different intensity randomization. Each subject experienced 12 total scans.

A battery-operated tDCS device (Soterix Medical Inc., New York, NY, USA) administered the stimulation. A previous study indicated that 5 min of tDCS over the primary motor cortex (M1) was sufficient to induce significant increases in MEP amplitude and that the effects of the stimulation returned to baseline after 10 min [16], and another study indicated significant blood flow changes from 4 min of motor cortex stimulation [17]. Thus, the tDCS parameters in this study included 5 min of stimulation at each intensity, each separated by ≥10 min, to avoid stimulation carry-over effects; this also allowed for sufficient time for the [^{15}O]water tracer to decay before the next injection. The stimulation was ramped up to the target intensity (1, 2, 3, or 4 mA) over 30 s, after which time the intensity was maintained for 5 min before ramping down to 0 mA over 30 s. The anode was over the left dorsolateral prefrontal cortex (dlPFC), and the cathode was over the contralateral supraorbital area. A dlPFC target helped to avoid potentially confounding M1 activity from counting or spontaneous movement during scanning.

Because significant excitability increases from 5 min of tDCS have been found at 1 min post-tDCS [16], imaging commenced 1 min after the ramp-down process was completed. A summed image of the 40 seconds immediate post-bolus transit was generated for each scan. The summed images were co-registered with the subject's T1-weighted magnetic resonance imaging (MRI). Individual, anatomically-based regions were defined by using the PNEURO tool of the PMOD Biomedical Image Quantification software package (PMOD Technologies, Ltd. Zürich, Switzerland. The mean global activity (gCBF) was calculated based on the volume-weighted average of all intracerebral regions. The CBF relative to the global activity was calculated for each region (rCBF = regional activity/gCBF; e.g., ratio 1.2 = rCBF 20% higher than gCBF) for each condition. Changes in gCBF and rCBF, focusing on regions under the electrodes, at the different tDCS intensities were investigated.

3. Results

The results revealed no notable differences in the gCBF or rCBF for the areas under the electrodes from the different tDCS intensities (Figure 2). We did not find any immediate effects (i.e., ~1–2 min post-stimulation) of tDCS on CBF in this functional [^{15}O]water PET study. None of the stimulation intensities (1, 2, 3, and 4 mA) over dlPFC were associated with changes in gCBF (not shown) or rCBF under the electrodes.

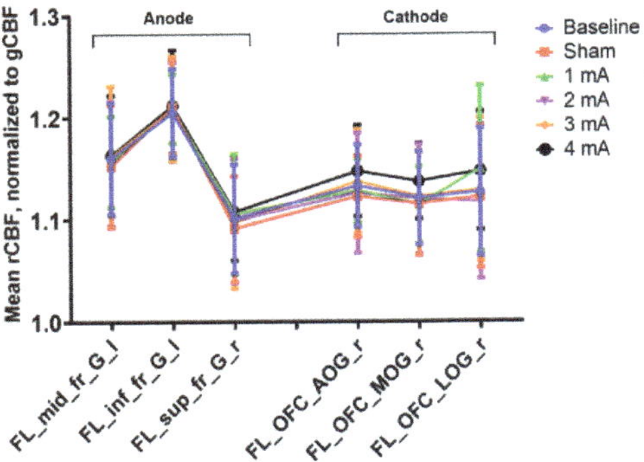

Figure 2. Mean tracer uptake (regional cerebral blood flow (rCBF)) relative to global tracer uptake (global CBF (gCBF)) by region and condition. Data are mean ± SEM. FL_mid_fr_G_l = left middle frontal gyrus, FL_inf_fr_G_l = left inferior frontal gyrus, FL_sup_fr_G_l = left superior frontal gyrus, FL_OFC_AOG_r = right anterior orbital gyrus, FL_OFC_MOG_r = right medial orbital gyrus, and FL_OFC_LOG_r = right lateral orbital gyrus.

4. Discussion

There are several possible explanations for this finding: (1) tDCS may not induce a sufficient current in the cortex to have a measurable effect on neural function. Vöröslakos et al. [18] suggested that a 1 V/m minimum voltage gradient was required for measurable online stimulation effects on neuronal spiking or membrane potentials in a rat model. To achieve 1 V/m at the cortex, the current applied at the scalp may need to be as high as 4–6 mA; however, the 4 mA that was used in this study did not elicit any CBF changes. (2) Only a small fraction of the transcranial current may reach the brain. One study [18] in rodents and human cadavers demonstrated that >75% of applied current is lost at the scalp, subcutaneous tissue, and muscle. These tissues serve as an effective shunt, resulting in at least a 50% reduction of current intensity. In addition, the serial resistance of the skull further reduces the current flow by 10%–25%, depending on skull thickness [18]. (3) Inter-individual variability in tDCS, with approximately 50% of people having poor or absent responses [19], may have biased our results. (4) Previous neuroimaging studies on the neurophysiological effects of tDCS detected immediate effects of tDCS on CBF by using [^{15}O]water PET [17,20]. However, these studies targeted M1 during a motor task, and diverse tDCS effects on different brain regions cannot be excluded. (5) Stimulating nerves on the scalp could also send signals to the brain or influence circulation, and/or other non-neuronal cells may react to the induced electrical fields and gradually alter brain function. Lastly, (6) the significant findings reported in the tDCS literature have assumed to have reflected one or more of a number of factors including placebo effects, influences from peripheral nerve stimulation, poor experimental designs, low statistical power, or inappropriate control conditions or analyses.

The sample size of this pilot study was small, and more subjects are recommended to confirm or refute these results. However, the study design, which included 12 injections that encompassed all conditions in duplicate, provided robust and remarkably consistent information, both between conditions and between subjects. Additionally, given that PET studies are costly and exposing the fewest subjects as possible to radiation is a major consideration [21], collecting images from additional subjects when there were no evident effects in these three PwMS was not justified. Furthermore, one of the purposes of reports like the present study is to encourage discussion and highlight the need to confirm the mechanistic effects of tDCS with neuroimaging (e.g., [^{15}O]water PET, [^{18}F]-fluorodeoxyglucose PET, functional magnetic resonance imaging (fMRI), and arterial spin labelling (ASL)). It should also be mentioned that multiple interventions over short periods might have complex and non-linear effects on the brain. Thus, any effect (or lack of effect) should to be interpreted with caution. Additionally, the absence of immediate tDCS effects on the cortex does not preclude longer-term effects. Furthermore, these findings were in a sample of PwMS, and different effects in healthy or patient populations cannot be excluded. Finally, this is a growing field for research, and a comprehensive understanding of how tDCS modulates the cortex is necessary. In humans, physiological changes associated with tDCS at different intensities and durations can be assessed with a variety of different neuroimaging methods (e.g., PET, fMRI, and ASL). All of these techniques have their limitations, but [^{15}O]water PET is considered to be the most "direct" measurement of cerebral blood flow. Thus, routinely combining neuroimaging with tDCS will help provide the necessary evidence that tDCS can affect neural function in humans.

5. Conclusions

In conclusion, we have demonstrated that 5 min of dlPFC tDCS at 1, 2, 3, or 4 mA did not result in immediate changes in CBF in PwMS. Thus, to achieve sufficient magnitudes of intracranial electrical fields without direct peripheral side effects, novel methods may be required.

Author Contributions: Conceptualization, C.D.W., T.R., and L.L.B.P.; methodology, C.D.W., T.R., and L.L.B.P.; data analysis, C.D.W. and L.L.B.P.; investigation, C.D.W. and L.L.B.P.; resources: T.R.; writing—original draft preparation: C.D.W. and T.R.; writing—review and editing; C.D.W., L.L.B.P., J.K., and T.R.; All authors approved the final version. All authors have read and agreed to the published version of the manuscript.

Funding: This research received no external funding.

Acknowledgments: The authors would like to thank all of the study participants for their time and effort during participation. In addition, the authors thank Lisa Dunnwald and Shannon Lehman for their assistance in data collection and study coordination.

Conflicts of Interest: The authors declare no conflict of interest.

References

1. Bishop, M.; Rumrill, P.D. Multiple sclerosis: Etiology, symptoms, incidence and prevalence, and implications for community living and employment. *Work* **2015**, *52*, 725–734. [CrossRef]
2. Solaro, C.; Tanganelli, P.; Messmer Uccelli, M. Pharmacological treatment of pain in multiple sclerosis. *Exp. Rev. Neurotherap.* **2007**, *7*, 1165–1174. [CrossRef]
3. Berra, E.; Bergamaschi, R.; De Icco, R.; Dagna, C.; Perrotta, A.; Rovaris, M.; Grasso, M.G.; Anastasio, M.G.; Pinardi, G.; Martello, F.; et al. The Effects of Transcutaneous Spinal Direct Current Stimulation on Neuropathic Pain in Multiple Sclerosis: Clinical and Neurophysiological Assessment. *Front. Hum. Neurosci.* **2019**, *13*, 31. [CrossRef]
4. Oveisgharan, S.; Karimi, Z.; Abdi, S.; Sikaroodi, H. The use of brain stimulation in the rehabilitation of walking disability in patients with multiple sclerosis: A randomized double-blind clinical trial study. *Iran J. Neurol.* **2019**, *18*, 57–63. [CrossRef]
5. Workman, C.D.; Kamholz, J.; Rudroff, T. Transcranial direct current stimulation (tDCS) for the treatment of a Multiple Sclerosis symptom cluster. *Brain Stimul.* **2020**, *13*, 263–264. [CrossRef]
6. Sanchez-Kuhn, A.; Perez-Fernandez, C.; Canovas, R.; Flores, P.; Sanchez-Santed, F. Transcranial direct current stimulation as a motor neurorehabilitation tool: An empirical review. *Biomed. Eng. Online* **2017**, *16*, 76. [CrossRef]
7. Horvath, J.C.; Forte, J.D.; Carter, O. Evidence that transcranial direct current stimulation (tDCS) generates little-to-no reliable neurophysiologic effect beyond MEP amplitude modulation in healthy human subjects: A systematic review. *Neuropsychologia* **2015**, *66*, 213–236. [CrossRef]
8. Krause, M.R.; Vieira, P.G.; Csorba, B.A.; Pilly, P.K.; Pack, C.C. Transcranial alternating current stimulation entrains single-neuron activity in the primate brain. *Proc. National Acad. Sci. USA* **2019**, *116*, 5747–5755. [CrossRef]
9. Sun, X.; Tanaka, M.; Kondo, S.; Okamoto, K.; Hirai, S. Clinical significance of reduced cerebral metabolism in multiple sclerosis: A combined PET and MRI study. *Ann. Nucl. Med.* **1998**, *12*, 89–94. [CrossRef]
10. Iadecola, C. Neurovascular regulation in the normal brain and in Alzheimer's disease. *Nat. Rev. Neurosci.* **2004**, *5*, 347–360. [CrossRef]
11. Herscovitch, P.; Markham, J.; Raichle, M.E. Brain blood flow measured with intravenous $H_2^{(15)}O$. I. Theory and error analysis. *J. Nucl. Med.* **1983**, *24*, 782–789.
12. Raichle, M.E.; Martin, W.R.; Herscovitch, P.; Mintun, M.A.; Markham, J. Brain blood flow measured with intravenous H2(15)O. II. Implementation and validation. *J. Nucl. Med.* **1983**, *24*, 790–798.
13. Giovannella, M.; Andresen, B.; Andersen, J.B.; El-Mahdaoui, S.; Contini, D.; Spinelli, L.; Torricelli, A.; Greisen, G.; Durduran, T.; Weigel, U.M.; et al. Validation of diffuse correlation spectroscopy against (15)O-water PET for regional cerebral blood flow measurement in neonatal piglets. *J. Cereb. Blood Flow Metab.* **2019**. [CrossRef]
14. Guo, J.; Gong, E.; Fan, A.P.; Goubran, M.; Khalighi, M.M.; Zaharchuk, G. Predicting (15)O-Water PET cerebral blood flow maps from multi-contrast MRI using a deep convolutional neural network with evaluation of training cohort bias. *J. Cereb. Blood Flow Metab.* **2019**. [CrossRef]
15. Puig, O.; Henriksen, O.M.; Vestergaard, M.B.; Hansen, A.E.; Andersen, F.L.; Ladefoged, C.N.; Rostrup, E.; Larsson, H.B.; Lindberg, U.; Law, I. Comparison of simultaneous arterial spin labeling MRI and (15)O-H2O PET measurements of regional cerebral blood flow in rest and altered perfusion states. *J. Cereb. Blood Flow Metab.* **2019**. [CrossRef]
16. Nitsche, M.A.; Paulus, W. Sustained excitability elevations induced by transcranial DC motor cortex stimulation in humans. *Neurology* **2001**, *57*, 1899–1901. [CrossRef]
17. Paquette, C.; Sidel, M.; Radinska, B.A.; Soucy, J.P.; Thiel, A. Bilateral transcranial direct current stimulation modulates activation-induced regional blood flow changes during voluntary movement. *J. Cereb. Blood Flow Metab.* **2011**, *31*, 2086–2095. [CrossRef]
18. Vöröslakos, M.; Takeuchi, Y.; Brinyiczki, K.; Zombori, T.; Oliva, A.; Fernández-Ruiz, A.; Kozák, G.; Kincses, Z.T.; Iványi, B.; Buzsáki, G.; et al. Direct effects of transcranial electric stimulation on brain circuits in rats and humans. *Nat. Commun.* **2018**, *9*, 483. [CrossRef]

19. Wiethoff, S.; Hamada, M.; Rothwell, J.C. Variability in response to transcranial direct current stimulation of the motor cortex. *Brain Stimul.* **2014**, *7*, 468–475. [CrossRef]

20. Lang, N.; Siebner, H.R.; Ward, N.S.; Lee, L.; Nitsche, M.A.; Paulus, W.; Rothwell, J.C.; Lemon, R.N.; Frackowiak, R.S. How does transcranial DC stimulation of the primary motor cortex alter regional neuronal activity in the human brain? *Eur. J. Neurosci.* **2005**, *22*, 495–504. [CrossRef]

21. Andreasen, N.C.; Arndt, S.; Cizadlo, T.; O'Leary, D.S.; Watkins, G.L.; Ponto, L.L.; Hichwa, R.D. Sample size and statistical power in [^{15}O]H$_2$O studies of human cognition. *J. Cereb. Blood Flow Metab.* **1996**, *16*, 804–816. [CrossRef]

Article

Streptococcus thermophilus ST285 Alters Pro-Inflammatory to Anti-Inflammatory Cytokine Secretion against Multiple Sclerosis Peptide in Mice

Narges Dargahi [1], John Matsoukas [2] and Vasso Apostolopoulos [1,*]

[1] Institute for Health and Sport, Victoria University, Melbourne VIC 3030, Australia; narges.dargahi@vu.edu.au
[2] NewDrug, Patras Science Park, 26500 Patras, Greece; imats1953@gmail.com
*   Correspondence: vasso.apostolopoulos@vu.edu.au; Tel.: +613-9919-2025

Received: 28 November 2019; Accepted: 21 February 2020; Published: 23 February 2020

Abstract: Probiotic bacteria have beneficial effects to the development and maintenance of a healthy microflora that subsequently has health benefits to humans. Some of the health benefits attributed to probiotics have been noted to be via their immune modulatory properties suppressing inflammatory conditions. Hence, probiotics have become prominent in recent years of investigation with regard to their health benefits. As such, in the current study, we determined the effects of *Streptococcus thermophilus* to agonist MBP_{83-99} peptide immunized mouse spleen cells. It was noted that *Streptococcus thermophilus* induced a significant increase in the expression of anti-inflammatory IL-4, IL-5, IL-10 cytokines, and decreased the secretion of pro-inflammatory IL-1β and IFN-γ. Regular consumption of *Streptococcus thermophilus* may therefore be beneficial in the management and treatment of autoimmune diseases such as multiple sclerosis.

Keywords: probiotics; *Streptococcus thermophilus*; ST285; MBP_{83-99} peptide; mannan; immune modulation; multiple sclerosis; agonist peptide

---

1. Introduction

There is an increasing trend in immune-mediated disorders across the world that is believed to be in part, a result of intestinal dysbiosis. The imbalance in the intestinal ecosystem can lead to a dysfunctional immune system that consequently causes immune disorders including autoimmune diseases (multiple sclerosis, MS) and other inflammatory disorders [1,2]. Probiotics have long been implicated for the overall improvement of health and the management of a number of health conditions including infection, constipation, allergies, and autoimmune diseases, and are either consumed in the form of different fermented foods and dairy products or taken as capsules. In either case, there is strong evidence that suggests that the ingestion of probiotics can alter intestinal dysbiosis and relieve dysfunctionality complications, with subsequent improvements to health [3].

Probiotic bacteria have been evolved inside the human intestinal tract (GIT), and through this co-evolution, the gut and its microbiome have developed a symbiotic relationship that is of mutual benefit. While the GIT microflora relies on the gut's warm habitat and food content, in return, it not only provides numerous unique bioactive components such as vitamins B and K, minerals, short chain fatty acids (SCF), and miosins to the host, but it also assists in modulating the immune system [3]. In fact, probiotics are able to modulate monocytes, macrophages, B cells, T helper (h)1, Th2, Th17, regulatory T cells (Treg), natural killer (NK) cells, and dendritic cells (DC) [3–6].

Chronic inflammation is the pathophysiological condition involved in neuro-degenerative disorders including MS, Parkinson's disease, and Alzheimer's disease [7,8]. There is cross-talk between the gut microbiota and the central nervous system (CNS) [8–10], known as the gut–brain axis. An insufficient or imbalanced GIT microflora can also lead to dysfunctions in the gut–brain axis and the pathogenesis of a number of diseases inside the GIT (such as inflammatory bowel disease, IBD) and outside the GIT (such as the CNS). Experimental autoimmune encephalomyelitis (EAE) is an animal model of human MS that has been used to study the effects of probiotic bacteria on CNS [11,12]. One of the safe and appropriate ways to modulate T cells in MS is to orally administer specific autoantigens [13,14]. Administration of Bifidobacteria or Lactobacteria probiotic strains to mice has been shown to increase Treg cells and tumor growth factor (TGF)-β levels and reduce the severity of EAE clinical symptoms, in parallel with improvement in the regeneration of myelin in the spinal cord compared to the control [15,16]. Administration of both Bifidobacteria and Lactobacteria strains induce an additional significant delay in the onset of EAE and related clinical symptoms, together with a substantial reduction of mononuclear infiltration into the CNS, and increased level of Treg cells of the $CD4^+CD25^+Foxp3^+$ phenotype in mouse spleen and lymph nodes.

In SJL/J mice, immunization with the MBP_{83-99} peptide mixed with mycobacterium stimulates autoimmune $CD4^+$ T cells in mice, and induces EAE [7,17]. Major histocompatibility complex (MHC) class II $H-2^s$ haplotype in the SJL/J mouse strain resembles many clinical, histopathological, and immunological characteristics of human MS, thus the SJL/J mouse is regularly used for immunization studies. Different peptides are immunogenic in different mouse strains however, in the SJL/J mouse strain, the peptide MBP_{81-100} binds to MHC class II $H-2^s$ with high affinity with the minimum epitope being MBP_{83-99} [7,17]. As such, the MBP_{83-99} epitope has been used as an agonist peptide to immunize mice for the activation of $CD4^+$ T cells [7,17]. We have shown that injection of the MBP_{83-99} peptide conjugated to the carrier mannan or mixed in complete Freund's adjuvant induces Th1 pro-inflammatory interferon-gamma (IFN γ-g) secreting $CD4^+$ T cells [18–25]. Studies have shown that there is a cross-reactivity between the MBP self-peptide and some microbial peptides (i.e., UL15, PMM) for Hy.1B11 T cell receptor (TCR), which has been isolated from a patient with MS. It has been highlighted that there are chemical interactions underlying the recognition mechanisms between TCR and the peptides presented by MHC proteins, as a critical constituent in adaptive immune responses to foreign antigens [26].

The Streptococcus genus constitutes over 100 species, amongst which *S. thermophilus* (ST) are non-pathogenic and food related bacteria that represent outstanding technological features in the food industry [27]. ST are commonly used as secondary starter cultures in dairy products to transform lactose into lactic acid and to acidify the pH of milk [27,28], contributing to both the fermentation and flavoring of dairy products [29]. Most probiotics belong to lactic acid bacteria (LAB); Gram-positive lactic acid producing bacteria that include lactobacilli, bifidobacterial, and enterococci [3]. As such, live LABs are not only used in foods for their health benefits, but exopolysaccharide-producing strains of ST such as ST1342, ST1275, and ST285 are generally used due to their beneficial properties (i.e., relieving lactose intolerance and suppressing acute conditions such as acute ulcerative colitis) [29]. Additionally, experimental studies designed to investigate the effect of VSL3 (Streptococcus, Bifidobacterium, and Lactobacillus species) on the peripheral immune system and the GIT microbiota in MS patients and healthy subjects showed improved abundance of many taxa with enriched taxa mainly consisting of Lactobacillus, Streptococcus, and Bifidobacterium. VSL3 also induced peripheral anti-inflammatory immune responses [30].

We recently showed that ST bacteria have anti-inflammatory properties [29]. U937 pro-monocytic cell line co-cultured with three ST bacteria (ST1342, ST1275 and ST285) induced an anti-inflammatory profile [29]. ST285 was further shown to have immune modulating effects via gene arrays to human peripheral blood mononuclear cells (PBMC) [31] and monocyte cells isolated from PBMC [32]. Herein, we immunized SJL/J mice with agonist MBP_{83-99} peptide conjugated to mannan three times, isolated spleen cells, and after re-stimulation of spleen cells with the MBP_{83-99} peptide, IFN-γ was secreted by spleen cells. Re-stimulation of spleen cells with the MBP_{83-99} peptide in the presence of ST285 probiotics was able to downregulate IFN-γ responses and stimulate the Th2, IL-4, IL-5, and IL-10 cytokine profile. These studies show that probiotics are able to modulate and alter the immune profile of MBP$_{83-99}$ specific cells to anti-inflammatory, which warrant in vivo EAE mouse experiments and hold promise as a therapeutic alternative approach to MS in human clinical trials.

2. Materials and Methods

2.1. Bacterial Strains

Pure bacterial cultures of *S. thermophilus* 285 (ST285) were obtained from the Victoria University culture collection (Werribee, Vic, Australia). Stock cultures were stored in cryobeads at −80 °C. Prior to each experiment, the cultures were propagated in M17 broth (Oxoid, Denmark) with 20 g/L lactose and incubated at 37 °C under aerobic conditions. Bacteria were also cultured on M17 agar (1.5% *w/v* agar) with 20 g/L lactose (Oxoid, Denmark) to assess the characteristics, morphology, purity, and Gram-positive confirmation [1].

2.2. Preparation of Live Bacterial Suspensions

Media were prepared and autoclaved at 121 °C for 15 min prior to the experiments. Bacterial cultures were grown three times in M17 broth with 20 g/L lactose at 37 °C aerobically for 18 h with a 1% inoculum transfer rate [33]. Cultures grow optimally at 37–42 °C for 24 h [29]. The growth period of cultures were consistent at 18 h (at the end of the exponential growth phase) and before the stationary growth phase to prevent cell lysis.

2.3. Enumeration of Bacterial Cells

For the actual experiment, bacteria were grown in broth media to the stationary phase at 37 °C aerobically, pelleted by centrifugation (6000× g) for 15 min at 4 °C, transferred, and resuspended in Dulbecco's phosphate-buffered saline, pH 7.4 (Invitrogen, Pty Ltd. Australia). The bacterial density in suspension was adjusted to 10^8 colony forming units (cfu)/mL for final concentration by determining the optical density at 600 nm, followed by two washes with Dulbecco's phosphate-buffered saline. These samples constituted the live-cell suspensions and were resuspended in the Roswell Park Memorial Institute (RPMI) 1640 culture media prior to co-culturing with spleen cells [1].

2.4. Mouse Experimental Procedures

2.4.1. Mice, Conjugates, and Immunization Schedule

Female SJL/J mice, aged 6–9 weeks, used in all experiments were purchased from the Animal Resources Center (ARC, Perth, Australia), and accommodated at the animal house (Victoria University, Werribee campus, Melbourne, Australia). All mice were ensured free access to water and food, and were housed in a temperature controlled room with a 12 h day 12 h night cycle. All immunizations were conducted according to the guidelines of the Australian Code of Practice for the Care and Use of Animals for Scientific Purposes and the study was approved by the Victoria University Animal Ethics Committee (AEC15/013) of Victoria University, Melbourne, Australia.

The MBP$_{83-99}$ agonist peptide of over 99% purity with (KG)$_5$ at the C-terminus was conjugated to mannan via a method previously described [34–38]. Briefly, 14 mg of mannan (Sigma, VIC Australia) was oxidized in sodium carbonate buffer and 0.1 M sodium periodate at 4 °C after which ethylene glycol was added and incubated for 30 min at 4 °C. Oxidized mannan comprising aldehyde groups was passed through a PD-10 column (Sigma, VIC Australia) pre-equilibrated in carbonate-bicarbonate buffer pH 9.0 and 2 mL of oxidized mannan was collected and 1 mg of MBP$_{83-99}$-(KG)$_5$ peptide was added and allowed to react overnight at room temperature in the dark. The resultant MBP$_{83-99}$-(KG)$_5$-mannan conjugate was used to immunize the mice.

The MBP$_{83-99}$ mannan peptide conjugate (50 µg/mouse) was injected in the SJL/J mice subcutaneously into the base of the tail, three times, every two weeks [17]. This conjugate has been shown to induce T cell proliferation and IFN-γ cytokine secretion to the agonist MBP$_{83-99}$ peptide in SJL/J mice [17,19,23,24]. Ten to fourteen days after the three injections, spleen cells were isolated, red blood cells were lysed using 0.73% NH_4Cl, and counted.

2.4.2. Isolation of Spleen Cells and In Vitro Stimulation with ST285

Spleen cells were resuspended in RPMI 1640 media supplemented with 10% heat-inactivated fetal bovine serum (Invitrogen, Pty Ltd. Australia), 1% antibiotic-antimycotic solution, and 2 mM L-glutamine in T75 cm^2 cell culture flasks. Mouse spleen cells (1×10^7) in RPMI media only was used as the negative control, 5 µg/mL recall agonist MBP_{83-99} peptide was used as the recall control, or 1×10^8 ST285 bacteria were added together with the MBP_{83-99} peptide, and cultured at 37 °C, 5% CO_2 for 24 h [29]. We previously showed that 24 h co-culture was adequate for the stimulation of the monocyte/macrophage cell line, human peripheral blood mononuclear cells, and human monocytes isolated from peripheral blood mononuclear cells [29,32]. At the end of the culture period, cells were transferred into falcon tubes, and centrifuged for 5 min at 1200 rpm to pellet the cells. All cell-free supernatants were collected and frozen at −20 °C until cytokine analysis.

2.5. Cytokine Production Analysis

Cytokine secretion of the spleen cell culture supernatants was analyzed by commercially available capture and detection antibodies in a Bioplex multiplex bead assay for a panel of nine mouse cytokines and chemokines using a 9-plex kit (BioRad, Melbourne Australia) to measure Interleukin (IL)-1β, IL-2, IL-4, IL-5, IL-6, IL-10, GM-CSF, TNF-α, and IFN-γ. Cell-free supernatants were collected and the assay procedures were performed according to the manufacturer's instructions. Briefly, a flat bottom 96-well plate was coated with 1× coupled beads and washed twice, followed by adding the standard serial dilutions, blanks, and samples to assigned wells. Post incubation at shaking at room temperature, plates were washed twice, adequate 1× detection antibody was added, and incubated at room temperature. Plates were washed three times and 1× Streptavidin Phycoerythrin (SA-PE) stop solution was added to each well, followed by incubating at room temperature and washing. Data collection was repeated twice, data were expressed as the mean cytokine response minus background (pg/mL) of each treatment from three replicate wells, plus or minus the standard error of the mean.

2.6. Statistical Analysis

Significant differences between all treatment groups were tested by analysis of variance (ANOVA) using the Statistical Package for the Social Sciences for Windows 25.0 (SPSS; IBM Corp), followed by a comparison between treatments performed by Tukey's honest significance test/degree and Fisher's least significant difference method, with a level of significance defined as $p < 0.05$.

3. Results

3.1. ST285 Reduces Pro-Inflammatory TNF-α and IFN-γ Production by MBP_{83-99} Primed Mouse Splenocytes

Interferon gamma (IFN-γ) is a pro-inflammatory Th1 cytokine involved in macrophage activation and cellular immunity. IFN-γ promotes Th1 cells and inhibits Th2 anti-inflammatory cells. In MS, IFN-γ is induced following CD4$^+$ T cell activation by agonist peptide MBP_{83-99}. SJL/J mice immunized with MBP_{83-99}–mannan conjugates induced IFN-γ responses by spleen cells, following overnight MBP_{83-99} peptide re-stimulation (Figure 1A, $p < 0.01$). Spleen cells re-stimulated with the agonist MBP_{83-99} peptide in the presence of ST285 reduced IFN-γ cytokine secretion (Figure 1A, $p < 0.05$). TNF-α, a Th1 cytokine, was not secreted by spleen cells from immunized mice wither by re-stimulation of the MBP_{83-99} peptide or MBP_{83-99} peptide plus ST285 (Figure 1B).

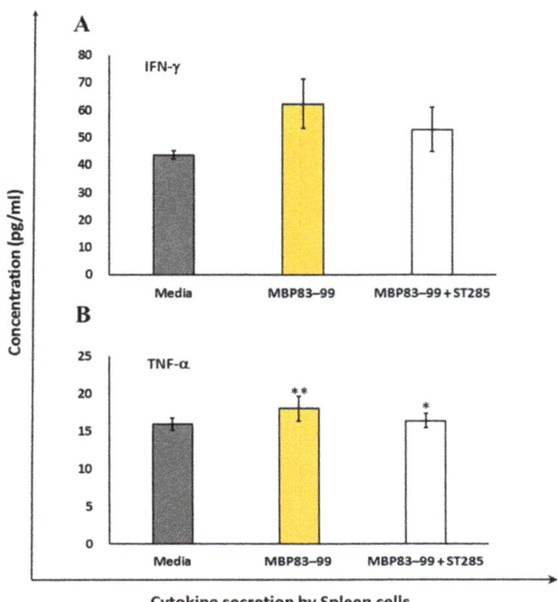

Figure 1. *S. thermophilus* 285 reduces pro-inflammatory cytokine production by mouse splenocytes. Spleen cells isolated from immunized mice ($n = 3$) were stimulated with *S. thermophilus* (ST) ST285 and recall agonist MBP_{83-99} peptide for 24 h and secretion of (**A**) IFN-γ and (**B**) TNF-α were measured. Recall MBP_{83-99} peptide was used as an internal positive control, and media refers to spleen cells from immunized mice ($n = 3$) without any additional recall peptide, or ST285 probiotic bacteria plus the MBP_{83-99} peptide. Means of two different readings of three replicate experiments were measured and analyzed. The means of readings for $n = 3$ mice were calculated and presented as plus or minus (±) the standard error of the mean. Symbols represent the p value for the Tukey test (one way ANOVA) where * $p < 0.05$ and ** $p < 0.01$.

3.2. ST285 Decreases Secretion of IL-1β, IL-2, and IL-6 by Mouse Spleen Cells

Secretion of IL-1β was slightly, but significantly reduced in immunized mouse spleen cells re-stimulated with the MBP_{83-99} peptide and ST285 compared to no re-stimulation, or the MBP_{83-99} peptide re-stimulation without ST285 ($p < 0.05$) (Figure 2A). IL-2 production was significantly increased in immunized spleen cells re-stimulated with the MBP_{83-99} peptide ($p < 0.01$), which was weakly but significantly decreased as a result of the co-stimulation of mouse spleen cells with the MBP_{83-99} peptide plus ST285 ($p < 0.05$) (Figure 2B). The production of IL-6 was profoundly increased by immunized mouse spleen cells upon co-culture of ST285 and the recall MBP_{83-99} peptide compared to the control media or MBP_{83-99} recall peptide ($p < 0.001$) (Figure 2C); spleen cells recalled with the MBP_{83-99} peptide alone also increased IL-6 secretion.

3.3. ST285 Induces Anti-Inflammatory Cytokine Profile by Mouse Splenocytes

Mice immunized with the MBP_{83-99} agonist peptide did not induce IL-4, IL-5, and IL-10 anti-inflammatory cytokines in the control (media alone) and recall agonist peptide MBP_{83-99} (Figure 3). However, the Th2 anti-inflammatory cytokine IL-4 was significantly ($p < 0.001$) increased by immunized mouse spleen cells when the MBP_{83-99} recall peptide was co-cultured with ST285 probiotic bacteria (Figure 3A). IL-5 was also increased by immunized spleen cells following co-culture with ST285 and the recall agonist MBP_{83-99} peptide (Figure 3B) ($p < 0.01$). The anti-inflammatory IL-10 cytokine was also significantly increased by immunized mouse spleen cells when co-cultured with ST285 and agonist recall MBP$_{83-99}$ peptide compared to the MBP$_{83-99}$ peptide alone or media control ($p < 0.001$) (Figure 3C).

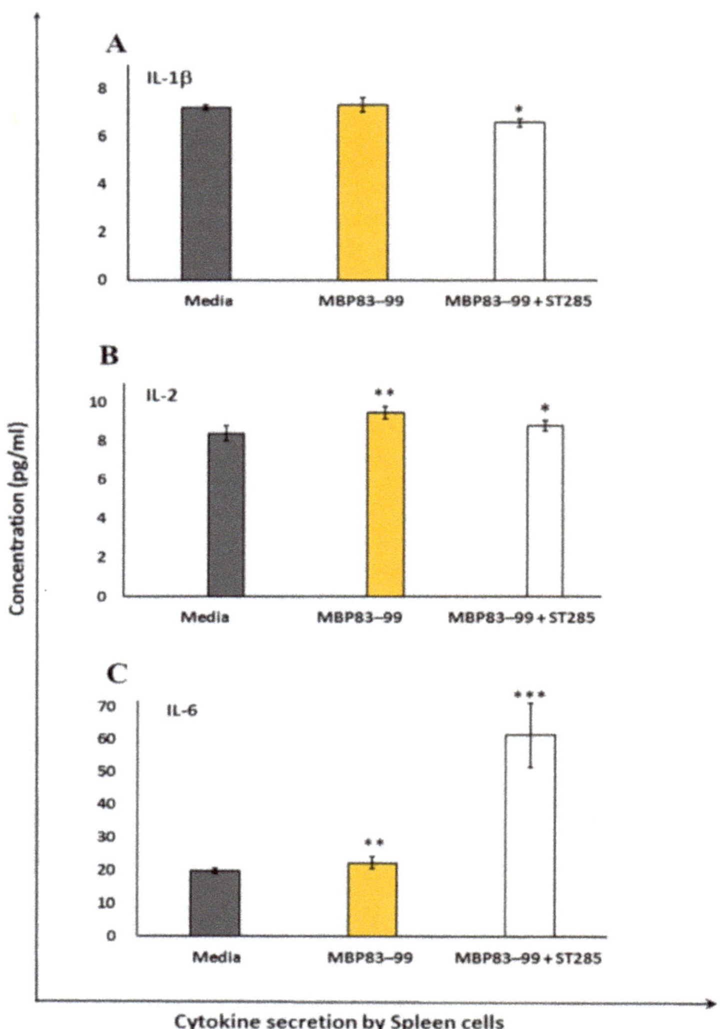

Figure 2. *S. thermophilus* 285 decreases expression of IL-1β, IL-2, and increases IL-6 by mouse spleen cells. Spleen cells isolated from immunized mice ($n = 3$) were stimulated with *S. thermophilus* (ST) ST285 and recall agonist MBP$_{83-99}$ peptide for 24 h and secretion of (**A**) IL-1β, (**B**) IL-2, and (**C**) IL-6 were measured. Recall MBP$_{83-99}$ peptide was used as the reference peptide, and media refers to spleen cells from immunized mice ($n = 3$) without any additional recall peptide or ST285 probiotic bacteria plus MBP$_{83-99}$ peptide. Means are shown as plus or minus ($\pm$) standard error of the means. Symbols represent the p value for the Tukey test (one way ANOVA) where * $p < 0.05$ and ** $p < 0.01$ and *** $p < 0.001$.

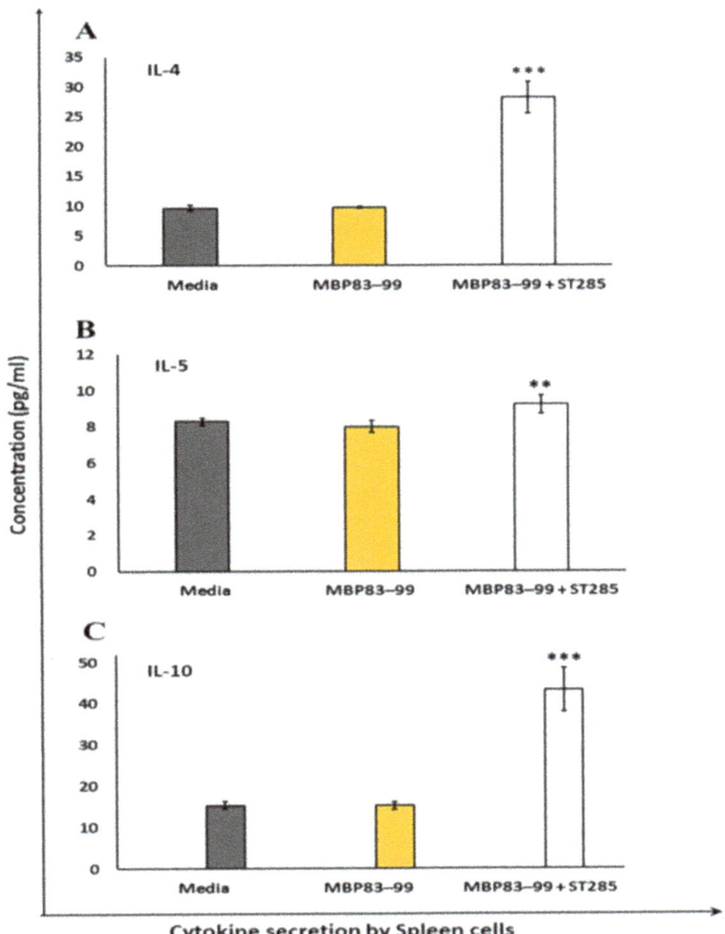

Figure 3. *S. thermophilus* 285 induces the anti-inflammatory cytokine profile by immunized mouse splenocytes. Spleen cells isolated from immunized mice ($n = 3$) were stimulated with *S. thermophilus* (ST) ST285 and the recall agonist MBP$_{83-99}$ peptide for 24 h and the secretion of (**A**) IL-4, (**B**) IL-5, and (**C**) IL-10 were measured. Recall MBP$_{83-99}$ peptide, media alone, or recall MBP$_{83-99}$ peptide plus ST285 are shown from immunized mice ($n = 3$). The means of readings for $n = 3$ mice were calculated and presented as plus or minus (±) the standard error of the mean. Symbols represent the p value for the Tukey test (one way ANOVA) where ** $p < 0.01$ and *** $p < 0.001$.

3.4. ST285 Does Not Alter the Secretion of Granulocyte-macrophage Colony-stimulating Factor by Mouse Spleen Cells

Secretion of granulocyte-macrophage colony-stimulating factor (GM-CSF) did not show any change by immunized mouse spleen cells upon co-culture with ST285 and agonist recall MBP$_{83-99}$ peptide compared to the negative control or MBP$_{83-99}$ peptide (Figure 4), despite significant upregulation of GM-CSF by ST285 on monocytes/macrophage cells [29].

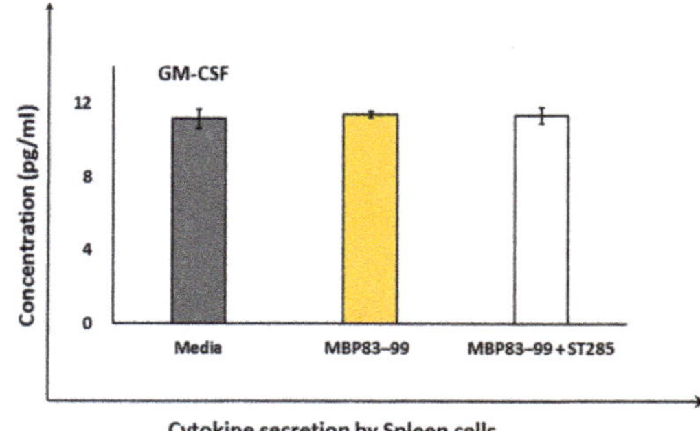

Figure 4. *S. thermophilus* 285 does not alter secretion of GM-CSF by mouse spleen cells. Spleen cells isolated from immunized mice ($n = 3$) were stimulated with *S. thermophilus* (ST) ST285 and the recall reference peptide for 24 hours and secretion of GM-CSF was measured. Recall MBP$_{83-99}$ peptide, media alone, or recall MBP$_{83-99}$ peptide plus ST285 are shown from immunized mice ($n = 3$). The means of readings for $n = 3$ mice were calculated and presented as plus or minus ($\pm$) the standard error of the mean.

4. Discussion

The Th1 pro-inflammatory cytokines IFN-γ and TNF-α are both involved in the defense against bacterial infections and in acute phase reactions. In MS, these two cytokines are implicated in the pathogenesis of disease by stimulating CD4$^+$ T cells against the myelin sheath. Mice immunized with the mannan MBP$_{83-99}$ peptide stimulated IFN-γ secretion, which was reduced in the presence of ST285. This reduction is very important in the context of inflamed situations such as autoimmune and inflammatory diseases, as any reduction in the amount of mediators that cause inflammation is imperative in the relief of symptoms. We previously noted that high levels of TNF-α and IFN-γ was secreted by the U937 monocytic cell line in the presence of ST285 [29]. However, the addition of ST285 to the MBP$_{83-99}$ recall peptide reduced IFN-γ secretion by mouse splenocytes. Spleen cells were populated with B, T, NK cells, macrophages, and monocytes, while the U937 cell line that we previously used were purely monocytic/macrophage cells. Additionally, the polarized inflammatory state of cytokines as a result of the immunization regimen and further exposure of spleen cells to the recall MBP$_{83-99}$ peptide that operate as inflammatory stimuli, compared to the U937 monoclonal cells only being exposed to ST285 bacteria, might give a clue as to the ability of ST285 probiotics to dampen the inflammatory immune response in the instance of exposure to polyclonal spleen cells.

Secretion of IL-1β by monocytes is involved in regulating the immune and inflammatory responses to infections and injuries; therefore, it has a role in innate immunity. IL-1β is also a major mediator in inflammatory responses associated with various cellular activities such as differentiation, proliferation, and apoptosis [39]. In addition, IL-1β is a regulator of inflammatory reactions and is involved in the stimulation of the central nervous system through cyclooxygenase-2 (PTGS2/COX2), which is involved in neurodegenerative disorders such as MS [33,40], Down's Syndrome, Alzheimer's disease, and HIV-associated dementia [41,42].

We noted the secretion of IL-1β by immunized mouse spleen cells was marginally, but significantly reduced in the presence of ST285 with the recall MBP$_{83-99}$ peptide. We previously noted that ST285 did not induce IL-1β cytokine to the U937 cell lines, however, significant upregulation of IL-1β mRNA was induced by human PBMC [31] and monocytes post co-culture with ST285 [32]. It is therefore clear that the immunized mouse spleens and the recall of T cells with the MBP$_{83-99}$ peptide in the presence of ST285 caused a reduction in IL-1β secretion. Likewise, IL-2 was marginally decreased in the presence of ST285 compared to the increased secretion caused by the MBP$_{83-99}$ peptide in the positive control. Co-culturing human PBMC with ST285 also downregulates IL-2 mRNA expression [31].

IL-6 is produced by activated immune cells including DC, B cells, and macrophages. Although IL-6 is associated with acute phase responses, it is also associated with a reduction of Th1 polarization, while promoting Th2 differentiation, B cell maturation, and macrophage differentiation. Proliferation and differentiation of Th2 cells changed the polarized Th1 environment and skewed the Th1/Th2 balance toward Th2, which is beneficial in relieving autoimmune conditions such as MS. IL-6 production was significantly higher (three times) in mouse splenocytes cultured with ST285 compared to the control, hence, it is likely that ST285 bacteria may potentially change the balance toward a healthier state in MS. We previously noted significant upregulation of IL-6 to human monocytes [manuscript submitted] and to bulk PBMC co-cultures [31] with ST285, which are also in accord with the increase in IL-6 levels by the U937 promonocytic cell line co-cultured with ST285 [29]. Likewise, the commercially used probiotic *L. paracasei* DG induces IL-6 cytokines to the THP-1 human monocyte cell line [43]. In contrast, ingestion of *B. bifidum* by mice did not increase the IL-6 levels, but boosted anti-oxidation activities in the spleen and thymus of mice and improved other immune functions by changing the gene expression of immune mediators [44].

It is likely that the constant-shifting in the equilibrium and the dynamics that exist between pro- and anti-inflammatory cytokines will lead to some controversy in the research findings regarding IL-6. On one hand, IL-6 may ease the autoimmune condition due to its downstream immunological effects. On the other hand, elevated levels of pro-inflammatory effector T cell cytokines such as IFN-γ, IL-17 as well as IL-6 are noted in patients with autoimmune myasthenia gravis and MS [45]. Thus, it might be likely that the role that cytokines such as IL-6 play may depend on their bio-environment and may be advantageous to the body, if probiotics such as ST285 are used for neutralization and/or reversing from a pro- to an anti-inflammatory state in the body.

IL-4 is one of the important cytokines required for anti-inflammatory responses against inflammatory conditions such as MS and allergies [29]. IL-4 production was significantly increased by mouse spleen cells in the presence of the recall MBP$_{83-99}$ peptide and ST285 compared to either the MBP$_{83-99}$ peptide alone or the negative control (media). Likewise, it was previously noted that ST285 induced U937 monocytic cells to produce IL-4, although no changes to mRNA expression levels of IL-4 were noted to human monocytes or to bulk human PBMC following co-cultures with ST285 [31]. In contrast, feeding BALB/c mice with *L. paracasei* BEJ01 alone or combined with aflatoxins B1 (AFB1) and fumonisin B1 (FB1) (known foodborne mycotoxins with immunomycotoxic effects on human health) was used to evaluate *L. paracasei* BEJ01 detoxification [46]. Assessing different splenic immunological factors indicated that exposure to these mycotoxins led to increased IL-4 mRNA levels, oxidative stress, and immunotoxicity in the spleen [46] whereas the combined LAB treatment with AFB1 or FB1 suppressed and normalized mRNA levels of IL-4, showing protective effects induced by LAB against AFB1 and FB1 via diminishing toxin adhesion and bioavailability [46]. In contrast, spleen cells isolated from BALB/c mice in vitro co-cultured individually with LAB strains (*L. casei* Lc2w (Lc), *L. plantarum* CCFM47 (Lp), and *L. acidophilus* CCFM137 (La)) showed reduced IL-4 production by spleen cells exposed to La only, while parallel animal studies displayed LAB-induced alleviation of inflammation post airway allergy for all strains through increased Treg cells and modulation of Th1/Th2 balance [47].

The anti-inflammatory cytokine IL-5 is produced by Th2 cells and mast cells. In the event of infection with helminth parasites, IL-5 leads to a lesser risk of autoimmune disorders, which is indirectly accredited to some therapeutic characteristics of IL-5 in autoimmune disorders. We noted a slight increase in the IL-5 production by spleen cells in response to ST285, whereas no changes to the mRNA expression levels of IL-5 were previously noted in ST285 co-cultures with human PBMC or human monocyte cells [manuscripts submitted]. A study showed that treating mice with *Fasciola hepatica* excretion/secretions (FHES) reduced EAE clinical signs due to a significant decrease in the infiltration of Th1 and Th17 cells into the brain and an increase in IL-5 (and IL-23) response, with subsequent increase in eosinophils [48]. It is likely that the small but significant increase of IL-5 may be beneficial to MS.

IL-10, an anti-inflammatory cytokine, is secreted by Th2 and Treg cells. Amongst all the anti-inflammatory cytokines and chemokines, anti-inflammatory properties of IL-10 are the most potent in suppressing inflammatory mediators by other activated immune cells (TNF-α, IFN-γ, IL-1, IL-17, and IL-23 cytokines) [49]. A significant amplification in the IL-10 levels secreted by the spleen cells in the presence of ST285 was noted, which was similarly shown for the U937 monocytic cell line in the presence of ST285 and to human PBMC, although no significant changes were shown in human monocyte cells [manuscripts submitted]. Likewise, oral administration of *L. reuteri* and *L. lactis* strains to mice stimulated the production of anti-inflammatory IL-10 and Treg cells [50,51]. In addition, sub-clinical studies of *L. salivarius* UCC118, *L. lactis* MG1363, and *L. plantarum* WCFS1 administered to mice and re-exposure of their isolated bone marrow cells to the three bacterial co-cultures showed all three strains differentially stimulated IL-10 production [52]. Correspondingly, when DC from spleen and mesenteric lymph nodes of mice were matured using *L. acidophilus* X37 and exposed to commensal gut *Bifidobacterium longum* Q46, *L. acidophilus* X37, and *Escherichia coli* Nissle 1917, increased IL-10 levels were noted [53]. Similarly, after BALB/c mice were fed with *L. paracasei* BEJ01 alone or combined with aflatoxins B1 and fumonisin B1, high IL-10 mRNA levels were induced [46]. In addition, mice fed with kefir-derived *Lactobacillus kefiri* CIDCA 8348 also increased IL-10 gene expression [54]. In the context of MS, the use of ST285 was shown to downregulate Th1 responses and upregulate Th2 responses, something of the utmost importance to patients with MS to alleviate MS symptoms and/or reversal of the disease [31].

5. Conclusions

Immunization of SJL/J mice with agonist MBP$_{83-99}$ peptide conjugated to mannan induces Th1 pro-inflammatory IFN-γ responses and no Th2 anti-inflammatory responses when spleen cells are co-cultured in vitro in the presence of the agonist recall MBP$_{83-99}$ peptide. However, stimulation of spleen cells with the recall MBP$_{83-99}$ peptide in the presence of ST285 significantly increased the secretion of IL-4, IL-6, and IL-10, along with mild upregulation in IL-2 and IL-5, suggesting a role for ST285 in the activation of immune response phenotypes toward a predominant anti-inflammatory profile, tolerance, and suppression of inflammation. In addition, ST285, downregulated the secretion of IL-1a and IFN-γ—the immune mediators involved in Th1 type responses—collectively pointing to a shift in immune responses from Th1 to a Th2 phenotype. More importantly, the significant increase of IL-10 could further contribute by the differentiation of naïve CD4$^+$ T cells and proliferation of Tregs, which can also drive the immune balance further toward a dominant anti-inflammatory phenotype. Additionally, given the drastic increase of GM-CSF in our previous studies of ST285 co-cultured with the U937 monocytic cell line, human PBMC, and human monocyte cells, and no change to the secretion of GM-CSF in spleen cells with GM-CSF being a major cytokine for proliferation and recruitment of the immune cells, this might indicate a deliberate and purposeful neutralization of GM-CSF by ST285. The effects of ST285 on the immune response could be used as a novel approach in modulating chronic inflammatory and autoimmune conditions such as MS. Further studies should involve the effects of ST285 in mice with EAE or be used to prevent EAE induction, which will pave the way for new modalities for the treatment of MS in human clinical trials.

Author Contributions: Conceptualization, N.D. and V.A.; Methodology, N.D and V.A.; Formal analysis, N.D.; Investigation, V.A and J.M.; Resources, V.A. and J.M.; Data curation, N.D.; Writing - original draft preparation, N.D.; Writing—review and editing, V.A and J.M.; Supervision, V.A.; Project administration, V.A.; Funding acquisition, V.A. All authors have read and agreed to the published version of the manuscript.

Funding: This research received no external funding.

Acknowledgments: V.A and N.D would like to thank the Immunology and Translational Research Group, the Mechanisms and Interventions in Health and Disease Program, and the Institute of Health and Sport, Victoria University, Australia for their support.

Conflicts of Interest: The authors declare no conflicts of interest.

References

1. De Palma, G.; Collins, S.M.; Bercik, P.; Verdu, E.F. The microbiota-gut-brain axis in gastrointestinal disorders: Stressed bugs, stressed brain or both? *J. Physiol.* **2014**, *592*, 2989–2997. [CrossRef] [PubMed]
2. Adamczyk-Sowa, M.; Medrek, A.; Madej, P.; Michlicka, W.; Dobrakowski, P. Does the Gut Microbiota Influence Immunity and Inflammation in Multiple Sclerosis Pathophysiology? *J. Immunol. Res.* **2017**, *2017*, 7904821. [CrossRef] [PubMed]
3. Dargahi, N.; Johnson, J.; Donkor, O.; Vasiljevic, T.; Apostolopoulos, V. Immunomodulatory effects of probiotics: Can they be used to treat allergies and autoimmune diseases? *Maturitas* **2019**, *119*, 25–38. [CrossRef] [PubMed]
4. Asarat, M.; Apostolopoulos, V.; Vasiljevik, T.; Donkor, O. Short-chain fatty acids produced by synbiotic mixtures in skim milk differentially regulate proliferation and cytokine production in peripheral blood mononuclear cells. *Int. J. Food Sci. Nutr.* **2015**, *66*, 755–765. [CrossRef]
5. Hardy, H.; Harris, J.; Lyon, E.; Beal, J.; Foey, A.D. Probiotics, prebiotics and immunomodulation of gut mucosal defences: Homeostasis and immunopathology. *Nutrients* **2013**, *5*, 1869–1912. [CrossRef]
6. Wolvers, D.; Antoine, J.M.; Myllyluoma, E.; Schrizenmeir, J.; Szajewska, H.; Rijkers, G.T. Guidance for substantiating the evidence for beneficial effects of probiotics: Prevention and management of infections by probiotics. *J. Nutr.* **2010**, *140*, 698S–712S. [CrossRef]
7. Dargahi, N.; Katsara, M.; Tselios, T.; Androutsou, M.E.; De Courten, M.; Matsoukas, J.; Apostolopoulos, V. Multiple sclerosis: Immunopathology and treatment update. *Brain Sci.* **2017**, *7*, 78. [CrossRef]
8. Mukherjee, A.; Biswas, A.; Das, S.K. Gut dysfunction in Parkinson's disease. *World J. Gastroenterol.* **2016**, *22*, 5742–5752. [CrossRef]
9. Catanzaro, R.; Anzalone, M.; Calabrese, F.; Milazzo, M.; Capuana, M.; Italia, A.; Occhipinti, S.; Marotta, F. The gut microbiota and its correlations with the central nervous system disorders. *Panminerva Med.* **2015**, *57*, 127–143.
10. Russo, R.; Cristiano, C.; Avagliano, C.; De Caro, C.; Rana, G.L.; Raso, G.M.; Canani, R.B.; Meli, R.; Calignano, A. Gut-brain axis: Role of lipids in the regulation of inflammation, pain and CNS diseases. *Curr. Med. Chem.* **2018**, *25*, 3930–3952. [CrossRef]
11. Kwon, H.K.; Kim, G.C.; Kim, Y.; Hwang, W.; Jash, A.; Sahoo, A.; Kim, J.E.; Nam, J.H.; Im, S.H. Amelioration of experimental autoimmune encephalomyelitis by probiotic mixture is mediated by a shift in T helper cell immune response. *Clin. Immunol.* **2013**, *146*, 217–227. [CrossRef] [PubMed]
12. Lavasani, S.; Dzhambazov, B.; Nouri, M.; Fak, F.; Buske, S.; Molin, G.; Thorlacius, H.; Alenfall, J.; Jeppsson, B.; Westrom, B. A novel probiotic mixture exerts a therapeutic effect on experimental autoimmune encephalomyelitis mediated by IL-10 producing regulatory T cells. *PLoS ONE* **2010**, *5*, e9009. [CrossRef] [PubMed]
13. Buerth, C.; Mausberg, A.K.; Heininger, M.K.; Hartung, H.P.; Kiesseier, B.C.; Ernst, J.F. Oral tolerance induction in experimental autoimmune encephalomyelitis with Candida utilis expressing the immunogenic MOG35-55 peptide. *PLoS ONE* **2016**, *11*, e0155082. [CrossRef] [PubMed]
14. Maassen, C.B.M.; Laman, J.D.; Van Holten-Neelen, C.; Hoogteijling, L.; Visser, L.; Schellekens, M.M.; Boersma, W.J.A.; Claassen, E. Reduced experimental autoimmune encephalomyelitis after intranasal and oral administration of recombinant lactobacilli expressing myelin antigens. *Vaccine* **2003**, *21*, 4685–4693. [CrossRef]
15. Consonni, A.; Cordiglieri, C.; Rinaldi, E.; Marolda, R.; Ravanelli, I.; Guidesi, E.; Eli, M.; Mantegaza, R.; Baggi, F. Administration of bifidobacterium and lactobacillus strains modulates experimental myasthenia gravis and experimental encephalomyelitis in Lewis rats. *Oncotarget* **2018**, *9*, 22269–22287. [CrossRef] [PubMed]
16. Salehipour, Z.; Haghmorad, D.; Sankian, M.; Nosratabdi, R.; Sotan Dallal, M.M.; Tabasi, N.; Khazee, M.; Nasirali, L.R.; Mahmoudi, M. Bifidobacterium animalis in combination with human origin of Lactobacillus plantarum ameliorate neuroinflammation in experimental model of multiple sclerosis by altering CD4+ T cell subset balance. *Biomed. Pharmacother.* **2017**, *95*, 1535–1548. [CrossRef] [PubMed]
17. Yannakakis, M.P.; Simal, C.; Tzoupis, H.; Rodi, M.; Dargahi, N.; Prakash, M.; Mouzaki, A.; Platts, J.A.; Apostolopoulos, V.; Tselios, T.V. Design and synthesis of non-peptide mimetics mapping the immunodominant myelin basic protein (MBP83–96) epitope to function as T-cell receptor antagonists. *Int. J. Mol. Sci.* **2017**, *18*, 1215. [CrossRef]

18. Katsara, M.; Apostolopoulos, V. Editorial: Multiple Sclerosis: Pathogenesis and Therapeutics. *Med. Chem.* **2018**, *14*, 104–105. [CrossRef]

19. Katsara, M.; Deraos, G.; Tselios, T.; Matsoukas, J.; Apostolopoulos, V. Design of novel cyclic altered peptide ligands of myelin basic protein MBP83-99 that modulate immune responses in SJL/J mice. *J. Med. Chem.* **2008**, *51*, 3971–3978. [CrossRef]

20. Katsara, M.; Deraos, G.; Tselios, T.; Matsoukas, M.T.; Friligou, I.; Matsoukas, J.; Apostolopoulos, V. Design and synthesis of a cyclic double mutant peptide (cyclo (87–99)[A91, A96] MBP87–99) induces altered responses in mice after conjugation to mannan: Implications in the immunotherapy of multiple sclerosis. *J. Med. Chem.* **2009**, *52*, 214–218. [CrossRef]

21. Katsara, M.; Deraos, S.; Tselios, T.; Pietersz, G.; Matsoukas, J.; Apostolopoulos, V. Immune responses of linear and cyclic PLP139-151 mutant peptides in SJL/J mice: Peptides in their free state versus mannan conjugation. *Immunotherapy* **2014**, *6*, 709–724. [CrossRef] [PubMed]

22. Katsara, M.; Matsoukas, J.; Deraos, G.; Apostolopoulos, V. Towards immunotherapeutic drugs and vaccines against multiple sclerosis. *Acta Biochim. Biophys. Sin.* **2008**, *40*, 636–642. [CrossRef] [PubMed]

23. Katsara, M.; Yuriev, E.; Ramsland, P.A.; Deraos, G.; Tselios, T.; Matsoukas, J.; Apostolopoulos, V. A double mutation of MBP83–99 peptide induces IL-4 responses and antagonizes IFN-γ responses. *J. Neuroimmunol.* **2008**, *200*, 77–89. [CrossRef] [PubMed]

24. Katsara, M.; Yuriev, E.; Ramsland, P.A.; Deraos, G.; Tselios, T.; Matsoukas, J.; Apostolopoulos, V. Mannosylation of mutated MBP83–99 peptides diverts immune responses from Th1 to Th2. *Mol. Immunol.* **2008**, *45*, 3661–3670. [CrossRef] [PubMed]

25. Katsara, M.; Yuriev, E.; Ramsland, P.A.; Tselios, T.; Deraos, G.; Lourbopoulos, A.; Grigoriadis, N.; Matsoukas, J.; Apostolopoulos, V. Altered peptide ligands of myelin basic protein (MBP87–99) conjugated to reduced mannan modulate immune responses in mice. *Immunology* **2009**, *128*, 521–533. [CrossRef] [PubMed]

26. Kumar, A.; Delogu, F. Dynamical footprint of cross-reactivity in a human autoimmune T-cell receptor. *Sci. Rep.* **2017**, *7*, 42496. [CrossRef] [PubMed]

27. Tarrah, A.; Treu, L.; Giaretta, S.; Duarte, V.; Corich, V.; Giacomini, A. Differences in Carbohydrates Utilization and Antibiotic Resistance Between Streptococcus macedonicus and Streptococcus thermophilus Strains Isolated from Dairy Products in Italy. *Curr. Microbiol.* **2018**, *75*, 1334–1344. [CrossRef]

28. Tarrah, A.; Noal, V.; Treu, L.; Giaretta, S.; da Silva Duarte, V.; Corich, V.; Giacomini, A. Short communication: Comparison of growth kinetics at different temperatures of Streptococcus macedonicus and Streptococcus thermophilus strains of dairy origin. *J. Dairy Sci.* **2018**, *101*, 7812–7816. [CrossRef]

29. Dargahi, N.; Johnson, J.; Donkor, O.; Vasiljevic, T.; Apostolopoulos, V. Immunomodulatory effects of Streptococcus thermophilus on U937 monocyte cell cultures. *J. Funct. Foods* **2018**, *49*, 241–249. [CrossRef]

30. Tankou, S.K.; Regev, K.; Healy, B.C.; Cox, L.M.; Tjon, E.; Kivisakk, P.; Vanande, I.P.; Cook, S.; Gandhi, R.; Glanz, B.; et al. Investigation of probiotics in multiple sclerosis. *Mult. Scler.* **2018**, *24*, 58–63. [CrossRef]

31. Dargahi, N.; Johnson, J.; Apostolopoulos, V. Streptococcus thermophilus alters the expression of genes associated with innate and adaptive immunity in human peripheral blood mononuclear cells. *PLoS ONE* **2020**, *15*, e0228531. [CrossRef]

32. Dargahi, N.; Johnson, J.; Apostolopoulos, V. Immune modulatory effects of probiotic Streptococcus thermophilus in human monocytes. 2020; in preparation.

33. McGuinness, M.C.; Powers, J.M.; Bias, W.B.; Schmeckpeper; Segal, A.H.; Gowda, V.C.; Wesselignh, S.L.; Berger, J.; Griffin, D.E.; Smith, K.D. Human leukocyte antigens and cytokine expression in cerebral inflammatory demyelinative lesions of X-linked adrenoleukodystrophy and multiple sclerosis. *J. Neuroimmunol.* **1997**, *75*, 174–182. [CrossRef]

34. Apostolopoulos, V.; Barnes, N.; Pietersz, G.A.; McKenzie, I.F. Ex vivo targeting of the macrophage mannose receptor generates anti-tumor CTL responses. *Vaccine* **2000**, *18*, 3174–3184. [CrossRef]

35. Apostolopoulos, V.; Pietersz, G.A.; Gordon, S.; Martinez-Pomares, L.; McKenzie, I.F. Aldehyde-mannan antigen complexes target the MHC class I antigen-presentation pathway. *Eur. J. Immunol.* **2000**, *30*, 1714–1723. [CrossRef]

36. Apostolopoulos, V.; Pietersz, G.A.; Loveland, B.E.; Sandrin, M.S.; McKenzie, I.F. Oxidative/reductive conjugation of mannan to antigen selects for T1 or T2 immune responses. *Proc. Natl. Acad. Sci. USA* **1995**, *92*, 10128–10132. [CrossRef]

37. Apostolopoulos, V.; Pietersz, G.A.; McKenzie, I.F. Cell-mediated immune responses to MUC1 fusion protein coupled to mannan. *Vaccine* **1996**, *14*, 930–938. [CrossRef]

38. Tselios, T.V.; Lamari, F.N.; Karathanasopoulos, I.; Katsara, M.; Apostolopoulos, V.; Pietersz, G.A.; Matsoukas, J.M.; Karamanos, N.K. Synthesis and study of the electrophoretic behavior of mannan conjugates with cyclic peptide analogue of myelin basic protein using lysine-glycine linker. *Anal. Biochem.* **2005**, *347*, 121–128. [CrossRef]

39. Masters, S.L.; Simon, A.; Aksentjevich, I.; Kastner, D.L. Horror autoinflammaticus: The molecular pathophysiology of autoinflammatory disease. *Annu. Rev. Immunol.* **2009**, *27*, 621–668. [CrossRef]

40. Paré, A.; Mailhot, B.; Levesque, S.A.; Juzwik, C.; Doss, P.M.I.A.; Lecuyer, M.A.; Prat, A.; Rangachari, M.; Fournier, A.; Lacroix, S. IL-1β enables CNS access to CCR2hi monocytes and the generation of pathogenic cells through GM-CSF released by CNS endothelial cells. *Proc. Natl. Acad. Sci. USA* **2018**, *115*, E1194–E1203. [CrossRef]

41. Griffin, W.S.T.; Stanley, L.C.; Ling, C.; White, L.; McLeod, V.; Perrot, L.J.; White Iii, C.L.; Araoz, C. Brain interleukin 1 and S-100 immunoreactivity are elevated in Down syndrome and Alzheimer disease. *Proc. Natl. Acad. Sci. USA* **1989**, *86*, 7611–7615. [CrossRef] [PubMed]

42. Shaftel, S.S.; Griffin, W.S.T.; Kerry, K.M. The role of interleukin-1 in neuroinflammation and Alzheimer disease: An evolving perspective. *J. Neuroinflamm.* **2008**, *5*, 7. [CrossRef] [PubMed]

43. Balzaretti, S.; Taverniti, S.; Guglielmetti, S.; Fiore, W.; Minuzzo, M.; Ngo, H.N.; Ngere, J.B.; Sadiq, S.; Humphreys, P.N.; Laws, A.P. A novel rhamnose-rich hetero-exopolysaccharide isolated from Lactobacillus paracasei DG activates THP-1 human monocytic cells. *Appl. Environ. Microbiol.* **2017**, *83*, e02702–e02716. [CrossRef] [PubMed]

44. Fu, Y.R.; Yi, Z.J.; Pei, J.L.; Guan, S. Effects of Bifidobacterium bifidum on adaptive immune senescence in aging mice. *Microbiol. Immunol.* **2010**, *54*, 578–583. [CrossRef] [PubMed]

45. Danikowski, K.M.; Jayaraman, S.; Prabhakar, B.S. Regulatory T cells in multiple sclerosis and myasthenia gravis. *J. Neuroinflamm.* **2017**, *14*, 117. [CrossRef] [PubMed]

46. Abbès, S.; Ben Salah-Abbes, J.; Jebali, R.; Younes, R.B.; Oueslati, R. Interaction of aflatoxin B1 and fumonisin B1 in mice causes immunotoxicity and oxidative stress: Possible protective role using lactic acid bacteria. *J. Immunotoxicol.* **2016**, *13*, 46–54. [CrossRef]

47. Ai, C.; Ma, N.; Zhang, Q.; Wang, G.; Liu, X.; Tian, F.; Chen, P.; Chen, W. Immunomodulatory effects of different lactic acid bacteria on allergic response and its relationship with in vitro properties. *PLoS ONE* **2016**, *11*, e0164697. [CrossRef]

48. Finlay, C.M.; Stefanska, A.M.; Walsh, K.P.; Kelly, P.J.; Boon, L.; Lavelle, E.C.; Walsh, P.T.; Mills, K.H.G. Helminth products protect against autoimmunity via innate type 2 cytokines IL-5 and IL-33, which promote eosinophilia. *J. Immunol.* **2016**, *196*, 703–714. [CrossRef]

49. Iyer, S.S.; Cheng, G. Role of interleukin 10 transcriptional regulation in inflammation and autoimmune disease. *Crit. Rev. Immunol.* **2012**, *32*, 23–63. [CrossRef]

50. Souza, B.M.; Preisser, T.M.; Pereira, V.B.; Zurita-Turk, M.; Castro, C.P.; Cunha, V.P.; Oliveira, R.P.; Gomes-Santos, A.C.; Faria, A.M.C.; Machado, D.C.C.; et al. Lactococcus lactis carrying the pValac eukaryotic expression vector coding for IL-4 reduces chemically-induced intestinal inflammation by increasing the levels of IL-10-producing regulatory cells. *Microb. Cell Fact.* **2016**, *15*, 150. [CrossRef]

51. Levkovich, T.; Poutahidis, T.; Smillie, C.; Varian, B.J.; Ibrahim, Y.M.; Lakritz, J.R.; Alm, E.J.; Erdman, S.E. Probiotic Bacteria Induce a 'Glow of Health'. *PLoS ONE* **2013**, *8*, e53867. [CrossRef] [PubMed]

52. Smelt, M.J.; de Haan, B.J.; Bron, P.A.; van Swam, I.; Meijerink, M.; Wells, J.M.; Faas, M.M.; de Vos, P.L. L. plantarum, L. salivarius, and L. lactis Attenuate Th2 Responses and Increase Treg Frequencies in Healthy Mice in a Strain Dependent Manner. *PLoS ONE* **2012**, *7*, e47244. [CrossRef] [PubMed]

53. Fink, L.N.; Frøkiær, H. Dendritic cells from Peyer's patches and mesenteric lymph nodes differ from spleen dendritic cells in their response to commensal gut bacteria. *Scand. J. Immunol.* **2008**, *68*, 270–279. [CrossRef] [PubMed]

54. Carasi, P.; Racedo, S.M.; Jacquot, C.; Romanin, D.E.; Serradell, M.A.; Urdaci, M.C. Impact of kefir derived lactobacillus kefiri on the mucosal immune response and gut microbiota. *J. Immunol. Res.* **2015**. [CrossRef]

Review

Microbiome in Multiple Sclerosis: Where Are We, What We Know and Do Not Know

Marina Kleopatra Boziki [1], Evangelia Kesidou [1], Paschalis Theotokis [1], Alexios-Fotios A. Mentis [2,3], Eleni Karafoulidou [1], Mikhail Melnikov [4,5], Anastasia Sviridova [4,5], Vladimir Rogovski [6], Alexey Boyko [4,5,*] and Nikolaos Grigoriadis [1,*]

[1] 2nd Neurological University Department, Aristotle University of Thessaloniki, AHEPA General Hospital, 54634 Thessaloniki, Greece; bozikim@auth.gr (M.K.B.); bioevangelia@yahoo.gr (E.K.); ptheotokis@gmail.com (P.T.); elenikarafoulidou95@hotmail.com (E.K.)

[2] Public Health Laboratories, Hellenic Pasteur Institute, Athens 11521, Greece; mentisaf@gmail.com

[3] Laboratory of Microbiology, University Hospital of Larissa, School of Medicine, University of Thessaly, 41110 Larissa, Greece

[4] Department of Neurology, Neurosurgery and Medical Genetics, Pirogov Russian National Research Medical University, Moscow 117997, Russia; medikms@yandex.ru (M.M.); anastasiya-ana@yandex.ru (A.S.)

[5] Department of Neuroimmunology, Federal Center of Cerebrovascular Pathology and Stroke, Moscow 117342, Russia

[6] Department of Molecular Pharmacology and Radiobiology, Pirogov Russian National Research Medical University, Moscow 117997, Russia; qwer555@mail.ru

[*] Correspondence: boykoan13@gmail.com (A.B.); ngrigoriadis@auth.gr (N.G.); Tel.: +30-231-099-4665 (N.G.)

Received: 19 March 2020; Accepted: 8 April 2020; Published: 14 April 2020

Abstract: An increase of multiple sclerosis (MS) incidence has been reported during the last decade, and this may be connected to environmental factors. This review article aims to encapsulate the current advances targeting the study of the gut–brain axis, which mediates the communication between the central nervous system and the gut microbiome. Clinical data arising from many research studies, which have assessed the effects of administered disease-modifying treatments in MS patients to the gut microbiome, are also recapitulated.

Keywords: gut microbiome; gut–brain axis; metagenomics; multiple sclerosis; disease-modifying treatments

1. Introduction

The prevalence of multiple sclerosis (MS) has reportedly increased over the last few decades, showing both higher absolute numbers of patients and a real increase in MS incidence [1,2]. Overall, the number of patients afflicted with MS may be increasing due to a prolongation of their life expectancy and of the disease duration. Moreover, the integration of MAGNIMS (magnetic resonance imaging in multiple sclerosis) consensus in the diagnostic criteria for MS and the universal application of these criteria, together with their constant re-evaluation in order to achieve optimal sensitivity and specificity, allow for more accurate and early diagnoses [3,4]. The development of novel disease-modifying treatments (DMTs) that are effective in controlling disease activity and delaying progression (even in cases with highly active disease [5]), as well as the increase in physician's awareness towards complications of the disease (such as spasticity, urinary disturbance, and chronic infections [6]), are also measures that increase life quality, and ultimately survival, for patients with MS. Moreover, the combined efforts of medical societies worldwide towards the development and application of universal registries and patient databases have led to improved case ascertainment that has also contributed to the observed increase of MS frequency [7]. In addition to the aforementioned advances in the health system and the medical services provided, a true increase in the incidence of MS has occurred in several ethnic populations over the last few decades. This is indicated by (i) the minimal number of ethnic populations that still remain free of the disease; (ii) a well-documented increase of the frequency of MS in previously low-incidence populations, such as in Asia, Southern and Eastern Europe; and (iii) the wider age window, i.e., younger than 16 and older than 50 years of age, in which the disease onset occurs [1,8,9]. According to the hygiene hypothesis, advanced civilization and technological progress in the recent past led to an improvement of the hygiene level of the overall life conditions for several ethnic populations and this improvement may be linked to increased MS frequency. In this respect, the observed alterations in MS incidence may be linked to an environmental shift towards a more MS-predisposing status. Another likely scenario is that the relative significance of the environmental factor with respect to MS pathogenesis has increased in the 21st Century.

This review aims to summarize the recent advances that have been achieved in the analysis of gut microbiota, an environmental factor with a well-described impact in autoimmunity, and to provide a critical assessment of the derived knowledge with respect to the role of gut microbiota in MS pathogenesis. Moreover, we attempt to form key questions in order to position the derived knowledge into a valuable context with respect to personalized medicine and patient-tailored therapeutic approaches.

2. The Environmental Factor in Autoimmune Disease

A complex interplay between genetic and environmental factors is necessary for the development of autoimmunity. In MS, genetically predisposing factors have been recognized, with specific polymorphisms of the major histocompatibility complex (MHC), namely the human leukocyte antigen (HLA) system, to be the factors accounting for the majority of cases [10]. For instance, beta chain of HLA (HLA-DRB1) and DQ beta 1 chain of HLA (HLA-DQB1)polymorphisms have been implicated in MS predisposition in Caucasians. In addition, more than 130 single nucleotide polymorphisms (SNPs) implicated in various responses of the innate and adaptive immune system, as well as in cell survival and/or pathways of cellular death, have been recognized. However, even by considering the cumulative effect of these polymorphisms, the effect of the genetic factor itself does not account for more than 30% of MS cases [10]. Environmental factors have long been implicated in MS pathogenesis, and they include lifestyle conditions, such as smoking and the level of physical exercise, as well as the type of overall diet (e.g., Western, Eastern, or Mediterranean) and/or specific dietary parameters, such as vitamin D and salt intake [11,12]. Recently, it became evident that dietary and lifestyle conditions may exert a profound impact on the gastrointestinal tract (GI) and, more specifically, the intestine. This is an organ that appears to pose a significant role in regulating several responses of the signaling systems of the human organism, namely the endocrine, the immune, and, more remotely, the central nervous system (CNS) [13,14].

More importantly, the genetic factor itself is the main factor that is present upon the prenatal and immediate postnatal stage that determines predisposition towards disease at a level that remains relatively constant throughout life. Nonetheless, its outcome is subject to the effect of several environmental factors that are, overall, actively present throughout life. Each factor acts for an individual period of time and possibly affects a specific stage of disease pathogenesis, namely the predisposition, onset, and/or course of the disease [15]. According to the classical paradigm of genetics–environment interplay, gene polymorphisms are constant for a given individual and exert an effect upon their phenotype that remains constant throughout life. Environmental factors, on the other hand, continue to exert a biological effect that may be cumulative for the time period that the factor is present, or they may act as triggers that induce the onset of disease. An environmental factor may, therefore (i) act before the biological onset of the disease, thus contributing towards predisposition; (ii) act upon disease onset (trigger); or (iii) be present during the disease course, according to the LEARn (Latent Early Life-Associated Regulation) model, an epigenetic model of disease development described by Lahiri et al. Similarly, the GERSMS (Genetic and Environmental Risk Score) has been proposed as a means to quantify a combined estimate of an individual's genetic burden and environmental exposures [16]. Special notice has been taken with respect to (i) the Western diet (ii) other lifestyle conditions, such as smoking, lack of physical excersise etc.; (iii) specific virus infections, such as the Epstein-Barr virus; (iv) the wide use of antibiotics; and (iv) the high sanitary level, as factors that promote pro-inflammatory responses; Several of these factors are present early (age <15 years old) in life [17].

3. Gut Microbiota and the Role of Intestinal Dysbiosis

The human GI tract is colonized by approximately 1014 different populations of microorganisms. Overall, gut microbiota are nowadays regarded as a separate organ in the human body, weighing approximately 2 kg and carrying information that is at least 100 times larger than the number of human genes for an individual [18]. Under steady-state conditions, these microorganisms are symbiotic, in the sense that they contribute to the homeostasis of the human organism. More specifically, gut microbiota (i) contribute to the maintenance of the motility and permeability of the gut; (ii) prevent colonization by pathogens; (iii) mediate nutrient metabolism; (iv) participate in the production of vitamins, such as vitamin B complex, vitamin K, and folate; and (v) promote intestinal epithelial functions, such as absorption and secretion [18]. Recently, gut microbiota have been shown to shape the immune responses of innate and adaptive immunity, both locally (at the level of the GI mucosa) and systemically, thus affecting remote organs [19]. Data stemming from two large metagenomic databases, i.e., the MetaHIT (Metagenomics of the Human Intestinal Tract) and the Human Microbiome Project, isolated 2172 species in humans that were classified into 12 different phyla, with 93.5% of them belonging to the Proteobacteria, Firmicutes, Actinobacteria, and Bacteroidetes [20]. Large fractions of the phyla Firmicutes and Bacteroidetes reportedly include the genera *Prevotella*, *Bacteroides*, and *Ruminococcus*, and these are followed in size by Actinobacteria [20]. Moreover, the relative composition of the gut microbiota does not appear to be constant throughout different parts of the GI tract. Rather, there appears to be some degree of regional specialization with respect to the exact microbes that colonize each part of the gut [21]. For instance, frequent *Lactobacilli* are present in the duodenum, whereas both *Lactobacilli* and *Streptococci* are abundant in the jejunum. A large diversity has been described for the colon with the caecum and the appendix; these are two areas that also bear larger burden of microorganisms, in terms of absolute numbers [21]. Similarly, the diversity in microbiota across the GI tract leads to differential profiles of the metabolites that are produced as a result of the various microbiota mediating nutrient absorption and metabolism: In the stomach and the duodenum, vitamin A and aryl hydrocarbon receptors (AHR) ligands are primarily produced, whereas in the colon, a gradual shift towards higher short-chain fatty acid (SCFA) production is evident [22]. The structural architecture of the GI tract, as well as the differences in cellular composition and the pH of the adjacent mucosa, account for the alterations in the microbial composition and in the associated metabolites across the GI tract. Disequilibrium in the relative composition of intestinal microbiota has recently been recognized as a common underlying condition in several autoimmune diseases. The alteration of the intestinal microbial community that might lead to either animal or human diseases is termed intestinal or gut dysbiosis. Intestinal microbiota have been proven to shape immune responses and to affect the neural and endocrine systems of the gut. All these pathways exert remote signaling in the human body and thus bear implications for systemic and organ-specific autoimmunity, as in the case of the CNS [19].

4. The Gut Microbiota in MS

4.1. Immunoregulation and the Gut–Brain Axis

The enteric nervous system has long been recognized as a second brain. More recently, the gut–brain axis has been recognized as a bi-directional communication system from the CNS to the gut and vice versa; this communication is mediated by neuronal connections, neuroendocrine signals, general humoral signals, and immune signaling [23]. The CNS regulates gut function by promoting gut motility via a dense innervation system and by orchestrating local immune responses through the high numbers of immune cells that are present in the gut. These humoral signals are delivered by the utilization of common molecular mediators, such as pro-inflammatory cytokines, neuropeptides (like cholecystokinin (CCK) and leptin), and neurotransmitters (like dopamine (DA), serotonin (5-HT), gamma-aminobutyric acid (GABA), acetylcholine (Ach), and glutamate [22]). Conversely, structures in immediate proximity to the microbiota—such as the intestinal epithelial cells and immune cells in gut-associated lymphatic tissue (GALT) and the enteric nervous system (ENS)—mediate the transmission of signaling pathways from the gut towards the CNS. In this respect, gut microbiota may modulate the host via several pathways that originate in parts of the neuroendocrine, neural, and immune systems [23].

For instance, structurally distinct lipopolysaccharide (LPS), a characteristic component of the outer envelope of many microbes, exhibits a differential immunogenic profile in terms of the associated cytokines that are produced as a response by the host [24]. Toll-like receptor (TLR) signaling, a part of the pattern-recognition receptor (PRR) signaling, appears to be a key mediator of the host's immune response towards bacteria, as it is the first-line sensing pathway that recognizes microbial structural patterns.

Moreover, the recognition of bacterial structures by the TLR system prevents microbial translocation towards the deep layers of the gut lumen, as demonstrated in myeloid differentiation primary response 88 (MyD88) -/- mice that lack the expression of epithelial MyD88-dependent TLR [25]. In the bi-directional communication between the microbes and the host, it is therefore evident that the host may also regulate microbial colonization by the early recruitment of sensing and defense mechanisms. For example, cluster of differentiation antigen (CD) 1d (CD1d)+ invariant natural killer T (iNKT) cells and $\gamma\delta$ intraepithelial lymphocytes ($\gamma\delta$ IELs) are T-cell subsets that respond to microbial antigens. These cells were shown to regulate bacterial colonization in the gut [26]. Local immunoglobulin A (IgA) production by B-cells is also regarded as a factor regulating gut microflora composition and density [27,28]. Conversely, germinal center formation and the production of IgA are shaped by activation of T-follicular helper cells; the latter is induced by microbes and mediated by programmed cell death protein 1 (PD-1) [28].

4.2. Gut Microbiota and Innate Immunity

Overall, microbiota are essential for priming the gastro-intestinal immune system to evoke specific immune responses: With respect to the innate immune system, several subsets of cells that participate antigen presentation respond to microbial stimuli by enhancing cytokine and chemokine production. The mucosa-associated invariant T (MAIT) cells, which express an invariant α T-cell receptor (TCR) chain and the non-classical MHC-I related protein located in mucosal tissues (e.g., intestinal lamina propria), produce diverse pro-inflammatory cytokines, such as interleukin (IL)-17, interferon gamma (IFNγ), granzyme B, or tumor necrosis factor alpha (TNFα) [29]. By expressing various chemokine receptors, MAIT cells exhibit a migratory capacity into remote tissues [29]. Natural killer (NK)-cells increase the expression of co-stimulatory molecules in response to microbial stimuli. NK cells are essential for the priming of other immune cells and the coordination of the overall host immune response by the production of IL-4, IL-13, and IFNγ, as well as the promotion of chemokine (C-X-C motif) ligand 16 (CXCL16) production by epithelial cells [29]. Dendritic cells and macrophages are, as is known, the classical antigen-presenting cells, and they play a key role in first-line host defense and the modulation of adaptive immunity. In so doing, they enhance the production of pro-IL-1β and its processing to bioactive IL-1β by caspase-1, thus discriminating between pathogenic and protective bacteria and dietary components [30].

4.3. Gut Microbiota and Adaptive Immunity

The adaptive immune system also exhibits the capacity for microbe-driven responses. T-helper (Th)-17 cells are prevalent in the intestine, and they are important for the gastro-intestinal host defense, as they secrete cytokines that are involved in the regulation of inflammation (IL-17A, IL-17F, and IL-22).

Specific microbes are capable of eliciting a differential T-effector phenotype (e.g., Th17 and Tregs) in the intestines and lymph nodes of mice that exhibit predisposition towards autoimmunity [31,32]. In germ-free mice or antibiotics-treated mice, the number of Th17 cells is reduced along with attenuated pro-inflammatory responses [33]. Moreover, mice that are resistant to autoimmunity exhibit the preferential sequestration of Th17 cells in the intestine, whereas peripheral blood Th17 repopulation by the administration of anti-(a4b7) integrin monoclonal antibodies rescues the autoimmune disease phenotype [34]. In this respect, the intestine appears to be a key regulating organ of immune system responses [34,35]. T regulatory (Treg) cells are two- to three-folds higher in abundance in the gastro-intestinal tract compared to other tissues. Mice with compromised gut microbiota, such as germ-free mice or antibiotic-treated mice, display a reduced frequency of Treg cells, and these Treg cells exhibit impaired anti-inflammatory cytokine-secretion, especially IL-10. In these mice, re-colonization by gut microbes promotes the function and frequency of Treg cells [32,36]. Moreover, bacterial antigens, such as LPS, are necessary for class-switch recombination in B cells towards IgA production. Additionally, B-cells primed by bacterial antigens have been shown to participate in antigen presentation and IgA selection in the germinal centers of the GALT (e.g., Peyer's patches) [37].

4.4. The Role of Microbial Metabolites

Apart from microbial structural components that may serve as antigens that shape immune responses in the intestine with implications for systemic disease, other molecular mediators also exhibit the capacity to induce pro- or anti-inflammatory reactions. Metabolites of microbial origin are present in the intestine, often as a by-product of nutrient degradation; these molecules may stimulate immune cells towards activated phenotype and cytokine production. SCFAs are metabolites produced by intestinal microbes that mediate a well-known anti-inflammatory effect. SCFAs inhibit histone deacetylases (HDACs) on Treg and microglia, a mechanism mediated by G-protein-coupled receptors (GPRs) [38]. Moreover, SCFAs may stimulate dendritic cells (DCs) towards the production of anti-inflammatory molecules, such as retinoic acid (RA) and transforming growth factor beta (TGFβ). Tryptophan metabolites evidently shape the phenotype of T cell subsets by promoting the production of either pro-inflammatory Th1 cytokines, such as IFN-γ and IL-2, or anti-inflammatory Th2 cytokines, such as IL-4 and IL-10 [38]. Tryptophan metabolites may also promote the Th17 pro-inflammatory phenotype by acting on AHR, a signaling pathway known to also affect astrocyte activation in the CNS as a response to microbial stimuli from the systemic circulation [38].

4.5. The Role of Intestinal Barrier

Clinical and experimental evidence on the role of the intestinal barrier and its structural and functional integrity has recently elucidated aspects of the interplay between intestinal microbes and the host. An impaired intestinal barrier is regarded a common underlying condition in several autoimmune diseases of the gut that exhibit systemic implications, such as inflammatory bowel disease (IBD). In this respect, impaired intestinal barrier integrity exposes the cells of the local and, more importantly, of the systemic immune system to stimuli of microbial origin with a potential to elicit immune responses, as suggested in the context of the leaky gut theory. Intestinal dysbiosis appears to mediate barrier dysfunction, as this microbiome-related process may induce changes in mucus composition, enterocyte apoptosis and tight junction dysfunction through the translocation of associated structural components, as well as bacterial translocation to the lamina propria [39]. These alterations lead to an increased homing of lymphocytes in the lamina propria, the immunological layer of the intestinal barrier, and, in doing such, they contribute in the host's predisposition towards local and systemic autoimmune responses. In the case of MS, such intestinal barrier alterations are linked to the presence of increased LPS and LPS-mediated signaling in the lamina propria, leading to chronic low-grade inflammation and endotoxemia [39]. Concomitant reduction in SCFAs associated to dysbiosis, that is, reduced microbial diversity, a condition frequently described in MS, results in compromized intestinal barrier and thus predisposes towards systemic pro-inflammatory reactions.

In the CNS, microglia and astrocytes respond to pro-inflammatory stimuli from systemic circulation and acquire activated phenotypes, thus further promoting pro-inflammatory milieu in the context of CNS autoimmunity [39]. In this respect, the intestinal barrier has recently emerged as a novel target of pharmacological intervention in MS. This is because the restoration of the intestinal barrier may reduce the exposure of the cellular components of the systemic immune system to microbial derivatives and the associated pro-inflammatory cascade.

4.6. Mechanisms of Immune-Modulation by Intestinal Microbiota—Experimental Evidence

Experimental evidence has dictated that the presence of intestinal microbiota is necessary in order for CNS autoimmunity to develop. In a myelin oligodendrocyte glycoprotein-specific t cell receptor (MOG-TCR) transgenic mouse model of spontaneous disease, experimental autoimmune encephalomyelitis (EAE) does not occur under germ-free (GF) conditions, whereas mice transferred from GF conditions into a conventional environment develop spontaneous EAE after few weeks of transfer [40]. Interestingly, MOG-TCR transgenic mice of a genetic background that is resistant towards autoimmunity, namely the B10.S mice, do not exhibit spontaneous EAE, even under conventional conditions [34]. In these mice, a preferential sequestration of Th17 pro-inflammatory T-cells in the intestine has been observed. The administration of anti-$\alpha4\beta7$, a monoclonal antibody (mAb) that blocks intestinal integrin, has been found to be able to repopulate peripheral blood with Th17 T-cells and to rescue the disease phenotype [34]. The intestine thus appears as an organ with the ability to control systemic autoimmune responses with implication towards CNS autoimmunity [41]. In a similar context, the modulation of gut microbiota, as achieved by antibiotic administration, reduces the severity of conventional EAE. In an experimental setting, specific immune responses have been linked with single bacteria, such as in the case of Clostridia and *Bacteroides fragilis* derived from human feces that have the potential to induce Foxp3+ T regulatory cells, thus ameliorating EAE [42]. The fecal transplantation of MOG-TCR transgenic mice with human feces stemming from twins discordant for MS has only been found to result in the development of spontaneous EAE in mice recipients for feces stemming from twins with MS. In contrast, mice recipients for feces stemming from healthy twins have not been found to develop the disease [43]. GF mice recipients for the human feces of MS patients have also been found to develop severe EAE, coupled with alterations in the peripheral immune profile [44]. More specifically, fecal transplantation with material provided by healthy adults has been found to result in the induction of T regulatory cells in the mesenteric lymph nodes of the recipient mice, thus, overall, exerting an immune-regulatory response [44]. Conversely, the administration of *Lactobacilli* has been repeatedly shown to protect form EAE by the induction of IL-4, IL-10, TGF-β1, and IL-27 [45,46] by mediating an increase in IL-10+ and Foxp3+ T regulatory cells [47–49].

However, findings often observed in EAE fail to be translated to human disease due to the differences that the model exhibits. With the exception of spontaneous models, EAE requires myelin peptide immunization with a strong adjuvant. This is a condition that exerts especially skewed immune responses towards inflammation and results in a monophasic disease of inflammatory origin with little demyelination compared to the human disease [50]. In TCR transgenic mice that exhibit spontaneous disease, more than 90% of the circulating T-cells bear transgenic, autoreactive TCRs [51,52]. This is a condition that also does not accurately depict CNS autoimmunity in humans. In this respect, a detailed profiling of human microbiota appears to be a necessary approach to elucidate human-specific mechanisms. These mechanisms stem from interactions between the gut microbiota and the host, and they show the potential to induce autoimmune responses with relevance to the CNS.

4.7. Mechanisms of Immune-Modulation by Intestinal Microbiota—Clinical Evidence

During the last five years, several clinical studies have provided evidence indicating that in MS, the gut microbiome is altered. In an approach similar to the experimental model, initial studies linked alterations in the relative abundance of Clostridia, in the context of gut dysbiosis with MS. However, the clinical relevance—with respect to whether these alterations contribute towards susceptibility for

MS or, instead, they exert a relative protective effect—remains controversial both for adult [53,54] and pediatric populations [55,56]. In 2016, two case control studies reported distinct patterns of gut microbiota composition by the use of 16S ribosomal ribonucleic acid rRNA metagenomics analysis [57,58] (for further discussion on the potential of metagenomic techniques as applied to MS, see [59,60]). These studies provided evidence of reduced diversity in the gut microbiome of MS patients compared to controls. Interestingly, this reduction was evident for patients with active MS, whereas patients in remission exhibited comparable diversity levels to the healthy population. Further studies verified this association of disease activity status with alterations of the relative abundance of microbes in the gut, with Firmicutes and Bacteroidetes exhibiting higher relative abundances reviewed by Kozhieva et al. [61]. These studies indicated the following: Though it is widely accepted that the gut microbiome in patients with MS is characterized by moderate dysbiosis, a clear and consistent multiple sclerosis microbiome phenotype has not been described. Moreover, given that a myriad of microbes have been implicated in MS, it is unlikely that, in the future, a single microbial organism will be isolated and characterized as an environmental trigger towards disease. This is in striking contrast with the paradigm stemming from mouse EAE. According to the latter, triggering of CNS autoimmunity by microbes provides mechanistic insight with respect to the molecular pathways that lead from the local immune responses in the gut to systemic inflammation and, eventually, to organ-specific autoimmunity towards the CNS [31,32].

Recently, a systematic review [62] of MS case-control studies with respect to gut microbiota composition concluded that, although differences in the diversity of microbiota were not reported by the majority of the included studies, several studies reported consistent patterns with respect to the taxonomic relative abundance. These findings further elucidated pattern alterations in the overall gut microbial composition of patients with MS compared to controls. Further prospective studies are necessary in order to establish a causative relation between these microbial pattern alterations and the disease pathogenesis and/or exacerbation. Notably, the majority of the reviewed studies referred to the Relapsing-remmitting type of MS(RRMS). A recent study addressed the differential gut microbiota profile in patients with primary progressive MS (PPMS), relatively to healthy controls [63]. As in the case of patients with RRMS, patients with PPMS have exhibited differences in a minority of a-diversity indices, whereas pattern differences have been observed at a taxonomic level [63].

In line with the observations described above, diet and dietary supplementation has recently emerged a major factor that affects gut microbiota's relative composition. It has been proposed that a diet that is rich in vegetables, complex carbohydrates (fibers) combined with probiotics, vitamin D supplementation, vitamin A supplementation, and lipoic acid promotes gut eubiosis. This is coupled with a concomitant increase in microbial diversity and microbe-associated anti-inflammatory mediators, such as SCFAs, microbial anti-inflammatory molecules (MAMs), histone deacetylase inhibitors, AHR receptor agonists, and an increase of the Treg/Th17 ratio [64]. Conversely, a Western diet rich in animal fat and trans-fatty acids, with a high sugar and salt intake, promotes gut dysbiosis and results in (i) an increased presence of pro-inflammatory mediators such as TNFa, IL-6, and IL-17; and (ii) increased gut barrier and blood–brain barrier (BBB) permeability with implications for systemic and CNS autoimmunity [64]. More specifically, gut dysbiosis predisposes one to intestinal inflammation, which is characterized by alterations in the immunological barrier layer of the lamina propria towards pro-inflammatory milieu in the GALT and an increase in the presence of endotoxin/LPS in the intestinal mucosa. The further translocation of LPS and other bacterial components, as well as whole bacteria in the deep layers of the intestinal wall and the local secondary lymphoid organs (such as the mesenteric lymph nodes) allows for the generation of circulating activated T-cells in the context of low-grade endotoxemia [39,64]. These systemic alterations (i) compromise the integrity of the BBB, (ii) allow for pro-inflammatory stimuli to cross the BBB towards the CNS, and (iii) affect microglia and astrocyte activation status, thus predisposing one towards neuroinflammation.

Clinical evidence of the possible causal relationship between the gut microbiota profile and the CNS autoimmunity stems from more interventional approaches that actively alter gut microflora composition; such approaches include fecal microbiota transplantation (FMT), an investigational method that has been used successfully to treat cases of severe enterocolitis [65,66] More specifically, prolonged antibiotic administration in certain individuals may cause expansion of *Clostridium difficile* (*C. difficile*) at the expense of symbiotic bacteria, thus serving as an example of intestinal dysbiosis and *C. difficile*-related severe enterocolitis. FMT protocols require that fecal material from a healthy donor, following careful donor screening and appropriate preparation procedures, is transferred to a patient, either via colonoscopy or via an oral route as capsule ingestion [67]. Due to risks linked with transplantation (i.e., possible transmittable disease) and colonoscopy procedures, FMT is reserved for cases that are refractory to the antibiotics that are typically prescribed against enterocolitis due to *C. difficile*. Recently, FMT has been advocated as an attractive therapeutic approach for several diseases that are linked to intestinal dysbiosis, either of the intestine, such as IBD [68,69] or systemic extragastric and CNS disease [70,71] (and reviewed in [72]). Isolated case reports have described the beneficial effects of FMT over MS disease course, and a clinical trial of FMT for patients with MS is currently underway [73].

5. Disease-Modifying Treatment (DMT) and Gut Microbiota

In the management of MS, DMTs serve as prophylactic treatments towards clinical and radiological disease activity, whereas other medications are prescribed for symptom management. The later are more often prescribed in the context of disability accumulation, such as gamma-Aminobutyric acid-type B GABA$_B$ receptor agonists (e.g., baclofen) for limb spasticity and a-adrenergic inhibitors for the control of overactive bladders. Several of these regimens are known to alter the profile of the gut microbiota [39]. Recently, several oral DMTs were shown to inhibit the growth of *Clostridium* in vitro. This feature has been proposed to contribute to the DMTs' overall anti-inflammatory mechanism of action [74,75]. In this study, fingolimod was proven to be bactericidal, whereas teriflunomide and dimethyl fumarate (DMF) exerted a bacteriostatic effect. Clinical data stemming from metagenomics analysis of gut microbiota alterations in patients receiving DMF and glatiramer acetate (GA) further verified that DMTs exert a profound effect on the relative composition of gut microbiota. The above could shed light into potential additional mechanisms of action [76].

5.1. Interferon-β

With respect to IFNβ, several lines of investigation have indicated that it may modify the immunological properties of the intestinal barrier. IFNβ is a member of the type 1 interferon (T1IFN) family, and it is considered a major cytokine that mediates local responses to viral, bacterial, and other antigen stimuli in the intestine (reviewed in [77]). In a mouse model of pneumococcal lung infection, IFNβ treatment led to the upregulation of tight junction proteins in lung epithelial and blood vessel endothelial layers, thus reducing lung–blood barrier permeability and preventing invasive pneumococcal infection [78]. Furthermore, IFNβ, produced by DCs following stimulation by gut commensal microbiota, was recently shown to mediate Treg proliferation in the intestine [79]. A clinical case-control study exploring the effect of IFNβ administration in patients with MS recently reported an increase of *Prevotella*, a known probiotic, in patients with MS treated with IFNβ. This increase was comparable to healthy controls, whereas untreated patients exhibited a reduced relative abundance of probiotics [80].

5.2. Glatiramer Acetate

GA is a myelin-basic protein (MBP) analog and a long-administered first-line DMT treatment for MS. In line with its anti-inflammatory properties, GA is known to ameliorate colonic injury in an experimental model of colitis by inducing a reduction in TNFa and a concomitant increase in Tregs, IL-10, and TGFb-producing cells [81]. Moreover, the role of GA in stabilizing the intestinal barrier and promoting tissue repair, as documented by analysis of syndecan-1 expression, was shown in a model of IBD [82]. In EAE, GA administration ameliorated the disease phenotype coupled with an increase in gut *Prevotella*, and the administration of GA combined with GI colonization with live *Prevotella* led to the further attenuation of disease [83]. GA treatment in patients with MS was shown to exert an effect in the relative abundance of gut microbiota, especially the Lachnospiraceae and Veillonellaceae families [76]. Similarly, another case-control study in patients with RRMS treated with GA reported alterations in the relative composition of the gut microbiome with respect to several *Clostridium* [84].

5.3. Dimethyl Fumarate

DMF is a long-prescribed DMT for psoriasis that was more recently approved as a first-line treatment for MS. In an experimental setting, DMF was shown to promote an increase in the relative abundance of probiotics and to stabilize the intestinal barrier. A concomitant increase in SCFA-producing microbes was shown to further promote the systemic anti-inflammatory effect of the drug [85]. Similarly, DMF administration in Lewis rats was shown to mediate (i) a reduction in the TLR-4 expression by the GALT, (ii) a reduction in IFNγ, and (iii) a concomitant increase in lamina propria's Foxp-3+ expression and the abundance of CD4+CD25+ Tregs in Peyer's patches [86]. A case-control study implementing metagenomics techniques reported an association between DMF treatment and decreased relative abundance of Firmicutes and Clostridia, as well as an increase of Bacteroidetes, relative to untreated patients [76].

5.4. Teriflunomide

Teriflunomide, another oral, first-line DMT approved as a prophylactic treatment for RRMS, has been shown to modify immune responses in the intestine by promoting the local proliferation of CD39+ Tregs. Thus, it exerted anti-inflammatory action that ameliorated CNS inflammation in a mouse model of EAE [87].

5.5. Natalizumab

Natalizumab (NTZ) is an injectable second-line DMT that has been approved for patients with highly active RRMS. NTZ is a monoclonal antibody targeting a4-integrin, a family that includes adhesion molecules expressed in T-cells. Integrins exhibit tissue specificity with a4b1 expressed in the CNS, whereas a4b7 is expressed in the intestine. NTZ is not selective; therefore, by inhibiting T-cell trafficking towards the CNS, it also blocks T-cell circulation in the gut. In addition to its beneficial effect in MS, NTZ has been shown to be effective in ameliorating the symptoms of IBD [88]. The administration of NTZ has been proposed as preferred treatment approach for patients with RRMS and IBD co-morbidity [88]. Furthermore, IBD is a well-described condition characterized by intestinal dysbiosis and local immune dysregulation. In this respect, also in patients with RRMS that do not exhibit signs of IBD, the amelioration of T-cell trafficking in the gut by NTZ may contribute to the drug's anti-inflammatory effect by inhibiting the circulation of activated T-cells in the gut. The gut has recently been proposed as a regulating organ with respect to the peripheral circulation of activated T-cells, with implications for CNS autoimmunity. In mice resistant to EAE that exhibit the preferential sequestration of autoreactive Th17 T-cells in the intestine, the administration of the a4b7 mAb led to the re-population of peripheral blood with Th17 T-cells and rescued the disease phenotype [34]. Interestingly, in this MOG-TCR transgenic mouse model, the selective accumulation of autoreactive T-cells in the intestine acts as a mechanism of immune tolerance that contributes in resistance towards EAE [34]. In the context of MS, it is reasonable to assume that the blocking of T-cell trafficking in the intestine may ameliorate the exposure of the immune system to stimuli of microbial origin. This amelioration could be performed by reducing antigen sampling and, subsequently, T-cell clonal expansion and activation in response to these antigens. As gut dysbiosis, and the associated local immune dysregulation, is frequently reported in patients with RRMS, this concomitant effect may serve as an additional mode of action for NTZ in RRMS.

5.6. Fingolimod

Fingolimod is an oral second-line DMT that is indicated as prophylactic treatment for highly active RRMS. Fingolimod is a sphingosine-1-phosphate (S1P) agonist, acting on four out of five S1P receptors, on various organs and cell types. Fingolimod ligation on S1P receptors leads to the downregulation of the S1P receptor expression; therefore, the drug is considered as a functional antagonist of S1P signaling. On a physiological level, S1P receptor expression is necessary for the lymphocytes, either naïve or activated, to egress from the secondary lymphoid organs, such as the peripheral lymph nodes. Due to this mechanism of action, fingolimod inhibits the egress of lymphocytes from lymph nodes to the peripheral blood stream. Thus, it ameliorates systemic immune responses and CNS-targeted autoimmunity. Similarly, other S1P ligands have been tested for intestinal autoimmune disease, such as IBD, as they exhibit the potential to ameliorate the transmigration of immune cells across the intestine [89]. Fingolimod has been shown to ameliorate experimental colitis [90], and two S1P ligands are currently being tested in terms of safety and efficacy in phase II and phase III clinical trials on colitis [88]. Interestingly, S1P signaling has been shown to regulate innate lymphoid cell (ILC) transmigration from intestinal lamina propria towards systemic circulation and other lymphoid organs, thus regulating infectious and inflammatory responses [91]. In a transgenic mouse model of enteric nervous system pathology resembling Parkinson's disease (PD) due to a-synuclein accumulation, fingolimod resulted in an enhanced gut motility and increased levels of brain-derived neurotrophic factor (BDNF) [92]. Moreover, S1P ligation has been shown to exert a stabilizing effect towards barrier function [93,94] and the BBB [95], with implications for EAE and MS [96–98]. Fingolimod may potentially exert an effect on the gut microbiome's relative composition, as it has been shown to regulate IgA plasmablasts' maturation from the intestinal Peyer's patches, a first-line defense mechanism of the host towards microbe colonization [99,100]. Moreover, fingolimod was shown to exert a direct anti-microbial effect by inhibiting the growth of *Clostridium* and the associated endotoxin production in vitro [74].

5.7. Alemtuzumab

Alemtuzumab is an anti-CD52 monoclonal antibody inducing T-cell and B-cell depletion in peripheral blood. It was originally used for the treatment of chronic B-cell lymphocytic leukemia. Recently, alemtuzumab has been approved for the treatment of highly active RRMS for patients with breakthrough disease that were previously exposed to other first- and/or second-line DMTs [101]. Apart from the obvious effect of alemtuzumab in depleting circulating primarily B- and T-cell lymphocytes, the long-term immunomodulating effect is mediated by alterations that the drug causes in the peripheral immune cell pool following repopulation. Some of these alterations are attributed to the homeostatic proliferation of mature lymphocytes that the drug promotes in peripheral tissues [102]. Due to this effect, the administration of alemtuzumab has been linked with an increased susceptibility towards autoimmune comorbidities, such as autoimmune thyroid disease, membranous glomerulonephritis, autoimmune hepatitis, and immune thrombocytopenic purpura [103]. Colitis due to *Clostridium* was the cause of fatal outcome in one patient with RRMS who received alemtuzumab [104], whereas another patient presented with pancolitis during the first course of alemtuzumab treatment [105]. Susceptibility towards infection was the assumed underlying cause in both cases. With respect to the second case, an immune-mediated mechanism contributing to sepsis has also been proposed [105]. Evidence stemming from a cynomolgus monkey model indicated that the intestinal barrier may be disrupted during alemtuzumab treatment [106]. In macaques monkeys, a single dose of alemtuzumab resulted in (i) intestinal epithelial cell loss, (ii) increased apoptosis in the villi, and (iii) an abnormal Paneth cell morphology [107]. Mouse anti-CD52 mAb also resulted in increased numbers of IELs undergoing apoptosis and disrupted intestinal barrier function in mice [108]. In a cynomolgus monkey model, alemtuzumab administration resulted in profound alterations in the relative composition of gut microbiota, namely Lactobacillales, Enterobacterales, and Clostridiales, as well as the genera *Prevotella* and *Faecalibacterium*. These alterations were primarily linked to alterations in the relative abundance of TCRαβ+ or TCRγδ+ T cells [109]. These data indicate that alemtuzumab administration exerts a profound effect on the intestinal homeostasis with respect to tissue integrity, barrier function, immune properties, and microbiome profile. However, it remains unknown whether these alterations are beneficial in the context of CNS autoimmunity, or, instead, if, in a proportion of patients, the overall beneficiary effect of alemtuzumab in subsiding disease activity is counterbalanced by a detrimental effect in intestinal function.

6. Conclusions: Treat the Microbiome—Treat MS?

The combined efforts of the scientific community in the field of MS have been focused on identifying strategies that may be implemented in order to either modify the peripheral immune responses or, as proposed by the less successful approach to date, to enhance neuroprotection and the endogenous regenerative capacity of the CNS. In addition to the classical paradigm of immune–brain interaction in the context of MS, the intestine has emerged as an additional regulating organ of responses that take place both in the immunological and the nervous (central and peripheral) counterparts. In this respect, the gut commensal microbiota may serve as environmental factors that shape the intestinal milieu. The modification of gut microbiota by either dietary (e.g., probiotic supplementation) or medicinal approaches (e.g., antibiotic administration) may serve as additional therapeutic strategies for MS prophylaxis [110]. More interventional approaches, such as FMT, have also been proposed. Moreover, the relative composition of gut microbiota may also serve as an indicator of reciprocal host–microorganism interactions. Further longitudinal studies that implement the profiling of intestinal microbiota during the pre-clinical phase and over the course of the disease are needed in order to elucidate this assumption. MS is a complex autoimmune disease with clinical variability. As such, the establishment of a causative role for intestinal microbiota towards disease pathogenesis requires combined efforts from the field of metagenomics and other "-omics" approaches [59,60,111] with the capacity for high throughput data production and the application of these data in the context of translational medicine.

With respect to future directions, we consider gut microbiota modulation as a promising intervention for the management of MS. Understanding the pathways that the gut microbiota implicate in order to shape host's immune responses may elucidate therapeutic targets, such as the induction of immune regulatory cell populations via the promotion of an "anti-inflammatory" gut microflora. Similar interventions, possibly in combination with DMTs, may contribute in promoting treatment efficacy and optimal response. As several newly available DMTs confer significant and potentially severe adverse effects, gut microbiota modification has emerged as a promising, and possibly less interventional, additional approach.

Author Contributions: M.K.B. conducted data collection, design, writing, overall preparation of the manuscript and synthesis. E.K. (Evangelia Kesidou), P.T., A.-F.A.M., E.K. (Eleni Karafoulidou), M.M., A.S., V.R. participated in data collection and preparation of the manuscript. A.B. and N.G. were responsible for conception and critical review of the manuscript. All authors have read and agreed to the published version of the manuscript.

Funding: This research received no external funding.

Conflicts of Interest: The authors declare no conflict of interest.

References

1. Global Health Metrics. Global, regional, and national incidence, prevalence, and years lived with disability for 328 diseases and injuries for 195 countries, 1990–2016: A systematic analysis for the Global Burden of Disease Study 2016. *Lancet* **2017**, *390*, 1211–1259. [CrossRef]
2. Grytten, N.; Torkildsen, O.; Myhr, K.M. Time trends in the incidence and prevalence of multiple sclerosis in Norway during eight decades. *Acta Neurol. Scand.* **2015**, *132*, 29–36. [CrossRef] [PubMed]
3. Filippi, M.; Rocca, M.A.; Ciccarelli, O.; De Stefano, N.; Evangelou, N.; Kappos, L.; Rovira, A.; Sastre-Garriga, J.; Tintore, M.; Frederiksen, J.L.; et al. MRI criteria for the diagnosis of multiple sclerosis: MAGNIMS consensus guidelines. *Lancet Neurol.* **2016**, *15*, 292–303. [CrossRef]

4. Thompson, A.J.; Banwell, B.L.; Barkhof, F.; Carroll, W.M.; Coetzee, T.; Comi, G.; Correale, J.; Fazekas, F.; Filippi, M.; Freedman, M.S.; et al. Diagnosis of multiple sclerosis: 2017 revisions of the McDonald criteria. *Lancet Neurol.* **2018**, *17*, 162–173. [CrossRef]

5. Tintore, M.; Vidal-Jordana, A.; Sastre-Garriga, J. Treatment of multiple sclerosis—Success from bench to bedside. *Nat. Rev. Neurol.* **2019**, *15*, 53–58. [CrossRef]

6. Henze, T.; Rieckmann, P.; Toyka, K.V. Symptomatic treatment of multiple sclerosis. Multiple Sclerosis Therapy Consensus Group (MSTCG) of the German Multiple Sclerosis Society. *Eur. Neurol.* **2006**, *56*. [CrossRef]

7. Glaser, A.; Stahmann, A.; Meissner, T.; Flachenecker, P.; Horakova, D.; Zaratin, P.; Brichetto, G.; Pugliatti, M.; Rienhoff, O.; Vukusic, S.; et al. Multiple sclerosis registries in Europe—An updated mapping survey. *Mult. Scler. Relat Disord.* **2019**, *27*, 171–178. [CrossRef]

8. McKay, K.A.; Hillert, J.; Manouchehrinia, A. Long-term disability progression of pediatric-onset multiple sclerosis. *Neurology* **2019**, *92*, e2764–e2773. [CrossRef] [PubMed]

9. Guillemin, F.; Baumann, C.; Epstein, J.; Kerschen, P.; Garot, T.; Mathey, G.; Debouverie, M. Older Age at Multiple Sclerosis Onset Is an Independent Factor of Poor Prognosis: A Population-Based Cohort Study. *Neuroepidemiology* **2017**, *48*, 179–187. [CrossRef] [PubMed]

10. Patsopoulos, N.A. Genetics of Multiple Sclerosis: An Overview and New Directions. *Cold Spring Harb. Perspect. Med.* **2018**, *8*. [CrossRef] [PubMed]

11. Belbasis, L.; Bellou, V.; Evangelou, E.; Ioannidis, J.P.; Tzoulaki, I. Environmental risk factors and multiple sclerosis: An umbrella review of systematic reviews and meta-analyses. *Lancet Neurol.* **2015**, *14*, 263–273. [CrossRef]

12. Zheng, C.; He, L.; Liu, L.; Zhu, J.; Jin, T. The efficacy of vitamin D in multiple sclerosis: A meta-analysis. *Mult. Scler. Relat. Disord.* **2018**, *23*, 56–61. [CrossRef] [PubMed]

13. Camara-Lemarroy, C.R.; Metz, L.M.; Yong, V.W. Focus on the gut-brain axis: Multiple sclerosis, the intestinal barrier and the microbiome. *World J. Gastroenterol.* **2018**, *24*, 4217–4223. [CrossRef] [PubMed]

14. Probstel, A.K.; Baranzini, S.E. The Role of the Gut Microbiome in Multiple Sclerosis Risk and Progression: Towards Characterization of the "MS Microbiome". *Neurotherapeutics* **2018**, *15*, 126–134. [CrossRef]

15. Lahiri, D.K.; Maloney, B. The "LEARn" (Latent Early-life Associated Regulation) model integrates environmental risk factors and the developmental basis of Alzheimer's disease, and proposes remedial steps. *Exp. Gerontol.* **2010**, *45*, 291–296. [CrossRef]

16. Xia, Z.; Steele, S.U.; Bakshi, A.; Clarkson, S.R.; White, C.C.; Schindler, M.K.; Nair, G.; Dewey, B.E.; Price, L.R.; Ohayon, J.; et al. Assessment of Early Evidence of Multiple Sclerosis in a Prospective Study of Asymptomatic High-Risk Family Members. *JAMA Neurol.* **2017**, *74*, 293–300. [CrossRef]

17. Freedman, S.N.; Shahi, S.K.; Mangalam, A.K. The "Gut Feeling": Breaking Down the Role of Gut Microbiome in Multiple Sclerosis. *Neurotherapeutics* **2018**, *15*, 109–125. [CrossRef]

18. Picca, A.; Fanelli, F.; Calvani, R.; Mule, G.; Pesce, V.; Sisto, A.; Pantanelli, C.; Bernabei, R.; Landi, F.; Marzetti, E. Gut Dysbiosis and Muscle Aging: Searching for Novel Targets against Sarcopenia. *Mediators Inflamm.* **2018**, *2018*, 7026198. [CrossRef]

19. Ochoa-Reparaz, J.; Magori, K.; Kasper, L.H. The chicken or the egg dilemma: Intestinal dysbiosis in multiple sclerosis. *Ann. Transl. Med.* **2017**, *5*, 145. [CrossRef]

20. Blum, H.E. The human microbiome. *Adv. Med. Sci.* **2017**, *62*, 414–420. [CrossRef]

21. Mowat, A.M.; Agace, W.W. Regional specialization within the intestinal immune system. *Nat. Rev. Immunol.* **2014**, *14*, 667–685. [CrossRef] [PubMed]

22. Ghaisas, S.; Maher, J.; Kanthasamy, A. Gut microbiome in health and disease: Linking the microbiome-gut-brain axis and environmental factors in the pathogenesis of systemic and neurodegenerative diseases. *Pharmacol. Ther.* **2016**, *158*, 52–62. [CrossRef] [PubMed]

23. Fleck, A.K.; Schuppan, D.; Wiendl, H.; Klotz, L. Gut-CNS-Axis as Possibility to Modulate Inflammatory Disease Activity-Implications for Multiple Sclerosis. *Int. J. Mol. Sci.* **2017**, *18*, 1526. [CrossRef] [PubMed]

24. Vatanen, T.; Kostic, A.D.; d'Hennezel, E.; Siljander, H.; Franzosa, E.A.; Yassour, M.; Kolde, R.; Vlamakis, H.; Arthur, T.D.; Hamalainen, A.M.; et al. Variation in Microbiome LPS Immunogenicity Contributes to Autoimmunity in Humans. *Cell* **2016**, *165*, 842–853. [CrossRef]

25. Frantz, A.L.; Rogier, E.W.; Weber, C.R.; Shen, L.; Cohen, D.A.; Fenton, L.A.; Bruno, M.E.; Kaetzel, C.S. Targeted deletion of MyD88 in intestinal epithelial cells results in compromised antibacterial immunity associated with downregulation of polymeric immunoglobulin receptor, mucin-2, and antibacterial peptides. *Mucosal. Immunol.* **2012**, *5*, 501–512. [CrossRef]

26. Nieuwenhuis, E.E.; Matsumoto, T.; Lindenbergh, D.; Willemsen, R.; Kaser, A.; Simons-Oosterhuis, Y.; Brugman, S.; Yamaguchi, K.; Ishikawa, H.; Aiba, Y.; et al. Cd1d-dependent regulation of bacterial colonization in the intestine of mice. *J. Clin. Investig.* **2009**, *119*, 1241–1250. [CrossRef]

27. Suzuki, K.; Meek, B.; Doi, Y.; Muramatsu, M.; Chiba, T.; Honjo, T.; Fagarasan, S. Aberrant expansion of segmented filamentous bacteria in IgA-deficient gut. *Proc. Natl. Acad. Sci. USA* **2004**, *101*, 1981–1986. [CrossRef]

28. Hapfelmeier, S.; Lawson, M.A.; Slack, E.; Kirundi, J.K.; Stoel, M.; Heikenwalder, M.; Cahenzli, J.; Velykoredko, Y.; Balmer, M.L.; Endt, K.; et al. Reversible microbial colonization of germ-free mice reveals the dynamics of IgA immune responses. *Science* **2010**, *328*, 1705–1709. [CrossRef]

29. Dias, J.; Leeansyah, E.; Sandberg, J.K. Multiple layers of heterogeneity and subset diversity in human MAIT cell responses to distinct microorganisms and to innate cytokines. *Proc. Natl. Acad. Sci. USA* **2017**, *114*, E5434–E5443. [CrossRef]

30. Franchi, L.; Kamada, N.; Nakamura, Y.; Burberry, A.; Kuffa, P.; Suzuki, S.; Shaw, M.H.; Kim, Y.G.; Nunez, G. NLRC4-driven production of IL-1beta discriminates between pathogenic and commensal bacteria and promotes host intestinal defense. *Nat. Immunol.* **2012**, *13*, 449–456. [CrossRef]

31. Ivanov, I.I.; Atarashi, K.; Manel, N.; Brodie, E.L.; Shima, T.; Karaoz, U.; Wei, D.; Goldfarb, K.C.; Santee, C.A.; Lynch, S.V.; et al. Induction of intestinal Th17 cells by segmented filamentous bacteria. *Cell* **2009**, *139*, 485–498. [CrossRef]

32. Atarashi, K.; Tanoue, T.; Shima, T.; Imaoka, A.; Kuwahara, T.; Momose, Y.; Cheng, G.; Yamasaki, S.; Saito, T.; Ohba, Y.; et al. Induction of colonic regulatory T cells by indigenous Clostridium species. *Science* **2011**, *331*, 337–341. [CrossRef] [PubMed]

33. Shaw, M.H.; Kamada, N.; Kim, Y.G.; Nunez, G. Microbiota-induced IL-1beta, but not IL-6, is critical for the development of steady-state TH17 cells in the intestine. *J. Exp. Med.* **2012**, *209*, 251–258. [CrossRef] [PubMed]

34. Berer, K.; Boziki, M.; Krishnamoorthy, G. Selective accumulation of pro-inflammatory T cells in the intestine contributes to the resistance to autoimmune demyelinating disease. *PLoS ONE* **2014**, *9*, e87876. [CrossRef]

35. Sano, T.; Huang, W.; Hall, J.A.; Yang, Y.; Chen, A.; Gavzy, S.J.; Lee, J.Y.; Ziel, J.W.; Miraldi, E.R.; Domingos, A.I.; et al. An IL-23R/IL-22 Circuit Regulates Epithelial Serum Amyloid A to Promote Local Effector Th17 Responses. *Cell* **2015**, *163*, 381–393. [CrossRef] [PubMed]

36. Tanoue, T.; Atarashi, K.; Honda, K. Development and maintenance of intestinal regulatory T cells. *Nat. Rev. Immunol.* **2016**, *16*, 295–309. [CrossRef]

37. Kim, M.; Kim, C.H. Regulation of humoral immunity by gut microbial products. *Gut. Microbes* **2017**, *8*, 392–399. [CrossRef] [PubMed]

38. Haase, S.; Haghikia, A.; Wilck, N.; Muller, D.N.; Linker, R.A. Impacts of microbiome metabolites on immune regulation and autoimmunity. *Immunology* **2018**, *154*, 230–238. [CrossRef] [PubMed]

39. Camara-Lemarroy, C.R.; Metz, L.; Meddings, J.B.; Sharkey, K.A.; Wee Yong, V. The intestinal barrier in multiple sclerosis: Implications for pathophysiology and therapeutics. *Brain* **2018**, *141*, 1900–1916. [CrossRef]

40. Berer, K.; Mues, M.; Koutrolos, M.; Rasbi, Z.A.; Boziki, M.; Johner, C.; Wekerle, H.; Krishnamoorthy, G. Commensal microbiota and myelin autoantigen cooperate to trigger autoimmune demyelination. *Nature* **2011**, *479*, 538–541. [CrossRef]

41. Yokote, H.; Miyake, S.; Croxford, J.L.; Oki, S.; Mizusawa, H.; Yamamura, T. NKT cell-dependent amelioration of a mouse model of multiple sclerosis by altering gut flora. *Am. J. Pathol.* **2008**, *173*, 1714–1723. [CrossRef] [PubMed]

42. Atarashi, K.; Tanoue, T.; Oshima, K.; Suda, W.; Nagano, Y.; Nishikawa, H.; Fukuda, S.; Saito, T.; Narushima, S.; Hase, K.; et al. Treg induction by a rationally selected mixture of Clostridia strains from the human microbiota. *Nature* **2013**, *500*, 232–236. [CrossRef] [PubMed]

43. Berer, K.; Gerdes, L.A.; Cekanaviciute, E.; Jia, X.; Xiao, L.; Xia, Z.; Liu, C.; Klotz, L.; Stauffer, U.; Baranzini, S.E.; et al. Gut microbiota from multiple sclerosis patients enables spontaneous autoimmune encephalomyelitis in mice. *Proc. Natl. Acad. Sci. USA* **2017**, *114*, 10719–10724. [CrossRef] [PubMed]

44. Cekanaviciute, E.; Yoo, B.B.; Runia, T.F.; Debelius, J.W.; Singh, S.; Nelson, C.A.; Kanner, R.; Bencosme, Y.; Lee, Y.K.; Hauser, S.L.; et al. Gut bacteria from multiple sclerosis patients modulate human T cells and exacerbate symptoms in mouse models. *Proc. Natl. Acad. Sci. USA* **2017**, *114*, 10713–10718. [CrossRef] [PubMed]

45. Lavasani, S.; Dzhambazov, B.; Nouri, M.; Fak, F.; Buske, S.; Molin, G.; Thorlacius, H.; Alenfall, J.; Jeppsson, B.; Westrom, B. A novel probiotic mixture exerts a therapeutic effect on experimental autoimmune encephalomyelitis mediated by IL-10 producing regulatory T cells. *PLoS ONE* **2010**, *5*, e9009. [CrossRef] [PubMed]

46. Kwon, H.K.; Kim, G.C.; Kim, Y.; Hwang, W.; Jash, A.; Sahoo, A.; Kim, J.E.; Nam, J.H.; Im, S.H. Amelioration of experimental autoimmune encephalomyelitis by probiotic mixture is mediated by a shift in T helper cell immune response. *Clin. Immunol.* **2013**, *146*, 217–227. [CrossRef] [PubMed]

47. Takata, K.; Kinoshita, M.; Okuno, T.; Moriya, M.; Kohda, T.; Honorat, J.A.; Sugimoto, T.; Kumanogoh, A.; Kayama, H.; Takeda, K.; et al. The lactic acid bacterium Pediococcus acidilactici suppresses autoimmune encephalomyelitis by inducing IL-10-producing regulatory T cells. *PLoS ONE* **2011**, *6*, e27644. [CrossRef]

48. Rezende, R.M.; Oliveira, R.P.; Medeiros, S.R.; Gomes-Santos, A.C.; Alves, A.C.; Loli, F.G.; Guimaraes, M.A.; Amaral, S.S.; da Cunha, A.P.; Weiner, H.L.; et al. Hsp65-producing Lactococcus lactis prevents experimental autoimmune encephalomyelitis in mice by inducing CD4+LAP+ regulatory T cells. *J. Autoimmun.* **2013**, *40*, 45–57. [CrossRef]

49. Mangalam, A.; Shahi, S.K.; Luckey, D.; Karau, M.; Marietta, E.; Luo, N.; Choung, R.S.; Ju, J.; Sompallae, R.; Gibson-Corley, K.; et al. Human Gut-Derived Commensal Bacteria Suppress CNS Inflammatory and Demyelinating Disease. *Cell Rep.* **2017**, *20*, 1269–1277. [CrossRef]

50. Gold, R.; Linington, C.; Lassmann, H. Understanding pathogenesis and therapy of multiple sclerosis via animal models: 70 years of merits and culprits in experimental autoimmune encephalomyelitis research. *Brain* **2006**, *129*, 1953–1971. [CrossRef]

51. Pollinger, B.; Krishnamoorthy, G.; Berer, K.; Lassmann, H.; Bosl, M.R.; Dunn, R.; Domingues, H.S.; Holz, A.; Kurschus, F.C.; Wekerle, H. Spontaneous relapsing-remitting EAE in the SJL/J mouse: MOG-reactive transgenic T cells recruit endogenous MOG-specific B cells. *J. Exp. Med.* **2009**, *206*, 1303–1316. [CrossRef] [PubMed]

52. Krishnamoorthy, G.; Wekerle, H. EAE: An immunologist's magic eye. *Eur. J. Immunol.* **2009**, *39*, 2031–2035. [CrossRef] [PubMed]

53. Rumah, K.R.; Linden, J.; Fischetti, V.A.; Vartanian, T. Isolation of Clostridium perfringens type B in an individual at first clinical presentation of multiple sclerosis provides clues for environmental triggers of the disease. *PLoS ONE* **2013**, *8*, e76359. [CrossRef]

54. Miyake, S.; Kim, S.; Suda, W.; Oshima, K.; Nakamura, M.; Matsuoka, T.; Chihara, N.; Tomita, A.; Sato, W.; Kim, S.W.; et al. Dysbiosis in the Gut Microbiota of Patients with Multiple Sclerosis, with a Striking Depletion of Species Belonging to Clostridia XIVa and IV Clusters. *PLoS ONE* **2015**, *10*, e0137429. [CrossRef] [PubMed]

55. Tremlett, H.; Fadrosh, D.W.; Faruqi, A.A.; Hart, J.; Roalstad, S.; Graves, J.; Spencer, C.M.; Lynch, S.V.; Zamvil, S.S.; Waubant, E. Associations between the gut microbiota and host immune markers in pediatric multiple sclerosis and controls. *BMC Neurol.* **2016**, *16*, 182. [CrossRef]

56. Tremlett, H.; Fadrosh, D.W.; Faruqi, A.A.; Zhu, F.; Hart, J.; Roalstad, S.; Graves, J.; Lynch, S.; Waubant, E. Gut microbiota in early pediatric multiple sclerosis: A case-control study. *Eur. J. Neurol.* **2016**, *23*, 1308–1321. [CrossRef] [PubMed]

57. Jangi, S.; Gandhi, R.; Cox, L.M.; Li, N.; von Glehn, F.; Yan, R.; Patel, B.; Mazzola, M.A.; Liu, S.; Glanz, B.L.; et al. Alterations of the human gut microbiome in multiple sclerosis. *Nat. Commun.* **2016**, *7*, 12015. [CrossRef]

58. Chen, J.; Chia, N.; Kalari, K.R.; Yao, J.Z.; Novotna, M.; Paz Soldan, M.M.; Luckey, D.H.; Marietta, E.V.; Jeraldo, P.R.; Chen, X.; et al. Multiple sclerosis patients have a distinct gut microbiota compared to healthy controls. *Sci Rep.* **2016**, *6*, 28484. [CrossRef]

59. Mentis, A.A.; Dardiotis, E.; Grigoriadis, N.; Petinaki, E.; Hadjigeorgiou, G.M. Viruses and Multiple Sclerosis: From Mechanisms and Pathways to Translational Research Opportunities. *Mol. Neurobiol* **2017**, *54*, 3911–3923. [CrossRef]

60. Mentis, A.A.; Dardiotis, E.; Grigoriadis, N.; Petinaki, E.; Hadjigeorgiou, G.M. Viruses and endogenous retroviruses in multiple sclerosis: From correlation to causation. *Acta Neurol. Scand.* **2017**, *136*, 606–616. [CrossRef]

61. Kozhieva, M.K.; Melnikov, M.V.; Rogovsky, V.S.; Oleskin, A.V.; Kabilov, M.R.; Boyko, A.N. Gut human microbiota and multiple sclerosis. *Zh Nevrol. Psikhiatr. Im S S Korsakova* **2017**, *117*, 11–19. [CrossRef] [PubMed]

62. Mirza, A.; Forbes, J.D.; Zhu, F.; Bernstein, C.N.; Van Domselaar, G.; Graham, M.; Waubant, E.; Tremlett, H. The multiple sclerosis gut microbiota: A systematic review. *Mult. Scler. Relat. Disord.* **2020**, *37*, 101427. [CrossRef] [PubMed]

63. Kozhieva, M.; Naumova, N.; Alikina, T.; Boyko, A.; Vlassov, V.; Kabilov, M.R. Primary progressive multiple sclerosis in a Russian cohort: Relationship with gut bacterial diversity. *BMC Microbiol.* **2019**, *19*, 309. [CrossRef]

64. Riccio, P.; Rossano, R. Diet, Gut Microbiota, and Vitamins D + A in Multiple Sclerosis. *Neurotherapeutics* **2018**, *15*, 75–91. [CrossRef] [PubMed]

65. Borody, T.J.; Brandt, L.J.; Paramsothy, S. Therapeutic faecal microbiota transplantation: Current status and future developments. *Curr. Opin. Gastroenterol.* **2014**, *30*, 97–105. [CrossRef]

66. Makkawi, S.; Camara-Lemarroy, C.; Metz, L. Fecal microbiota transplantation associated with 10 years of stability in a patient with SPMS. *Neurol. Neuroimmunol. Neuroinflamm.* **2018**, *5*, e459. [CrossRef]

67. Cammarota, G.; Ianiro, G.; Tilg, H.; Rajilic-Stojanovic, M.; Kump, P.; Satokari, R.; Sokol, H.; Arkkila, P.; Pintus, C.; Hart, A.; et al. European consensus conference on faecal microbiota transplantation in clinical practice. *Gut* **2017**, *66*, 569–580. [CrossRef]

68. Chin, S.M.; Sauk, J.; Mahabamunuge, J.; Kaplan, J.L.; Hohmann, E.L.; Khalili, H. Fecal Microbiota Transplantation for Recurrent Clostridium difficile Infection in Patients with Inflammatory Bowel Disease: A Single-Center Experience. *Clin. Gastroenterol. Hepatol.* **2017**, *15*, 597–599. [CrossRef]

69. Paramsothy, S.; Kamm, M.A.; Kaakoush, N.O.; Walsh, A.J.; van den Bogaerde, J.; Samuel, D.; Leong, R.W.L.; Connor, S.; Ng, W.; Paramsothy, R.; et al. Multidonor intensive faecal microbiota transplantation for active ulcerative colitis: A randomised placebo-controlled trial. *Lancet* **2017**, *389*, 1218–1228. [CrossRef]

70. Jayasinghe, T.N.; Chiavaroli, V.; Holland, D.J.; Cutfield, W.S.; O'Sullivan, J.M. The New Era of Treatment for Obesity and Metabolic Disorders: Evidence and Expectations for Gut Microbiome Transplantation. *Front. Cell Infect. Microbiol.* **2016**, *6*, 15. [CrossRef]

71. Bajaj, J.S.; Kassam, Z.; Fagan, A.; Gavis, E.A.; Liu, E.; Cox, I.J.; Kheradman, R.; Heuman, D.; Wang, J.; Gurry, T.; et al. Fecal microbiota transplant from a rational stool donor improves hepatic encephalopathy: A randomized clinical trial. *Hepatology* **2017**, *66*, 1727–1738. [CrossRef]

72. Choi, H.H.; Cho, Y.S. Fecal Microbiota Transplantation: Current Applications, Effectiveness, and Future Perspectives. *Clin. Endosc.* **2016**, *49*, 257–265. [CrossRef] [PubMed]

73. Lawson Health Research Institute. Fecal Microbial Transplantation in Relapsing Multiple Sclerosis Patients. Available online: https://clinicaltrials.gov/ct2/show/NCT03183869 (accessed on 5 April 2020).

74. Rumah, K.R.; Vartanian, T.K.; Fischetti, V.A. Oral Multiple Sclerosis Drugs Inhibit the In vitro Growth of Epsilon Toxin Producing Gut Bacterium, Clostridium perfringens. *Front. Cell Infect. Microbiol.* **2017**, *7*, 11. [CrossRef] [PubMed]

75. Linden, J.R.; Ma, Y.; Zhao, B.; Harris, J.M.; Rumah, K.R.; Schaeren-Wiemers, N.; Vartanian, T. Clostridium perfringens Epsilon Toxin Causes Selective Death of Mature Oligodendrocytes and Central Nervous System Demyelination. *MBio* **2015**, *6*, e02513. [CrossRef] [PubMed]

76. Katz Sand, I.; Zhu, Y.; Ntranos, A.; Clemente, J.C.; Cekanaviciute, E.; Brandstadter, R.; Crabtree-Hartman, E.; Singh, S.; Bencosme, Y.; Debelius, J.; et al. Disease-modifying therapies alter gut microbial composition in MS. *Neurol. Neuroimmunol. Neuroinflamm.* **2019**, *6*, e517. [CrossRef] [PubMed]

77. Giles, E.M.; Stagg, A.J. Type 1 Interferon in the Human Intestine-A Co-ordinator of the Immune Response to the Microbiota. *Inflamm. Bowel Dis.* **2017**, *23*, 524–533. [CrossRef]

78. LeMessurier, K.S.; Hacker, H.; Chi, L.; Tuomanen, E.; Redecke, V. Type I interferon protects against pneumococcal invasive disease by inhibiting bacterial transmigration across the lung. *PLoS Pathog.* **2013**, *9*, e1003727. [CrossRef]

79. Nakahashi-Oda, C.; Udayanga, K.G.; Nakamura, Y.; Nakazawa, Y.; Totsuka, N.; Miki, H.; Iino, S.; Tahara-Hanaoka, S.; Honda, S.; Shibuya, K.; et al. Apoptotic epithelial cells control the abundance of Treg cells at barrier surfaces. *Nat. Immunol.* **2016**, *17*, 441–450. [CrossRef]

80. Castillo-Alvarez, F.; Perez-Matute, P.; Oteo, J.A.; Marzo-Sola, M.E. The influence of interferon beta-1b on gut microbiota composition in patients with multiple sclerosis. *Neurologia* **2018**. [CrossRef]

81. Aharoni, R.; Sonego, H.; Brenner, O.; Eilam, R.; Arnon, R. The therapeutic effect of glatiramer acetate in a murine model of inflammatory bowel disease is mediated by anti-inflammatory T-cells. *Immunol. Lett.* **2007**, *112*, 110–119. [CrossRef]

82. Yablecovitch, D.; Shabat-Simon, M.; Aharoni, R.; Eilam, R.; Brenner, O.; Arnon, R. Beneficial effect of glatiramer acetate treatment on syndecan-1 expression in dextran sodium sulfate colitis. *J. Pharmacol. Exp. Ther.* **2011**, *337*, 391–399. [CrossRef] [PubMed]

83. Shahi, S.K.; Freedman, S.N.; Murra, A.C.; Zarei, K.; Sompallae, R.; Gibson-Corley, K.N.; Karandikar, N.J.; Murray, J.A.; Mangalam, A.K. Prevotella histicola, A Human Gut Commensal, Is as Potent as COPAXONE(R) in an Animal Model of Multiple Sclerosis. *Front. Immunol.* **2019**, *10*, 462. [CrossRef] [PubMed]

84. Cantarel, B.L.; Waubant, E.; Chehoud, C.; Kuczynski, J.; DeSantis, T.Z.; Warrington, J.; Venkatesan, A.; Fraser, C.M.; Mowry, E.M. Gut microbiota in multiple sclerosis: Possible influence of immunomodulators. *J. Investig. Med.* **2015**, *63*, 729–734. [CrossRef] [PubMed]

85. Ma, N.; Wu, Y.; Xie, F.; Du, K.; Wang, Y.; Shi, L.; Ji, L.; Liu, T.; Ma, X. Dimethyl fumarate reduces the risk of mycotoxins via improving intestinal barrier and microbiota. *Oncotarget* **2017**, *8*, 44625–44638. [CrossRef] [PubMed]

86. Pitarokoili, K.; Bachir, H.; Sgodzai, M.; Gruter, T.; Haupeltshofer, S.; Duscha, A.; Pedreiturria, X.; Motte, J.; Gold, R. Induction of Regulatory Properties in the Intestinal Immune System by Dimethyl Fumarate in Lewis Rat Experimental Autoimmune Neuritis. *Front. Immunol.* **2019**, *10*, 2132. [CrossRef]

87. Ochoa-Reparaz, J.; Colpitts, S.L.; Kircher, C.; Kasper, E.J.; Telesford, K.M.; Begum-Haque, S.; Pant, A.; Kasper, L.H. Induction of gut regulatory CD39(+) T cells by teriflunomide protects against EAE. *Neurol Neuroimmunol. Neuroinflamm.* **2016**, *3*, e291. [CrossRef] [PubMed]

88. Biswas, S.; Bryant, R.V.; Travis, S. Interfering with leukocyte trafficking in Crohn's disease. *Best Pract. Res. Clin. Gastroenterol.* **2019**, *38–39*, 101617. [CrossRef]

89. Kunisawa, J.; Kurashima, Y.; Higuchi, M.; Gohda, M.; Ishikawa, I.; Ogahara, I.; Kim, N.; Shimizu, M.; Kiyono, H. Sphingosine 1-phosphate dependence in the regulation of lymphocyte trafficking to the gut epithelium. *J. Exp. Med.* **2007**, *204*, 2335–2348. [CrossRef]

90. Deguchi, Y.; Andoh, A.; Yagi, Y.; Bamba, S.; Inatomi, O.; Tsujikawa, T.; Fujiyama, Y. The S1P receptor modulator FTY720 prevents the development of experimental colitis in mice. *Oncol. Rep.* **2006**, *16*, 699–703. [CrossRef]

91. Huang, Y.; Mao, K.; Chen, X.; Sun, M.A.; Kawabe, T.; Li, W.; Usher, N.; Zhu, J.; Urban, J.F., Jr.; Paul, W.E.; et al. S1P-dependent interorgan trafficking of group 2 innate lymphoid cells supports host defense. *Science* **2018**, *359*, 114–119. [CrossRef]

92. Vidal-Martinez, G.; Vargas-Medrano, J.; Gil-Tommee, C.; Medina, D.; Garza, N.T.; Yang, B.; Segura-Ulate, I.; Dominguez, S.J.; Perez, R.G. FTY720/Fingolimod Reduces Synucleinopathy and Improves Gut Motility in A53T Mice: CONTRIBUTIONS OF PRO-BRAIN-DERIVED NEUROTROPHIC FACTOR (PRO-BDNF) AND MATURE BDNF. *J. Biol. Chem.* **2016**, *291*, 20811–20821. [CrossRef] [PubMed]

93. Bonitz, J.A.; Son, J.Y.; Chandler, B.; Tomaio, J.N.; Qin, Y.; Prescott, L.M.; Feketeova, E.; Deitch, E.A. A sphingosine-1 phosphate agonist (FTY720) limits trauma/hemorrhagic shock-induced multiple organ dysfunction syndrome. *Shock* **2014**, *42*, 448–455. [CrossRef] [PubMed]

94. Garcia, J.G.; Liu, F.; Verin, A.D.; Birukova, A.; Dechert, M.A.; Gerthoffer, W.T.; Bamberg, J.R.; English, D. Sphingosine 1-phosphate promotes endothelial cell barrier integrity by Edg-dependent cytoskeletal rearrangement. *J. Clin. Investig.* **2001**, *108*, 689–701. [CrossRef]

95. Wang, X.; Maruvada, R.; Morris, A.J.; Liu, J.O.; Wolfgang, M.J.; Baek, D.J.; Bittman, R.; Kim, K.S. Sphingosine 1-Phosphate Activation of EGFR As a Novel Target for Meningitic *Escherichia coli* Penetration of the Blood-Brain Barrier. *PLoS Pathog.* **2016**, *12*, e1005926. [CrossRef] [PubMed]

96. Cruz-Orengo, L.; Daniels, B.P.; Dorsey, D.; Basak, S.A.; Grajales-Reyes, J.G.; McCandless, E.E.; Piccio, L.; Schmidt, R.E.; Cross, A.H.; Crosby, S.D.; et al. Enhanced sphingosine-1-phosphate receptor 2 expression underlies female CNS autoimmunity susceptibility. *J. Clin. Investig.* **2014**, *124*, 2571–2584. [CrossRef] [PubMed]

97. Choi, J.W.; Gardell, S.E.; Herr, D.R.; Rivera, R.; Lee, C.W.; Noguchi, K.; Teo, S.T.; Yung, Y.C.; Lu, M.; Kennedy, G.; et al. FTY720 (fingolimod) efficacy in an animal model of multiple sclerosis requires astrocyte sphingosine 1-phosphate receptor 1 (S1P1) modulation. *Proc. Natl. Acad. Sci. USA* **2011**, *108*, 751–756. [CrossRef]

98. Colombo, E.; Di Dario, M.; Capitolo, E.; Chaabane, L.; Newcombe, J.; Martino, G.; Farina, C. Fingolimod may support neuroprotection via blockade of astrocyte nitric oxide. *Ann. Neurol.* **2014**, *76*, 325–337. [CrossRef]

99. Gohda, M.; Kunisawa, J.; Miura, F.; Kagiyama, Y.; Kurashima, Y.; Higuchi, M.; Ishikawa, I.; Ogahara, I.; Kiyono, H. Sphingosine 1-phosphate regulates the egress of IgA plasmablasts from Peyer's patches for intestinal IgA responses. *J. Immunol.* **2008**, *180*, 5335–5343. [CrossRef]

100. Kunisawa, J.; Kurashima, Y.; Gohda, M.; Higuchi, M.; Ishikawa, I.; Miura, F.; Ogahara, I.; Kiyono, H. Sphingosine 1-phosphate regulates peritoneal B-cell trafficking for subsequent intestinal IgA production. *Blood* **2007**, *109*, 3749–3756. [CrossRef]

101. Coles, A.J.; Cohen, J.A.; Fox, E.J.; Giovannoni, G.; Hartung, H.P.; Havrdova, E.; Schippling, S.; Selmaj, K.W.; Traboulsee, A.; Compston, D.A.S.; et al. Alemtuzumab CARE-MS II 5-year follow-up: Efficacy and safety findings. *Neurology* **2017**, *89*, 1117–1126. [CrossRef]

102. Havrdova, E.; Horakova, D.; Kovarova, I. Alemtuzumab in the treatment of multiple sclerosis: Key clinical trial results and considerations for use. *Ther. Adv. Neurol. Disord.* **2015**, *8*, 31–45. [CrossRef] [PubMed]

103. Holmoy, T.; Fevang, B.; Olsen, D.B.; Spigset, O.; Bo, L. Adverse events with fatal outcome associated with alemtuzumab treatment in multiple sclerosis. *BMC Res. Notes* **2019**, *12*, 497. [CrossRef] [PubMed]

104. Baker, D.; Giovannoni, G.; Schmierer, K. Marked neutropenia: Significant but rare in people with multiple sclerosis after alemtuzumab treatment. *Mult. Scler. Relat. Disord.* **2017**, *18*, 181–183. [CrossRef] [PubMed]

105. Vijiaratnam, N.; Rath, L.; Xu, S.S.; Skibina, O. Pancolitis a novel early complication of Alemtuzumab for MS treatment. *Mult. Scler. Relat. Disord.* **2016**, *7*, 83–84. [CrossRef] [PubMed]

106. Qu, L.L.; Lyu, Y.Q.; Jiang, H.T.; Shan, T.; Zhang, J.B.; Li, Q.R.; Li, J.S. Effect of alemtuzumab on intestinal intraepithelial lymphocytes and intestinal barrier function in cynomolgus model. *Chin. Med. J. (Engl.)* **2015**, *128*, 680–686. [CrossRef] [PubMed]

107. Li, Q.; Zhang, Q.; Wang, C.; Jiang, S.; Li, N.; Li, J. The response of intestinal stem cells and epithelium after alemtuzumab administration. *Cell Mol. Immunol.* **2011**, *8*, 325–332. [CrossRef]

108. Qu, L.; Li, Q.; Jiang, H.; Gu, L.; Zhang, Q.; Wang, C.; Li, J. Effect of anti-mouse CD52 monoclonal antibody on mouse intestinal intraepithelial lymphocytes. *Transplantation* **2009**, *88*, 766–772. [CrossRef]

109. Li, Q.R.; Wang, C.Y.; Tang, C.; He, Q.; Li, N.; Li, J.S. Reciprocal interaction between intestinal microbiota and mucosal lymphocyte in cynomolgus monkeys after alemtuzumab treatment. *Am. J. Transplant.* **2013**, *13*, 899–910. [CrossRef]

110. Metz, L.M.; Li, D.K.B.; Traboulsee, A.L.; Duquette, P.; Eliasziw, M.; Cerchiaro, G.; Greenfield, J.; Riddehough, A.; Yeung, M.; Kremenchutzky, M.; et al. Trial of Minocycline in a Clinically Isolated Syndrome of Multiple Sclerosis. *N. Engl. J. Med.* **2017**, *376*, 2122–2133. [CrossRef]

111. Mentis, A.A.; Pantelidi, K.; Dardiotis, E.; Hadjigeorgiou, G.M.; Petinaki, E. Precision Medicine and Global Health: The Good, the Bad, and the Ugly. *Front. Med. (Lausanne)* **2018**, *5*, 67. [CrossRef]

Editorial

The Long Road of Immunotherapeutics against Multiple Sclerosis [†]

Vasso Apostolopoulos [1], Abdolmohamad Rostami [2] and John Matsoukas [3,*]

[1]   Institute for Health and Sport, Victoria University, Melbourne, VIC 3030, Australia; vasso.apostolopoulos@mail.com

[2]   Neurology Department, Thomas Jefferson University, Philadelphia, PA 19107, USA; A.M.Rostami@jefferson.edu

[3]   NewDrug, Patras Science Park, 26500 Patras, Greece

*   Correspondence: imats1953@gmail.com; Tel.: +30-2610-911546

†   This Editorial is dedicated to Elizabeth Matsoukas who was the inspiration of the research.

Received: 24 April 2020; Accepted: 26 April 2020; Published: 11 May 2020

Abstract: This commentary highlights novel immunomodulation and vaccine-based research against multiple sclerosis (MS) and reveals the amazing story that triggered this cutting-edge MS research in Greece and worldwide. It further reveals the interest and solid support of some of the world's leading scientists, including sixteen Nobel Laureates who requested from European leadership to take action in supporting Greece and its universities in the biggest ever financial crisis the country has encountered in the last decades. This support endorsed vaccine-based research on MS, initiated in Greece and Australia, leading to a worldwide network aiming to treat or manage disease outcomes. Initiatives by bright and determined researchers can result in frontiers science. We shed light on a unique story behind great research on MS which is a step forward in our efforts to develop effective treatments for MS.

Keywords: multiple sclerosis; MS; vaccine; immunomodulation; carriers

---

1. Introduction

Nobel Laureates Taking Action to Support Research in Greece

It was realized and clearly understood by the governments in Greece five years ago, and especially now, during this period of COVID-19 pandemic, the necessity for research, as first priority in their policies, for innovation, development and growth. Greece has suffered a lot the last decade from recession. Initiatives by eminent scientists were taken to support research in Greece, with remarkable positive outcomes. Fifteen Nobel Laureates cosigned the "Support for Greece" petition that was addressed by Nobel Laureate Professor Harald zur Hausen, on 14 December 2015, to the European leadership (Jean-Claude Juncker, Martin Schulz and Donald Tusk), pleading for the support for research in universities and to the country. The first to sign was the DNA discoverer Nobel Laureate Professor James Watson, who also sent a letter to the then President of the USA, Mr. Barack Obama, urging him to support Greece [1]. This petition to support research and universities in Greece led to the Hellenic Foundation for Research and Innovation (HFRI) to spur economic development. The European Investment Bank co-financed the creation of the HFRI fund with the Ministry of Finance. Professor Costas Fotakis, Alternate Minister of Research and Technology then, has greatly contributed to the establishment of HFRI. The HFRI fund launches regular calls for all scientists at all stages in support of their research. The "Support for Greece" petition, which was co-signed by the Nobel Laureates and led to the HFRI fund, was a joint initiative between Professor John Matsoukas from the University of Patras Greece and Nobel Laureate Professor Harald zur Hausen from the German Cancer

Research Center in Heidelberg, Germany. This is the second petition after the first in 2012 co-signed by twenty-one Nobel Laureates and addressed again by Professor Harald zur Hausen [2,3]. The second petition worked out successfully.

2. The Sparkle of Immunotherapeutics MS Research

Nobel Laureates Professors James Watson (Cold Spring Harbor Laboratory, New York, NY, USA), Harald zur Hausen (German Cancer Research Center, Heidelberg, Germany) and Andrew Schally (University of Miami, FL, USA) were attracted by the excellent research in Greece and have stated in particular that MS research in Greece is world-class research that is worthy of support. This research had its reason and sparkle. Myelin based immunotherapeutics research for MS in Greece was triggered by Dr. Elizabeth Matsoukas, a Biologist, who has been struck by the disease. That happened to her in 1982, at the age of 30. Following her diagnosis, she dedicated her life to promote research for MS. Her PhD dissertation from the National Hellenic Research Foundation in Athens identified and evaluated myelin epitopes, in particular myelin basic protein (MBP) epitopes, which are implicated in the pathogenesis of the disease [4–7]. Now these epitopes are the tools and the core for developing therapeutics and vaccines for the treatment of MS.

3. The First EAE Experiment in Pennsylvania

In 1994, Professor John Matsoukas, brother of Elizabeth, decided to introduce into his drug discovery research program at the University of Patras the design, synthesis and development of drugs, mimetics and immunotherapeutics, using MBP epitopes against MS. The first experiment, "experimental allergic encephalomyelitis" (EAE), an animal model of the disease, was run that year, at the University of Pennsylvania, in Professor Abdolmohamad Rostami's lab, at that time professor of neurology at the University of Pennsylvania (currently Professor and Chairman of the Department of Neurology at Thomas Jefferson University, Philadelphia USA) [8–10]. Elizabeth had visited Professor Rostami earlier that year in his Pennsylvania clinic for diagnosis and prescription of an interferon drug for her case which was not possible yet at that time in Europe. First experiments were carried out, using the guinea pig epitope MBP_{72-85}, as suggested by Elizabeth [10]. The research for new therapies was part of her curiosity to determine the mechanisms of disease and, based on that, to pursue treatment of disease. Her first EAE experiments in Pennsylvania were successful and paved the way for further research to identify new peptide immunomodulators, which resulted in research based on the other myelin epitopes, primarily MPB_{83-99}, MOG_{35-55} and $PLP_{139-151}$ [4–7,11–21]. This research was quickly spread to the research community in the field, all over the world.

4. Development of a Worldwide MS Consortium

A multi-institutional and multidisciplinary consortium was established in 1999, by Professor John Matsoukas (from the University of Patras; currently Head of NewDrug P.C, Patras Science Park, Greece) and Professor Vasso Apostolopoulos (from the Austin Research Institute Australia/Scripps Research Institute USA); currently Pro Vice-Chancellor, Research Partnerships at Victoria University Australia). The consortium comprised over 15 top universities and research Institutions worldwide (Europe, USA, Canada and Australia), and over 50 researchers have taken part in the consortium over time. Consortium members/collaborators were included, as each had expertise in various disciplines including, chemistry, structural biology, crystallography, molecular dynamics, nuclear magnetic resonance, protein chemistry, cell biology, biochemistry, molecular biology, immunology, neuropathology, animal research, clinical research and neurology clinicians. Each team approached the MS immunotherapeutics research program, using their specialist discipline areas, which together resulted in novel findings and potential new immunotherapeutics against MS. In addition, Professor John Matsoukas (organic chemist) and his team pioneered, through rational design, cyclic constraints of myelin peptides of architectural beauty, which were evaluated for efficacy and stability by members of the consortium. In addition, his team developed novel altered peptide ligands of myelin peptides.

The linear and cyclic peptides, native or as altered peptide ligands, were evaluated for stability in vitro, binding affinities to major histocompatibility complex class II, efficacy in mice and rats and to human peripheral blood mononuclear cells from patients with MS by various consortium groups [6,12–14,22–28].

5. Optimizing Immunotherapeutics and Vaccines against MS, Using a Novel Delivery System

Professor Vasso Apostolopoulos (immunologist and crystallographer), who had developed a novel antigen delivery system against breast and ovarian cancer [29–43], which were translated into human clinical trials [44–48], applied her insights into MS research [26,49–54]. The delivery system specifically targets dendritic cells and, when applied to myelin peptides (cyclic, linear and altered peptide ligands), was able to modulate immune responses from pro-inflammatory to anti- inflammatory, with protection and reversal of EAE in animal models and altered cytokine profile in peripheral blood mononuclear cells isolated from patients with MS [26,49–54]. Over 10 candidate immunotherapeutics have been developed and are justified for their use in phase I human clinical trials.

6. Awarding the Inspiration and the Pioneers of This MS Research

Dr. Elizabeth Matsoukas can justifiably be proud of what she has achieved. Her pain was translated into promising global research to fight the disease. Her dream to see novel immunotherapeutics development against the disease is very close to being materialized. At last, she has seen research due to her case flourish globally. Elizabeth was honored by the Greek Academy of Athens for her dissertation and in 2018, by His Excellency, the President of the Hellenic Democracy, Mr. Prokopis Pavlopoulos, for her initial research and for being the inspiration, the spur and the motivating power of this research. In a special ceremony on 22 September 2018, in Amaliada (province of Ilida), celebrating 20 years of Medicinal Chemistry excellence in Greece, she was awarded by the president with a DNA-inspired plague made by famous sculptor Eustathios Leontis. Standing-ovation applause for her contribution was an emotional moment. In this special ceremony, the protagonists of this MS research, Professors Apostolopoulos, Rostami and Matsoukas, were also awarded, as well Professor Harald zur Hausen, for his contribution to science and society. In addition, Professor Vasso Apostolopoulos and Nobel Laureate Professor Harald zur Hausen received an award for career excellence by His Excellency, the President of the Hellenic Democracy, Mr. Prokopis Pavlopoulos. Last year, Professor John Matsoukas and Professor Vasso Apostolopoulos, were each independently awarded the Salus Index Award, from New Times Publishing, for outstanding career achievements, including their work on MS.

7. Conclusions

The development of drugs, immunotherapeutics and vaccines against diseases is a long process, often taking researchers a lifetime. Researchers often work in silos, limiting their research output; as such, the breaking down of silos would improve research outcomes. Here, we provided an insight of a multi-institutional and multidisciplinary consortium which was developed over 20 years ago that has led to the identification and development of over 10 candidate immunotherapeutics against MS. Today, most research-funding bodies, require multi-institutional and multidisciplinary teams in order to be successful in grant applications. Most importantly, alliances are required to get to the target of the research.

Author Contributions: Conceptualization, J.M. and V.A.; writing, review and editing, J.M., A.R. and V.A. All authors have read and agreed to the published version of the manuscript.

Funding: The writing of this commentary received no external funding.

Acknowledgments: V.A. would like to thank the Institute for Health and Sport, Victoria University, for supporting her current efforts into MS research. J.M. would like to thank the General Secretariat for Research and Technology (GSRT) for supporting his MS research.

Conflicts of Interest: The authors declare no conflict of interest.

References

1. Nobel Laureates "Support for Greece". Available online: http://www.elenapanaritis.com/22-nobelprize-winners-publish-support-greece-letter-science/ (accessed on 4 May 2020).
2. Syriza May Have Lost the Election, but Greece's Research Reforms Deserve to Stay. Available online: https://www.nature.com/articles/d41586-019-02323-y (accessed on 4 May 2020).
3. Hausen, H.Z. Support for Greece. *Science* **2012**, *336*, 978–979. [CrossRef]
4. Tselios, T.; Daliani, I.; Deraos, S.; Thymianou, S.; Matsoukas, E.; Troganis, A.; Gerothanassis, I.; Mouzaki, A.; Mavromoustakos, T.; Probert, L.; et al. Treatment of experimental allergic encephalomyelitis (EAE) by a rationally designed cyclic analogue of myelin basic protein (MBP) epitope 72–85. *Bioorg. Med. Chem. Lett.* **2000**, *10*, 2713–2717. [CrossRef]
5. Tselios, T.; Daliani, I.; Probert, L.; Deraos, S.; Matsoukas, E.; Roy, S.; Pires, J.; Moore, G.; Matsoukas, J. Treatment of experimental allergic encephalomyelitis (EAE) induced by guinea pig myelin basic protein epitope 72–85 with a Human MBP 87–99 analogue and effects of cyclic peptides. *Bioorg Med. Chem.* **2000**, *8*, 1903–1909. [CrossRef]
6. Tselios, T.; Probert, L.; Daliani, I.; Matsoukas, E.; Troganis, A.; Gerothanassis, I.P.; Mavromoustakos, T.; Moore, G.J.; Matsoukas, J.M. Design and Synthesis of a Potent Cyclic Analogue of the Myelin Basic Protein Epitope MBP72-85: Importance of the Ala81Carboxyl Group and of a Cyclic Conformation for Induction of Experimental Allergic Encephalomyelitis. *J. Med. Chem.* **1999**, *42*, 1170–1177. [CrossRef] [PubMed]
7. Tselios, T.; Probert, L.; Kollias, G.; Matsoukas, E.; Roumelioti, P.; Alexopoulos, K.; Moore, G.; Matsoukas, J. Design and synthesis of small semi-mimetic peptides with immunomodulatory activity based on Myelin Basic Protein (MBP). *Amino Acids* **1998**, *14*, 333–341. [CrossRef]
8. Rostami, A.; Gregorian, S.K. Peptide 53–78 of myelin P2 protein is a T cell epitope for the induction of experimental autoimmune neuritis. *Cell. Immunol.* **1991**, *132*, 433–441. [CrossRef]
9. Rostami, A.; Gregorian, S.K.; Brown, M.J.; Pleasure, D.E. Induction of severe experimental autoimmune neuritis with a synthetic peptide corresponding to the 53–78 amino acid sequence of the myelin P2 protein. *J. Neuroimmunol.* **1990**, *30*, 145–151. [CrossRef]
10. Tselios, T.; Deraos, S.; Matsoukas, E.; Panagiotopoulos, D.; Matsoukas, J.; Moore, G.J.; Probert, L.; Kollias, G.; Hilliard, B.; Rostami, A.; et al. Myelin Basic Protein Peptides: Induction and Inhibition of Experimental Allergic Encephalomyelitis. *Rev. Clin. Pharmacol. Pharmacokinetics* **1997**, *11*, 60–64.
11. Dargahi, N.; Katsara, M.; Tselios, T.; Androutsou, M.-E.; De Courten, M.P.J.; Matsoukas, J.; Apostolopoulos, V. Multiple Sclerosis: Immunopathology and Treatment Update. *Brain Sci.* **2017**, *7*, 78. [CrossRef]
12. Day, S.; Tselios, T.; Androutsou, M.-E.; Tapeinou, A.; Frilligou, I.; Stojanovska, L.; Matsoukas, J.; Apostolopoulos, V. Mannosylated Linear and Cyclic Single Amino Acid Mutant Peptides Using a Small 10 Amino Acid Linker Constitute Promising Candidates Against Multiple Sclerosis. *Front. Immunol.* **2015**, *6*. [CrossRef]
13. Deraos, G.; Kritsi, E.; Matsoukas, M.-T.; Christopoulou, K.; Kalbacher, H.; Zoumpoulakis, P.; Apostolopoulos, V.; Matsoukas, J. Design of Linear and Cyclic Mutant Analogues of Dirucotide Peptide (MBP82-98) against Multiple Sclerosis: Conformational and Binding Studies to MHC Class II. *Brain Sci.* **2018**, *8*, 213. [CrossRef] [PubMed]
14. Deraos, G.; Rodi, M.; Kalbacher, H.; Chatzantoni, K.; Karagiannis, F.; Synodinos, L.; Plotas, P.; Papalois, A.; Dimisianos, N.; Papathanasopoulos, P.; et al. Properties of myelin altered peptide ligand cyclo(87-99)(Ala91,Ala96)MBP87-99 render it a promising drug lead for immunotherapy of multiple sclerosis. *Eur. J. Med. Chem.* **2015**, *101*, 13–23. [CrossRef] [PubMed]
15. Laimou, D.; Lazoura, E.; Troganis, A.N.; Matsoukas, M.-T.; Deraos, S.N.; Katsara, M.; Matsoukas, J.; Apostolopoulos, V.; Tselios, T. Conformational studies of immunodominant myelin basic protein 1–11 analogues using NMR and molecular modeling. *J. Comput. Mol. Des.* **2011**, *25*, 1019–1032. [CrossRef] [PubMed]

16. Matsoukas, J.; Apostolopoulos, V.; Kalbacher, H.; Papini, A.M.; Tselios, T.; Chatzantoni, K.; Biagioli, T.; Lolli, F.; Deraos, S.; Papathanassopoulos, P.; et al. Design And Synthesis of a Novel Potent Myelin Basic Protein Epitope 87–99 Cyclic Analogue: Enhanced Stability and Biological Properties of Mimics Render Them a Potentially New Class of Immunomodulators†. *J. Med. Chem.* **2005**, *48*, 1470–1480. [CrossRef] [PubMed]

17. Matsoukas, J.; Apostolopoulos, V.; Mavromoustakos, T. Designing peptide mimetics for the treatment of multiple sclerosis. *Mini-Rev. Med. Chem.* **2001**, *1*, 273–282. [CrossRef] [PubMed]

18. Tapeinou, A.; Androutsou, M.-E.; Kyrtata, K.; Vlamis-Gardikas, A.; Apostolopoulos, V.; Matsoukas, J.; Tselios, T.; Vlamis, A. Conjugation of a peptide to mannan and its confirmation by tricine sodium dodecyl sulfate–polyacrylamide gel electrophoresis. *Anal. Biochem.* **2015**, *485*, 43–45. [CrossRef]

19. Tselios, T.; Apostolopoulos, V.; Daliani, I.; Deraos, S.; Grdadolnik, S.G.; Mavromoustakos, T.; Melachrinou, M.; Thymianou, S.; Probert, L.; Mouzaki, A.; et al. Antagonistic Effects of Human Cyclic MBP87-99 Altered Peptide Ligands in Experimental Allergic Encephalomyelitis and Human T-Cell Proliferation. *J. Med. Chem.* **2002**, *45*, 275–283. [CrossRef]

20. Tselios, T.V.; Lamari, F.; Karathanasopoulou, I.; Katsara, M.; Apostolopoulos, V.; Pietersz, G.A.; Matsoukas, J.M.; Karamanos, N.K. Synthesis and study of the electrophoretic behavior of mannan conjugates with cyclic peptide analogue of myelin basic protein using lysine-glycine linker. *Anal. Biochem.* **2005**, *347*, 121–128. [CrossRef]

21. Tzakos, A.; Kursula, P.; Troganis, A.; Theodorou, V.; Tselios, T.; Svarnas, C.; Matsoukas, J.; Apostolopoulos, V.; Gerothanassis, I. Structure and Function of the Myelin Proteins: Current Status and Perspectives in Relation to Multiple Sclerosis. *Curr. Med. Chem.* **2005**, *12*, 1569–1587. [CrossRef]

22. Deraos, G.; Chatzantoni, K.; Matsoukas, M.-T.; Tselios, T.; Deraos, S.; Katsara, M.; Papathanasopoulos, P.; Vynios, D.; Apostolopoulos, V.; Mouzaki, A.; et al. Citrullination of Linear and Cyclic Altered Peptide Ligands from Myelin Basic Protein (MBP87–99) Epitope Elicits a Th1 Polarized Response by T Cells Isolated from Multiple Sclerosis Patients: Implications in Triggering Disease. *J. Med. Chem.* **2008**, *51*, 7834–7842. [CrossRef]

23. Katsara, M.; Deraos, G.; Tselios, T.; Matsoukas, J.; Apostolopoulos, V. Design of Novel Cyclic Altered Peptide Ligands of Myelin Basic Protein MBP83–99 That Modulate Immune Responses in SJL/J Mice. *J. Med. Chem.* **2008**, *51*, 3971–3978. [CrossRef] [PubMed]

24. Katsara, M.; Matsoukas, J.; Deraos, G.; Apostolopoulos, V. Towards immunotherapeutic drugs and vaccines against multiple sclerosis. *Acta Biochim. Biophys. Sin.* **2008**, *40*, 636–642. [CrossRef] [PubMed]

25. Matsoukas, J.; Apostolopoulos, V.; Lazoura, E.; Deraos, G.; Matsoukas, M.-T.; Katsara, M.; Tselios, T.; Deraos, S. Round and round we go: Cyclic peptides in disease. *Curr. Med. Chem.* **2006**, *13*, 2221–2232. [CrossRef]

26. Katsara, M.; Yuriev, E.; Ramsland, P.A.; Deraos, G.; Tselios, T.; Matsoukas, J.; Apostolopoulos, V. A double mutation of MBP83–99 peptide induces IL-4 responses and antagonizes IFN-? responses. *J. Neuroimmunol.* **2008**, *200*, 77–89. [CrossRef] [PubMed]

27. Lourbopoulos, A.; Deraos, G.; Matsoukas, M.-T.; Touloumi, O.; Giannakopoulou, A.; Kalbacher, H.; Grigoriadis, N.; Apostolopoulos, V.; Matsoukas, J. Cyclic MOG 35 – 55 ameliorates clinical and neuropathological features of experimental autoimmune encephalomyelitis. *Bioorganic Med. Chem.* **2017**, *25*, 4163–4174. [CrossRef]

28. Lourbopoulos, A.; Matsoukas, M.-T.; Katsara, M.; Deraos, G.; Giannakopoulou, A.; Lagoudaki, R.; Grigoriadis, N.; Matsoukas, J.; Apostolopoulos, V. Cyclization of PLP139-151 peptide reduces its encephalitogenic potential in experimental autoimmune encephalomyelitis. *Bioorganic Med. Chem.* **2018**, *26*, 2221–2228. [CrossRef]

29. Apostolopoulos, V.; Barnes, N.; Pietersz, G.A.; McKenzie, I.F. Ex vivo targeting of the macrophage mannose receptor generates anti-tumor CTL responses. *Vaccine* **2000**, *18*, 3174–3184. [CrossRef]

30. Apostolopoulos, V.; Pietersz, G.A.; Gordon, S.; Martínez-Pomares, L.; McKenzie, I.F. Aldehyde-mannan antigen complexes target the MHC class I antigen-presentation pathway. *Eur. J. Immunol.* **2000**, *30*, 1714–1723. [CrossRef]

31. Apostolopoulos, V.; Pietersz, G.A.; Loveland, B.E.; Sandrin, M.S.; McKenzie, I.F. Oxidative/reductive conjugation of mannan to antigen selects for T1 or T2 immune responses. *Proc. Natl. Acad. Sci. USA* **1995**, *92*, 10128–10132. [CrossRef]

32. Apostolopoulos, V. Cell-mediated immune responses to MUC1 fusion protein coupled to mannan. *Vaccine* **1996**, *14*, 930–938. [CrossRef]

33. Apostolopoulos, V.; Pietersz, G.A.; Tsibanis, A.; Tsikkinis, A.; Drakaki, H.; Loveland, B.E.; Piddlesden, S.J.; Plebanski, M.; Pouniotis, D.S.; Alexis, M.N.; et al. Pilot phase III immunotherapy study in early-stage breast cancer patients using oxidized mannan-MUC1 [ISRCTN71711835]. *Breast Cancer Res.* **2006**, *8*, R27. [CrossRef] [PubMed]

34. Apostolopoulos, V.; Pietersz, G.A.; Tsibanis, A.; Tsikkinis, A.; Stojanovska, L.; McKenzie, I.F.; Vassilaros, S. Dendritic cell immunotherapy: Clinical outcomes. *Clin. Transl. Immunol.* **2014**, *3*, e21. [CrossRef] [PubMed]

35. Karanikas, V.; Hwang, L.A.; Pearson, J.; Ong, C.S.; Apostolopoulos, V.; Vaughan, H.; Xing, P.X.; Jamieson, G.; Pietersz, G.; Tait, B.; et al. Antibody and T cell responses of patients with adenocarcinoma immunized with mannan-MUC1 fusion protein. *J. Clin. Investig.* **1997**, *100*, 2783–2792. [CrossRef] [PubMed]

36. Lofthouse, S.A.; Apostolopoulos, V.; Pietersz, G.A.; Li, W.; McKenzie, I.F. Induction of T1 (cytotoxic lymphocyte) and/or T2 (antibody) responses to a mucin-1 tumour antigen. *Vaccine* **1997**, *15*, 1586–1593. [CrossRef]

37. Loveland, B.E.; Zhao, A.; White, S.; Gan, H.; Hamilton, K.; Xing, P.X.; Pietersz, G.A.; Apostolopoulos, V.; Vaughan, H.; Karanikas, V.; et al. Mannan-MUC1-Pulsed Dendritic Cell Immunotherapy: A Phase I Trial in Patients with Adenocarcinoma. *Clin. Cancer Res.* **2006**, *12*, 869–877. [CrossRef]

38. Sheng, K.-C.; Kalkanidis, M.; Pouniotis, D.S.; Esparon, S.; Tang, C.K.; Apostolopoulos, V.; Pietersz, G.A. Delivery of antigen using a novel mannosylated dendrimer potentiates immunogenicityin vitro andin vivo. *Eur. J. Immunol.* **2008**, *38*, 424–436. [CrossRef]

39. Sheng, K.-C.; Kalkanidis, M.; Pouniotis, D.S.; Wright, M.D.; Pietersz, G.A.; Apostolopoulos, V. The adjuvanticity of a mannosylated antigen reveals TLR4 functionality essential for subset specialization and functional maturation of mouse dendritic cells. *J. Immunol.* **2008**, *181*, 2455–2464. [CrossRef]

40. Sheng, K.-C.; Pouniotis, D.S.; Wright, M.D.; Tang, C.K.; Lazoura, E.; Pietersz, G.A.; Apostolopoulos, V. Mannan derivatives induce phenotypic and functional maturation of mouse dendritic cells. *Immunology* **2006**, *118*, 372–383. [CrossRef]

41. Tang, C.K.; Lodding, J.; Minigo, G.; Pouniotis, D.S.; Plebanski, M.; Scholzen, A.; McKenzie, I.F.C.; Pietersz, G.A.; Apostolopoulos, V. Mannan-mediated gene delivery for cancer immunotherapy. *Immunology* **2007**, *120*, 325–335. [CrossRef]

42. Tang, C.-K.; Sheng, K.-C.; Pouniotis, D.S.; Esparon, S.; Son, H.-Y.; Kim, C.-W.; Pietersz, G.A.; Apostolopoulos, V. Oxidized and reduced mannan mediated MUC1 DNA immunization induce effective anti-tumor responses. *Vaccine* **2008**, *26*, 3827–3834. [CrossRef]

43. Vassilaros, S.; Tsibanis, A.; Tsikkinis, A.; Pietersz, G.A.; McKenzie, I.F.; Apostolopoulos, V. Up to 15-year clinical follow-up of a pilot Phase III immunotherapy study in stage II breast cancer patients using oxidized mannan–MUC1. *Immunotherapy* **2013**, *5*, 1177–1182. [CrossRef]

44. Apostolopoulos, V.; Lofthouse, S.A.; Popovski, V.; Chelvanayagam, G.; Sandrin, M.S.; McKenzie, I.F.C. Peptide mimics of a tumor antigen induce functional cytotoxic T cells. *Nat. Biotechnol.* **1998**, *16*, 276–280. [CrossRef]

45. Apostolopoulos, V.; Osinski, C.; McKenzie, I.F. MUC1 cross-reactive Galα(1,3)Gal antibodies in humans switch immune responses from cellular to humoral. *Nat. Med.* **1998**, *4*, 315–320. [CrossRef]

46. Karanikas, V.; Lodding, J.; Maino, V.C.; McKenzie, I.F. Flow cytometric measurement of intracellular cytokines detects immune responses in MUC1 immunotherapy. *Clin. Cancer Res.* **2000**, *6*, 829–837.

47. Karanikas, V.; Thynne, G.; Mitchell, P.; Ong, C.-S.; Gunawardana, D.; Blum, R.; Pearson, J.; Lodding, J.; Pietersz, G.; Broadbent, R.; et al. Mannan Mucin-1 Peptide Immunization: Influence of Cyclophosphamide and the Route of Injection. *J. Immunother.* **2001**, *24*, 172–183. [CrossRef] [PubMed]

48. Mitchell, P.; Quinn, M.; Grant, P.T.; Allen, D.G.; Jobling, T.W.; White, S.; Zhao, A.; Karanikas, V.; Vaughan, H.; Pietersz, G.; et al. A phase 2, single-arm study of an autologous dendritic cell treatment against mucin 1 in patients with advanced epithelial ovarian cancer. *J. Immunother. Cancer* **2014**, *2*, 16. [CrossRef] [PubMed]

49. Dargahi, N.; Matsoukas, J.; Apostolopoulos, V. Streptococcusthermophilus ST285 Alters Pro-Inflammatory to Anti-Inflammatory Cytokine Secretion against Multiple Sclerosis Peptide in Mice. *Brain Sci.* **2020**, *10*, 126. [CrossRef] [PubMed]

50. Katsara, M.; Deraos, G.; Tselios, T.; Matsoukas, M.-T.; Friligou, I.; Matsoukas, J.; Apostolopoulos, V. Design and Synthesis of a Cyclic Double Mutant Peptide (cyclo(87–99)[A91,A96]MBP87–99) Induces Altered Responses in Mice after Conjugation to Mannan: Implications in the Immunotherapy of Multiple Sclerosis. *J. Med. Chem.* **2009**, *52*, 214–218. [CrossRef] [PubMed]
51. Katsara, M.; Deraos, S.; Tselios, T.; Pietersz, G.; Matsoukas, J.; Apostolopoulos, V. Immune responses of linear and cyclic PLP139-151 mutant peptides in SJL/J mice: Peptides in their free state versus mannan conjugation. *Immunotherapy* **2014**, *6*, 709–724. [CrossRef]
52. Katsara, M.; Yuriev, E.; Ramsland, P.A.; Deraos, G.; Tselios, T.; Matsoukas, J.; Apostolopoulos, V. Mannosylation of mutated MBP83–99 peptides diverts immune responses from Th1 to Th2. *Mol. Immunol.* **2008**, *45*, 3661–3670. [CrossRef]
53. Katsara, M.; Yuriev, E.; Ramsland, P.A.; Tselios, T.; Deraos, G.; Lourbopoulos, A.; Grigoriadis, N.; Matsoukas, J.; Apostolopoulos, V. Altered peptide ligands of myelin basic protein (MBP87–99) conjugated to reduced mannan modulate immune responses in mice. *Immunology* **2009**, *128*, 521–533. [CrossRef] [PubMed]
54. Tseveleki, V.; Tselios, T.; Kanistras, I.; Koutsoni, O.; Karamita, M.; Vamvakas, S.-S.; Apostolopoulos, V.; Dotsika, E.; Matsoukas, J.; Lassmann, H.; et al. Mannan-conjugated myelin peptides prime non-pathogenic Th1 and Th17 cells and ameliorate experimental autoimmune encephalomyelitis. *Exp. Neurol.* **2015**, *267*, 254–267. [CrossRef] [PubMed]

Review

Molecular Interventions towards Multiple Sclerosis Treatment

Athanasios Metaxakis [1], Dionysia Petratou [1] and Nektarios Tavernarakis [1,2,*

[1] Institute of Molecular Biology and Biotechnology, Foundation for Research and Technology Hellas, Nikolaou Plastira 100, 70013 Heraklion, Greece; thanos_metaxakis@imbb.forth.gr (A.M.); dipetratou@imbb.forth.gr (D.P.)

[2] Department of Basic Sciences, Faculty of Medicine, University of Crete, 71110 Heraklion, Greece

* Correspondence: tavernarakis@imbb.forth.gr; Tel.: +30-2810-391066

Received: 28 April 2020; Accepted: 12 May 2020; Published: 15 May 2020

Abstract: Multiple sclerosis (MS) is an autoimmune life-threatening disease, afflicting millions of people worldwide. Although the disease is non-curable, considerable therapeutic advances have been achieved through molecular immunotherapeutic approaches, such as peptides vaccination, administration of monoclonal antibodies, and immunogenic copolymers. The main aims of these therapeutic strategies are to shift the MS-related autoimmune response towards a non-inflammatory T helper 2 (Th2) cells response, inactivate or ameliorate cytotoxic autoreactive T cells, induce secretion of anti-inflammatory cytokines, and inhibit recruitment of autoreactive lymphocytes to the central nervous system (CNS). These approaches can efficiently treat autoimmune encephalomyelitis (EAE), an essential system to study MS in animals, but they can only partially inhibit disease progress in humans. Nevertheless, modern immunotherapeutic techniques remain the most promising tools for the development of safe MS treatments, specifically targeting the cellular factors that trigger the initiation of the disease.

Keywords: B cell receptor; delivery methods; immunotherapy; monoclonal antibodies; multiple sclerosis; T cell receptor; tolerance; vaccine

1. Introduction

Multiple sclerosis (MS) is the commonest inflammatory autoimmune disorder of the central nervous system (CNS), progressively leading to demyelination, neurodegeneration, and neuronal disability [1–3]. MS globally affects more than 2.5 million people and it often afflicts young people, mainly women [4,5]. Despite the availability of a large arsenal of putative therapeutic approaches, numerous studies in animal model systems, and clinical trials, MS is still non-curable. As a result, the average life expectancy of MS patients is shorter by 5 to 10 years [6].

Inflammatory lesions at the CNS, generated by autoreactive lymphocytes, are suggested to underlie the pathophysiology of the disease, which results in neuronal demyelination and damage. Genetic and environmental factors influence MS susceptibility: Family history, single nucleotide polymorphisms, Epstein–Barr virus (EBV) infection, smoking, obesity, and vitamin D shortage are associated with MS development [7–11]. Patients experience relapsing-remitting phases of the disease, which are followed, even years later, by a progressive phase, accompanied by neurodegeneration [12,13]. MS symptomatology largely varies among patients, including sensory disturbances, cognitive defects, loss of vision, weakness, bladder dysfunction and neurological disability among others [14,15].

Therapeutic strategies against MS have been mainly relied on immune function suppressors, such as glucocorticoids, methotrexate, and antihistamines, which non-specifically reduce immune activity. These strategies have been enforced in recent years by the usage of antibodies against proinflammatory mediators [16]. However, this approach has severe side effects and dangers for patients, since the general inhibition of immune responses risks the development of infections and tumors. Hence, modern therapeutic approaches must aim at disease-modifying interventions that will counteract specifically the excessive immune response against self-antigens. Administration of self-antigens, an intervention that has been successfully applied in other autoimmune diseases and has been shown to eliminate the autoimmune response, is a widely accepted methodology to achieve this [17]. A major drawback of this technique is the poor targeting of CNS by the exogenously supplied antigens, for their inability to cross the brain–blood barrier and increased degradation. As such, the improvement of delivery methods used to protect and adequately transfer self-antigens to the inflammation sites has been an intriguing research field [18]. Nevertheless, a prerequisite for the success of this approach is that the epitope of the self-antigen is known. This is not true in the case of MS yet, although proteins within the myelin sheath have been suggested to be promising candidates [19,20]. Consequently, much research effort must be invested before modern immunomodulatory approaches can assure the cure of MS.

Recent experimental studies and clinical trials show that modern immunotherapeutic techniques have the potential to treat MS with less or no side effects in the future. Extensive work in mammalian model organisms has given insights into the mechanisms of the disease development and efficiency of several drugs in animals and humans. Indeed, novel drugs, such as Glatiramer acetate (Copaxone), a random sequence of four synthetic polypeptides with similar immunogenic properties to myelin protein, are currently being used against MS with very promising results [21]. In this review, we discuss antigen-specific and cell-specific immunotherapeutic approaches, applications of monoclonal antibodies against MS, anti-inflammatory strategies, peptide delivery methodologies and biological mechanisms that can serve as targets for the development of adjunctive MS treatments.

2. Immunotherapeutic Approaches

2.1. Antigen-Specific Immunotherapy (ASI)

Antigen-specific immunotherapy (ASI) is a promising strategy to treat MS with the least possible side effects. It was firstly introduced several decades ago, when Leonard Noon suppressed conjunctival sensitivity to grass pollen through prophylactic inoculation with grass pollen extracts [22]. His work paved the way for the first clinical trial of allergen immunotherapy a few decades later [23,24]. Allergen immunotherapy is based on the prevention of immune over-reaction against an allergen when repetitive doses of the latest are supplied to the organism. Repeated exposure to increasing amounts of an allergen results in altered cytokine production and shifts the immune response from a T helper 2 (Th2) to a T helper 1 (Th1) response, and also in the activation of regulatory T cells (Tregs) that secrete interleukin (IL)-10 and transforming growth factor (TGF)-β [25].

Contrary to allergic responses, where Th2 immune responses prevail, in autoimmune diseases, the prevalent responses are Th1 and Th17 against self-antigens. ASI for MS aims to induce Tregs in order to promote autoantigen-specific tolerance. The elimination of pathogenic Th1 and Th17 cells or the inhibition of the autoantigen-specific T cells-induced immune response might be the treatment for MS. Through repeated exposure to antigens, both allergen immunotherapy and ASI aim to promote self-tolerance [26].

Inspired by the progress in allergen immunotherapy, researchers have aimed at treating MS through the administration of self-peptides, which are expected to mimic the immunogenicity of self-antigens. This technique is called 'peptide vaccination' and promises to eliminate the antigen-specific attack without diminishing the organism's immune capacity against other threats. The most successful peptide vaccines applied so far are fractions of myelin proteins, such as myelin basic protein (MBP), myelin oligodendrocyte glycoprotein (MOG), and proteolipid protein (PLP) [27]. These antigens have been used to induce autoimmune encephalomyelitis (EAE) in mouse models, a widely accepted inflammatory model used to study MS. Several trials of myelin self-antigen peptide vaccines have cured EAE to a lesser or greater extent. Vaccination of an immunodominant epitope of myelin basic protein (MBP) (peptide 87–99), shown to be recognized and attacked by the T cell receptor (TCR), prevented and treated EAE, while it reduced tumor necrosis factor (TNF)-alpha and interferon (IFN)-gamma production, two determinant cytokines in the pathogenesis of EAE and MS [28]. More MBP peptides are shown to be immunogenic, and upon vaccination, they can mildly or strongly counteract EAE pathogenesis [29]. Myelin PLP (peptide 139–151) peptides can also prevent or treat EAE in animals [30,31]. A peptide from another myelin protein, the myelin oligodendrocyte glycoprotein (MOG) (peptide 35–55), can inhibit EAE development in mice [32,33], similarly to peptides derived from proteolipid protein (PLP) [34,35]. Hence, promising results from animal model systems have recommended peptide vaccination as a featured strategy to counteract MS.

In humans, two promising vaccination-based clinical trials with myelin peptides were safe and well tolerated by MS patients. Moreover, vaccination suppressed autoreactive responses and IFN-gamma production, while it significantly improved clinical disease measures. The activation of Langerhans cells and generation of IL-10-secreting cells are suggested to underlie these effects [36,37]. *Chataway et al.* showed that a mixture of peptides derived from MBP (peptide ATX-MS-1467) was safe and well tolerated by MS patients, while it improved radiographic activity in magnetic resonance imaging (MRI) [38]. *Crowe et al.* used a fragment of MBP (peptide 83–99) to induce immune responses and enhance anti-inflammatory cytokine secretion from T lymphocytes that cross-react with MBP [39]. Similarly, subcutaneous administration of a mixture of three MBP peptides (peptides 46-64, 124–139, and 147-170), termed Xemys, in MS patients was safe, while treatment decreased the cytokines monocyte chemoattractant protein-1, macrophage inflammatory protein-1β, and IL-7 and -2 levels, thus indicating reduced inflammation. However, clinical parameters were not significantly changed in patients [40]. In another scheme, researchers vaccinated MS patients with autologous peripheral blood mononuclear cells, chemically coupled with seven myelin peptides. Administration of antigen-coupled cells did not cause adverse effects, it was well tolerated and patients exhibited decreased antigen-specific T cell responses after treatment [41].

Contrary to the above, some studies show that peptide vaccination can have severe side effects and few clinical trials have not been completed for safety reasons. In two studies, MBP peptide 83–99 not only did not improve the disease state of MS [42], but even aggravated it, with few patients having exacerbations of MS [20]. Furthermore, administration of myelin epitopes has raised safety concerns of anaphylaxis [43–45]. In conclusion, specific attention should be paid to the adverse effects of peptides vaccination and future studies must identify the factors underlying the diversity of evoked responses in MS patients. Genomic profiling of MS patients that develop such effects can indicate factors that underlie the toxicity of this approach and indicate complementary treatments to reduce side effects. Moreover, trials with novel immunogenic peptides and further experimentation on the timing and dosage of vaccination can improve the efficiency and reduce the adverse effects of peptides vaccination.

Another immunotherapy technique that has been applied to induce self-tolerance in MS patients is the administration of genetically engineered DNA that encodes human MBP protein (BHT-3009). Experiments with animals clearly highlighted the potential of DNA vaccination as a safe and efficient technique at inducing regulatory T cells and EAE inhibition in animals. Its application in MS patients was safe and well tolerated, thus offering an alternative to peptide vaccination in terms of safety. Moreover, it decreased the proliferation of IFN-gamma-producing myelin-reactive T cells, the number of myelin-specific autoantibodies in the cerebrospinal fluid, and MRI-measured disease activity, while it increased the antigen-specific tolerance to myelin-specific B and T cells [46–49]. Nevertheless, no significant clinical improvements in the disease development were observed in these trials.

2.2. Cell-specific Immunotherapy

T cell vaccination is another immunotherapeutic approach, which is aimed at reducing or inactivating pathogenic T cells that maintain an autoimmune attack on myelin in MS. T cells' reaction is believed to be the initial step that drives the pathogenesis of MS [50]. In this technique, autologous myelin-reactive T cells are isolated and inactivated prior to their administration to MS patients.

Initial trials clearly showed safety and encouraging effects from T cell vaccination [51]. In a matched trial, MS patients were vaccinated with irradiated MBP-reactive T cells. Vaccinated patients with relapsing-remitting disease phases experienced a remarkable decrease in disease exacerbations and a five-fold lower increase in brain lesion size, compared to controls [52]. In three cases, however, T cell vaccine aggravated brain lesions and worsened relapses, a condition accompanied by reactivation of circulating MBP-reactive T cells. *Zhang et al.* showed that inhibition of MBP-reactive T cells was correlated with a 40% reduction in the rate of disease relapses, while brain lesion activity in vaccinated patients was stabilized [53]. This trial revealed that repetitive T cell vaccinations are needed to hamper the reappearance of myelin-reactive T cell clones.

Alternative T cell vaccination schemes use mixtures of inactivated autoreactive T cells, selected with more than one myelin peptides. In one trial, T cells activated with synthetic MBP and MOG peptides were administrated in MS patients, with no adverse effects being reported. Patients exhibited stabilized neurological symptoms and vaccination reduced active brain lesions both in number and size [54,55]. Tcelna (formerly known as Tovaxin) is a T cell vaccine containing T cell populations selected with peptides derived from MBP, PLP, and MOG. In a double-blind trial involving a restricted number of MS patients, vaccination did not cause adverse effects and showed mild clinical efficacy [56]. More studies are required to properly evaluate the potency of Tcelna to treat MS.

Another suggested methodology to inhibit the autoimmune response in MS is via the elimination of dendritic cells, which play a major role in inflammation induction. Dendritic cells are the most efficient antigen-presenting cells (APCs) of the immune system and they have a particular role in the stimulation of naïve T cells. They regulate T cell differentiation and priming, secrete proinflammatory cytokines, orchestrate the immune response against self-antigens, and initiate chronic inflammation and loss of tolerance [57]. Dendritic cells respond occasionally to a specific antigen, in a manner dependent on the tissue environment. Tolerance-inducing (Tolerogenic) dendritic cells are dendritic cells with immunosuppressive properties, elicited by the induction of T cell anergy, T cell apoptosis, regulatory T cell activity, and production of anti-inflammatory cytokines [58]. In vitro treatment of monocyte-derived dendritic cells with vitamin D3 causes T cell hyporesponsiveness to myelin [19,59]. MOG 40–55 peptide-treated tolerogenic cells that were administrated in mice preventively or after EAE induction reduced incidence of the disease or improved its clinical features, respectively [60]. Several trials in humans show that the technique is safe in patients with other autoimmune diseases [19]. Recently, engineered dendritic cells, loaded with specific antigens, were used to induce tolerance in MS patients. Therapy was safe and well tolerated; it increased IL-10 levels and the number of regulatory T cells, indicating that antigen-specific tolerance can be, at least partially, induced with this approach [61].

2.3. Cell Receptor-Specific Immunotherapy

A similar approach to cell-specific immunotherapy is T cell receptor-specific immunotherapy. Here, fragments of the T cell receptor (TCR) from pathogenic T cell clones are used as peptide vaccines, in order to activate immune responses against TCR-expressing T cells. TCR is a protein complex that recognizes antigens bound to major histocompatibility complex (MHC) molecules. Different TCRs can be specific for the same antigen, while more than one antigen peptides can be recognized by the same TCR [62].

Vaccination of rats with a synthetic TCR V-region peptide conferred resistance to subsequent induction of EAE [63]. According to the study, T cells specific for the TCR peptide weakened the immune attack to the encephalitogenic epitope. Furthermore, *Offner et al.* showed that TCR vaccination can not only prevent EAE but also cure it. When a TCR-V beta 8-39-59 peptide was injected into rats with EAE, disease symptoms were alleviated and recovery from the disease was fast [64].

To test safety and immunogenicity of TCR vaccines in humans, *Bourdette et al.* intradermally injected MS patients with two synthetic TCR peptides (TCR peptides V beta 5.2, 39-59 and V beta 6.1, 39-59). Low doses of the TCR vaccine caused no side effects, restricted spectrum immunosuppression, generated TCR peptide-specific T cells, and reduced MBP-specific T cells [65]. In a subsequent trial,

TCR vaccination enhanced TCR-reactive T cells, reduced the MBP response against MBP antigen, stabilized clinical features, and caused no adverse effects to MS patients [66]. In support, TCR-specific Th2 cells inhibit the MBP-specific Th1 response in vitro through the release of IL-10, and a triplicate TCR vaccine (BV5S2, BV6S5, and BV13S1 peptides) increases the numbers of circulating IL-10-secreting T cells, reactive to the TCR peptides, in MS patients [67].

Together with pathogenic T cells, autoreactive B cells are involved in MS induction. Hence, the B cell receptor (BCR) can be used as a vaccine as well. Single-cell sequencing and phage display libraries of B cells derived from MS patients have been performed to identify BCR structures involved in MS autoimmunity [68–70]. *Gabibov et al.* showed that, antibodies induced against Epstein–Barr virus latent membrane protein 1 (LMP1) potentially react with MBP. This suggests that natural molecular reactivity might underlie MS induction and raises questions about the causal link between virus infection and MS development. Recently, antibody engineering techniques have allowed for the targeting of BCR with toxins, resulting in the cell death of pathogenic B cells [29,71,72]. This makes BCR-specific immunotherapy an alternative, although still at a preliminary state, approach to treat MS.

2.4. Monoclonal Antibodies (MABs)

The usage of monoclonal antibodies is another encouraging molecular therapy against MS, for their high specificity and high efficacy. Several ones have been approved for MS treatment [73,74]. Natalizumab, an adhesion molecule inhibitor, was the first MAB to be approved in 2004 [75]. It is a recombinant humanized MAB that binds integrin α-4 on the surface of activated inflammatory lymphocytes and monocytes. This inhibits the interaction of integrin a-4 with vascular cell adhesion molecule-1 (VCAM-1) on endothelial cells and consequently circulation into the CNS. Clinical trials show that it is safe, well tolerated, and efficient, since it reduces the risk of sustained progression of disability and MS relapses [76]. Ocrelizumab and Rituximab are MABs that target CD20 protein on B lymphocytes. They have been shown to reduce the rates of disease activity and disease progression [77, 78]. Ofatumumab also binds on CD20, albeit at a different epitope, and its administration in MS patients reduces new MRI-detected lesions by 99% [79]. Another MAB, Opicinumab, has been designed to repair and enhance re-myelination of lesions in MS patients. Opicinumab is a fully humanized MAB that targets and inactivates leucine rich repeat and immunoglobin-like domain-containing protein 1 (LINGO-1), a transmembrane signaling protein that inhibits the differentiation of oligodendrocytes and myelination. Hence, it is potentially a promising tool to induce re-myelination in MS patients and alleviate disease symptoms. It has been tested in mice and in humans, where it increases myelination and re-myelination in MS patients [80,81]. Alemtuzumab is a humanized monoclonal antibody, approved in several countries for the treatment of relapsing-remitting MS. It targets CD52 antigen on lymphocytes, resulting in their depletion [82]. Hence, monoclonal antibodies are very promising tools for MS therapy for their safety, specificity, and efficacy but also for the various cellular procedures they can target to reduce autoimmunity and its clinical consequences.

2.5. HLA Antagonistic Co-polymers

Synthetic materials (copolymers) can mimic the immunogenic properties of endogenous proteins and compete with them for binding to HLA class II molecules. Glatiramer acetate (Copaxone or GA) is a random polymer of four amino acids (L-alanine, L-glutamic acid, L-lysine, and L-tyrosine) that effectively treats experimental encephalomyelitis and reduces relapses in MS patients [83–85]. GA is suggested to specifically inhibit the production of myelin-reactive antibodies, by directly acting on APCs. This modifies them into non-inflammatory type II cells. APCs-mediated presentation of GA to CD8+ and CD4+ T cells results in the generation of CD4+ regulatory T cells and immune response deviation towards Th2 responses [86,87]. A second generation of polymers has been synthesized with stronger binding activities on HLA molecules compared to GA. They have been successfully used to suppress EAE in mice [88]. In transgenic mice with human HLA-DR-TCR, poly(VWAK)n copolymers are shown to induce T cells' anergy, while poly(FYAK)n copolymers induce Th2 cells that secrete anti-inflammatory cytokines [29]. Hence, they can serve as alternative tools for shifting the immune response towards Th2 activation in MS patients.

3. Delivery Methods of Immunotherapeutic Factors

A key point for the successful implementation of immunotherapy treatment is the efficacy of the delivery methodology. Oral, skin, parenteral, intramuscular, intravenous, and intra-peritoneal routes are mainly used with various delivery vehicles. These vehicles must enhance the tolerance of immunomodulatory molecules against the harsh intra-organismal environment and advance their efficacy to overcome the brain–blood barrier. Synthetic polymers, such as poly lactide-co-glycolide (PLGA), polyethylene glycol (PEG), and polymethylmethacrylate (PMMA), are easily synthesized and modified, capable of transferring sufficient amounts of immunotherapeutic molecules and facilitating their gradual release [18]. Permeability is decreased when electrically charged nanoparticles are used, such as orally administrated polyethylene imine-based nanoparticles and thiol-modified Eudragit polymers (polymethacrylates) [89,90]. Transgenic plant delivery is another technique that takes advantage of the protective effect of the plant cell wall, especially for delivery through the gastrointestinal tract [91,92]. Nanoemulsions, small colloidal particles, provide a high encapsulation efficiency [93], while phosphatidylserine-liposomes have been efficiently used to reduce EAE severity in mice [94]. Much attention has been paid to lipid-based nanocarriers, such as nanoemulsions, nanoliposomes, solid lipid nanoparticles (SLNs), and nanostructured lipid carriers (NLCs), which are suggested to be efficient for brain targeting. NLCs have been reported to be very safe and stable, with a high encapsulation efficiency [95,96]. A major challenge in the field of immunotherapy treatment is the improvement of delivery methods so that immunotherapeutic molecules can be transferred more efficiently through the brain–blood barrier. This will improve the therapeutic efficiency, reduce side effects, and decrease the number of administration procedures. More selective delivery to the CNS can be achieved through the covalent tethering of delivery molecules with ligands capable of overcoming the brain–blood barrier, the use of fusion antibodies that target specific lymphocytes, and of liposomes that intrinsically tend to reach inflammation sites.

Therapeutic treatments for MS target lymphocyte subpopulations, specific for autoreactive response towards the myelin sheath. Tolerogenic DCs, myelin peptide and DNA vaccines, TCR peptides and GA lead to the activation of Th2 cells, through Tregs. Subsequent release of IL-10 leads to the inhibition of Th1 cells. DMF acts on HCAR2, found on dendritic cells, to induce Th2 cells. Toxins targeting BCRs lead to the elimination of pathogenic B cells. Fingolimod blocks the circulation of mature lymphocytes through S1PR, and Teriflunomide and Mitoxantrone inhibit T and B cell proliferation. Anti-CD 20 and anti-CD 52 antibodies deplete CD 20+ and CD 52+ lymphocytes. Tolerogenic TCs block MBP-reactive T and B cells. Natalizumab binds to α 4 β 1 integrin on activated T and B cells and prevent their interaction with VCAM-1. Opicinumab promotes the differentiation of oligodendrocyte precursor cells by inactivating LINGO-1. Abbreviations: Antigen Presenting Cell (APC), Blood–Brain Barrier (BBB), B Cell Receptor (BCR), cluster of differentiation 52/20 (CD52/20), Dendritic Cells (DCs), DMF (Dimethyl Fumarate), Glatiramer Acetate (GA), hydroxycarboxylic acid receptor 2 (HCAR2), Interferon (IFN), Interleukin (IL), Immunoglobin-like domain-containing protein 1 (LINGO-1), MBP (Myelin Binding Protein), MMF (Monomethyl Fumarate), Multiple Sclerosis (MS), Sphingosine-1-phosphate receptor (S1PR), TCR (T cell Receptor), T helper 2 cell (Th2), T helper 1 cell (Th1), T Cell Receptor (TCR), T regulatory cells (Tregs), TCs (T cells), vascular cell adhesion molecule-1 (VCAM-1). (Table 1).

Table 1. Overview of medical treatments for multiple sclerosis.

Treatment	Mode of Action	MS Type	Study Format (Number of Participants)	Clinical Outcomes	Adverse Effects	Administration Route	References
Interferons							
Interferon-β1a *	reduces immature-transitional B cell subset/plasmablasts ratio, increases CD27$^-$ and CD27$^+$IgM$^+$ memory B cell subsets, enhances Tregs	RRMS	case-control study/multicenter, open-label, prospective clinical trial, phase 4 (96)	reduction in relapse rates, reduction in MRI measurement of disease, well tolerated	flu-like symptoms, asthenia, fever, malaise, fatigue, local pain at the injection site	intramuscular injection	[97,98]
Interferon-β1b *	reduces neuron inflammation	RRMS	multicenter, randomized, double-blind, placebo-controlled trial (372)	reduced ARR, and MRI lesions	lymphopenia, skin reactions to injection, flu-like symptoms, fever, chills, myalgia, sweating, malaise	subcutaneous injection	[99,100]
Peptides							
Peptide loaded cells							
Myelin peptides (MOG1-20, MOG35-55, MBP13-32, MBP83-99, MBP111-129, MBP146-170, PLP139-154)	myelin peptide coupled autologous peripheral blood mononuclear cells, slightly increase T regulatory cells	RRMS SPMS	open-label, single-center, dose-escalation study, phase 1 trial (9)	safe and well tolerated	metallic flavor during infusion and IARs (diarrhea, headache, diverticulitis of sigma, neck pain, vision disturbance, dysesthesia, cold, gastric pain)	infusion	[41]
Peptide vaccines							
NBI-5788	altered MBP83-99 peptide, induces Th2-like cells APL-reactive	PPMS SPMS RRMS	multicenter phase 1 trial (11)	induced NBI-5788 responsive T cells, no clinical exacerbations	-	subcutaneous infusion	[39]
Xemys	mannosylated liposomes encapsulating MBP peptides, increases TNF-α, cytokine's levels normalization	RRMS SPMS	phase 1 trial (18)/phase 1, open-label, dose-escalating, proof-of-concept study (20)	increased TNF-α serum levels, safe and well tolerated	injection site reaction, rhinitis, general weakness	subcutaneous infusion	[40,101]
peptides MBP85-99, MOG35-55, and PLP139-155	induce T regs producing IL-10, reduce IFN-γ and TGF-β	RRMS	double-blind, placebo-controlled cohort study (30)	reduced GdE lesions and ARR	local skin reaction (redness, itching), upper respiratory tract infection, lacrimation	transdermally, with skin patch	[36,37]
ATX-MS-1467	peptide mixture of MBP derived epitopes, induces MBP tolerance and IL-10 secreting T regs	RMS	multicenter, phase 1b (43), phase 2a, multicenter, single-arm trial (37)	reduced GdE lesions	erythema, induration, pain, pruritus, hemorrhage, alopecia, diarrhea	intradermal/ subcutaneous injection	[38,102]
DNA vaccine							
BHT-3009	decreases T cells	RRMS	randomized, multicenter, double-blind, placebo-controlled dose escalation, phase 1/2 trial (30)/randomized, placebo-controlled, phase 2 trial (289)	reduced GdE lesions, reduced myelin-specific autoantibodies, safe and well tolerated	infections, musculoskeletal, urinary, gastrointestinal psychiatric, respiratory effects (IARs)	intramuscular injections	[47,48]

Table 1. *Cont.*

Treatment	Mode of Action	MS Type	Study Format (Number of Participants)	Clinical Outcomes	Adverse Effects	Administration Route	References
TCR vaccines							
TCR V beta 5.2, 39-59 and V beta 6.1, 39-59	induce T regs	PMS	dose escalation study (11)	induced T cell immunity to synthetic peptides, safe	skin hypersensitivity reaction to the injection, no side effects or broad immunosuppression	intradermal injection	[65,103]
vβ5.2-38-58	induce Th2 cells and inhibits MBP-specific Th1 cells	PMS	double-blind (23)	induced T cell immunity to synthetic peptides, attenuated disease progression	no side effects or broad immunosuppression	intradermal injection	[66]
BV5S2, BV6S5 and BV13S1	induce IL-10 secreting T cells	RRM PMS	single-arm, open-label study (23)	induced T cell immunity to synthetic peptides, stabilized disease, improved FoxP3 expression, safe	no side effects	intramuscular injection	[67]
Monoclonal antibodies							
Natalizumab *	anti-a4-integrin Ab, prevents leukocytes crossing BBB	early RRMS	controlled, non-randomized trial (34)/multicenter, observational, open-label, single-arm, phase 4 study (222)	reduced relapse rates, MRI lesions and progression of disability, improvement in information processing speed, NEDA, SDMT and MSIS-29 physical, psychological and quality-of-life	suicide attempt, acute kidney injury, anaphylactic reactions, bronchial obstruction, clostridium difficile colitis, conversion disorder, hydronephrosis, hyperkaliemia, hypotension, ileus, melanoma recurrent, migraine	intravenous infusion	[104,105]
		SPMS	randomized, double-blind, placebo-controlled, phase 3 trial (889), open-label extension (291)	reduced progression of disability, improved ARR and MRI measurements, well tolerated	urinary tract infection, nasopharyngitis, fall, MS relapse, headache, fatigue, upper respiratory tract infection, back pain, arthralgia, pain in hands and feet, muscular weakness (IARs)	intravenous infusion	[106]
Opicinumab	anti-LINGO-1 Ab, allows oligodendricy maturation	RRMS SPMS	double-blind, dose-ranging, proof-of-concept, phase 2b study (418)/phase 1, randomized, multiple ascending dose study	primary endpoint was not met, inverted U-shaped dose-response	unaffected immune function	intravenous infusion	[81,107]
Alemtuzumab*	anti-CD52 IgG Ab, depletes circulating T and B lymphocytes	RRMS	rater-masked, randomized, controlled phase 3 trial (667)	reduced ARR, stabilized disability levels, improved clinical and MRI outcomes, reduced brain volume loss	infections, thyroid-associated adverse events, thrombocytopenia IARs (headache, pyrexia, rash, bradycardia, insomnia, erythema, nausea, Urticaria, pruritus, abdominal pain, fatigue, dyspnea, flushing)	intravenous infusion	[108]
Ofatumumab	anti-CD20, cytotoxic to B lymphocytes	RRMS	randomized, double-blind, placebo-controlled, phase 2 study (36)/randomized, double-blind, phase 2b study (232)	decreased new MRI lesions, safe	rash, erythema, upper respiratory tract infection, viral infection, throat irritation, headache, fatigue, back pain, flushing, injection related reactions	subcutaneous injection	[79,109]

Table 1. *Cont.*

Treatment	Mode of Action	MS Type	Study Format (Number of Participants)	Clinical Outcomes	Adverse Effects	Administration Route	References
Rituximab	selective depletion of CD20+ B lymphocytes	PMS	single-center, open-label trial (8)/retrospective, uncontrolled, observational, multicenter study (822)	reduced peripheral B cells, CSF B cells and CXCL-13 levels, increased BAFF levels/ lower EDSS score, delayed CDP	IARs (lower extremity paresthesia), lower extremity spasticity or weakness, fatigue, fever, rigors/ infections (respiratory, intestinal), disorders (cardiac, respiratory, neuronal, immune) and IARs (malaise, headache, chills, nausea)	intrathecal infusion	[110,111]
		RRMS	blind, single-center, phase 2 trial (30)	reduced relapses and GdE lesions	IARs (fever, chills, flushing, itching of body or throat, and/or diarrhea, shortness of breath), urinary tract infections, thigh pain, upper respiratory tract infection, bronchitis, hand tendonitis, dizziness	intravenous infusion	[112]
		PPMS SPMS	multicenter, prospective, open-label phase 1b trial (23)/randomized, double-blind, placebo-controlled, multicenter, phase 2/3 trial (439)	well tolerated and feasible, reduced GdE lesions, delayed CDP	IARs (vertigo, nausea), infections, paresthesia, fall, nervous system disorders, fever, fatigue, meningitis/IARs (nausea, fatigue, chills, pyrexia, headache, dizziness, throat irritation, pharyngolaryngeal pain, pruritus, rash, flushing, hypotension), pneumonia, bronchitis	intravenous or intrathecal infusion	[113,114]
Ocrelizumab*	anti-CD20 Ab, depletes circulating CD20+ B cells	RMS PPMS	randomized, double-blind, active-controlled, phase 3 trials (1651), randomized, parallel-group, double-blind, placebo- controlled, phase 3 study (725)	reduced new and GdE lesions, improved ARR, disability progression, and MRI outputs	IARs (pruritus, rash, throat irritation, flushing, urticaria, oropharyngeal pain, headache, tachycardia, pyrexia, nausea, hypo-, hyper-tension, myalgia, dizziness, fatigue)	intravenous infusion	[115,116]
		PPMS	randomized, double-blind, placebo-controlled, phase 3 trial (732)	reduced risk of Upper Extremity disability progression, enhanced NEPAD, reduced brain volume loss	IARs (upper respiratory tract infections, oral herpes infections, pruritus, rash, throat irritation, flushing)	intravenous infusion	[117,118]
HLA antagonistic co-polymers							
Glatiramer acetate *	increases Tregs to suppress inflammatory response	RRMS	randomized, placebo-controlled, double-blind study (251), open-label (208)	reduced relapse rate, reduced GdE and new lesions	IARs (flushing, anxiety, dyspnea)	subcutaneous injection	[119]

Table 1. *Cont.*

Treatment	Mode of Action	MS Type	Study Format (Number of Participants)	Clinical Outcomes	Adverse Effects	Administration Route	References
Sphingosine-1-phosphate receptor modulators							
Fingolimod *	structural analogue of sphingosine, anti-inflammatory, impairs cytotoxic CD8 T cells function	RRMS	prospective observational study (60)	higher retention rate, increased satisfaction at MSQ, reduced dGM volume loss, ARR and EDSS	influenza-like illness, pain in extremity, headache, anxiety, depression, nasopharyngitis, hypoesthesia, arthralgia, dizziness, fatigue, rash, urinary tract infection, abdominal pain, hypertension, lymphopenia	oral	[120,121]
Other inhibitors							
Teriflunomide *	DHODH inhibitor, reduces proliferation of T- and B-cells	RMS	prospective, single-arm, open-label, phase 4 real-world study (1000)/randomized, double-blind, placebo-controlled, phase 3 trial (168)/multicenter, multinational, randomized, double-blind, parallel-group, placebo-controlled, phase 3 study (2251)	well tolerated, improved MRI outcomes, reduced ARR and CDW, improved TSQM scores, stabilized disability measures, improved cognition and quality of life measures	neutropenia, hair thinning, diarrhea, nausea, headache, urinary tract infection, increased alanine aminotransferase, nasopharyngitis, fatigue, paresthesia	oral	[122–124]
T cell vaccination							
MBP-reactive T cells	deplete circulating MBP-reactive T cells.	RRMSSPMS	pilot, controlled (8)/preliminary open label study (54)	safe and well tolerated, improved MRI outcome, reduced relapse rates	no adverse effects, skin infection	subcutaneous injection	[52,53]
MBB-, MOG-reactive T cells	deplete circulating MBP-, MOG-reactive T cells.	RRMS	20	improved MRI outcome	no adverse effects, skin infection	S subcutaneous injection	[55]
MBP-, MOG-, PLP-reactive T cells/ Tovaxin	deplete circulating MBP-, MOG-, PLP-reactive T cells.	RRMSSPMS	open-label dose escalation study (16)/randomized, double-blind trial, phase 2 study (26)	well tolerated, reduced EDSS, ARR and 10 min walking time, stabilized MRI lesions, improved EDSS and MSIS-29	relapse of MS, pain in extremity, IARs (injection site pain, erythema, inflammation, pruritus), unrelated to TCV administration (anemia, intestinal obstruction, pneumonia, carpal tunnel syndrome, headache, respiratory distress, infections)	Subcutaneous injection	[54,56]

Table 1. *Cont.*

Treatment	Mode of Action	MS Type	Study Format (Number of Participants)	Clinical Outcomes	Adverse Effects	Administration Route	References
Dendritic cell vaccination							
peptide loaded cells	increase T regulatory cells and IL-10 levels	RRMSSPMSPPMS	open-label, single-center, multiple ascending-dose, phase 1b trial (12)	well tolerated, stabilized disease progress	headache, leg pain, cold, palpitations, influenza (and unrelated to TCV administration)	intravenous	[61]
Esters							
Dimethyl Fumarate *	fumaric acid ester, modulates CD4(+) cells, M2 monocytes and B-cells, induction of antioxidant response	RRMS	randomized, double-blind, placebo controlled, phase 3 trial (213)/open-label, observational, phase 4 study (1105)	decreased EDSS, GdE and new lesions, reduced ARR, improved treatment satisfaction and quality of life measures	flushing, nausea, abdominal pain, diarrhea, gastrointestinal events, nasopharyngitis, infections, cardiovascular, skin and hepatic events, pruritus, rash, headache, fall, lymphopenia, breast cancer, MS relapse	oral delayed release	[125,126]
Other Immunomodulators							
Mitoxantrone *	a synthetic anthracenedione, inhibits T-cell, B-cell and macrophage proliferation	SPMS RRMS PRMS	multicenter, prospective, open-label, observational, phase 4 study (509)	reduced GdE lesions and relapse rate, improved EDSS	congestive heart failure, leukemia, amenorrhea, decreased ejection fraction, urinary tract infection	intravenous infusion	[127]

Table 1. The main MS treatments are summarized. Some of them are approved while others are still under clinical trial. Their mode of action and outcomes of some indicative clinical trials are tabulated. With asterisk (*) are indicated the MS medications approved by the FDA. Abbreviations: Antibody (Ab), Altered Peptide Ligand (APL), Annualized Relapse Rate (ARR), B-cell Activating Factor (BAFF), Blood-Brain Barrier (BBB), Confirmed Disability Progression (CDP), Confirmed Disability Worsening (CDW), CerebroSpinal Fluid (CSF), C-X-C motif chemokinebinding Ligand-13 (CXCL-13), DiHydro-Orotate DeHydrogenase (DHODH), deep Gray Matter (dGM), Expanded Disability Status Scale (EDSS), Gadolinium-Enhanced (GdE), Infusion-Associated Reactions (IARs), InterFeroN (IFN), InterLeukin (IL), Leucine rich repeat and Immunoglobin-like domain-containing protein 1 (LINGO-1), Myelin Basic Protein (MBP), myelin oligodendrocyte glycoprotein (MOG), Modified Fatigue Impact Scale (MFIS), Mental Health Inventory (MHI), Medication Satisfaction Questionnaire (MSQ), Multiple Sclerosis (MS), No Evidence of Disease Activity (NEDA), No Evidence of Progression or active Disease (NEPAD), proteolipid protein (PLP), Primary Progressive Multiple Sclerosis (PPMS), Relapsing Multiple Sclerosis (RMS), Relapsing-Remitting Multiple Sclerosis (RRMS), Sphingosine-1-phosphate receptor (S1PR), Symbol Digit Modalities Test (SDMT), Secondary Progressive Multiple Sclerosis (SPMS), T-helper-2 cell (Th2), T Cell Receptor (TCR), Transforming Growth Factor beta (TGF-β), T regulatory cells (Tregs), Treatment Satisfaction Questionnaire for Medication Version 1.4 (TSQM 1.4).

4. Conclusions

Researchers in the field of MS treatment have been trying to cure the disease via the elimination of CNS inflammation, elicited by the MS-related autoimmune response. Different applied strategies include the deviation of the immune response towards non-inflammatory Th2 activation, inactivation or amelioration of cytotoxic autoreactive T cells, induction of anti-inflammatory cytokines' secretion, inhibition of inflammatory cytokines, blockage of autoreactive-lymphocytes' recruitment to the CNS, and enhancement of myelination mechanisms (Figure 1). Several drugs have been tested so far in clinical trials, some of which can reduce relapses and symptoms in MS patients (Table 1), thus significantly improving their quality of life. However, none of them can cure MS. Despite the success of allergen immunotherapy in treating allergies, ASI has not displayed great achievements so far as a putative MS treatment. Reasons underlying this might be the difficulty in the identification of the self-antigens that trigger autoimmunity, the inability of regulatory T cells to suppress cytokine production under inflammatory conditions, the different immune players participating in allergies compared to MS (e.g., IgE antibodies, Th2 responses), and also the route, dosage, and timing used for ASI treatments [128]. Nevertheless, more than 10 drugs are currently being used against the secondary progressive form of MS, characterized by the relapsing-remitting phases, significantly reducing the frequency of relapses and disease symptoms [14]. These drugs are either immunosuppressants (such as Natalizumab, Ocrelizumab, Fingolimod, Alemtuzumab) or immunomodulatory (such as Interferon beta, GA, Teriflunomide, Mitoxantrone, Dimethylfumarate). Fingolimod reduces the number of circulating mature lymphocytes [129], Teriflunomide and Mitoxantrone are inhibitors of lymphocytes proliferation and the secretion of cytokines [130,131], while Dimethylfumarate (DMF), used for psoriasis treatment, shifts the Th1 and Th17 immune responses to Th2 [132]. However, these drugs do not cure the primary progressive form of MS, they must be repetitively supplied to the MS patients, and they can have adverse effects. As such, more selective and efficient drugs are required to assure safe treatment of MS in the future.

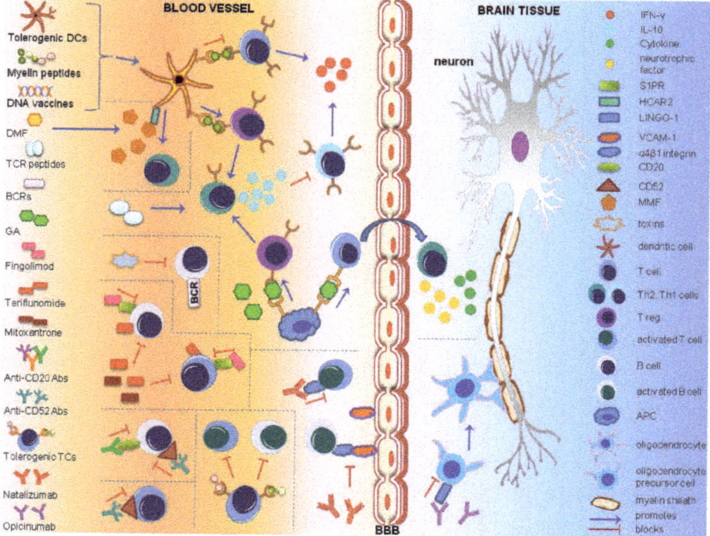

Figure 1. Mechanism of action of immunomodulatory treatments for multiple sclerosis.

Basic research on the mechanisms that underlie MS can reveal novel targets for monoclonal antibodies, identify the specific self-antigens that trigger autoimmunity, and characterize the types of lymphocytes that participate in the inflammatory reaction, so that antigen and cell-specific immunotherapies expand and become more precise. In addition, the identification of novel carriers or ligands that, upon conjugation, will lead these immunotherapeutic molecules to the CNS inflammatory sites can improve the efficiency of treatments. It is also important to clarify the role of Epstein–Barr virus infection on MS development and their possible association, which might give further insights into the disease etiology and treatment. Improved delivery of therapeutic molecules is another challenge of research in MS, which can be achieved through the generation of fusions between the therapeutic molecules and peptide leaders that will efficiently guide them to the inflammation sites in the brain [133]. Recently, a fusion protein of an NOD-like receptor family member X1 (NLRX1) and blood–brain barrier-permeable peptide dNP2 treated experimental autoimmune encephalomyelitis in mice [134] and a peptide that selectively recognizes the CNS was used for targeted drug delivery to the CNS in mice [135]. Genome-wide DNA sequencing analysis of MS patients is another approach that can advance our knowledge on the disease etiology and on MS patients' responses to medical treatments; it can reveal genes that make people more susceptible to MS and identify the reasons why specific drug treatments have adverse effects in some patients. In this case, the proper therapy could be administrated to patients that have certain genetic profiles, so that adverse effects of MS therapy could be minimized. Furthermore, drugs that enhance myelination, such as metformin [136], growth factors shown to regulate inflammation [137], and hormones known to affect autoimmunity [138] can offer new perspectives into the development of novel complementary treatments of MS in the future.

Author Contributions: A.M. and D.P. summarized the literature, wrote the paper and created the initial figure and table. N.T. edited the paper, revised the figures and contributed to the writing. All authors have read and agreed to the published version of the manuscript.

Funding: A.M. is supported by the project "BIOIMAGING-GR" (MIS5002755), which is implemented under the Action "Reinforcement of the Research and Innovation Infrastructure", funded by the Operational Programme "Competitiveness, Entrepreneurship and Innovation" (NSRF 2014-2020). D.P. is funded by grants from the European Research Council (ERC-GA695190-MANNA, ERC-GA737599-NeuronAgeScreen).

Conflicts of Interest: The authors declare no conflict of interest.

References

1. Compston, A.; Coles, A. Multiple sclerosis. *Lancet* **2008**, *372*, 1502–1517. [CrossRef]
2. Dendrou, C.A.; Fugger, L.; Friese, M.A. Immunopathology of multiple sclerosis. *Nat. Rev. Immunol.* **2015**, *15*, 545–558. [CrossRef]
3. Grigoriadis, N.; van Pesch, V. A basic overview of multiple sclerosis immunopathology. *Eur. J. Neurol.* **2015**, *22* (Suppl. 2), 3–13. [CrossRef]
4. Koch-Henriksen, N.; Sorensen, P.S. The changing demographic pattern of multiple sclerosis epidemiology. *Lancet. Neurol.* **2010**, *9*, 520–532. [CrossRef]
5. Kurtzke, J.F. Epidemiology of multiple sclerosis. Does this really point toward an etiology? Lectio doctoralis. *Neurol. Sci.* **2000**, *21*, 383–403. [CrossRef]
6. GBD 2015 Disease and Injury Incidence and Prevalence Collaborators. Global, regional, and national incidence, prevalence, and years lived with disability for 310 diseases and injuries, 1990–2015: A systematic analysis for the global burden of disease study 2015. *Lancet* **2016**, *388*, 1545–1602. [CrossRef]
7. Sintzel, M.B.; Rametta, M.; Reder, A.T. Vitamin d and multiple sclerosis: A comprehensive review. *Neurol. Ther.* **2018**, *7*, 59–85. [CrossRef]
8. Oksenberg, J.R. Decoding multiple sclerosis: An update on genomics and future directions. *Expert Rev. Neurother.* **2013**, *13*, 11–19. [CrossRef]
9. Ascherio, A. Environmental factors in multiple sclerosis. *Expert Rev. Neurother.* **2013**, *13*, 3–9. [CrossRef]
10. Ramagopalan, S.V.; Dobson, R.; Meier, U.C.; Giovannoni, G. Multiple sclerosis: Risk factors, prodromes, and potential causal pathways. *Lancet. Neurol.* **2010**, *9*, 727–739. [CrossRef]
11. Hedstrom, A.K.; Baarnhielm, M.; Olsson, T.; Alfredsson, L. Tobacco smoking, but not swedish snuff use, increases the risk of multiple sclerosis. *Neurology* **2009**, *73*, 696–701. [CrossRef]
12. Bjartmar, C.; Wujek, J.R.; Trapp, B.D. Axonal loss in the pathology of ms: Consequences for understanding the progressive phase of the disease. *J. Neurol. Sci.* **2003**, *206*, 165–171. [CrossRef]

13. Lublin, F.D.; Reingold, S.C. Defining the clinical course of multiple sclerosis: Results of an international survey. National multiple sclerosis society (USA) advisory committee on clinical trials of new agents in multiple sclerosis. *Neurology* **1996**, *46*, 907–911. [CrossRef]

14. Huang, W.J.; Chen, W.W.; Zhang, X. Multiple sclerosis: Pathology, diagnosis and treatments. *Exp. Ther. Med.* **2017**, *13*, 3163–3166. [CrossRef]

15. de Sa, J.C.; Airas, L.; Bartholome, E.; Grigoriadis, N.; Mattle, H.; Oreja-Guevara, C.; O'Riordan, J.; Sellebjerg, F.; Stankoff, B.; Vass, K.; et al. Symptomatic therapy in multiple sclerosis: A review for a multimodal approach in clinical practice. *Ther. Adv. Neurol. Disord.* **2011**, *4*, 139–168. [CrossRef]

16. Ransohoff, R.M.; Hafler, D.A.; Lucchinetti, C.F. Multiple sclerosis-a quiet revolution. *Nat. Rev. Neurol.* **2015**, *11*, 134–142. [CrossRef]

17. Critchfield, J.M.; Racke, M.K.; Zuniga-Pflucker, J.C.; Cannella, B.; Raine, C.S.; Goverman, J.; Lenardo, M.J. T cell deletion in high antigen dose therapy of autoimmune encephalomyelitis. *Science* **1994**, *263*, 1139–1143. [CrossRef]

18. Shakya, A.K.; Nandakumar, K.S. Antigen-specific tolerization and targeted delivery as therapeutic strategies for autoimmune diseases. *Trends Biotechnol.* **2018**, *36*, 686–699. [CrossRef]

19. Willekens, B.; Cools, N. Beyond the magic bullet: Current progress of therapeutic vaccination in multiple sclerosis. *CNS Drugs* **2018**, *32*, 401–410. [CrossRef]

20. Bielekova, B.; Goodwin, B.; Richert, N.; Cortese, I.; Kondo, T.; Afshar, G.; Gran, B.; Eaton, J.; Antel, J.; Frank, J.A.; et al. Encephalitogenic potential of the myelin basic protein peptide (amino acids 83–99) in multiple sclerosis: Results of a phase ii clinical trial with an altered peptide ligand. *Nat. Med.* **2000**, *6*, 1167–1175. [CrossRef]

21. Schrempf, W.; Ziemssen, T. Glatiramer acetate: Mechanisms of action in multiple sclerosis. *Autoimmun. Rev.* **2007**, *6*, 469–475. [CrossRef]

22. Noon, L. Prophylactic inoculation against hay fever. *Int. Arch. Allergy Appl. Immunol.* **1953**, *4*, 285–288. [CrossRef]

23. Frankland, A.W.; Augustin, R. Prophylaxis of summer hay-fever and asthma: A controlled trial comparing crude grass-pollen extracts with the isolated main protein component. *Lancet* **1954**, *266*, 1055–1057. [CrossRef]

24. Freeman, J. " Rush " inoculation, with special reference to hay-fever treatment. *Lancet* **1930**, *215*, 744–747. [CrossRef]

25. Hochfelder, J.L.; Ponda, P. Allergen immunotherapy: Routes, safety, efficacy, and mode of action. *Immunotargets Ther.* **2013**, *2*, 61–71. [CrossRef]

26. Pozsgay, J.; Szekanecz, Z.; Sarmay, G. Antigen-specific immunotherapies in rheumatic diseases. *Nat. Rev. Rheumatol.* **2017**, *13*, 525–537. [CrossRef]

27. Hohlfeld, R.; Wekerle, H. Autoimmune concepts of multiple sclerosis as a basis for selective immunotherapy: From pipe dreams to (therapeutic) pipelines. *Proc. Natl. Acad. Sci. USA* **2004**, *101* (Suppl. 2), 14599–14606. [CrossRef]

28. Karin, N.; Mitchell, D.J.; Brocke, S.; Ling, N.; Steinman, L. Reversal of experimental autoimmune encephalomyelitis by a soluble peptide variant of a myelin basic protein epitope: T cell receptor antagonism and reduction of interferon gamma and tumor necrosis factor alpha production. *J. Exp. Med.* **1994**, *180*, 2227–2237. [CrossRef]

29. Stepanov, A.; Lomakin, Y.; Gabibov, A.; Belogurov, A. Peptides against autoimmune neurodegeneration. *Curr. Med. Chem.* **2017**, *24*, 1761–1771. [CrossRef]

30. Puentes, F.; Dickhaut, K.; Hofstatter, M.; Falk, K.; Rotzschke, O. Active suppression induced by repetitive self-epitopes protects against eae development. *PLoS ONE* **2013**, *8*, e64888. [CrossRef]

31. Metzler, B.; Wraith, D.C. Inhibition of experimental autoimmune encephalomyelitis by inhalation but not oral administration of the encephalitogenic peptide: Influence of mhc binding affinity. *Int. Immunol.* **1993**, *5*, 1159–1165. [CrossRef] [PubMed]

32. Tselios, T.; Aggelidakis, M.; Tapeinou, A.; Tseveleki, V.; Kanistras, I.; Gatos, D.; Matsoukas, J. Rational design and synthesis of altered peptide ligands based on human myelin oligodendrocyte glycoprotein 35–55 epitope: Inhibition of chronic experimental autoimmune encephalomyelitis in mice. *Molecules* **2014**, *19*, 17968–17984. [CrossRef] [PubMed]

33. Yeste, A.; Nadeau, M.; Burns, E.J.; Weiner, H.L.; Quintana, F.J. Nanoparticle-mediated codelivery of myelin antigen and a tolerogenic small molecule suppresses experimental autoimmune encephalomyelitis. *Proc. Natl. Acad. Sci. USA* **2012**, *109*, 11270–11275. [CrossRef] [PubMed]

34. Nicholson, L.B.; Greer, J.M.; Sobel, R.A.; Lees, M.B.; Kuchroo, V.K. An altered peptide ligand mediates immune deviation and prevents autoimmune encephalomyelitis. *Immunity* **1995**, *3*, 397–405. [CrossRef]

35. Kuchroo, V.K.; Greer, J.M.; Kaul, D.; Ishioka, G.; Franco, A.; Sette, A.; Sobel, R.A.; Lees, M.B. A single tcr antagonist peptide inhibits experimental allergic encephalomyelitis mediated by a diverse t cell repertoire. *J. Immunol.* **1994**, *153*, 3326–3336.

36. Walczak, A.; Siger, M.; Ciach, A.; Szczepanik, M.; Selmaj, K. Transdermal application of myelin peptides in multiple sclerosis treatment. *JAMA Neurol.* **2013**, *70*, 1105–1109. [CrossRef]

37. Jurynczyk, M.; Walczak, A.; Jurewicz, A.; Jesionek-Kupnicka, D.; Szczepanik, M.; Selmaj, K. Immune regulation of multiple sclerosis by transdermally applied myelin peptides. *Ann. Neurol.* **2010**, *68*, 593–601. [CrossRef]

38. Chataway, J.; Martin, K.; Barrell, K.; Sharrack, B.; Stolt, P.; Wraith, D.C. Effects of atx-ms-1467 immunotherapy over 16 weeks in relapsing multiple sclerosis. *Neurology* **2018**, *90*, e955–e962. [CrossRef]

39. Crowe, P.D.; Qin, Y.; Conlon, P.J.; Antel, J.P. Nbi-5788, an altered mbp83–99 peptide, induces a t-helper 2-like immune response in multiple sclerosis patients. *Ann. Neurol.* **2000**, *48*, 758–765. [CrossRef]

40. Lomakin, Y.; Belogurov, A., Jr.; Glagoleva, I.; Stepanov, A.; Zakharov, K.; Okunola, J.; Smirnov, I.; Genkin, D.; Gabibov, A. Administration of myelin basic protein peptides encapsulated in mannosylated liposomes normalizes level of serum tnf-alpha and il-2 and chemoattractants ccl2 and ccl4 in multiple sclerosis patients. *Mediat. Inflamm.* **2016**, *2016*, 2847232. [CrossRef]

41. Lutterotti, A.; Yousef, S.; Sputtek, A.; Sturner, K.H.; Stellmann, J.P.; Breiden, P.; Reinhardt, S.; Schulze, C.; Bester, M.; Heesen, C.; et al. Antigen-specific tolerance by autologous myelin peptide-coupled cells: A phase 1 trial in multiple sclerosis. *Sci. Transl. Med.* **2013**, *5*, 188ra175. [CrossRef] [PubMed]

42. Kappos, L.; Comi, G.; Panitch, H.; Oger, J.; Antel, J.; Conlon, P.; Steinman, L. Induction of a non-encephalitogenic type 2 t helper-cell autoimmune response in multiple sclerosis after administration of an altered peptide ligand in a placebo-controlled, randomized phase ii trial. The altered peptide ligand in relapsing ms study group. *Nat. Med.* **2000**, *6*, 1176–1182. [CrossRef] [PubMed]

43. Smith, C.E.; Eagar, T.N.; Strominger, J.L.; Miller, S.D. Differential induction of ige-mediated anaphylaxis after soluble vs. Cell-bound tolerogenic peptide therapy of autoimmune encephalomyelitis. *Proc. Natl. Acad. Sci. USA* **2005**, *102*, 9595–9600. [CrossRef]

44. Warren, K.G.; Catz, I.; Wucherpfennig, K.W. Tolerance induction to myelin basic protein by intravenous synthetic peptides containing epitope p85 vvhffknivtp96 in chronic progressive multiple sclerosis. *J. Neurol. Sci.* **1997**, *152*, 31–38. [CrossRef]

45. Weiner, H.L.; Mackin, G.A.; Matsui, M.; Orav, E.J.; Khoury, S.J.; Dawson, D.M.; Hafler, D.A. Double-blind pilot trial of oral tolerization with myelin antigens in multiple sclerosis. *Science* **1993**, *259*, 1321–1324. [CrossRef]

46. Fissolo, N.; Montalban, X.; Comabella, M. DNA-based vaccines for multiple sclerosis: Current status and future directions. *Clin. Immunol.* **2012**, *142*, 76–83. [CrossRef]

47. Stuve, O.; Cravens, P.D.; Eagar, T.N. DNA-based vaccines: The future of multiple sclerosis therapy? *Expert Rev. Neurother.* **2008**, *8*, 351–360. [CrossRef]

48. Garren, H.; Robinson, W.H.; Krasulova, E.; Havrdova, E.; Nadj, C.; Selmaj, K.; Losy, J.; Nadj, I.; Radue, E.W.; Kidd, B.A.; et al. Phase 2 trial of a DNA vaccine encoding myelin basic protein for multiple sclerosis. *Ann. Neurol.* **2008**, *63*, 611–620. [CrossRef]

49. Bar-Or, A.; Vollmer, T.; Antel, J.; Arnold, D.L.; Bodner, C.A.; Campagnolo, D.; Gianettoni, J.; Jalili, F.; Kachuck, N.; Lapierre, Y.; et al. Induction of antigen-specific tolerance in multiple sclerosis after immunization with DNA encoding myelin basic protein in a randomized, placebo-controlled phase 1/2 trial. *Arch. Neurol.* **2007**, *64*, 1407–1415. [CrossRef]

50. Friese, M.A.; Schattling, B.; Fugger, L. Mechanisms of neurodegeneration and axonal dysfunction in multiple sclerosis. *Nat. Rev. Neurol.* **2014**, *10*, 225–238. [CrossRef]

51. Zhang, J.; Raus, J. T cell vaccination in multiple sclerosis: Hopes and facts. *Acta Neurol. Belg.* **1994**, *94*, 112–115.

52. Medaer, R.; Stinissen, P.; Truyen, L.; Raus, J.; Zhang, J. Depletion of myelin-basic-protein autoreactive t cells by t-cell vaccination: Pilot trial in multiple sclerosis. *Lancet* **1995**, *346*, 807–808. [CrossRef]

53. Zhang, J.Z.; Rivera, V.M.; Tejada-Simon, M.V.; Yang, D.; Hong, J.; Li, S.; Haykal, H.; Killian, J.; Zang, Y.C. T cell vaccination in multiple sclerosis: Results of a preliminary study. *J. Neurol.* **2002**, *249*, 212–218. [CrossRef]

54. Loftus, B.; Newsom, B.; Montgomery, M.; Von Gynz-Rekowski, K.; Riser, M.; Inman, S.; Garces, P.; Rill, D.; Zhang, J.; Williams, J.C. Autologous attenuated t-cell vaccine (tovaxin) dose escalation in multiple sclerosis relapsing-remitting and secondary progressive patients nonresponsive to approved immunomodulatory therapies. *Clin. Immunol.* **2009**, *131*, 202–215. [CrossRef]

55. Achiron, A.; Lavie, G.; Kishner, I.; Stern, Y.; Sarova-Pinhas, I.; Ben-Aharon, T.; Barak, Y.; Raz, H.; Lavie, M.; Barliya, T.; et al. T cell vaccination in multiple sclerosis relapsing-remitting nonresponders patients. *Clin. Immunol.* **2004**, *113*, 155–160. [CrossRef]

56. Karussis, D.; Shor, H.; Yachnin, J.; Lanxner, N.; Amiel, M.; Baruch, K.; Keren-Zur, Y.; Haviv, O.; Filippi, M.; Petrou, P.; et al. T cell vaccination benefits relapsing progressive multiple sclerosis patients: A randomized, double-blind clinical trial. *PLoS ONE* **2012**, *7*, e50478. [CrossRef]

57. Agrawal, A.; Agrawal, S.; Gupta, S. Role of dendritic cells in inflammation and loss of tolerance in the elderly. *Front. Immunol.* **2017**, *8*, 896. [CrossRef]

58. Raker, V.K.; Domogalla, M.P.; Steinbrink, K. Tolerogenic dendritic cells for regulatory t cell induction in man. *Front. Immunol.* **2015**, *6*, 569. [CrossRef] [PubMed]

59. Lee, W.P.; Willekens, B.; Cras, P.; Goossens, H.; Martinez-Caceres, E.; Berneman, Z.N.; Cools, N. Immunomodulatory effects of 1,25-dihydroxyvitamin d3 on dendritic cells promote induction of t cell hyporesponsiveness to myelin-derived antigens. *J. Immunol. Res.* **2016**, *2016*, 5392623. [CrossRef]

60. Mansilla, M.J.; Selles-Moreno, C.; Fabregas-Puig, S.; Amoedo, J.; Navarro-Barriuso, J.; Teniente-Serra, A.; Grau-Lopez, L.; Ramo-Tello, C.; Martinez-Caceres, E.M. Beneficial effect of tolerogenic dendritic cells pulsed with mog autoantigen in experimental autoimmune encephalomyelitis. *CNS Neurosci. Ther.* **2015**, *21*, 222–230. [CrossRef]

61. Zubizarreta, I.; Florez-Grau, G.; Vila, G.; Cabezon, R.; Espana, C.; Andorra, M.; Saiz, A.; Llufriu, S.; Sepulveda, M.; Sola-Valls, N.; et al. Immune tolerance in multiple sclerosis and neuromyelitis optica with peptide-loaded tolerogenic dendritic cells in a phase 1b trial. *Proc. Natl. Acad. Sci. USA* **2019**, *116*, 8463–8470. [CrossRef] [PubMed]

62. Sewell, A.K. Why must t cells be cross-reactive? *Nat. Rev. Immunol.* **2012**, *12*, 669–677. [CrossRef] [PubMed]

63. Vandenbark, A.A.; Hashim, G.; Offner, H. Immunization with a synthetic t-cell receptor v-region peptide protects against experimental autoimmune encephalomyelitis. *Nature* **1989**, *341*, 541–544. [CrossRef] [PubMed]

64. Offner, H.; Hashim, G.A.; Vandenbark, A.A. T cell receptor peptide therapy triggers autoregulation of experimental encephalomyelitis. *Science* **1991**, *251*, 430–432. [CrossRef]

65. Bourdette, D.N.; Whitham, R.H.; Chou, Y.K.; Morrison, W.J.; Atherton, J.; Kenny, C.; Liefeld, D.; Hashim, G.A.; Offner, H.; Vandenbark, A.A. Immunity to tcr peptides in multiple sclerosis. I. Successful immunization of patients with synthetic v beta 5.2 and v beta 6.1 cdr2 peptides. *J. Immunol.* **1994**, *152*, 2510–2519.

66. Vandenbark, A.A.; Chou, Y.K.; Whitham, R.; Mass, M.; Buenafe, A.; Liefeld, D.; Kavanagh, D.; Cooper, S.; Hashim, G.A.; Offner, H. Treatment of multiple sclerosis with t-cell receptor peptides: Results of a double-blind pilot trial. *Nat. Med.* **1996**, *2*, 1109–1115. [CrossRef]

67. Vandenbark, A.A.; Culbertson, N.E.; Bartholomew, R.M.; Huan, J.; Agotsch, M.; LaTocha, D.; Yadav, V.; Mass, M.; Whitham, R.; Lovera, J.; et al. Therapeutic vaccination with a trivalent t-cell receptor (tcr) peptide vaccine restores deficient foxp3 expression and tcr recognition in subjects with multiple sclerosis. *Immunology* **2008**, *123*, 66–78. [CrossRef]

68. Gabibov, A.G.; Belogurov, A.A., Jr.; Lomakin, Y.A.; Zakharova, M.Y.; Avakyan, M.E.; Dubrovskaya, V.V.; Smirnov, I.V.; Ivanov, A.S.; Molnar, A.A.; Gurtsevitch, V.E.; et al. Combinatorial antibody library from multiple sclerosis patients reveals antibodies that cross-react with myelin basic protein and ebv antigen. *FASEB J.* **2011**, *25*, 4211–4221. [CrossRef]

69. Lambracht-Washington, D.; O'Connor, K.C.; Cameron, E.M.; Jowdry, A.; Ward, E.S.; Frohman, E.; Racke, M.K.; Monson, N.L. Antigen specificity of clonally expanded and receptor edited cerebrospinal fluid b cells from patients with relapsing remitting ms. *J. Neuroimmunol.* **2007**, *186*, 164–176. [CrossRef]

70. Yu, X.; Gilden, D.H.; Ritchie, A.M.; Burgoon, M.P.; Keays, K.M.; Owens, G.P. Specificity of recombinant antibodies generated from multiple sclerosis cerebrospinal fluid probed with a random peptide library. *J. Neuroimmunol.* **2006**, *172*, 121–131. [CrossRef]

71. Stepanov, A.V.; Belogurov, A.A., Jr.; Ponomarenko, N.A.; Stremovskiy, O.A.; Kozlov, L.V.; Bichucher, A.M.; Dmitriev, S.E.; Smirnov, I.V.; Shamborant, O.G.; Balabashin, D.S.; et al. Design of targeted b cell killing agents. *PLoS ONE* **2011**, *6*, e20991. [CrossRef]

72. Madhumathi, J.; Verma, R.S. Therapeutic targets and recent advances in protein immunotoxins. *Curr. Opin. Microbiol.* **2012**, *15*, 300–309. [CrossRef] [PubMed]

73. Voge, N.V.; Alvarez, E. Monoclonal antibodies in multiple sclerosis: Present and future. *Biomedicines* **2019**, *7*, 20. [CrossRef]

74. Wootla, B.; Watzlawik, J.O.; Stavropoulos, N.; Wittenberg, N.J.; Dasari, H.; Abdelrahim, M.A.; Henley, J.R.; Oh, S.H.; Warrington, A.E.; Rodriguez, M. Recent advances in monoclonal antibody therapies for multiple sclerosis. *Expert Opin. Biol. Ther.* **2016**, *16*, 827–839. [CrossRef]

75. Yaldizli, O.; Putzki, N. Natalizumab in the treatment of multiple sclerosis. *Ther. Adv. Neurol. Disord.* **2009**, *2*, 115–128. [CrossRef]

76. Polman, C.H.; O'Connor, P.W.; Havrdova, E.; Hutchinson, M.; Kappos, L.; Miller, D.H.; Phillips, J.T.; Lublin, F.D.; Giovannoni, G.; Wajgt, A.; et al. A randomized, placebo-controlled trial of natalizumab for relapsing multiple sclerosis. *New Engl. J. Med.* **2006**, *354*, 899–910. [CrossRef]

77. Hauser, S.L.; Waubant, E.; Arnold, D.L.; Vollmer, T.; Antel, J.; Fox, R.J.; Bar-Or, A.; Panzara, M.; Sarkar, N.; Agarwal, S.; et al. B-cell depletion with rituximab in relapsing-remitting multiple sclerosis. *New Engl. J. Med.* **2008**, *358*, 676–688. [CrossRef]

78. Hauser, S.L.; Bar-Or, A.; Comi, G.; Giovannoni, G.; Hartung, H.P.; Hemmer, B.; Lublin, F.; Montalban, X.; Rammohan, K.W.; Selmaj, K.; et al. Ocrelizumab versus interferon beta-1a in relapsing multiple sclerosis. *New Engl. J. Med.* **2017**, *376*, 221–234. [CrossRef]

79. Sorensen, P.S.; Lisby, S.; Grove, R.; Derosier, F.; Shackelford, S.; Havrdova, E.; Drulovic, J.; Filippi, M. Safety and efficacy of ofatumumab in relapsing-remitting multiple sclerosis: A phase 2 study. *Neurology* **2014**, *82*, 573–581. [CrossRef]

80. Mi, S.; Miller, R.H.; Lee, X.; Scott, M.L.; Shulag-Morskaya, S.; Shao, Z.; Chang, J.; Thill, G.; Levesque, M.; Zhang, M.; et al. Lingo-1 negatively regulates myelination by oligodendrocytes. *Nat. Neurosci.* **2005**, *8*, 745–751. [CrossRef]

81. Mellion, M.; Edwards, K.R.; Hupperts, R.; Drulović, J.; Montalban, X.; Hartung, H.P.; Brochet, B.; Calabresi, P.A.; Rudick, R.; Ibrahim, A.; et al. Efficacy results from the phase 2b synergy study: Treatment of disabling multiple sclerosis with the anti-lingo-1 monoclonal antibody opicinumab (s33.004). *Neurology* **2017**, *88* (Suppl. 16).

82. Havrdova, E.; Horakova, D.; Kovarova, I. Alemtuzumab in the treatment of multiple sclerosis: Key clinical trial results and considerations for use. *Ther. Adv. Neurol. Disord.* **2015**, *8*, 31–45. [CrossRef] [PubMed]

83. Ford, C.; Goodman, A.D.; Johnson, K.; Kachuck, N.; Lindsey, J.W.; Lisak, R.; Luzzio, C.; Myers, L.; Panitch, H.; Preiningerova, J.; et al. Continuous long-term immunomodulatory therapy in relapsing multiple sclerosis: Results from the 15-year analysis of the us prospective open-label study of glatiramer acetate. *Mult. Scler.* **2010**, *16*, 342–350. [CrossRef] [PubMed]

84. Johnson, K.P.; Brooks, B.R.; Cohen, J.A.; Ford, C.C.; Goldstein, J.; Lisak, R.P.; Myers, L.W.; Panitch, H.S.; Rose, J.W.; Schiffer, R.B. Copolymer 1 reduces relapse rate and improves disability in relapsing-remitting multiple sclerosis: Results of a phase iii multicenter, double-blind placebo-controlled trial. The copolymer 1 multiple sclerosis study group. *Neurology* **1995**, *45*, 1268–1276. [CrossRef]

85. Arnon, R.; Teitelbaum, D.; Sela, M. Suppression of experimental allergic encephalomyelitis by cop1–relevance to multiple sclerosis. *Isr. J. Med Sci.* **1989**, *25*, 686–689.

86. Racke, M.K.; Lovett-Racke, A.E. Glatiramer acetate treatment of multiple sclerosis: An immunological perspective. *J. Immunol.* **2011**, *186*, 1887–1890. [CrossRef]

87. Ponomarenko, N.A.; Durova, O.M.; Vorobiev, I.I.; Belogurov, A.A., Jr.; Kurkova, I.N.; Petrenko, A.G.; Telegin, G.B.; Suchkov, S.V.; Kiselev, S.L.; Lagarkova, M.A.; et al. Autoantibodies to myelin basic protein catalyze site-specific degradation of their antigen. *Proc. Natl. Acad. Sci. USA* **2006**, *103*, 281–286. [CrossRef]

88. Fridkis-Hareli, M.; Santambrogio, L.; Stern, J.N.; Fugger, L.; Brosnan, C.; Strominger, J.L. Novel synthetic amino acid copolymers that inhibit autoantigen-specific t cell responses and suppress experimental autoimmune encephalomyelitis. *J. Clin. Investig.* **2002**, *109*, 1635–1643. [CrossRef]

89. Salvioni, L.; Fiandra, L.; Del Curto, M.D.; Mazzucchelli, S.; Allevi, R.; Truffi, M.; Sorrentino, L.; Santini, B.; Cerea, M.; Palugan, L.; et al. Oral delivery of insulin via polyethylene imine-based nanoparticles for colonic release allows glycemic control in diabetic rats. *Pharmacol. Res.* **2016**, *110*, 122–130. [CrossRef]

90. Sajeesh, S.; Vauthier, C.; Gueutin, C.; Ponchel, G.; Sharma, C.P. Thiol functionalized polymethacrylic acid-based hydrogel microparticles for oral insulin delivery. *Acta Biomater.* **2010**, *6*, 3072–3080. [CrossRef]

91. Posgai, A.L.; Wasserfall, C.H.; Kwon, K.C.; Daniell, H.; Schatz, D.A.; Atkinson, M.A. Plant-based vaccines for oral delivery of type 1 diabetes-related autoantigens: Evaluating oral tolerance mechanisms and disease prevention in nod mice. *Sci. Rep.* **2017**, *7*, 42372. [CrossRef] [PubMed]

92. Ma, S.; Huang, Y.; Yin, Z.; Menassa, R.; Brandle, J.E.; Jevnikar, A.M. Induction of oral tolerance to prevent diabetes with transgenic plants requires glutamic acid decarboxylase (gad) and il-4. *Proc. Natl. Acad. Sci. USA* **2004**, *101*, 5680–5685. [CrossRef] [PubMed]

93. Ma, Y.; Liu, D.; Wang, D.; Wang, Y.; Fu, Q.; Fallon, J.K.; Yang, X.; He, Z.; Liu, F. Combinational delivery of hydrophobic and hydrophilic anticancer drugs in single nanoemulsions to treat mdr in cancer. *Mol. Pharm.* **2014**, *11*, 2623–2630. [CrossRef] [PubMed]

94. Pujol-Autonell, I.; Mansilla, M.J.; Rodriguez-Fernandez, S.; Cano-Sarabia, M.; Navarro-Barriuso, J.; Ampudia, R.M.; Rius, A.; Garcia-Jimeno, S.; Perna-Barrull, D.; Martinez-Caceres, E.; et al. Liposome-based immunotherapy against autoimmune diseases: Therapeutic effect on multiple sclerosis. *Nanomed. (Lond.)* **2017**, *12*, 1231–1242. [CrossRef]

95. Teixeira, M.I.; Lopes, C.M.; Amaral, M.H.; Costa, P.C. Current insights on lipid nanocarrier-assisted drug delivery in the treatment of neurodegenerative diseases. *Eur. J. Pharm. Biopharm.* **2020**, *149*, 192–217. [CrossRef]

96. Agrawal, M.; Saraf, S.; Dubey, S.K.; Puri, A.; Patel, R.J.; Ajazuddin; Ravichandiran, V.; Murty, U.S.; Alexander, A. Recent strategies and advances in the fabrication of nano lipid carriers and their application towards brain targeting. *J. Control. Release* **2020**, *321*, 372–415. [CrossRef]

97. Ebrahimimonfared, M.; Ganji, A.; Zahedi, S.; Nourbakhsh, P.; Ghasami, K.; Mosayebi, G. Characterization of regulatory t-cells in multiple sclerosis patients treated with interferon beta-1a. *CNS Neurol. Disord. Drug Targets* **2018**, *17*, 113–118. [CrossRef]

98. Fernandez, O.; Arbizu, T.; Izquierdo, G.; Martinez-Yelamos, A.; Gata, J.M.; Luque, G.; de Ramon, E. Clinical benefits of interferon beta-1a in relapsing-remitting ms: A phase iv study. *Acta Neurol. Scand.* **2003**, *107*, 7–11. [CrossRef]

99. The IFNB multiple sclerosis study group. Interferon beta-1b is effective in relapsing-remitting multiple sclerosis, I. Clinical results of a multicenter, randomized, double-blind, placebo-controlled trial. *Neurology* **1993**, *43*, 655–661. [CrossRef]

100. Paty, D.W.; Li, D.K.; Ubc ms/mri study group; the ifnb multiple sclerosis study group. Interferon beta-1b is effective in relapsing-remitting multiple sclerosis. II. Mri analysis results of a multicenter, randomized, double-blind, placebo-controlled trial. *Neurology* **1993**, *43*, 662–667. [CrossRef]

101. Belogurov, A., Jr.; Zakharov, K.; Lomakin, Y.; Surkov, K.; Avtushenko, S.; Kruglyakov, P.; Smirnov, I.; Makshakov, G.; Lockshin, C.; Gregoriadis, G.; et al. Cd206-targeted liposomal myelin basic protein peptides in patients with multiple sclerosis resistant to first-line disease-modifying therapies: A first-in-human, proof-of-concept dose-escalation study. *Neurotherapeutics* **2016**, *13*, 895–904. [CrossRef]

102. De Souza, A.L.S.; Rudin, S.; Chang, R.; Mitchell, K.; Crandall, T.; Huang, S.; Choi, J.K.; Okitsu, S.L.; Graham, D.L.; Tomkinson, B.; et al. Atx-ms-1467 induces long-term tolerance to myelin basic protein in (dr2 x ob1)f1 mice by induction of il-10-secreting itregs. *Neurol. Ther.* **2018**, *7*, 103–128. [CrossRef] [PubMed]

103. Vandenbark, A.A. Tcr peptide vaccination in multiple sclerosis: Boosting a deficient natural regulatory network that may involve tcr-specific cd4+cd25+ treg cells. *Curr. Drug Targets. Inflamm. Allergy* **2005**, *4*, 217–229. [CrossRef]

104. Rorsman, I.; Petersen, C.; Nilsson, P.C. Cognitive functioning following one-year natalizumab treatment: A non-randomized clinical trial. *Acta Neurol. Scand.* **2018**, *137*, 117–124. [CrossRef]

105. Perumal, J.; Fox, R.J.; Balabanov, R.; Balcer, L.J.; Galetta, S.; Makh, S.; Santra, S.; Hotermans, C.; Lee, L. Outcomes of natalizumab treatment within 3 years of relapsing-remitting multiple sclerosis diagnosis: A prespecified 2-year interim analysis of strive. *BMC Neurol.* **2019**, *19*, 116. [CrossRef]

106. Kapoor, R.; Ho, P.R.; Campbell, N.; Chang, I.; Deykin, A.; Forrestal, F.; Lucas, N.; Yu, B.; Arnold, D.L.; Freedman, M.S.; et al. Effect of natalizumab on disease progression in secondary progressive multiple sclerosis (ascend): A phase 3, randomised, double-blind, placebo-controlled trial with an open-label extension. *Lancet. Neurol.* **2018**, *17*, 405–415. [CrossRef]

107. Ranger, A.; Ray, S.; Szak, S.; Dearth, A.; Allaire, N.; Murray, R.; Gardner, R.; Cadavid, D.; Mi, S. Anti-lingo-1 has no detectable immunomodulatory effects in preclinical and phase 1 studies. *Neurol. (R) Neuroimmunol. Neuroinflammation* **2018**, *5*, e417. [CrossRef] [PubMed]

108. Coles, A.J.; Twyman, C.L.; Arnold, D.L.; Cohen, J.A.; Confavreux, C.; Fox, E.J.; Hartung, H.P.; Havrdova, E.; Selmaj, K.W.; Weiner, H.L.; et al. Alemtuzumab for patients with relapsing multiple sclerosis after disease-modifying therapy: A randomised controlled phase 3 trial. *Lancet* **2012**, *380*, 1829–1839. [CrossRef]

109. Bar-Or, A.; Grove, R.A.; Austin, D.J.; Tolson, J.M.; VanMeter, S.A.; Lewis, E.W.; Derosier, F.J.; Lopez, M.C.; Kavanagh, S.T.; Miller, A.E.; et al. Subcutaneous ofatumumab in patients with relapsing-remitting multiple sclerosis: The mirror study. *Neurology* **2018**, *90*, e1805–e1814. [CrossRef]

110. Bhargava, P.; Wicken, C.; Smith, M.D.; Strowd, R.E.; Cortese, I.; Reich, D.S.; Calabresi, P.A.; Mowry, E.M. Trial of intrathecal rituximab in progressive multiple sclerosis patients with evidence of leptomeningeal contrast enhancement. *Mult. Scler. Relat. Disord.* **2019**, *30*, 136–140. [CrossRef]

111. Salzer, J.; Svenningsson, R.; Alping, P.; Novakova, L.; Bjorck, A.; Fink, K.; Islam-Jakobsson, P.; Malmestrom, C.; Axelsson, M.; Vagberg, M.; et al. Rituximab in multiple sclerosis: A retrospective observational study on safety and efficacy. *Neurology* **2016**, *87*, 2074–2081. [CrossRef] [PubMed]

112. Naismith, R.T.; Piccio, L.; Lyons, J.A.; Lauber, J.; Tutlam, N.T.; Parks, B.J.; Trinkaus, K.; Song, S.K.; Cross, A.H. Rituximab add-on therapy for breakthrough relapsing multiple sclerosis: A 52-week phase ii trial. *Neurology* **2010**, *74*, 1860–1867. [CrossRef] [PubMed]

113. Bergman, J.; Burman, J.; Gilthorpe, J.D.; Zetterberg, H.; Jiltsova, E.; Bergenheim, T.; Svenningsson, A. Intrathecal treatment trial of rituximab in progressive ms: An open-label phase 1b study. *Neurology* **2018**, *91*, e1893–e1901. [CrossRef] [PubMed]

114. Hawker, K.; O'Connor, P.; Freedman, M.S.; Calabresi, P.A.; Antel, J.; Simon, J.; Hauser, S.; Waubant, E.; Vollmer, T.; Panitch, H.; et al. Rituximab in patients with primary progressive multiple sclerosis: Results of a randomized double-blind placebo-controlled multicenter trial. *Ann. Neurol.* **2009**, *66*, 460–471. [CrossRef]

115. Mayer, L.; Kappos, L.; Racke, M.K.; Rammohan, K.; Traboulsee, A.; Hauser, S.L.; Julian, L.; Kondgen, H.; Li, C.; Napieralski, J.; et al. Ocrelizumab infusion experience in patients with relapsing and primary progressive multiple sclerosis: Results from the phase 3 randomized opera i, opera ii, and oratorio studies. *Mult. Scler. Relat. Disord.* **2019**, *30*, 236–243. [CrossRef]

116. Turner, B.; Cree, B.A.C.; Kappos, L.; Montalban, X.; Papeix, C.; Wolinsky, J.S.; Buffels, R.; Fiore, D.; Garren, H.; Han, J.; et al. Ocrelizumab efficacy in subgroups of patients with relapsing multiple sclerosis. *J. Neurol.* **2019**, *266*, 1182–1193. [CrossRef]

117. Fox, E.J.; Markowitz, C.; Applebee, A.; Montalban, X.; Wolinsky, J.S.; Belachew, S.; Fiore, D.; Pei, J.; Musch, B.; Giovannoni, G. Ocrelizumab reduces progression of upper extremity impairment in patients with primary progressive multiple sclerosis: Findings from the phase iii randomized oratorio trial. *Mult. Scler.* **2018**, *24*, 1862–1870. [CrossRef]

118. Montalban, X.; Hauser, S.L.; Kappos, L.; Arnold, D.L.; Bar-Or, A.; Comi, G.; de Seze, J.; Giovannoni, G.; Hartung, H.P.; Hemmer, B.; et al. Ocrelizumab versus placebo in primary progressive multiple sclerosis. *New Engl. J. Med.* **2017**, *376*, 209–220. [CrossRef]

119. Johnson, K.P.; Brooks, B.R.; Ford, C.C.; Goodman, A.; Guarnaccia, J.; Lisak, R.P.; Myers, L.W.; Panitch, H.S.; Pruitt, A.; Rose, J.W.; et al. Sustained clinical benefits of glatiramer acetate in relapsing multiple sclerosis patients observed for 6 years. Copolymer 1 multiple sclerosis study group. *Mult. Scler.* **2000**, *6*, 255–266. [CrossRef]

120. Ouspid, E.; Razazian, N.; Moghadasi, A.N.; Moradian, N.; Afshari, D.; Bostani, A.; Sariaslani, P.; Ansarian, A. Clinical effectiveness and safety of fingolimod in relapsing remitting multiple sclerosis in western iran. *Neurosciences (Riyadh)* **2018**, *23*, 129–134. [CrossRef]

121. Ntranos, A.; Hall, O.; Robinson, D.P.; Grishkan, I.V.; Schott, J.T.; Tosi, D.M.; Klein, S.L.; Calabresi, P.A.; Gocke, A.R. Fty720 impairs cd8 t-cell function independently of the sphingosine-1-phosphate pathway. *J. Neuroimmunol.* **2014**, *270*, 13–21. [CrossRef] [PubMed]

122. Coyle, P.K.; Khatri, B.; Edwards, K.R.; Meca-Lallana, J.E.; Cavalier, S.; Rufi, P.; Benamor, M.; Poole, E.M.; Robinson, M.; Gold, R. Teriflunomide real-world evidence: Global differences in the phase 4 teri-pro study. *Mult. Scler. Relat. Disord.* **2019**, *31*, 157–164. [CrossRef] [PubMed]

123. Miller, A.E.; Xu, X.; Macdonell, R.; Vucic, S.; Truffinet, P.; Benamor, M.; Thangavelu, K.; Freedman, M.S. Efficacy and safety of teriflunomide in asian patients with relapsing forms of multiple sclerosis: A subgroup analysis of the phase 3 tower study. *J. Clin. Neurosci.* **2019**, *59*, 229–231. [CrossRef] [PubMed]

124. Freedman, M.S.; Wolinsky, J.S.; Comi, G.; Kappos, L.; Olsson, T.P.; Miller, A.E.; Thangavelu, K.; Benamor, M.; Truffinet, P.; O'Connor, P.W. The efficacy of teriflunomide in patients who received prior disease-modifying treatments: Subgroup analyses of the teriflunomide phase 3 temso and tower studies. *Mult. Scler.* **2018**, *24*, 535–539. [CrossRef]

125. Saida, T.; Yamamura, T.; Kondo, T.; Yun, J.; Yang, M.; Li, J.; Mahadavan, L.; Zhu, B.; Sheikh, S.I. A randomized placebo-controlled trial of delayed-release dimethyl fumarate in patients with relapsing-remitting multiple sclerosis from east asia and other countries. *BMC Neurol.* **2019**, *19*, 5. [CrossRef]

126. Berger, T.; Brochet, B.; Brambilla, L.; Giacomini, P.S.; Montalban, X.; Vasco Salgado, A.; Su, R.; Bretagne, A. Effectiveness of delayed-release dimethyl fumarate on patient-reported outcomes and clinical measures in patients with relapsing-remitting multiple sclerosis in a real-world clinical setting: Protec. *Mult. Scler. J. Exp. Transl. Clin.* **2019**, *5*, 2055217319887191. [CrossRef]

127. Rivera, V.M.; Jeffery, D.R.; Weinstock-Guttman, B.; Bock, D.; Dangond, F. Results from the 5-year, phase iv renew (registry to evaluate novantrone effects in worsening multiple sclerosis) study. *BMC Neurol.* **2013**, *13*, 80. [CrossRef]

128. Hirsch, D.L.; Ponda, P. Antigen-based immunotherapy for autoimmune disease: Current status. *Immunotargets Ther.* **2015**, *4*, 1–11. [CrossRef]

129. Chiba, K.; Yanagawa, Y.; Masubuchi, Y.; Kataoka, H.; Kawaguchi, T.; Ohtsuki, M.; Hoshino, Y. Fty720, a novel immunosuppressant, induces sequestration of circulating mature lymphocytes by acceleration of lymphocyte homing in rats. I. Fty720 selectively decreases the number of circulating mature lymphocytes by acceleration of lymphocyte homing. *J. Immunol.* **1998**, *160*, 5037–5044.

130. Claussen, M.C.; Korn, T. Immune mechanisms of new therapeutic strategies in ms: Teriflunomide. *Clin. Immunol.* **2012**, *142*, 49–56. [CrossRef]

131. Fox, E.J. Mechanism of action of mitoxantrone. *Neurology* **2004**, *63*, S15–S18. [CrossRef] [PubMed]

132. Linker, R.A.; Gold, R. Dimethyl fumarate for treatment of multiple sclerosis: Mechanism of action, effectiveness, and side effects. *Curr. Neurol. Neurosci. Rep.* **2013**, *13*, 394. [CrossRef] [PubMed]

133. Lim, S.; Kim, W.J.; Kim, Y.H.; Lee, S.; Koo, J.H.; Lee, J.A.; Yoon, H.; Kim, D.H.; Park, H.J.; Kim, H.M.; et al. Dnp2 is a blood-brain barrier-permeable peptide enabling ctctla-4 protein delivery to ameliorate experimental autoimmune encephalomyelitis. *Nat. Commun.* **2015**, *6*, 8244. [CrossRef]

134. Koo, J.H.; Kim, D.H.; Cha, D.; Kang, M.J.; Choi, J.M. Lrr domain of nlrx1 protein delivery by dnp2 inhibits t cell functions and alleviates autoimmune encephalomyelitis. *Theranostics* **2020**, *10*, 3138–3150. [CrossRef] [PubMed]

135. Acharya, B.; Meka, R.R.; Venkatesha, S.H.; Lees, J.R.; Teesalu, T.; Moudgil, K.D. A novel cns-homing peptide for targeting neuroinflammatory lesions in experimental autoimmune encephalomyelitis. *Mol. Cell. Probes* **2020**, *51*, 101530. [CrossRef] [PubMed]

136. Sanadgol, N.; Barati, M.; Houshmand, F.; Hassani, S.; Clarner, T.; Shahlaei, M.; Golab, F. Metformin accelerates myelin recovery and ameliorates behavioral deficits in the animal model of multiple sclerosis via adjustment of ampk/nrf2/mtor signaling and maintenance of endogenous oligodendrogenesis during brain self-repairing period. *Pharmacol. Rep.* **2019**. [CrossRef]

137. Harada, M.; Kamimura, D.; Arima, Y.; Kohsaka, H.; Nakatsuji, Y.; Nishida, M.; Atsumi, T.; Meng, J.; Bando, H.; Singh, R.; et al. Temporal expression of growth factors triggered by epiregulin regulates inflammation development. *J. Immunol.* **2015**, *194*, 1039–1046. [CrossRef]

138. Severa, M.; Zhang, J.; Giacomini, E.; Rizzo, F.; Etna, M.P.; Cruciani, M.; Garaci, E.; Chopp, M.; Coccia, E.M. Thymosins in multiple sclerosis and its experimental models: Moving from basic to clinical application. *Mult. Scler. Relat. Disord.* **2019**, *27*, 52–60. [CrossRef]

Article

Kappa Free Light Chains and IgG Combined in a Novel Algorithm for the Detection of Multiple Sclerosis

Monika Gudowska-Sawczuk [1,*]**, Joanna Tarasiuk** [2]**, Alina Kułakowska** [2]**, Jan Kochanowicz** [2] **and Barbara Mroczko** [1,3]

1   Department of Biochemical Diagnostics, Medical University of Bialystok, Waszyngtona 15A St., 15-269 Bialystok, Poland; mroczko@umb.edu.pl
2   Department of Neurology, Medical University of Bialystok, M. Skłodowskiej—Curie 24A St., 15-276 Bialystok, Poland; amirtarasiuk@wp.pl (J.T.); alakul@umb.edu.pl (A.K.); kochanowicz@vp.pl (J.K.)
3   Department of Neurodegeneration Diagnostics, Medical University of Bialystok, Waszyngtona 15A St., 15-269 Bialystok, Poland
*   Correspondence: monika.gudowska-sawczuk@umb.edu.pl; Tel.: +48-85-831-8703

Received: 29 April 2020; Accepted: 24 May 2020; Published: 27 May 2020

Abstract: Background: It is well known that the cerebrospinal fluid (CSF) concentrations of free light chains (FLC) and immunoglobulin G (IgG) are elevated in multiple sclerosis patients (MS). Therefore, in this study we aimed to develop a model based on the concentrations of free light chains and IgG to predict multiple sclerosis. We tried to evaluate the diagnostic usefulness of the novel κIgG index and λIgG index, here presented for the first time, and compare them with the κFLC index and the λFLC index in multiple sclerosis patients. Methods: CSF and serum samples were obtained from 76 subjects who underwent lumbar puncture for diagnostic purposes and, as a result, were divided into two groups: patients with multiple sclerosis ($n = 34$) and patients with other neurological disorders (control group; $n = 42$). The samples were analyzed using turbidimetry and isoelectric focusing. The κIgG index, λIgG index, κFLC index, and λFLC index were calculated using specific formulas. Results: The concentrations of CSF κFLC, CSF λFLC, and serum κFLC and the values of κFLC index, λFLC index, and κIgG index were significantly higher in patients with multiple sclerosis compared to controls. CSF κFLC concentration and the values of κFLC index, λFLC index, and κIgG index differed in patients depending on their pattern type of oligoclonal bands. κFLC concentration was significantly higher in patients with pattern type 2 and type 3 in comparison to those with pattern type 1 and type 4. The κFLC index, λFLC index, and κIgG index were significantly higher in patients with pattern type 2 in comparison to those with pattern type 4. The κFLC index and κIgG index were significantly higher in patients with pattern type 2 in comparison to those with pattern type 1, and in patients with pattern type 3 compared to those with pattern type 4. The κIgG index was markedly elevated in patients with pattern type 3 compared to those with pattern type 1. In the total study group, κFLC, λFLC, κFLC index, λFLC index, κIgG index, and λIgG index correlated with each other. The κIgG index showed the highest diagnostic power (area under the curve, AUC) in the detection of multiple sclerosis. The κFLC index and κIgG index showed the highest diagnostic sensitivity, and the κIgG index presented the highest ability to exclude multiple sclerosis. Conclusion: This study provides novel information about the diagnostic significance of four markers combined in the κIgG index. More investigations in larger study groups are needed to confirm that the κIgG index can reflect the intrathecal synthesis of immunoglobulins and may improve the diagnosis of multiple sclerosis.

Keywords: multiple sclerosis; diagnostic markers; immunoglobulins; kappa; free light chains

---

1. Introduction

Multiple sclerosis (MS) is a common neuroinflammatory and neurodegenerative disorder of the central nervous system (CNS) [1]. The etiology of multiple sclerosis is still unknown. However, the major pathology is mediated by an auto-reactive immune process of multifocal myelin destruction throughout the CNS. Prompt and accurate diagnosis is particularly important for the clinical management of patients, since disease-modifying therapies are the most effective at the early stage of the disease [2,3]. A perfect biomarker should allow the early diagnosis of a disease, aid in determining the prognosis of a disease, and be rapid and easily testable. Currently, there is no specific test for the diagnosis of multiple sclerosis. According to the 2017 revisions of the McDonald diagnostic criteria for MS, the diagnosis of this disease is based on clinical symptoms, imaging by MRI technology, and laboratory testing including cerebrospinal fluid (CSF) examination [4].

The main feature of multiple sclerosis consists of abnormalities of the cellular and humoral immune system. Combined actions of B cells and T cells play a role in the full development of demyelination and in the secretion of immunoglobulins. Therefore, in more than 90% of patients, an elevated level of immunoglobulins synthesized in the intrathecal space can be observed, and IgG oligoclonal bands (OCBs) are detected in the CSF. However, there is a proportion of subjects, i.e., patients presenting with their first episode of multiple sclerosis, whose results of oligoclonal bands are negative. On the other hand, increased intrathecal immunoglobulin synthesis may occur also in other inflammatory CNS disorders, and therefore, this test is not specific for MS [5–8].

It is well known that human immunoglobulins are composed of two heavy and two light chains. There are two types of light chains, kappa (κ) and lambda (λ), that are produced by B lymphocytes during the synthesis of immunoglobulins. Physiologically, an excess of light chains is normally produced. These light chains that are not combined with heavy chains are called free light chains (FLC). It has been proven that B cell abnormalities are associated with disorders leading to an abnormal concentration of free light chains [9,10]. Therefore, in this study we aimed to develop a model based on free light chains and other available laboratory data to predict multiple sclerosis. We tried to evaluate the diagnostics usefulness of the novel κIgG index and λIgG index and compare them to the already known κFLC index and λFLC index used for the assessment of patients with MS.

2. Material and Methods

2.1. Subjects

This study was approved by the Bioethical Committee of the Medical University of Bialystok. Informed consent was obtained from all individuals included in the study. The patients were admitted to the Department of Neurology at the Medical University of Bialystok and underwent lumbar puncture for diagnostic purposes. Paired CSF and serum samples from the patients were collected between 2018 and 2020. The tested group consisted of 76 patients with neurological disorders who were divided into 2 subgroups: relapsing–remitting MS patients ($n = 34$) and a control group ($n = 42$) (Figure 1). All MS patients included in the study were in the process of receiving an MS diagnosis. They had a history of one clinical attack, and there was no evidence of dissemination in time according to magnetic resonance imaging (MRI). Finally, after CSF analysis which revealed OCBs presence, they were diagnosed with relapsing–remitting multiple sclerosis according to MacDonald criteria 2017 [4]. The degree of neurological impairment in patients diagnosed with multiple sclerosis from whom CSF was obtained was evaluated using the expanded disability status scale [11]. All evaluations rated between 1 and 2 points, indicating an early stage of the disease. All MS patients were not treated with any disease-modifying drugs or glucocorticosteroids at the time of lumbar puncture. The control group (29 females and 13 males; age range: 18–78 years) included patients eventually diagnosed with multifocal vascular lesions of the CNS ($n = 18$), discopathy ($n = 6$), idiopathic cephalgia ($n = 9$), dementia ($n = 3$), idiopathic (Bell's) facial nerve palsy ($n = 3$), epilepsy ($n = 1$), herpetic encephalitis ($n = 1$), hydrocephalus ($n = 1$). Out of 34 patients with multiple sclerosis, 31 had OCBs in the CSF but not in serum (pattern type 2), and 3 had OCBs in CSF and serum, with additional OCBs in the CSF (pattern type 3). Out of 42 patients in the control group, 21 had no bands in CSF and serum (pattern type 1), 4 had pattern type 3, 16 had identical OCBs in CSF and serum (pattern type 4), and 1 had monoclonal bands in CSF and serum (pattern type 5).

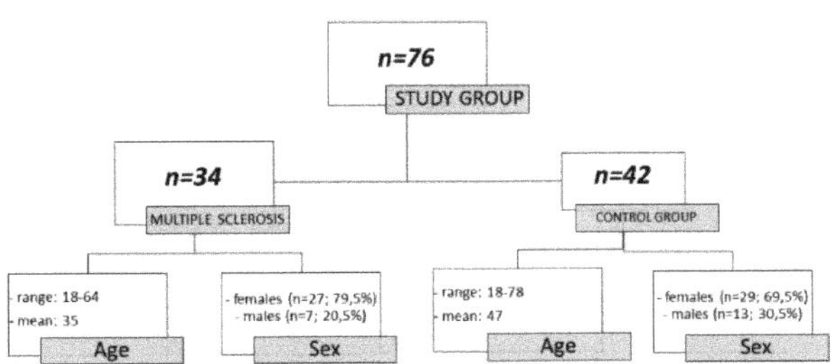

Figure 1. Characteristics of the study group.

2.2. Sample Collection

CSF specimens were collected from each patient by lumbar puncture. The samples were collected into polypropylene tubes, centrifuged, aliquoted, and frozen at −80 °C until assayed. Venous blood samples were collected and centrifuged to separate the serum. The serum samples were aliquoted and frozen at −80 °C until assayed.

IgG oligoclonal bands determination in human CSF and serum was performed at the time of diagnosis using isoelectric focusing on agarose gel. Each patient's serum and CSF samples were analyzed in parallel, in order to compare the IgG distribution. According to the manufacturer's instructions, the assay includes two steps. Firstly, we performed isoelectrofocusing on agarose gel to fractionate the proteins in the CSF and serum med. Secondly, we carried out immunofixation with peroxidase-labelled anti-IgG antiserum to detect IgG oligoclonal bands and demonstrate the distribution of IgG in both fluids (Hydragel 3 CSF Isofocusing; Hydrasys; Sebia). The concentrations of κFLC, λFLC, albumin, IgG, IgM, and IgA in CSF and serum were measured according to the turbidimetric method (Optilite; The Binding Site). The κIgG index, λIgG index, κFLC index, and λFLC index were calculated according to the following formulas: $\dfrac{\text{CSF }\kappa\text{FLC }\left(\frac{mg}{L}\right)/\text{serum }\kappa\text{FLC}\left(\frac{mg}{L}\right)}{\text{CSF IgG }\left(\frac{mg}{L}\right)/\text{serum IgG}\left(\frac{g}{L}\right)} \times 100$, $\dfrac{\text{CSF }\lambda\text{FLC }\left(\frac{mg}{L}\right)/\text{serum }\lambda\text{FLC}\left(\frac{mg}{L}\right)}{\text{CSF IgG }\left(\frac{mg}{L}\right)/\text{serum IgG}\left(\frac{g}{L}\right)} \times 100$, $\dfrac{\text{CSF }\kappa\text{FLC}\left(\frac{mg}{L}\right)/\text{serum }\kappa\text{FLC}\left(\frac{mg}{L}\right)}{\text{CSF albumin}\left(\frac{mg}{L}\right)/\text{serum albumin}\left(\frac{mg}{L}\right)}$ and $\dfrac{\text{CSF }\lambda\text{FLC}\left(\frac{mg}{L}\right)/\text{serum }\lambda\text{FLC}\left(\frac{mg}{L}\right)}{\text{CSF albumin}\left(\frac{mg}{L}\right)/\text{serum albumin}\left(\frac{mg}{L}\right)}$, respectively. In cases of FLCs concentrations below the lower limit of detection, we used the corresponding detection limit (CSF κFLC, 0.30 mg/L, CSF λFLC, 0.65 mg/L). Intrathecal synthesis was also evaluated using albumin, IgG, IgA, and IgM quotients (Q_{Alb}, Q_{IgG}, Q_{IgA}, Q_{IgM}, respectively).

2.3. Statistical Analysis

Data were stored and analyzed in Statistica 13.3. Differences between the multiple sclerosis and the control group were evaluated by Mann–Whitney U test. To test the hypothesis about the differences between subgroups, ANOVA rank Kruskal–Wallis test was performed. The post-hoc test was applied to determine which groups were different. We considered *p*-values < 0.05 as statistically significant. The diagnostic performance of each test was calculated as sensitivity, specificity, positive predictive value (PPV), negative predictive value (NPV), and accuracy (ACC). We used the area under the receiver operating characteristic (AUC ROC) curve to determine the optimal cut-off value and to calculate the diagnostic performance of the tests.

3. Results

The results of routine laboratory tests for patients with MS and the control group are presented in Table 1. Statistically significant differences between MS and controls in the Mann–Whitney U test were observed for the concentration of serum albumin and serum and CSF IgM and the values of Q_{IgM} and Q_{IgG} ($p = 0.010$; $p = 0.047$; $p = 0.003$; $p = 0.002$; $p = 0.002$, respectively).

3.1. CSF and Serum Concentrations of κFLC and λFLC

We determined the concentrations of κFLC and λFLC in the CSF and serum. κFLC and λFLC concentrations in the CSF and serum κFLC concentration were markedly elevated in MS patients (3.050 mg/L, 2.050 mg/L, 13.480 mg/L, respectively) compared to controls (0.310 mg/L, $p < 0.001$; 0.720 mg/L, $p = 0.017$; 16.265 mg/L, $p = 0.019$, respectively), while the concentration of serum λFLC did not differ between MS patients (11.715 mg/L) and controls (13.220 mg/L. $p = 0.066$). Furthermore, the concentrations of κFLC in the CSF differed depending on the types of OCB patterns (ANOVA rang Kruskal–Wallis test: $p < 0.001$, H = 36.472). Post-hoc analysis revealed that the CSF concentrations of κFLC were significantly lower in patients with pattern type 1 (0.300 mg/L) and type 4 (0.936 mg/L) of OCBs in comparison with those with pattern type 2 (2.905 mg/L; $p < 0.001$, $p = 0.002$, respectively) and type 3 (4.400 mg/L; $p = 0.002$, $p = 0.030$, respectively). There were no significant differences in CSF κFLC concentrations between patients with OCB pattern type 2 and type 3 ($p = 1.000$). The concentrations of serum κFLC and λFLC as well as CSF λFLC were similar in all patients irrespective of their OCB pattern type.

3.2. Values of κFLC Index, λFLC Index, κIgG Index, and λIgG Index

The values of κFLC index, λFLC index, κIgG index, and λIgG index are presented in Table 2. The values of κFLC index, λFLC index, and κIgG index were significantly higher in patients with multiple sclerosis compared to controls, but there were no differences in the λIgG index between the tested groups (Figure 2). The values of κFLC index, λFLC index, and κIgG index differed depending on the OCB pattern type ($p < 0.001$, H = 25.593; $p = 0.010$, H = 11.355; $p < 0.001$, H = 29.608). Post-hoc analysis revealed that the values of the κFLC index and κIgG index were significantly higher in patients with pattern type 2 (median: 58.551, 5.063) in comparison with those with pattern type 1 (5.933, 0.987; $p < 0.001$ for both) and type 4 (4.166, 0.636; $p < 0.001$ for both). The λFLC index was significantly elevated in patients with pattern type 2 (35.065) in comparison with those with pattern type 4 (7.208, $p = 0.013$). There were also differences in the κFLC index and κIgG index values between patients with pattern type 3 (56.172; 4.503) and those with pattern type 4 ($p = 0.034$; $p = 0.029$, respectively). In addition, the κIgG index was markedly elevated in patients with pattern type 3 compared with those with pattern type 1 ($p = 0.033$). There were no significant differences in the λIgG index between patients with different OCB types ($p = 0.106$, H = 6.123).

Table 1. Results of laboratory tests for patients with multiple sclerosis and the control group.

| | Variable Tested Median (Min–Max Values) | | | | | | | | | | | |
	Albumin S [g/L]	Albumin CSF [mg/L]	Q_{Alb}	IgG S [g/L]	IgG CSF [mg/L]	Q_{IgG}	IgM S [g/L]	IgM CSF [mg/L]	Q_{IgM}	IgA S [g/L]	IgA CSF [mg/L]	Q_{IgA}
Multiple Sclerosis (*n* = 34)	43.90 * (33.70–57.40)	187.95 (20.60–487.70)	4.80 (2.77–16.31)	10.72 (6.35–1320.00)	43.19 (3.37–20.47)	5.06 * (2.41–18.60)	1.57 * (0.57–360.00)	1.53 * (0.31–9.43)	1.05 (0.3–4.41)	2.18 (0.81–263.00)	3.47 (0.92–24.20)	1.67 (0.72–15.28)
Control group (*n* = 42)	40.00 (17.8–53.90)	196.45 (16.5–815.00)	5.95 (2.15–20.99)	10.00 (4.95–1150.00)	26.65 (2.14–151.72)	3.12 (1.07–19.50)	1.19 (0.35–249.00)	0.58 (0.11–103.00)	0.52 (0.12–8.95)	2.28 (0.02–434.00)	3.84 (0.88–37.20)	1.59 (0.45–15.44)
***p*–value**	0.010 *	0.368	0.123	0.541	0.071	0.002 *	0.047 *	0.004 *	0.002 *	0.965	0.952	0.673

S, serum; CSF, cerebrospinal fluid; *, significant differences in comparison to the controls.

Table 2. Values of κFLC index, λFLC index, κIgG index, and λIgG index in multiple sclerosis patients and control group and their diagnostic significance.

	Median	Min	Max	Cut-off from the ROC	Sensitivity [%]	Specificity [%]	PPV [%]	NPV [%]	ACC	AUC
				κFLC-index						
MS (*n* = 34)	59.338 *	4.466	623.565	9.417	93.50	68.30	79.20	69.00	79.20	0.866
C (*n* = 42)	6.196	0.912	91.081							
				λFLC-index						
MS (*n* = 34)	35.070 *	6.336	792.533	21.446	71.90	64.30	60.50	75.00	67.60	0.693
C (*n* = 42)	14.450	1.015	157.741							
				κIgG-index						
MS (*n* = 34)	5.660 *	0.751	16.400	1.929	90.30	80.50	77.80	91.70	84.70	0.871
C (*n* = 42)	0.956	0.216	9.581							
				λIgG-index						
MS (*n* = 34)	3.571	0.330	10.374	3.161	65.6	71.4	63.6	73.2	68.9	0.632
C (*n* = 42)	1.974	0.241	35.665							

MS, multiple sclerosis; C, control group; FLC, free light chains; ROC, receiver operating characteristic; PPV, positive predictive value; NPV, negative predictive value; ACC, accuracy; AUC, area under the ROC; *, significant differences in comparison to the control group.

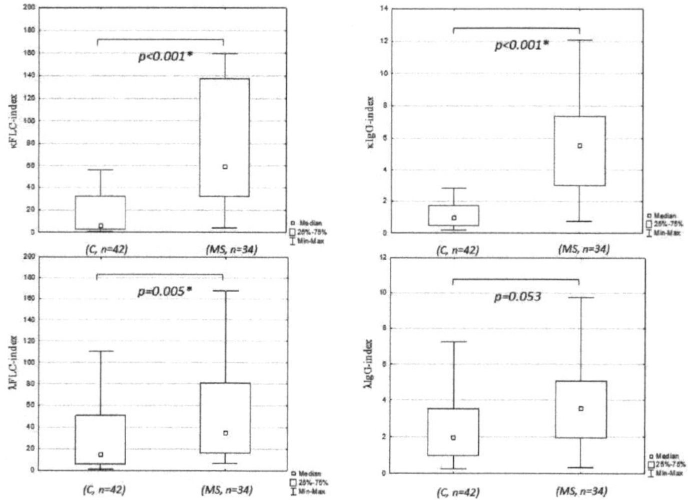

Figure 2. κFLC index, λFLC index, κIgG index, and λIgG index in the study groups. C, control group; MS, multiple sclerosis; *, significant differences in comparison to the controls.

3.3. Correlations of CSF κFLC, CSF λFLC, κFLC Index, λFLC Index, κIgG Index, and λIgG Index with Other Parameters Reflecting Pathological Processes in the CNS

The correlations between CSF κFLC, CSF λFLC, κFLC index, λFLC index, κIgG index, and λIgG index with other parameters reflecting pathological processed in the CNS are presented in Table 3. Spearman's rank correlation test demonstrated that in the total study group, κFLC, λFLC, κFLC-index, λFLC index, κIgG index, and λIgG index correlated with each other. The CSF concentrations of κFLC and the values of the λIgG index were significantly associated with Q_{IgG}. CSF κFLC, CSF λFLC, and κFLC index correlated with Q_{IgM} values, while Q_{IgA} was associated with the values of κIgG index and λIgG index. Furthermore, we observed a negative correlation of Q_{Alb} and patients' age with κFLC index, λFLC index, κIgG index, and λIgG index.

Table 3. Spearman's correlations between tested variables in the total study group.

Total Study Group (n = 76)	Age	Q_{Alb}	Q_{IgG}	Q_{IgM}	Q_{IgA}	CSF κ	CSF λ	κFLC–Index	λFLC–Index	κIgG–Index	λIgG–Index
Q_{Alb}											
r	0.403		0.648	0.268	0.778	−0.120	0.049	−0.253	−0.244	−0.472	−0.433
p	<0.005 *		<0.005 *	0.030 *	<0.005 *	0.316	0.678	0.032 *	0.036 *	<0.005 *	<0.005 *
Q_{IgG}											
r	0.01	0.648		0.553	0.678	0.405	0.208	−0.197	−0.028	0.034	−0.296
p	0.936	<0.005 *		<0.005 *	<0.005 *	<0.005 *	0.076	0.097	0.81	0.776	0.010 *
Q_{IgM}											
r	−0.121	−0.268	0.553		0.547	0.425	0.302	0.333	0.164	0.231	0.015
p	0.331	0.030 *	<0.005 *		<0.005 *	0.005 *	0.013 *	0.007 *	0.185	0.064	0.907
Q_{IgA}											
r	0.253	0.778	0.678	0.547		0.01	0.078	−0.034	−0.101	−0.273	−0.335
p	0.031 *	<0.005 *	<0.005 *	<0.005 *		0.936	0.512	0.779	0.397	0.021 *	0.004 *
CSF κ											
r	−0.309	−0.120	0.405	0.424	0.01		0.661	0.802	0.515	0.843	0.372
p	0.007	0.316	<0.005 *	<0.005 *	0.936		<0.005 *	<0.005 *	<0.005 *	<0.005 *	0.001 *
CSF λ											
r	−0.138	0.049	0.208	0.302	0.078	−0.126		0.536	0.72	0.557	0.686
p	0.236	0.678	0.08	0.013	0.512	<0.005 *		<0.005 *	<0.005 *	<0.005 *	<0.005 *
κFLC index											
r	−0.459	−0.253	0.207	0.333	−0.034	0.802	0.536		0.784	0.866	0.495
p	<0.005 *	0.032	0.081	0.007 *	0.779	<0.005 *	<0.005 *		<0.005 *	<0.005 *	<0.005 *
λFLC index											
r	−0.369	−0.244	−0.030	0.164	−0.101	0.371	0.72	0.784		0.659	0.809
p	<0.005 *	0.040 *	0.81	0.185	0.397	<0.005 *	<0.005 *	<0.005 *		<0.005 *	<0.005 *
κIgG index											
r	−0.472	−0.472	0.034	0.231	−0.273	0.843	0.557	0.867	0.659		0.647
p	<0.005 *	<0.005 *	0.776	0.064	0.020 *	<0.005 *	<0.005 *	<0.005 *	<0.005 *		<0.005 *
λIgG index											
r	−0.388	0.432	−0.296	0.015	−0.335	0.372	0.686	0.495	0.809	0.647	
p	<0.005 *	<0.005 *	−0.010 *	0.907	<0.005 *	0.001 *	<0.005 *	<0.005 *	<0.005 *	<0.005 *	

* Statistically significant ($p < 0.05$).

3.4. Diagnostic Power of κFLC Index, λFLC Index, κIgG Index, and λIgG Index

The diagnostic usefulness of κFLC index, λFLC index, κIgG index, and λIgG index in multiple sclerosis is presented in Table 2. The κFLC index and κIgG index showed a very high ability to detect MS (sensitivity > 90.00% for both) in comparison to the λFLC index (sensitivity, 71.90%) and the λIgG index (sensitivity, 65.60%). The κIgG index showed the highest ability to exclude multiple sclerosis, with 80.50% specificity and 91.70% negative predictive value. The κIgG index presented the highest diagnostic power (AUC) in the detection of multiple sclerosis in comparison to the λIgG index, κFLC index, and λFLC index (Figure 3).

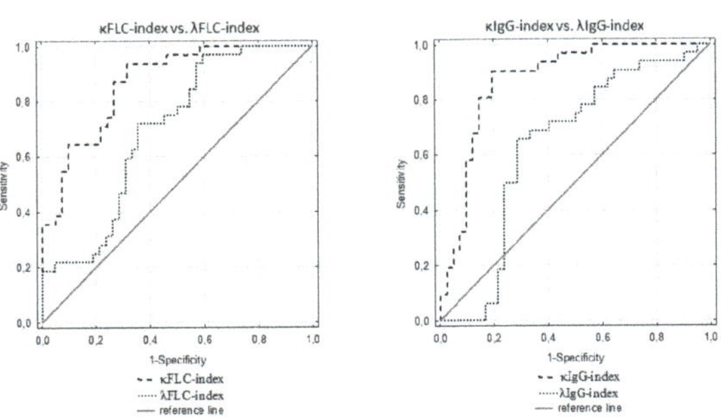

Figure 3. ROC curves for κFLC index, λFLC index, κIgG index, and λIgG index in multiple sclerosis.

4. Discussion

Multiple sclerosis is an inflammatory neurodegenerative disease characterized by intrathecal IgG synthesis. The detection by isoelectric focusing methods of oligoclonal IgG bands in parallel cerebrospinal fluid and serum samples is actually the gold standard for multiple sclerosis diagnosis [3,12]. However, there are some limitations of OCBs detection, such as still indefinite number of bands in the CSF without corresponding bands in serum defining positive results [13]. OCBs determination is not specific for multiple sclerosis, because the elevated intrathecal synthesis of IgG may occur in other CNS disorders [14]. In addition, OCBs are found in the CSF of about 90% of patients with multiple sclerosis, which means that there is always a group of MS patients without CSF bands [15]. Also, another problem is that isoelectric focusing methods are laborious and often difficult [16]. Taking all this into account, we believe that there is a need to find an additional indicator that can be used to diagnose multiple sclerosis. Therefore, in this study, we tried to define a novel diagnostic model using routinely available laboratory test results to predict multiple sclerosis in patients with symptoms of neurological disorders.

Firstly, we showed that the mean concentrations of κFLC and λFLC in the CSF and of serum κFLC are markedly elevated in patients with multiple sclerosis. Clearly, these changes in free light chains concentrations may originate from increased synthesis of immunoglobulins, a phenomenon firstly observed in the 1970s–1980s [17,18], or from the fact that light chains are synthesized at a speed more than twice higher compared to fully formed A, M, and G immunoglobulins [19]. Our results are totally consistent with the results obtained by other researchers [20–25]. Additionally, our study revealed that the CSF concentrations of κFLC were significantly increased in patients with OCB pattern types 2 and 3, which confirmed intrathecal immunoglobulins synthesis. This may suggest that the concentrations of FLCs in the CSF are highly sensitive and specific for the diagnosis of multiple sclerosis. Our findings of increased free light chains are consistent with those of other studies and support the inclusion of free light chains in our algorithm.

Many studies on the prediction of multiple sclerosis have been published in the past few years. Some studies have proposed a model based on κFLCs and albumin concentrations [16,17,22,23]. Presslauer et al. were the first scientists who developed a formula for the κFLC index and tried to evaluate its diagnostic significance. An index using a cut-off value ≥ 5.9 showed higher sensitivity for the diagnosis of multiple sclerosis than OCBs (96% vs. 80%, respectively) [16]. In our study, using the cut-off proposed by Presslauer et al., the κFLC index showed identical sensitivity with that previously reported, but the specificity for our patients' group was lower (46.3% vs. 86.0%). Therefore, for further analysis, we decided to use the best cut-off form the ROC. When we used a κFLC index value ≥ 9.4, we achieved a similar sensitivity, but the specificity was still lower than in Presslauer et al. study and equaled 68%. On the other hand, this index value was lower than the cut-off published by Menendez-Valladares et al., which was >10.62 and associated with higher specificity [21]. Despite the differences in the cut-off values and specificity, authors unanimously say that the κFLC index has high sensitivity and probably would avoid OCBs determination in most of patients with suspected multiple sclerosis.

It is well known that the CSF concentrations of FLCs and IgG are increased in patients with multiple sclerosis. The concentrations of free light chains and IgG have been used for the diagnosis of multiple sclerosis but never combined in a single algorithm. Our study was conducted to develop a new simple model for MS diagnosis using routine laboratory tests to predict this disease in a group of patients with neurological disorders. In our study, these variables were used together for the first time to create the novel κIgG index and λIgG index. We compared the already investigated κFLC index and λFLC index with panels named κIgG-index and λIgG-index combined of FLCs and IgG concentrations. The findings of our study confirmed significant differences in the values of κIgG index and λIgG index between multiple sclerosis patients and individuals with other neurological disorders. We denoted about a 9,5-fold difference of median κIgG index and a 2,4-fold difference of median λIgG index in multiple sclerosis patients in comparison to controls. Moreover, it is important to recognize that our model was developed considering different types of OCBs. Differentiation according to OCBs was chosen because clinically, patients with OCB pattern type 2 are almost always classified as multiple sclerosis patients. We observed that the κIgG index was significantly higher in patients with pattern type 2 in comparison with those with pattern type 1 and type 4. Additionally, only the values of the κIgG index were markedly higher in patients with pattern type 3 than in those with pattern type 4 and type 1, which does not exclude multiple sclerosis. In general, the κIgG index showed higher diagnostic significance compared with the λIgG index. The main factor causing this is probably the dominance of κ free light chains in humans (the normal total κFLC/λFLC ratio is approximately 2:1) [26]. These results indicate that the algorithm combining κFLC with IgG is more valid to evaluate the intrathecal synthesis of immunoglobulins in patients with neurological system disorders than other known algorithms.

5. Conclusions

In conclusion, we showed that a novel, simple κIgG index consisting of four variables combined together (serum κFLC, CSF κFLC, serum IgG, and CSF IgG) can predict the intrathecal synthesis of immunoglobulins and may serve as an additional, potential diagnostic marker for the diagnosis of multiple sclerosis, with a high degree of diagnostic sensitivity and accuracy. The main strength of our study is the use of readily available routine laboratory diagnostics tests. In addition, we examined a group of well-characterized patients including 45% multiple sclerosis patients and 55% controls. The control group in this study was highly heterogeneous; however, the purpose of this study was to determine the value of the κIgG index in the differentiation of multiple sclerosis from other neurological disorders. It is very important to differentiate multiple sclerosis from other neurological diseases, because they often require different treatments. While this study provides novel information about the diagnostic significance of four combined markers in the κIgG-index, in the context of practicality, further studies are required to determine the appropriateness of using the κIgG index as a diagnostic tool for multiple sclerosis in a clinical setting. Studies on larger samples should be performed to validate the quality and precision of the κIgG index. To our knowledge, there are no other studies combining FLCs with IgG concentrations, but we cautiously suggest that, in the future, this parameter could be determined as a complementary diagnostic element to oligoclonal bands determination.

Author Contributions: M.G.-S. and B.M. produced the idea of the study. M.G.-S. and B.M. contributed to research design and measurement of the tested proteins. J.T., A.K., and J.K. were involved in sample collection. All authors analyzed the data. M.G.-S. and B.M. coordinated project funding. All authors have read and agreed to the published version of the manuscript.

Funding: This research was funded by the Medical University of Białystok, Poland, and received external funding from The Binding Site and Biokom (ZEW/1/BD/18/001/1198).

Acknowledgments: M.G.-S. has received consultation honorarium from Roche. B.M. has received consultation and/or lecture honoraria from Abbott, Wiener, Roche, Cormay, and Biameditek.

Conflicts of Interest: The authors declare no conflict of interest.

References

1. Amato, M.P.; Prestipino, E.; Bellinvia, A.; Niccolai, C.; Razzolini, L.; Pastò, L.; Fratangelo, R.; Tudisco, L.; Fonderico, M.; Goretti, B.; et al. Cognitive impairment in multiple sclerosis: An exploratory analysis of environmental and lifestyle risk factors. *PLoS ONE* **2019**, *14*, e0222929. [CrossRef] [PubMed]

2. Lutton, J.D.; Winston, R.; Rodman, T.C. Multiple sclerosis: Etiological mechanisms and future directions. *Exp. Biol. Med.* **2004**, *229*, 12–20. [CrossRef] [PubMed]

3. Wootla, B.; Eriguchi, M.; Rodriguez, M. Is multiple sclerosis an autoimmune disease? *Autoimmune Dis.* **2012**, *2012*, 969657. [CrossRef] [PubMed]

4. Thompson, A.J.; Banwell, B.L.; Barkhof, F. Diagnosis of multiple sclerosis: 2017 revisions of the McDonald criteria. *Lancet Neurol.* **2018**, *17*, 162–173. [CrossRef]

5. Link, H.; Huang, Y.M. Oligoclonal bands in multiple sclerosis cerebrospinal fluid: An update on methodology and clinical usefulness. *J. Neuroimmunol.* **2006**, *180*, 17–28. [CrossRef]

6. Giles, P.D.; Wroe, S.J. Cerebrospinal fluid oligoclonal IgM in multiple sclerosis: Analytical problems and clinical limitations. *Ann. Clin. Biochem.* **1990**, *27*, 199–207. [CrossRef]

7. Puthenparampil, M.; Altinier, S.; Stropparo, E.; Zywicki, S.; Poggiali, D.; Cazzola, C.; Toffanin, E.; Ruggero, S.; Gallo, P.; Plebani, M.; et al. Intrathecal K free light chain synthesis in multiple sclerosis at clinical onset associates with local IgG production and improves the diagnostic value of cerebrospinal fluid examination. *Mult. Scler. Relat. Disord.* **2018**, *25*, 241–245. [CrossRef]

8. McLaughlin, K.A.; Wucherpfennig, K.W. B cells and autoantibodies in the pathogenesis of multiple sclerosis and related inflammatory demyelinating diseases. *Adv. Immunol.* **2008**, *98*, 121–149. [CrossRef]

9. Wang, X.; Xu, K.; Chen, S.; Li, Y.; Li, M. Role of Interleukin-37 in Inflammatory and Autoimmune Diseases. *Iran J. Immunol.* **2018**, *15*, 165–174. [CrossRef]

10. Janeway, C. Chapter 3. Antigen Recognition by B-cell and T-cell Receptors. In *Immunobiology: The Immune System in Health and Disease*; Garland Science: New York, NY, USA, 2001.

11. Kurtzke, J.F. Rating neurologic impairment in multiple sclerosis: An expanded disability status scale (EDSS). *Neurology* **1983**, *33*, 1444–1452. [CrossRef]

12. Trbojevic-Cepe, M. Detection of Oligoclonal Ig Bands: Clinical Significance and Trends in Methodological Improvement. *EJIFCC* **2004**, *15*, 86–94. [PubMed]

13. Hegen, H.; Zinganell, A.; Auer, M.; Deisenhammer, F. The clinical significance of single or double bands in cerebrospinal fluid isoelectric focusing. A retrospective study and systematic review. *PLoS ONE* **2019**, *14*, e0215410. [CrossRef] [PubMed]

14. Deisenhammer, F.; Zetterberg, H.; Fitzner, B.; Zettl, U.K. The Cerebrospinal Fluid in Multiple Sclerosis. *Front. Immunol.* **2019**, *10*, 726. [CrossRef] [PubMed]

15. Imrell, K.; Landtblom, A.M.; Hillert, J.; Masterman, T. Multiple sclerosis with and without CSF bands: Clinically indistinguishable but immunogenetically distinct. *Neurology* **2006**, *67*, 1062–1064. [CrossRef] [PubMed]

16. Cornell, F.N. Isoelectric focusing, blotting and probing methods for detection and identification of monoclonal proteins. *Clin. Biochem. Rev.* **2009**, *30*, 123–130.

17. Bracco, F.; Gallo, P.; Menna, R.; Battistin, L.; Tavolato, B. Free light chains in the CSF in multiple sclerosis. *J. Neurol.* **1987**, *234*, 303–307. [CrossRef]

18. Vandvik, B. Oligoclonal IgG and free light chains in the cerebrospinal fluid of patients with multiple sclerosis and infectious diseases of the central nervous system. *Scand. J. Immunol.* **1977**, *6*, 913–922. [CrossRef]

19. Zemana, D.; Hradilek, P.; Kusnierova, P. Oligoclonal free light chains in cerebrospinal fluid as markers of intrathecal inflammation. Comparison with oligoclonal IgG. *Biomed. Pap. Med. Fac. Univ. Palacky Olomouc Czech Repub.* **2015**, *15*, 104–113. [CrossRef]

20. Presslauer, S.; Milosavljevic, D.; Brücke, T.; Bayer, P.; Hübl, W. Elevated levels of kappa free light chains in CSF support the diagnosis of multiple sclerosis. *J. Neurol.* **2008**, *255*, 1508–1514. [CrossRef]

21. Menéndez-Valladares, P.; García-Sánchez, M.I.; Cuadri Benítez, P.; Lucas, M.; Adorna Martínez, M.; Carranco Galán, V.; García De Veas Silva, J.L.; Bermudo Guitarte, C.; Izquierdo Ayuso, G. Free kappa light chains in cerebrospinal fluid as a biomarker to assess risk conversion to multiple sclerosis. *Mult. Scler. J. Exp. Transl. Clin.* **2015**, *1*. [CrossRef]

22. Duranti, F.; Pieri, M.; Centonze, D.; Buttari, F.; Bernardini, S.; Dessi, M. Determination of kappa FLC and kappa index in cerebrospinal fluid: A valid alternative to assess intrathecal immunoglobulin synthesis. *J. Neuroimmunol.* **2013**, *263*, 116–120. [CrossRef] [PubMed]

23. Presslauer, S.; Milosavljevic, D.; Huebl, W.; Parigger, S.; Schneider-Koch, G.; Bruecke, T. Kappa free light chains: Diagnostic and prognostic relevance in MS and CIS. *PLoS ONE* **2014**, *9*, e89945. [CrossRef] [PubMed]

24. Voortman, M.M.; Stojakovic, T.; Pirpamer, L.; Jehna, M.; Langkammer, C.; Scharnagl, H.; Reindl, M.; Ropele, S.; Enzinger, C.; Fuchs, S.; et al. Prognostic value of free light chains lambda and kappa in early multiple sclerosis. *Mult. Scler.* **2017**, *23*, 1496–1505. [CrossRef] [PubMed]

25. Sáez, M.S.; Rojas, J.I.; Lorenzón, M.V.; Sánchez, F.; Patrucco, L.; Míguez, J.; Azcona, C.; Sorroche, P.; Cristiano, E. Validation of CSF free light chain in diagnosis and prognosis of multiple sclerosis and clinically isolated syndrome: Prospective cohort study in Buenos Aires. *J. Neurol.* **2019**, *266*, 112–118. [CrossRef]

26. Katzmann, J.A.; Clark, R.J.; Abraham, R.S.; Bryant, S.; Lymp, J.F.; Bradwell, A.R.; Kyle, R.A. Serum reference intervals and diagnostic ranges for free kappa and free lambda immunoglobulin light chains: Relative sensitivity for detection of monoclonal light chains. *Clin. Chem.* **2002**, *48*, 1437–1444. [CrossRef]

Review

Recent Advances in Antigen-Specific Immunotherapies for the Treatment of Multiple Sclerosis

Olga Kammona [1] and Costas Kiparissides [1,2,*]

[1]   Chemical Process and Energy Resources Institute, Centre for Research and Technology Hellas, P.O. Box 60361, 57001 Thessaloniki, Greece; kammona@cperi.certh.gr
[2]   Department of Chemical Engineering, Aristotle University of Thessaloniki, 54124 Thessaloniki, Greece
*   Correspondence: costas.lpre@cperi.certh.gr

Received: 5 May 2020; Accepted: 26 May 2020; Published: 29 May 2020

Abstract: Multiple sclerosis (MS) is an autoimmune disease of the central nervous system and is considered to be the leading non-traumatic cause of neurological disability in young adults. Current treatments for MS comprise long-term immunosuppressant drugs and disease-modifying therapies (DMTs) designed to alter its progress with the enhanced risk of severe side effects. The Holy Grail for the treatment of MS is to specifically suppress the disease while at the same time allow the immune system to be functionally active against infectious diseases and malignancy. This could be achieved via the development of immunotherapies designed to specifically suppress immune responses to self-antigens (e.g., myelin antigens). The present study attempts to highlight the various antigen-specific immunotherapies developed so far for the treatment of multiple sclerosis (e.g., vaccination with myelin-derived peptides/proteins, plasmid DNA encoding myelin epitopes, tolerogenic dendritic cells pulsed with encephalitogenic epitopes of myelin proteins, attenuated autologous T cells specific for myelin antigens, T cell receptor peptides, carriers loaded/conjugated with myelin immunodominant peptides, etc.), focusing on the outcome of their recent preclinical and clinical evaluation, and to shed light on the mechanisms involved in the immunopathogenesis and treatment of multiple sclerosis.

Keywords: multiple sclerosis; autoimmune diseases; antigen-specific immunotherapies; tolerogenic vaccines; tolerance induction; central nervous system; myelin peptides; myelin basic protei; proteolipid protein; myelin oligodendrocyte glycoprotein

---

1. Introduction

Multiple sclerosis (MS) is a chronic inflammatory disease of the central nervous system (CNS) caused by genetically-predisposed hosts by infectious and environmental factors which induce complex autoimmune responses in the CNS resulting in degeneration of the myelin sheath and axonal loss in the brain and spinal cord [1–14] It is the most prominent demyelinating disease leading to progressive clinical disability in MS patients [5,6,15] due to ineffective remyelination [13,15]. More than 2 million people worldwide suffer from MS and it is considered as the leading non-traumatic cause of neurological disability in young adults with a disease onset commonly around 20 and 40 years of age [4,6,15,16]. High prevalence of the disease is reported in North America and Europe [15].

MS exhibits a vastly heterogeneous clinical course [6,17] which varies from a benign disease course that doesn't lead to serious disability, demonstrated by 10–15% of MS patients, to aggressive forms of the disease leading to severe disability and even paralysis. The increased heterogeneity of the disease severity strongly affects the design and duration of therapeutic schemes administered to MS patients [17].

MS features the following stages: a pre-clinical stage, namely, a radiologically-isolated syndrome (RIS), which is then demonstrated as a clinically-isolated syndrome (CIS) [2,3], followed by a relapsing remitting stage (RRMS) which may later advance into secondary progressive disease (SPMS) [2,4,6,16,18]. It should be noted that a minority of MS patients (e.g., 10–15% [3,6,16]) exhibit progressive MS from the disease onset, known as primary progressive MS (PPMS) [2,4,6,18] (Figure 1). The aforementioned classification corresponds to the inflammatory image of MS which can be detected via magnetic resonance imaging (MRI) [2,16].

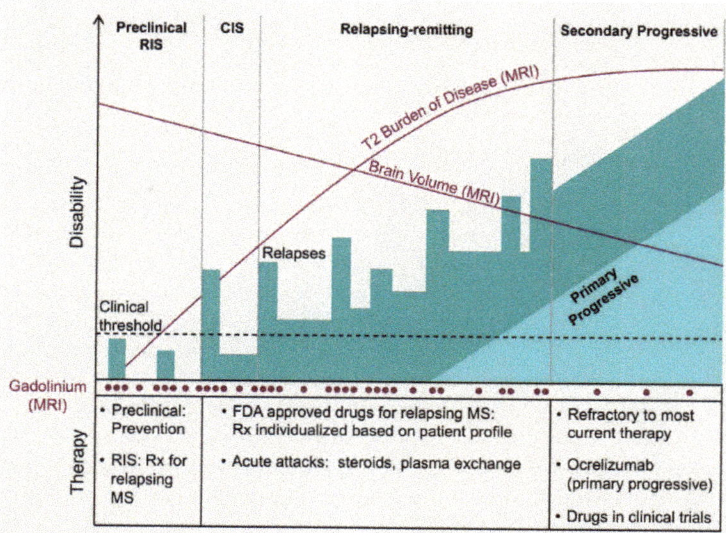

Figure 1. Stages of multiple sclerosis (MS). RIS: radiologically isolated syndrome; CIS: clinically isolated syndrome; FDA: U.S. food and drug administration (with the permission of [2]).

RRMS affects approximately 85% of MS patients [3,6,19] of whom women are twice as many as men [6]. It is characterized by periods of relapses (i.e., episodes of neurologic dysfunction, such as sensory disturbances, optic neuritis, or disturbances of motor/cerebellar function) followed by remission periods (i.e., periods of partial or full clinical recovery) [2,3,6,14,16]. Relapses coincide with CNS inflammation/demyelination visualized by MRI as lesions found mainly in the white matter [3]. In the majority of patients, RRMS advances to SPMS [16] within 10–20 years after diagnosis [3,6].

RRMS involves the movement of immune cells from the peripheral sites to the CNS (mainly in the white matter, even though extensive number of demyelinated plaques can be located in the grey matter [20]) resulting in the formation of localized inflammatory sites. Inflammatory processes in these sites induce killing of oligodendrocytes, myelin damage, and axon injury and loss, resulting in impaired neurological function [20]. On the other hand, the progressive disease implicates the generation of a pathological process within the brain [2]. Thus, the characteristic feature of SPMS is no longer the inflammatory lesions but an atrophic brain attributed to enhanced loss of axons, cortical demyelination, activation of microglia, and inefficient remyelination [2,3]. SPMS patients demonstrate progressive neurological dysfunction resulting in enhanced physical disability (e.g., inability to walk) [2,3].

PPMS is also characterized by gradual neurological decline without relapses [3,6]. In comparison with RRMS, the disease onset for PPMS is usually ten years later and it does not exhibit female predominance [6]. To date, clinical evidence shows significant differences between RRMS and progressive MS [21], reflected by the diverse response to currently existing treatments, but not between SPMS and PPMS. [18].

Currently, there is no cure for MS. Some existing treatments appear to be beneficial for patients with RRMS. However, there is still a lack of effective therapies for the progressive forms of MS [2].

The present paper aims to extensively review the different, recently developed myelin antigen-specific strategies (e.g., myelin peptide based vaccination, vaccination with plasmid DNA encoding myelin epitopes, tolerogenic dendritic cells pulsed with encephalitogenic epitopes of myelin proteins, vaccination with attenuated autologous T cells specific for myelin antigens, T cell receptor vaccination, carrier-aided administration of myelin immunodominant peptides, etc.) for the prevention/treatment of MS, especially with respect to their in vivo and clinical evaluation outcomes and the challenges they face in order to be translated to MS patients. It also seeks to unravel the mechanisms involved in the immunopathogenesis of the relapsing remitting and progressive MS, as well as the mechanisms of action of the developed tolerance-inducing vaccines.

The different antigen-specific immunotherapies are analytically presented in a comparative manner in separate tables providing detailed information about the selected myelin antigen, the vaccination strategy (e.g., prophylactic, preclinical, therapeutic), the administration route (e.g., intravenous, subcutaneous, intraperitoneal, epicutaneous, intradermal, oral, nasal, pulmonary) and the administered dose, the cell type (e.g., tolerogenic dendritic cells, T cells, hematopoietic stem cells, bone marrow cells) and the inductive agent, the carrier type (e.g., polymer particles, soluble antigen arrays, immune polyelectrolyte multilayers, inorganic particles, pMHC-NPs, mannan-conjugated myelin peptides, liposomes, exosomes, antigen-presenting yeast cells), and its characteristics (e.g., size, zeta potential, antigen loading), as well as the vaccination outcome.

The review paper is based on a systematic search of PubMed using the following search terms: multiple sclerosis, antigen-specific immunotherapies, tolerogenic vaccines, nanocarriers, nanomedicine, DNA vaccination, cell-based vaccination, clinical trials. The search covered the time period from 1 January 2000 till today. Publications addressing pre-clinical and clinical evaluation of antigen-specific immunotherapies for multiple sclerosis were selected for inclusion.

2. Immunopathogenesis of MS

Successful preclinical studies and clinical trials for MS which target cells and molecules of the immune system support the idea that the latter has a dominant role in the pathogenesis of MS. These studies have proposed that cells of the adaptive immune system like B cells and various effector T cells, combined with cells of the innate immune system such as natural killer cells and microglia, uniquely contribute to the disease [2]. However, it should be mentioned that while the peripheral adaptive immune system (T lymphocytes) is the primary driver of RRMS, the innate immune system (microglia and astrocytes) together with B lymphocytes is considered to drive progressive MS [2]. The CNS of MS patients has been also found to exhibit infiltration of activated T cells, B cells, plasma cells, dendritic cells (DCs), and macrophages indicating the contribution of both cellular and humoral (i.e., antibody-mediated) immune responses as well as of various immunopathological effector mechanisms to the damage of CNS tissue [22,23].

It has been suggested that two independent types of inflammation, developing in parallel, can occur in multiple sclerosis patients. The first one is related with the focal invasion of T and B cells through BBB leakage, giving rise to classic active demyelinated plaques in the white matter. The second one deals with a slow accumulation of T and B lymphocytes without profound BBB damage in the perivascular Virchow Robin spaces and the meninges, where they form cellular aggregates resembling, in most severe cases, tertiary lymph follicles. The latter can be linked with the development of demyelinated lesions in the cerebral and cerebellar cortex, slow expansion of existing lesions in the white matter, and diffuse neurodegeneration in normal-appearing white and/or grey matter [18]. The presence of the lymphoid follicle-like structures (follicle-like ectopic germinal centers) in the inflamed cerebral meninges of some SPMS patients could indicate that B-cell maturation is sustained locally in the CNS and contributes to the induction of a compartmentalized humoral immune response [2,22].

The role of the various immune cells and the immunopathological effector mechanisms contributing to the development of MS are discussed below.

The ability of the human immune system to respond to an enormous number of encountered antigens comes with the risk that some T cells will be able to recognize self-antigens, such as CNS (e.g., myelin) antigens. Most autoreactive T lymphocytes are usually deleted in the thymus via a process known as negative selection (central tolerance). However, a number of these T cells escape from the thymus to peripheral sites where they are normally kept under control by mechanisms of peripheral tolerance. If these mechanisms fail, due to reduced action of regulatory T cells and/or enhanced resistance of effector T and B lymphocytes to suppression, autoreactive T cells recognizing CNS antigens are activated in the peripheral lymphoid system to become effector cells, via molecular mimicry (i.e., activation by a viral peptide having sufficient sequence similarity [24] or otherwise sharing an immunologic epitope [25] with the CNS antigen), recognition of CNS proteins released in the periphery, presentation of new autoantigens and bystander activation (i.e., T cell receptor (TCR)-independent and cytokine-dependent activation probably due to viral infection [26]). Then the activated T cells (CD8+ T cells, and CD4+ T cells differentiate to T helper 1 (Th1) and Th17 cells) together with B cells and monocytes (cells of the innate immune system) infiltrate the CNS by crossing the blood-brain barrier (BBB) leading to inflammation. There, they are reactivated via encountered resident antigen presenting cells, APCs (e.g., microglial cells) and infiltrating APCs (e.g., dendritic cells, macrophages) presenting CNS autoantigens on the major histocompatibility complex, MHC (also known as human leucocyte antigen, HLA, in humans [11]) molecules. Specifically, CD4+ T cells interact with MHC II expressing cells, like dendritic cells, macrophages and B cells, whereas CD8+ T cells directly interact with MHC I/antigen-expressing cells, like neurons and oligodendrocytes. It should be noted that MHC class II is adequately expressed only on professional APCs, while MHC class I is expressed by all cell types in the CNS inflammatory milieu. Therefore, CD4+ T cells are mainly found in perivascular cuffs, and meninges, whereas CD8+ T cells additionally infiltrate the parenchyma of the irritated lesions. Upon contact with their cognate antigen, CD4+ T cells are thought to secrete cytokines and immune mediators resulting in the attraction of resident immune cells like microglia, macrophages and astrocytes, secretion of proinflammatory cytokines, enhanced APC function, and increased production of reactive oxygen and nitrogen species (ROS/RNS). On the other hand, apart from secreting inflammatory mediators, CD8+ T cells directly attack oligodendrocytes and neurons, thus causing oligodendrocyte death (e.g., via secretion of granzymes and perforin leading to pore formation and stimulation of programmed cell death [2]) and neuronal damage (e.g., release of cytolytic granules leading to axonal dissection [2]) (Figure 2). The above result in inflammation, myelin loss, and axonal injury. This inflammatory cascade leads to the recruitment of monocytes and macrophages into the lesion resulting in the release of more CNS antigens and their presentation to potentially autoreactive T cells. It should be mentioned that epitope spreading could result in a broader autoimmune response involving additional autoantigens [1–3,11,27–33].

CD4+ T cells are considered to have a paramount role in the immunopathogenesis of MS due to the secretion of interferon gamma (IFNγ) and IL-17 [2,20,34]. However, it has been lately revealed that CD8+ T cells are also responsible for the initiation of human MS pathogenesis where, contrary to experimental autoimune encephalomyelitis (EAE), CD8+ T cells are the predominant T lymphocyte infiltrate in acute and chronic MS lesions [1,2]. Compared with CD4+ T cells, CD8+ T cells can be found more frequently in the white matter and in the cortical demyelinating lesions in the grey matter, and their density can be closely correlated with axonal damage [1,3]. Epitope spreading, assisted by cross-presentation of antigens by monocyte-derived DCs, has been found to activate myelin-specific CD8+ T cells also in an EAE model [3]. It has been suggested that CD8+ T cells remain in the CNS (e.g., brain and spinal cord) as tissue-resident cells, and upon re-encounter of their cognate antigen, focally propagate neuroinflammation [18].

Despite the fact that MS is considered a T lymphocyte-mediated disease [35], the important results of anti-CD20 therapy (e.g., rituximab, ocrelizumab) in MS indicate a significant role for B cells in its pathogenesis. B cells can have either a pro- or an anti-inflammatory role, based on their subtype and context. Their pro-inflammatory functions, comprise critical antigen presentation in the context of MHC class II molecules to Th17 and Th1 cells, secretion of pro-inflammatory cytokines (e.g., tumor necrosis factor alpha, TNF-α, interleukin-6 (IL-6) and granulocyte-macrophage colony-stimulating factor, GM-CSF) that promote CNS inflammation and propagate demyelination and neurodegeneration, and production of antibodies [36]. B lymphocytes can traffic out of the CNS to the cervical lymph nodes where they can undergo affinity maturation and then re-enter the CNS and promote further damage [3].

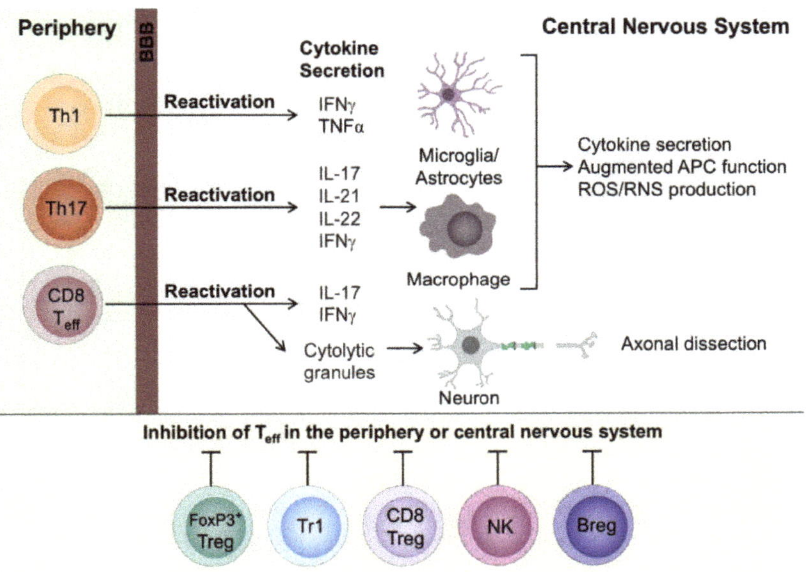

Figure 2. Effector T cells in multiple sclerosis (with the permission of [2]).

B cells are considered a unique population of APCs since, in contrast to other APCs which recognize various exogenous and endogenous antigens, B cells are highly selective (i.e., they specifically recognize only the antigens that are bound to their unique surface B cell receptor). Studies with the EAE model have indicated that some autoantigens, like the highly immunogenic myelin oligodendrocyte glycoprotein (MOG), require their presentation by B cells to activate CD4+ T cells. Accordingly, it can be speculated that the antigen(s) which trigger human MS are likewise B cell dependent [36]. Furthermore, active genes in B cells represent a major component of more than 200 variants known to increase the risk for developing MS. Remarkably, the gene that encodes the MHC class II DR β chain, which is known to be critical for APC function, is considered, genome-wide, the strongest MS predisposition signal. Probably, the net effect of this genetic burden is biased biology of B cells towards a pro-inflammatory phenotype, which promotes the presentation of self-antigens to effector T cells or augments the autoimmune responses through the production of cytokines and other immune mediators [36].

Regulatory T cells (CD4 FoxP3+ Tregs, CD4+ Tr1 regulatory cells, CD8 Tregs), regulatory B cells (Breg) cells and natural killer cells (NK cells) can achieve regulation of effector T cells in the peripheral lymphoid tissue or in the CNS. CD4 FoxP3+ Tregs (<4% of circulating CD4 T cells) express the transcription factor Forkhead box protein 3 (FoxP3) along with numerous inhibitory checkpoint molecules on their surface. They are activated by self-antigens and they suppress the activation of other cell types through a mechanism that requires cell contact [37]. CD4+ Tr1 regulatory cells impede cell proliferation mainly via the secretion of IL-10 [38]. Both Tregs are considered important in MS due to the exhibition of unique characteristics. Subsets of CD8+ Tregs that have been indicated to suppress immune responses and disease progression via distinct mechanisms have been identified by a unique expression of molecules like CD122, CD28, CD102 and HLA-G [2,39,40]. In addition, Th2 cells secreting cytokines like IL-4, IL-5, and IL-13, are considered to be able to downregulate the activity of pro-inflammatory cells [27]. B cells can also regulate various B and T cell mediated effector immune functions via secretion of regulatory cytokines IL-10 and IL-35, transforming growth factor beta (TGF-β), or programmed death-ligand 1 (PD-L1). Specifically, IL-10 secreting B-regs inhibit pro-inflammatory T cell responses, partly mediated via IFNγ and IL17 [2,3,36]. Finally, NK cells are known to suppress immune responses via killing activated, possibly pathogenic, CD4+ T cells.

Immune-modulatory networks are triggered in parallel with the deleterious activity of effector T cells, in order to limit CNS inflammation and initiate tissue repair, resulting in partial remyelination. The modulation of immune activation can be associated with clinical remission. However, it should be mentioned that in the absence of treatment, suppression of autoimmunity cannot be fully achieved. Consequently, additional attacks will normally lead to the progressive form of MS [2]. The action of autoreactive T and B cells in MS could be owed to the defective function of regulatory cells. Disease-associated HLA class II variants might skew the selection in the thymus so that the regulatory T cells which are released into the peripheral sites cannot adequately suppress autoreactive effector T cells [3].

3. MS Therapies

3.1. Disease-Modifying Therapies

Current treatments for MS can be categorized into long-term immunosuppressant drugs, which have significant risks for various infections and cancer, and disease-modifying therapies (DMTs) designed to alter the progress of the disease via interference with B and T cells activity, and reduction of BBB disruption. For example, the more recently engineered monoclonal antibodies (mAbs) act via blocking α4 integrin interactions (e.g., natalizumab) or lysing immune cells exhibiting surface markers like CD20 (ocrelizumab, ofatumumab) [41] or CD52 (alemtuzumab). Due to their different mechanisms of action (Figure 3), DMTs' efficacy and safety profiles [42] vary significantly. Presently, there exist more than 10 FDA (U.S. Food and Drug Administration) approved DMTs for RRMS aiming to reduce relapse level and severity of inflammation in CNS. DMTs can be classified based on the administration route as intravenous, self-injectable and oral formulations (Table 1) [16,23,31,43–49].

Among the FDA-approved DMTs, ocrelizumab, alemtuzumab and natalizumab seem to have the highest anti-inflammatory effect and to efficiently reduce relapses as proven by MRI scans [2,50]. Another approach for the treatment of MS involves the use of low-dose interleukin 2 (IL-2). This treatment is based on the weak in vivo response of effector T cells to low-dose IL-2 compared with Foxp3+ Treg cells which proliferate due to the expression of the high-affinity IL-2 receptor (CD25). This treatment has been shown to be well tolerated but, since non-specific expansion of the Foxp3+ Treg population cannot be excluded, it may effect susceptibility to infections and malignancies in some patients [51]. Interestingly, it has been shown that the more aggressive and less selective targeting of immune cells leads to more effective disease suppression, though at the cost of enhanced risk of side effects like infections and neoplasms due to decreased normal immune surveillance [27].

Despite the noteworthy advancements in the treatment of MS, the observed rates of progressive disability as well as of early mortality are still bothersome. Accordingly, there exists a need for safer, well tolerated and highly efficient treatments. This need is even higher for therapies capable of stopping or slowing the progression, and improving the disability in progressive MS [14,16,52–54]. Till now, only one therapy (ocrelizumab) appeared to be beneficial for the treatment of PPMS [14,16].

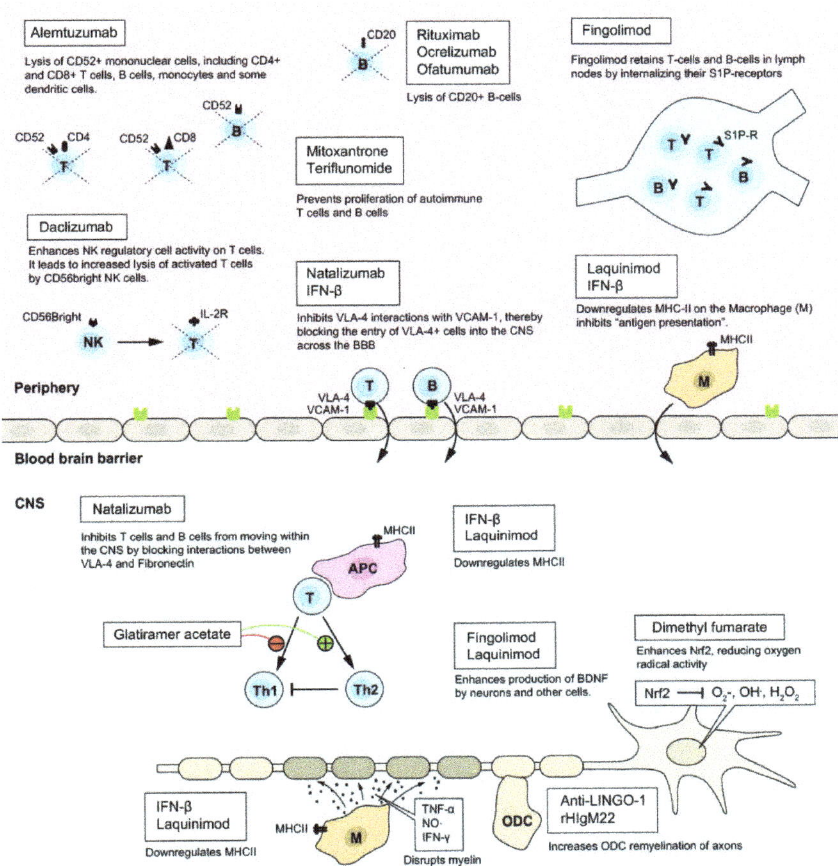

Figure 3. Suggested mechanism of action of several disease-modifying therapies (DMTs) (with the permission of [47]).

Table 1. Disease-modifying-therapies for RRMS (based on [16,23,43,45]).

Therapeutic Molecule	Commercial Name	Year of Approval	Admin. Route	Admin. Frequency	Mode of Action	Side Effects
IFN-β1a	Avonex® Rebif®	1993	i.m. s.c.	Once a week Three times a week	Decrease of proinflammatory and increase of anti-inflammatory cytokines; decreased migration of inflammatory cells across the BBB; decrease of Th17 cells; modulation of T and B cells.	Symptoms similar to those of flu; leukopenia; liver damage.
pegIFN-β1a	Plegridy®		s.c.	Once per two weeks	Decrease of proinflammatory and increase of anti-inflammatory cytokines; decreased migration of inflammatory cells across the BBB; decrease of Th17 cells; modulation of T and B cells	Symptoms similar to those of flu; leukopenia; liver damage.
IFN-β1b	Betaseron® Extavia®	1993	s.c.	Once per two days	Decrease of proinflammatory and increase of anti-inflammatory cytokines; decreased migration of inflammatory cells across the BBB; decrease of Th17 cells; modulation of T and B cells; down regulation of MHC expression on APCs.	Symptoms similar to those of flu; leukopenia; liver damage.
Glatiramer acetate	Copaxone®	1996	s.c.	-	Decrease of proinflammatory and increase of anti-inflammatory cytokines; decrease of Th17 cells; increase of Th2 cells and Tregs; blocking of pMHC.	Erythema; induration; heart palpitations; dyspnea; tightness of chest; flushes/anxiety.
Dimethyl fumarate	Tecfidera®	2013	oral	Twice or three times per day	Anti-inflammatory-Increase of Th2 cells; anti-oxidative stress; neuroprotection through activation of Nrf-2 pathway.	Flushes; vomit; diarrhea; nausea; decrease of WBC.
Teriflunomide	Aubagio®	2012	oral	Once per day	Inhibition of dihydroorotate dehydrogenase; inhibition of T and B cells;	Lymphopenia; nausea; hypertension; fatigue; headache; diarrhea; peripheral neuropathy; acute renal failure; alopecia.
Fingolimod	Glenya®	2010	oral	Once per day	S1P receptor modulator; preventing the circulation of lymphocytes in non-lymphoid tissues including the CNS.	Weakening of heart rate; hypertension; macular edema; increased liver enzymes; decreased lymphocyte levels.
Siponimod [55] Ozanimod [56]	Mayzent® Zeposia®	2019 2020 USA	oral oral		Binding to S1P-1 and S1P-5 S1P receptor agonist	
Laquinimod			Oral		Immunomodulation of T cells, DCs and monocytes; neuroprotection of astrocytes; decrease of proinflammatory and increase of anti-inflammatory cytokines; reduced infiltration of cells into the CNS.	No severe cardiac adverse effects were detected during Phase III clinical trials.
Cladribine [57]	Mavenclad®	2017 EU 2019 USA			Reduction of circulating T and B cells.	Risk of cancer

Table 1. *Cont.*

Therapeutic Molecule	Commercial Name	Year of Approval	Admin. Route	Admin. Frequency	Mode of Action	Side Effects
Mitoxantrone	Novatrone®	2000 USA	i.v.	Once per three months	Cytotoxic for B and T cells; reduction of Th1 cytokines; inhibition of type II topoisomerase.	Cardiotoxicity; leukemia
Methylprednisolone			i.v.	-	Immunosuppression; anti-inflammatory effects.	Risk of infections; retention of sodium; glucose intolerance; mood disturbances.
Dalfampridine	Ampyra®		oral	Twice per day	Blocking of potassium channel; improvement of motor symptoms.	
Natalizumab	Tysabr®	2004	i.v.	Once per 28 days	Targeting α4-integrin	Progressive multifocal leukoencephalopathy.
Ofatumumab	Arzerra®		i.v.	Once per two weeks	Targeting CD20	
Ocrelizumab	Ocrevus®		i.v.	Once per six months	Targeting CD20	
Alemtuzumab	Lemtrada®	2013 EU	i.v.	Once a year	Targeting CD52	High risk of infections Graves' disease
Daclizumab	Zinbryta®		s.c.	Once per month	Targeting CD25	
Rituximab	Rituxan®		i.v.	-	Targeting CD20	Chills; nausea; hypotension
Obinutuzumab	Gazyva®		i.v.	-	Direct cell death	Risk of infections; nausea; thrombocytopenia; neutropenia

IFN: interferon; i.m.: intramuscular; s.c.: subcutaneous; BBB: blood-brain barrier; MHC: major histocompatibility complex; APCs: antigen presenting cells; Nrf-2: nuclear factor erythroid-2; WBC: white blood cell; CNS: central nervous system; i.v.: intravenous.

3.2. Antigen-Specific Immunotherapies

The Holy Grail for the treatment of MS is to specifically suppress the disease while at the same time allow the immune system to be functionally active against infectious diseases and malignancy. This could be achieved via the development of immunotherapies designed to specifically suppress immune responses to self-antigens [43,51,58–60]. Even though the detailed mechanisms of MS induction have not been fully clarified, a dominant hypothesis is that the loss of immune tolerance to myelin proteins like myelin basic protein (MBP), proteolipid protein (PLP) and myelin oligodendrocyte glycoprotein (MOG) leads to the recruitment of myelin-specific CD4+ T cells, resulting in myelin damage [14,61].

Antigen-specific immunotherapies are based on the introduction of self-antigens to APCs in the absence or presence of very low levels of costimulatory molecules (i) acting directly via TCR on effector T cells resulting in immunological anergy and deletion of pathogenic T cell clones (passive tolerance), and (ii) through activation, expansion, and differentiation of antigen-specific regulatory T cells which secrete anti-inflammatory cytokines (active tolerance) [62,63] (Figure 4).

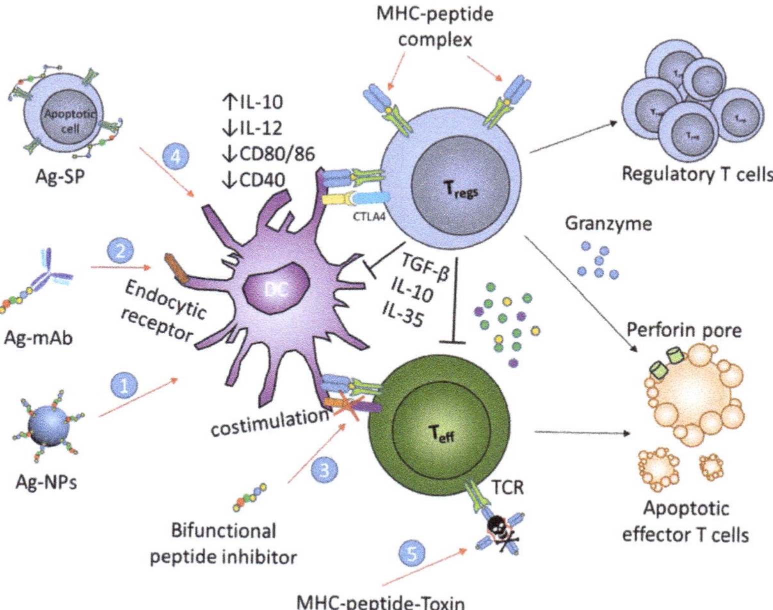

Figure 4. Bioconjugate-based approaches for the induction of Ag-specific tolerance in autoimmune diseases. The engineered bioconjugates target autoantigens and tolerogenic molecules to DCs (1); to facilitate antigen-processing via endocytic receptors (2); to hinder costimulation (3); to link to apoptotic cells for tolerogenic presentation (4); and to deliver toxin to autoantigen-specific T cells (5). These strategic approaches lead to peripheral tolerance as a consequence of anergy and deletion of cognate T cells, and/or induction of Tregs (with permission of [62]).

More specifically, an immunological synapse is established between APCs and T cells that is based on the formation of a trimolecular complex (signal 1) comprising the HLA class II molecule on the APC, the antigen (e.g., immunodominant epitope of a myelin protein) bound to this molecule and the TCR [64,65]. The establishment of the immunological synapse is the most vital process for the activation of effector T cells. In the absence of costimulatory molecules (signal 2), T cells become unresponsive to the antigen stimulation, a state known as anergy [65,66]. The presence of a costimulatory molecule exhibiting inhibitory properties could result to clonal deletion via apoptosis of the T cells. Autoreactivity of T lymphocytes can be also suppressed by the induction of regulatory T cells resulting in stable and long-term immune tolerance [59,65]. In vivo experiments have revealed that antigen-specific regulatory T cells are more effective than polyclonal Tregs regarding the control of organ-specific autoimmune diseases [67]. Finally, immune tolerance can be achieved via cytokine induced immune deviation, i.e., skewing of effector T cell subsets from Th1 and Th17 (proinflammatory phenotype) towards Th2 and Tr1 (anti-inflammatory phenotype) [59,65].

Antigen-specific therapies can be categorized according to the nature of the tolerogen (e.g., peptides derived from MBP, PLP, or MOG, mixtures of myelin derived peptides; altered peptide ligands; plasmids encoding myelin derived peptides, peptides related to TCR regions, attenuated myelin-specific T cells, tolerogenic DCs, antigen-coupled cells), the administration route (e.g., intravenous, subcutaneous, intraperitoneal, mucosal, epicutaneous, infusion of Ag-coupled cells) [14,43,51,59,65] and the antigen dose [68]. Since, antigen-specific therapies are thought to combine maximal efficiency with minimal side effects, they could be considered especially appealing [14]. On the other hand, they need to overcome major challenges in order to be efficiently used for the treatment of MS.

The first challenge is that the target antigens in MS are not known and remain to be identified [14,27,65]. The disease is largely heterogeneous. It involves multiple autoantigens (contrary for example to neuromyelitis optica that involves reactivity to Aquaporin-4, AQP4) that can vary between patients depending on genetic characteristics, age, environmental and/or triggering factors, and duration of the disease [2,27,69,70]. It has been assumed that myelin targets like MBP, PLP and MOG are relevant, but this is mainly based on EAE models and not on MS patients. Furthermore, therapeutic efficiency in EAE cannot always be translated in MS. Accordingly, the interpretation of the above remains a crucial challenge for the translation of antigen-specific therapies from bench to bedside [27].

Furthermore, it should be noted that the clinical/neuropathological features of MS change noticeably with time [5,70]. Thus, not all patients will necessarily have similar responses to myelin antigen-specific immunotherapies [5]. Additionally, in chronic MS, the pattern of recognized autoantigens progressively increases during the course of the disease, due to a spread of the adaptive immunity to related self-antigens, a phenomenon recognized as epitope spreading [69,70]. Epitope spreading has been defined as the broadening of epitope specificity from the initial immunodominant epitope-specific immune response to other subdominant protein epitopes [71]. Epitope spreading can be categorized as "intra-molecular" related to shifting of immune responses between different epitopes of the same protein (e.g., MBP) and "intermolecular" related to the shifting of immune responses between two proteins (e.g., MBP and PLP) [27,72]. The hierarchy of immunodominant and cryptic epitopes is supposed to be dependent on a combination of peptide processing and presentation by various APCs, and also on the availability of epitope-specific T lymphocytes, taking into account the mechanisms of central and peripheral tolerance [71]. Accordingly, identifying the autoantigens that should be included in the therapeutic formulation can be rather challenging. This problem might be partially overcome via tolerance spreading, i.e., a gradual spread of the tolerance to the administered autoantigens also to other self-antigens which are involved in autoimmunity [70]. Elucidation of the cellular and molecular mechanisms involved in epitope spreading in MS is very important in order to design efficient antigen-specific immunotherapies for MS patients [71]. In this respect, therapeutic strategies targeting a broader array of epitopes may need to be pursued. Furthermore, since immune reactivity broadens with disease duration, antigen-specific immunotherapies should ideally be delivered early in the course of the disease when epitope spreading has not yet occurred, according to an optimized dosage and frequency schedule [14,27,65,73]. An alternative approach could be to achieve bystander suppression (i.e., modulation of the responses to one target antigen leads to modulation of the responses to neighboring target antigens). However, limiting evidence exists for such therapies [27].

Finally, another challenge regarding the translation of antigen-specific immunotherapies from bench to bedside is that the administration of tolerogenic vaccines to MS patients with inapparent infections could be immunogenic and worsen the course of the disease due to its presentation in the immune system in a pro-inflammatory environment. This has been the case in clinical trials with APL [74]. Thus, a crucial test for tolerogenic vaccines could be the in vivo assessment of their delivery in a proinflammatory environment, either after EAE onset, or by co-delivery of adjuvants and/or pro-inflammatory stimuli during EAE immunization [63].

Continuing research efforts towards the development of effective and safe antigen-specific therapies for MS gave rise to the epicutaneous administration of antigens (e.g., dermal patch loaded with myelin derived peptides) for the establishment of skin-induced immune tolerance in MS. The ability of skin DCs to induce myelin-specific tolerance has already been demonstrated in both in vivo experiments (Table 2) and early clinical trials [28,58]. Finally, oral tolerance has appeared to be efficient regarding the prevention of EAE, but significantly less efficient concerning the therapy of ongoing EAE and MS [75].

4. In Vivo Assessment of Tolerance-Inducing Vaccination in MS

4.1. Animal Model of MS

The typically used animal model of MS is that of the experimental autoimmune encephalomyelitis (EAE) [3,4,18,76–80]. EAE is an acute or chronic neuro-inflammatory brain and spinal cord disease [18] which can be induced in various animal strains such as mice, rats, guinea pigs, rabbits, and even primates [7], via immunization with spinal cord homogenate or with various myelin proteins (e.g., MBP, PLP, MOG) emulsified in complete Freund's adjuvant (active EAE) [7,78,81]. EAE can be also transferred to naïve mice via adoptive transfer of T cells specific for myelin [8,78]. In EAE, myelin peptides are presented on MHC class II molecules to autoreactive T cells, together with costimulatory molecules (e.g., CD80 and CD86), resulting in activation of the T lymphocytes and, consequently, in an autoimmune attack on the myelin sheath [79]. EAE is principally mediated by myelin specific CD4+ T cells [20,78,82,83]. The clinical course of EAE varies based on the immunized animal species and the encephalitogenic antigen used for the inoculation. Usually the animals experience either an acute monophasic, progressive or not, disease, or a chronic relapsing-remitting disease. Ataxia, weight loss, sagging hind limb and paralysis are among the typical clinical signs of EAE [78]. Interestingly, various effective RRMS therapies (e.g., anti-inflammatory, immunomodulatory therapies) have been developed with the aid of EAE models. However, to date, no EAE model exists, that is capable of reproducing the specific features (e.g., clinical and neuropathological) of progressive MS. Therefore, despite the undeniable value of EAE for basic research concerning the mechanisms of brain inflammation and immune mediated CNS tissue damage, its value as model for MS is limited [18].

4.2. Myelin Peptide-Based Vaccination

4.2.1. Immunodominant Myelin Petides

Myelin is a multilaminar sheath around nerve fibers comprising lipid bilayers and different proteins. The major myelin proteins are MBP and PLP which represent more than 75% of the total myelin protein. Additionally, myelin contains MOG [84] representing ~0.05% of the myelin proteins [7], myelin-associated oligodendrocyte basic protein (MOBP), oligodendrocyte-specific protein (OSP), myelin-associated glycoprotein (MAG), and Nogo-A [85].

While the etiology of MS is not clear yet, a favored hypothesis supported by experimental evidence indicates that the cross-reactive immune response between myelin derived epitopic peptides and viral or bacterial components can be considered as an important factor that contributes to the development of autoimmune T cells which initiate a demyelinating inflammatory response. Thus, the determination of the main epitopes of the encephalitogenic myelin and/or neuronal proteins that are implicated in MS is considered of major significance both for the development of antigen-specific therapies for MS and the elucidation of MS pathophysiology and etiology [85].

In recent decades, extensive studies have been performed aiming to identify the immunodominant epitopes recognized by T lymphocytes in MS. These studies have revealed that only the myelin proteins MBP, PLP, MOG, MOBP, and OSP can induce clinical EAE in laboratory animals and that autoimmune T cells against these proteins can be detected in MS patients. Other myelin proteins, like MAG and Nogo-A have been also identified as encephalitogenic proteins. Finally, some neuronal components (e.g., β-Synuclein, Neurofilament) have been found to exhibit encephalitogenic potential [85]. Antigen recognition takes place in the setting of a trimolecular complex formed by HLA, myelin peptide and TCR [64,86,87]. The immunodominant PLP epitopes which can be processed by human APCs lie within the PLP regions 30–60 and 180–230. Similarly, the PLP epitopes that activate T lymphocytes in EAE are within the 40–70, 90–120 and 180–230 regions of the protein [5]. Immunodominant epitopes of MOG that are recognized by encephalitogenic T cells in MS as foreign antigens are MOG_{1-22}, MOG_{35-55} and MOG_{92-106} with the 35–55 epitope being the major immunodominant region of MOG [86]. Analysis of T-cell responses to MOBP in SJL/J mice indicated $MOBP_{15-36}$ as the main encephalitogenic epitope of MOBP [85].

A cyclic analogue of MBP_{87-99} has been designed by Matsoukas and coworkers taking into consideration HLA (His^{88}, Phe^{90}, Ile^{93}) and T-cell (Phe^{89}, Lys^{91}, Pro^{96}) contact side-chain information. cyclo(87-99)MBP_{87-99} was shown to induce EAE, bind HLA-DR4, and enhance CD4+ T-cell proliferation, similarly to the linear MBP_{87-99} peptide [83]. Additionally, peptide analogues derived from the encephalitogenic peptide MBP_{82-98}, the altered peptide ligand MBP_{82-98} (Ala^{91}) and their cyclic analogues were synthesized by Deraos and coworkers and assessed regarding their binding to HLA-DR2 and HLA-DR4 alleles involved in the presentation of myelin epitopes to T cells. The cyclic MBP_{82-98} was shown to bind strongly to HLA-DR2 and to have a lower affinity to the HLA-DR4 allele. Both the cyclic and APL analogues of MBP_{82-98} were found to be promising and were selected to be further evaluated regarding their ability to modulate the responses of autoreactive T cells in MS [88]. In addition to the abovementioned studies, Tapeinou and coworkers developed a peptide compound comprising the MBP_{85-99} immunodominant epitope coupled to an anthraquinone derivative (AQ) via a disulfide (S-S) and six amino hexanoic acid (Ahx) residues. AQ-S-S-(Ahx)6MBP_{85-99} was found to bind reasonably to HLA II DRB1*-1501 antigen indicating the possibility of eliminating encephalitogenic T lymphocytes through generation of a toxic, thiol-containing moiety (AQ-SH) [89].

Yannakakis and coworkers used molecular dynamic simulations to study the interactions of the MOG epitope MOG_{35-55} with the HLA and TCR receptors during the formation of the trimolecular complex TCR-hMOG_{35-55}-HLA DR2 [64]. They also used robust computational methods (e.g., molecular dynamics, pharmacophore modeling, molecular docking) to rationally design non-peptide mimetic molecules capable of binding with enhanced affinity to the T-cell receptor and not to the MHC-peptide complex, thus impeding the formation of the trimolecular complex [90].

To date various studies have assessed different myelin epitopes, as single peptides or mixtures of them, regarding their ability to induce antigen-specific tolerance in EAE animal models (Table 2).

4.2.2. Altered Peptide Ligands (APLs)

Altered peptide analogues (APLs) of the immunodominant myelin protein epitopes have been successfully synthesized and applied in antigen-specific immunotherapies in vivo (Table 2). They are molecules where one or more amino acids in the sequence of the native immunodominant peptides, crucial for the interaction with the TCR, have been substituted. Depending on the substitutions, APLs can induce protective or therapeutic immune responses against EAE [91]. APLs can change agonist peptides into antagonist ones. Antagonistic peptides participating in the trimolecular complex MHC-peptide-TCR and causing suppression of EAE exhibit loss of their side chain interactions with the complementarity determining region 3 (CDR3) loop of the TCR. Substitution of large side chains interacting with the TCR with small side chain amino acids (e.g., Ala) causes antagonism and, therefore, inhibition of EAE symptoms. Moreover, APLs can switch Th1 cell response towards Th2 thus leading to disease suppression. Finally, APLs might activate regulatory T cells capable of antagonizing the deleterious actions of encephalitogenic cells in the CNS [83,87]. Accordingly, mutant cyclic peptides of MBP87-99 (e.g., cyclo(91-99)[Ala96]MBP87-99 and cyclo(87-99)[Arg91Ala96]MBP87-99) were shown to suppress the proliferation of a CD4 T-cell line from a MS patient, bind to HLA-DR4 and exhibit an increased Th2/Th1 cytokine ratio in peripheral BMCs derived from MS patients [83].

Molecular dynamics were applied by Mantzourani and coworkers to study the interactions of the MBP_{87-99} epitope and its antagonistic APLs (e.g., [Arg^{91}, Ala^{96}] MBP_{87-99} and [$Ala^{91,96}$] MBP_{87-99}) with the receptor HLA-DR2b [92].

4.2.3. Y-MSPc

Kaushansky and coworkers [93,94] pursued a "multi-epitope-targeting" approach aiming to simultaneously neutralize T lymphocytes reactive against various major encephalitogenic epitopes. In this respect, they designed a recombinant synthetic protein comprising multiple epitopes of the human myelin protein (Y-MSPc). Y-MSPc was shown to efficiently inhibit the development of EAE induced in mice by a single epitope of myelin protein (classical EAE) or by a cocktail of five different encephalitogenic peptides (complex EAE) and suppress its progression, outperforming the single disease-specific epitope and the.mixture of peptides (Table 2).

4.2.4. Cytokine-Neuroantigen (NAg) Fusion Proteins

Fusion proteins consisting of a cytokine (N-terminal domain) fused with or without an appropriate linker to a neuroantigen (C-terminal domain) represent an emerging platform for antigen-specific vaccination [95,96]. Regarding their mechanism of action, the cytokine domain of the vaccine exhibits high affinity binding to specific surface cytokine receptors on certain subsets of APCs. This results in highly efficient uptake of the neuroantigen domain by these APCs, and its processing and presentation on MHC class II molecules to NAg-specific T lymphocytes. NAg tolerogenic presentation is assumed to induce regulatory responses and results in the establishment of antigen-specific immunological tolerance (Figure 5) [96,97].

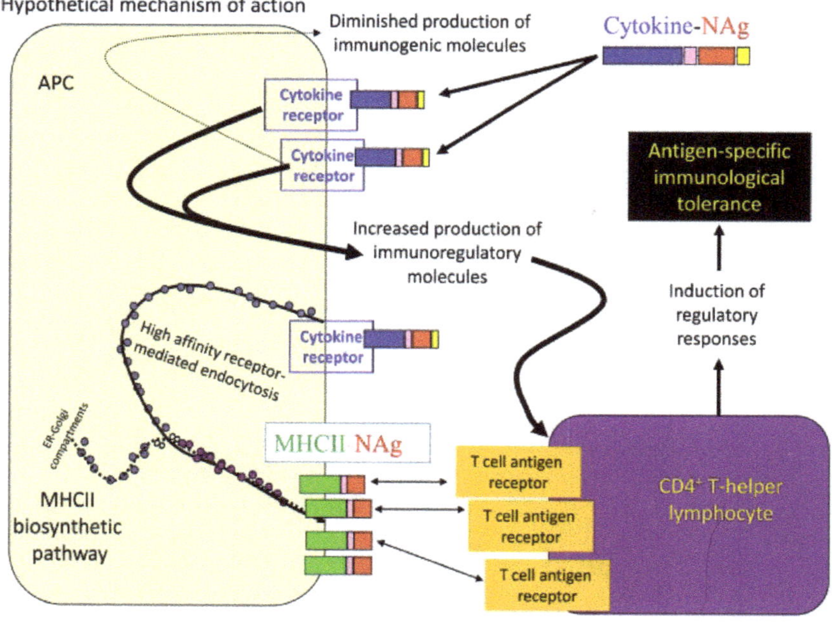

Figure 5. Mechanism of action of cytokine-NAg fusion proteins [96].

Various single-chain cytokine-neuroantigen (NAg) fusion proteins (e.g., granulocyte-macrophage colony-stimulating factor (GMCSF)-NAg, IFNβ-NAg, IL16-NAg, IL2-NAg), where NAg comprises self-myelin epitopes, have been examined as potential tolerogenic and/or therapeutic antigen-specific vaccines in EAE mouse models (Table 2). The developed fusion proteins have been found to target APCs and to effectively prevent the induction of EAE when administered prophylactically as well as to suppress pre-developed EAE. Due to their combined preventive and therapeutic activities, the cytokine-NAg vaccines were characterized as both tolerogenic and therapeutic. The ranking order with respect to their inhibitory activity was the following: GMCSF-NAg, IFNβ -NAg > NAgIL16 > IL2-NAg > MCSF-NAg, IL4-NAg, IL-13-NAg, IL1RA-NAg. [96].

Apart from the aforementioned cytokine-NAg fusion proteins, the macrophage colony stimulating factor (MCSF)-NAg fusion protein was used in order to increase the presentation of NAg by macrophages. However, it was found to be less tolerogenic than GMCSF-Nag, thus indicating the latter fusion protein as the most suitable for antigen-specific vaccination [95,98]. Additionally, it was revealed that GMCSF-MOG does not require a non-inflammatory quiescent environment to effectively prevent the development of EAE which contradicts the previous knowledge regarding tolerogenic vaccines [95,98].

4.2.5. Antibodies Coupled with Myelin Peptides

The dendritic and epithelial cell receptor with molecular weight equal to 205 kDa (DEC205) is expressed by DCs and enables antigen presentation. Injection of antigens (Ags) coupled to antibodies (Abs) specific for DEC205 into mice, at a low dose (e.g., $\leq$ 0.1 µg of fusion mAb [99]) leads to Ag presentation by nonactivated DCs, resulting in induction of regulatory T lymphocytes. In this respect, fusion of αDEC-205 Abs with MOG$_{35-55}$ [100] and PLP$_{139-151}$ [101] ameliorated EAE in mice. Similarly, Ring and coworkers synthesized single chain fragment variables (scFv) specific for DEC205. scFvs were subsequently fused with MOG (scFvDEC:MOG) and administered to mice both before and after induction of EAE. Significant prevention of EAE was observed by vaccination with scFv DEC:MOG before immunization. In addition, administration of scFv DEC:MOG post immunization led to substantial alleviation of the clinical symptoms of the disease [102]. On the other hand, Tabansky and coworkers targeted the dendritic cell inhibitory receptor 2 (DCIR2) receptor with αDCIR2 Abs fused to PLP$_{139-151}$ and observed significant alleviation of EAE clinical symptoms [79]. In another approach, Kasagy and co-workers demonstrated that administration of anti-CD4 and anti-CD8 Abs followed by injection of PLP$_{139-151}$ resulted in substantially lower EAE scores and reduced rate of relapses in chronic disease in mice [103] (Table 2).

4.2.6. Recombinant T-cell Receptor Ligands (RTLs)

Antigen-specific immunosuppression can be induced via the utilization of MHC-peptide complexes as specific TCR ligands interacting with autoimmune T cells in the absence of co-stimulatory molecules. A recombinant TCR ligand (RTL) typically comprises a single polypeptide chain encoding the β1 and α1 domains of MHC class II molecules linked to a self-antigen [104] and represents the minimal interactive surface with antigen-specific TCR. RTLs fold in a similar manner to native four-domain MHC/peptide complexes but they deliver qualitatively different, suboptimal signals which cause a "cytokine change" to anti-inflammatory factors in targeted autoreactive T cells. Treatment with RTLs could reverse the clinical/histological signs of EAE in different experimental cases (e.g., MBP-induced monophasic disease, MOG peptide-induced chronic EAE, PLP-induced relapsing remitting EAE) and even promote recovery of myelin and axons in mice with chronic disease [105–107] (Table 2).

Alternatively, RTLs could involve natural or recombinant $\alpha_1\alpha_2$ and $\beta_1\beta_2$ MHC class II domains covalently or noncovalently linked with encephalitogenic or other pathogenic peptides. These specific RTLs could bind both to the TCR and the CD4 molecule on the T cells surface via the β_2 MHC domain and were shown to hinder the activation of T cell and thus prevent EAE in rodents [108].

Table 2. Myelin protein/peptide-based vaccination.

Vaccine	Antigen	Targeting Ligand/Drug	Vaccination Type	Admin. Route	Admin. Dose	Animal Model	Vaccination Outcome
Myelin Proteins/Peptides							
MBP [112]	Guinea pig MBP	-	Prophylactic: seven days b.i.	e.c.		SJLxB10.PL female mice (6–8 weeks old) with EAE induced with MBP	Protection from RR form of EAE Reduction of disease incidence to 58%
MBP [113]	Guinea pig MBP	-	Prophylactic: seven and three days b.i. Therapeutic: at initial signs of EAE and after four days	e.c.		B10.PL female mice (6–8 weeks old) with EAE induced with MBP	Prophylactic vaccine: protection from EAE Therapeutic vaccine: suppression of EAE
MBP [114]	Guinea pig MBP	-	Prophylactic: seven and three days b.i.	e.c.		B10.PL and SJLxB10.PL female mice (6–8 weeks old) with acute or RR EAE respectively, induced with MBP Knock out mice: TCRδ-/-, CD1d-/- and β_2m-/- on H-2^u background.	Vaccination with MBP prior to EAE induction prevented the development of the disease (incidence reduction by 50%) and reduced the severity of the clinical symptoms in the mice that developed EAE. Experiments with knock out mice showed that the disease could not be completely suppressed only in β_2m-/- mice.
MOG$_{35-55}$ [115]	MOG$_{35-55}$	-	Preclinical/Therapeutic: 3, 5, and 7 days p.i.	i.v.		C57BL/6 female mice (8–10 weeks old) with EAE induced with MOG$_{35-55}$	Dramatic suppression of EAE development
c-MOG$_{35-55}$ [116]	MOG$_{35-55}$ and cyclic-MOG$_{35-55}$	-	Preclinical/Therapeutic on the same day with immunization and seven days p.i.	s.c.		C57BL/6 female mice (6–10 weeks old) with EAE induced with MOG$_{35-55}$	Amelioration of EAE clinical course and pathology. Reduction of clinical severity of acute phase of EAE and reduction of overall EAE burden.
ATX-MS-1467 [117]	Mixture of MBP$_{30-44}$, MBP$_{131-145}$, MBP$_{140-154}$, MBP$_{83-99}$	-	Prophylactic Preclinical/Therapeutic	s.c.	100 µL of ATX-MS-1467 twice a week	(ObxDR2)F1 mice with EAE induced with spinal cord homogenate	ATX-MS-1467 was shown to effectively prevent and treat EAE. The inhibition of the disease was found to be dose-dependent.
Pool of MBP peptides [118]	MBP$_{68-86}$ and MBP$_{87-99}$	-	Therapeutic: secen and 11 days p.i.	i.n.	500 µg of each MBP peptide /rat	Lewis female rats (9 weeks old) with EAE induced with MBP$_{68-86}$	Tolerization to a pool of MBP peptides was found to result in amelioration of clinical symptoms of EAE.
MOG$_{35-55}$ [119]	MOG$_{35-55}$	-	Prophylactic: every other day, for 10 days b.i.	oral	200 µg of MOG$_{35-55}$	C57BL/6 male mice (6–8 weeks old) with EAE induced with MOG$_{35-55}$.	Oral vaccination with MOG$_{35-55}$ was found capable of efficiently suppressing pathogenic cells.
MBP [120]	MBP	-	Prophylactic: one day b.i.	oral	100 mg of MBP	Euthymic and adult thymectomized Tg mice with EAE induced with MBP.	Euthymic Tg mice were shown to be protected from EAE after oral administration of MBP contrary to thymectomized mice, thus indicating the key role of thymus in oral tolerance induction.

Table 2. *Cont.*

Vaccine	Antigen	Targeting Ligand/Drug	Vaccination Type	Admin. Route	Admin. Dose	Animal Model	Vaccination Outcome
Altered peptide ligands (APLs)							
APL [121]	P1: MBP$_{87-99}$, P2: (Ala^{91},Ala^{96})MBP$_{87-99}$ P3: cyclo(87–99) (Ala^{91},Ala^{96})MBP$_{87-99}$	-	Prophylactic: on the day of immunization	s.c.		Female Lewis rats (6–8 weeks old) with EAE induced with MBP$_{74-85}$	Suppression of EAE was detected 8 days post P2 and P3 administration. P1 was not found to suppress EAE. P2 was shown to suppress EAE between 8–16 days whereas P3 suppressed EAE until the end of the experiment (e.g., day 18 or 20).
APL [87]	[Ala^{41}]MOG$_{35-55}$, [Ala^{41,46}]MOG$_{35-55}$ and [TyrOMe^{40}]MOG$_{35-55}$ cyclo(46–55)MOG$_{35-55}$ and cyclo(41–55)MOG$_{35-55}$	-	Prophylactic: on the day of immunization.	s.c.		C57BL/6 female mice (12–18 weeks old) with EAE induced with rat MOG$_{35-55}$	Significant reduction of EAE incidence and symptons with the administration of [Ala^{41,46}]MOG$_{35-55}$ or [Ala^{41}]MOG$_{35-55}$ as compared with the delivery of [TyrOMe^{40}]MOG$_{35-55}$, cyclo(46–55)MOG$_{35-55}$ and cyclo(41–55)MOG$_{35-55}$
Y-MSPc							
Y-MSPc [94]	MOG$_{34-56}$ MBP$_{89-104}$ OSP$_{55-80}$ OSP$_{179-201}$ MOBP$_{15-36}$ PLP$_{139-151}$ PLP$_{178-191}$	-	Preclinical/Therapeutic: 3, 5, 7, and 21 days p.i.	i.v.	75 µg of Y-MSPc/mouse	SJL/J female mice (2–3 months old) with EAE induced with PLP$_{139-151}$	Y-MSPc was revealed to be more efficient in inhibiting the development of the disease and suppressing its progression in comparison with a single encephalitogenic peptide or a cocktail of peptides.
Y-MSPc [93]	OSP$_{55-74}$ MOBP$_{55-77}$ MOBP$_{15-36}$ MOG$_{34-56}$ PLP$_{175-194}$ PLP$_{139-151}$ MBP$_{89-104}$		Preclinical/Therapeutic: administration post immunization	i.v.	75 µg of Y-MSPc/mouse	(C57Bl/6J6SJL/J)F1 mice with EAE induced with PLP$_{139-151}$ or rhMOG (active classical EAE), or a mixture of hMOG$_{34-56}$, hPLP$_{139-151}$, hMOBP$_{15-36}$, hMBP$_{89-104}$, hOSP$_{55-80}$ (active complex EAE), or via transfer of line T cells specific for phMOG$_{34-56}$ or phPLP$_{139-151}$ (passive EAE)	Y-MSPc was shown to be more efficient in inhibiting the development of classical or complex EAE, suppressing the disease course and reversing the chronic disease, compared with a single encephalitogenic peptide or a cocktail of peptides. Additionally, Y-MSPc appeared to be more effective in suppressing passive EAE.

Table 2. *Cont.*

Vaccine	Antigen	Targeting Ligand/Drug	Vaccination Type	Admin. Route	Admin. Dose	Animal Model	Vaccination Outcome
Cytokine-neuroantigen (NAg) fusion proteins							
GMCSF-NAg and MCSF-NAg [60]	Guinea pig MBP$_{69-87}$	GM-CSF M-CSF cytokines	Therapeutic: Exp.1: 9, 10, 12, and 14 days p.i.; exp. 2: 10, 11, and 13 days p.i.; exp. 3: eight and 11 days p.i.	s.c.	1 nmol of fusion protein(s) per injection (exp. 1 and 2), 4 nmol on day 8 and 1 nmol on day 11 (exp. 3)	Lewis rats with EAE induced with DHFR-NAg fusion protein	GMCSF-NAg was found to potently target MBP$_{69-87}$ to subsets of myeloid APCs and to successfully induce antigen-specific tolerance.
GMCSF-NAg MCSF-NAg [98]	MBP$_{69-87}$	GMC-SF MCSF	Prophylactic: 21, 1,4 and 7 days b.i. Therapeutic: 9, 10, 12 and 14 days p.i. (exp. 1), or 10, 11, and 13 days p.i. (exp. 2), or eight and 11 days p.i. (exp. 3)	s.c.	Prophylactic: 4 nmol of fusion protein(s) per injection Therapeutic: 1 nmol (exp. 1 & 2), 4 nmol on day 8 and 1 nmol on day 11 (exp. 3)	Lewis rats with EAE induced with DHFR-NAg fusion protein	Prophylactic vaccination with GMCSF-NAg resulted in attenuation of EAE severity. Furthermore, treatment with GMCSF-NAg successfully inhibited EAE progression to more severe stages.
GMCSF-NAg [122]	MOG$_{35-55}$	GM-CSF	Preclinical/Therapeutic: p.i.	s.c.	2 or 1 nmol of GMCSF-NAg	C57BL/6 mice with EAE induced with MOG$_{35-55}$ (active EAE) or with activated MOG-specific Th1 T cells (passive EAE). SJL mice with EAE induced with PLP$_{139-151}$. B cell deficient, CD4-deficient, IFN-γR1-deficient, and 2D2	GMCSF-NAg was shown to suppress the established disease especially in passive EAE models. It also proved to be an efficient therapy for *Cd4*-defficient mice and to exhibit tolerogenic activity in B cell deficient mice.
Cytokine-NAg [97]	MOG$_{35-55}$ PLP$_{139-151}$	GM-CSF	Prophylactic: 21, 14 and 7 days b.i. Therapeutic: 13, 15, 17, and 20 days p.i.	s.c.	Prophylactic: 2 nmol of cytokine-NAg Therapeutic: 4 nmol on days 9 and 11, and 2 nmol on day 14 p.i.	C57BL/6 with EAE induced with MOG$_{35-55}$ (active EAE) or with transfer of activated MOG$_{35-55}$-specific T lymphocytes. In order to provoke another bout of EAE on day 42, mice were challenged with MOG$_{35-55}$. SJL mice with EAE induced with PLP$_{139-151}$.	Fusion of GM-CSF with myelin protein epitopes was found to lead to efficient antigen uptake by myeloid APCs resulting in blocking of the development and progression of EAE.
Cytokine-NAg [96]	MBP$_{69-87}$ MBP$_{73-87}$ PLP$_{139-151}$ MOG$_{35-55}$	GMCSF IFN-β IL16 IL2	Prophylactic: 21, 14, and 7 days b.i. Therapeutic: 13, 15, 17, and 20 days p.i. or alternatively after the onset of paralysis	s.c.		C57BL/6 mice with EAE induced with MOG$_{35-55}$. SJL mice with RR EAE induced with PLP$_{139-151}$. Lewis rats with EAE (acute monophasic form) induced with MBP$_{73-87}$	The developed cytokine-NAg fusion proteins were shown to target APCs and to successfully prevent the induction of EAE when administered prophylactically as well as to suppress on-going EAE.
Cytokine-NAg [123]	Guinea pig MBP	rat IL-2 or IL-4	Prophylactic: 21, 14 and 7 days b.i. Preclinical/Therapeutic: five days p.i. and on every other day through days 9, 11, or 13 p.i.	s.c.	Prophylactic: 0.5-1 nmol per injection	Lewis rats with EAE induced with guinea pig MBP fusion protein	Prophylactic or therapeutic vaccination with IL-2/NAg resulted in attenuation of EAE course, whereas administration of IL4-NAg indicated lack of tolerogenic activity.

Table 2. *Cont.*

Vaccine	Antigen	Targeting Ligand/Drug	Vaccination Type	Admin. Route	Admin. Dose	Animal Model	Vaccination Outcome
GMCSF-NAg [95]	MOG_{35-55}	GM-CSF	C57BL/6 mice: Prophylactic 21, 14, and 7 days b.i. 2D2-FIG mice: Preclinical/Therapeutic: 0, 7, and 14 days, or 7 and 14 days, or 14 days p.i.	C57BL/6 mice: s.c. 2D2-FIG mice: i.v.	C57BL/6 mice: 2 nmol $GMCSF\text{-}MOG_{35-55}$ per injection 2D2-FIG mice: 4 nmol per injection	C57BL/6 mice with EAE induced with MOG_{35-55} 2D2-FIG mice with a transgenic MOG-specific repertoire of T cells and a GFP reporter of FOXP3 expression	The pretreatment with the GMCSF-MOG fusion protein elicited CD25+ Tregs which were required for the induction of tolerance. Vaccination of 2D2-FIG with GMCSF-MOG elicited circulating FOXP3+ Tregs the number of which was maintained with multiple boosters.
$MOG_{35-55}/\text{I-A}^b$ dimer [107]	MOG_{35-55}	I-A^b dimer	Therapeutic: nine days p.i. (treatment duration: four days).	i.p.	12 nM $MOG_{35-55}/\text{I-A}^b$ dimer (1 µg/mouse/day)	C57BL/6 female mice (6–8 weeks old) with EAE induced with MOG_{35-55}	The administration of $MOG_{35-55}/\text{I-A}^b$ dimer resulted in the reduction of antigen-specific T cells and amelioration of EAE symptoms.

Antibodies coupled with myelin peptides

Vaccine	Antigen	Targeting Ligand/Drug	Vaccination Type	Admin. Route	Admin. Dose	Animal Model	Vaccination Outcome
α-receptor–MOGp mAbs [100]	DNA for MOG_{29-59} (MOGp)	α-DEC mAbs α-Langerin mAb	Prophylactic: transfer of MOG-specific CD4+ T cells 15 days b.i. and admin. of α-receptor–MOGp mAbs 14 days b.i.	s.c.	3 µg of α-receptor mAbs	C57BL/6 (B6) mice with EAE induced with MOG_{35-55}	Prophylactic vaccination with α-DEC- and a-Langerin–MOGp mAbs led to reduction of disease incidence, onset delay and amelioration of clinical scores.
αDEC205-PLP$_{139-151}$ mAb [Stern et al., 2010]	PLP$_{139-151}$	anti-DEC205	Prophylactic: 10 or 15 days b.i.	i.p.	1 µg of fusion mAb	SJL/J female mice (6–10 weeks old) with EAE induced with PLP$_{139-151}$	Administration of αDEC205-PLP$_{139-151}$ mAb was found to alleviate the disease symptoms.
scFv DEC:MOG [102]	MOG	scFv specific for DEC205	Prophylactic: seven and three days b.i. Therapeutic: oje and four days after disease onset, signified by a clinical score equal to 1	i.v.	10 µg of scFvDEC:MOG	C57/Bl6 mice with EAE induced with WSCH	Almost complete prevention of EAE (90% of mice) was observed by administration of scFv DEC:MOG b.i. Moreover, vaccination with scFv DEC:MOG p.i. resulted in significant alleviation of the clinical symptoms in 90% of the mice.
αDCIR2-PLP$_{139-151}$ fusion mAb [79]	PLP$_{139-151}$	αDCIR2	Prophylactic: 10 days b.i.	i.p.	1 µg of fusion mAbs	SJL/J female mice (6–10 weeks old) with EAE induced with PLP$_{139-151}$ (active EAE) or via adoptive transfer of splenocytes from αDCIR2-PLP$_{139-151}$-treated mice (passive EAE)	Vaccination with αDCIR2+-PLP$_{139-151}$ fusion mAb was shown to decrease the severity of the disease and to delay its onset. Mice receiving splenocytes from αDCIR2-PLP$_{139-151}$-treated mice exhibited substantially lower clinical scores in comparison to those receiving cells from αDCIR2 mAb-treated mice.

Table 2. *Cont.*

Vaccine	Antigen	Targeting Ligand/Drug	Vaccination Type	Admin. Route	Admin. Dose	Animal Model	Vaccination Outcome
αCD4/CD8+PLP$_{139-151}$ [121]	PLP$_{139-151}$	Anti-CD4, anti-CD8a Ab	Prophylactic: admin. of mAb 21 days b.i. followed by PLP$_{139-151}$ delivery every other day for 16 days. Therapeutic: Mice treated with αCD4/CD8 Abs on day 11 p.i. were injected with αCD4/CD8+PLP$_{139-151}$ every other day from day 12–26.	i.p.	100 µg of CD4-/mouse) 100 µg of CD8a-/mouse 25 µg PLP$_{139-151}$ per injection	SJL female mice (seven weeks old) with EAE induced with PLP$_{139-151}$	αCD4/CD8+PLP$_{139-151}$-treated mice exhibited substantially lower EAE scores and reduced rate of relapses in chronic disease
Recombinant T-cell receptor ligands (RTLs)							
RTL342M [124]	MOG$_{35-55}$	HLA-DR2 peptide-binding domains	Therapeutic (s.c. or i.v.): admin. on the day that the clinical score for each mouse was ≥ 2. Daily admin. for mice receiving multiple doses. Prophylactic (s.c.): admin. of 4, 9, or 14 doses within 15 days. EAE was induced 2 days after the admin. of the final dose.	i.v. s.c.	50 µg of RTL342M	HLA-DR2 positive male/female mice (8–12 weeks old) with EAE induced with MOG$_{35-55}$	RTL treatment was revealed to be more efficient in reducing paralysis when administered in the form of multiple doses instead of a single dose, independently of the administration mode. Furthermore, the treatment with RTL342M could treat or prevent relapses. Pretreatment with RTL342M was shown to prevent the disease.
RTL401 [125]	PLP$_{139-151}$	α1 and β1 domains of the I-A^s class II molecule	Upon EAE onset, daily i) i.v. admin. for 3–4 days and ii) s.c. admin. for 8 days.	i.v. s.c.	100 µg of RTL401	SJL mice (6–7 weeks of age) with EAE induced with PLP$_{139-151}$ or PLP$_{178-191}$ or MBP$_{84-104}$. C57BL/6 X SJL) F1 mice (6–7 weeks of age) with EAE induced with MOG$_{35-55}$ or PLP$_{139-151}$.	i.v. or s.c. vaccination with RTL401 resulted in prevention of relapses and long-term reduction of clinical severity only in SJL mice and C57BL/6 X SJL) F1 mice with EAE induced with PLP$_{139-151}$.
RTL401 [126]	PLP$_{139-151}$	α1 and β1 domains of the I-A^s class II molecule	Upon EAE onset, daily (i) i.v. admin. for five days and (ii) s.c. for eight days.	i.v. s.c.	100 µL of 1 mg/mL RTL401	SJL female mice (7–8 weeks old) with EAE induced with PLP$_{139-151}$ (active EAE) or via transfer of activated PLP$_{139-150}$-specific T cells (passice EAE)	i.v. or s.c. vaccination with RTL401 was shown to effectively discontinue passive EAE progression, reverse its clinical severity and reduce the infiltration of cells into the CNS, as in the treatment of active EAE. Injury to axons was also prevented.
RTL551 [127]	MOG$_{35-55}$	α1 and β1 domains of the I-A^b class II molecule	Upon EAE onset (days 12–14 for active EAE and days 7–12 for passive EAE), daily i.v. admin. for five days.	i.v.	100 µL of 1 mg/mL RTL551	C57BL/6 male mice (6–7 weeks of age) with EAE induced with MOG$_{35-55}$ (active EAE) or via transfer of activated cells (passive EAE).	RTL551 treatment of actively or passively induced EAE resulted in significant reduction of clinical symptoms and spinal cord lesions.

Table 2. *Cont.*

Vaccine	Antigen	Targeting Ligand/Drug	Vaccination Type	Admin. Route	Admin. Dose	Animal Model	Vaccination Outcome
RTL401, RTL402, RTL403 [128]	$PLP_{139-151}$ $PLP_{178-191}$ MBP_{84-104}	$\alpha 1$ and $\beta 1$ domains of the I-A^s class II molecule	At EAE onset (days 10–11), when the clinical score was ≥ 2, daily s.c. admin. for 8 days.	s.c.	100 µL of 1 mg/mL RTL	SJL/J female mice (7–8 weeks old) with EAE induced with WSCH or with a mixture of $PLP_{139-151}$ and $PLP_{178-191}$.	A single RTL was found capable of successfully treating ongoing disease induced with a mixture of encephalitogenic epitopes as long as the cognate T cell specificity was present.
RTL551 [106]	rhMOG, $hMOG_{35-55}$, $mMOG_{35-55}$	$\alpha 1$ and $\beta 1$ domains of the I-A^b class II molecule	At EAE onset (days 10–13), when the clinical score was ≥ 2, daily i.v. admin. for eight days.	i.v.	100 µL of 1 mg/mL RTL551	C57BL/6 male mice (7–8 weeks old) with EAE induced with rhMOG or $mMOG_{35-55}$.	Vaccination with RTL551 could reverse the progression of EAE, reduce demyelination and damage of axons without however induce suppression of anti-MOG Ab response.
RTL401 [129]	$PLP_{139-151}$	$\alpha 1$ and $\beta 1$ domains of the I-A^s class II molecule	Upon EAE onset (days 10–11), daily admin. for 1, 2, or 5 days.	s.c.	100 µL of 1 mg/mL RTL401	SJL/J female mice (7–8 weeks old) with EAE induced with $PLP_{139-151}$ (active EAE) or via transfer of activated cells (passive EAE). TCR Tg 5B6 mice with EAE induced with $PLP_{139-151}$ B cell deficient (µMT knock-out, KO) mice on C57BL/6 background (7–8 weeks old) with EAE induced with MOG_{35-55}.	A new interaction between cells was revealed via which the RTL-equipped myeloid APCs reverse EAE progression by transferring tolerogenic signals to cognate T lymphocytes. It was also found that splenocytes incubated with RTL401 exhibited reduced ability to passively transfer EAE. Finally, it was shown that EAE can be treated by RTL551 in the absence of B cells.
VG312, VG303, VG311 [108]	MOG_{35-55}, MBP_{85-99}, CABL	$\alpha 1$ and $\beta 1$ domains of DR2	Therapeutic: i.v. administration for eight consecutive days, 2–4 days after the disease onset.	i.v.	100 µL of VG312, VG303, VG311	Tg HLA-DR2 male and female mice (8–12 weeks old) with EAE induced with MOG_{35-55}	Vaccination with VG312 led to peptide- and dose-dependent induction of long-term tolerance to the encephalitogenic epitope MOG_{35-55} and reversal of the clinical/histological symptoms of EAE
RTL401 [130]	$PLP_{139-151}$	$\alpha 1$ and $\beta 1$ domains of the I-A^s class II molecule	Therapeutic: (i) i.v. admin. for five consecutive days (days 20–24) and (ii) s.c. admin. for 3 days (days 32–34).	i.v. s.c.	100 µg of RTL401	SJL/J female mice (7–8 weeks old) with EAE induced with $PLP_{139-151}$.	Administration of RTL401 post the relapsing EAE peak resulted in prevention of disease relapses, reduction of demyelination and axonal damage.

Table 2. *Cont.*

Vaccine	Antigen	Targeting Ligand/Drug	Vaccination Type	Admin. Route	Admin. Dose	Animal Model	Vaccination Outcome
Bifunctional peptide inhibitor (BPI)							
PLP-B7AP [131]	PLP$_{139-151}$	B7 antisense peptide (AP) derived from CD28 receptor	Prophylactic 11, 8, and 5 days b.i. Preclinical/Therapeutic: 4, 7, and 10 days p.i.	s.c.	Prophylactic: 50 or 100 nmol PLP-B7AP/injection Therapeutic: 100 nmol PBI/injection	SJL/J female mice (5–7 weeks old) with EAE induced with PLP$_{139-151}$	Both prophylactic and therapeutic vaccination with PLP-B7AP resulted in efficient suppression of EAE. Mice treated with PLP-B7AP exhibited significantly low demyelination.
PLP-LABL [132]	PLP$_{139-151}$	LABL	Prophylactic: 11, 8, and 5 days b.i.	s.c.	100 nmol/injection/day	SJL/J female mice (5–7 weeks old) with EAE induced with PLP	The vaccination with PLP-LABL inhibited the inflammatory response resulting in prevention of BBB disruption and thus inhibition of EAE onset and progression.
PLP-LABL derivatives [110]	PLP$_{139-151}$	LABL	Therapeutic: admin. on disease onset, signified by a clinical score $\geq$1, and for three consecutive days until the score was <1)	i.v.	100 nmol/mouse	SJL/J (H-2S) female mice (5–7 weeks old)	Vaccination with the synthesized BPI derivatives was shown to efficiently inhibit EAE severity, and incidence.
PLP-LABL [133]	PLP$_{139-151}$	LABL	Preclinical/Therapeutic: 4, 7, 10, and 14 days p.i.	i.v.	100 mol/mouse	SJL/J female mice (5–7 weeks old) with EAE induced with PLP$_{139-151}$	Low disease scores and incidence could be observed in mice vaccinated with PLP-LABL.
PLP-LABL derivatives [134]	PLP$_{139-151}$	LABL	Therapeutic: admin. on disease onset, signified by a clinical score $\geq$1, and for three consecutive days until the score was <1)	i.v.	100 nmol/mouse	SJL/J female mice (5–7 weeks old) with EAE induced with PLP$_{139-151}$	The synthesized BPI derivatives were revealed to suppress EAE progression after intravenous administration more efficiently in comparison with unmodified BPI.
BPI-Fc fusion peptides LABL-Fc-ST-PLP and LABL-Fc-ST-MOG [109]	PLP$_{139-151}$ MOG$_{38-50}$	LABL-Fc-ST	Preclinical/Therapeutic: four and seven days p.i.	i.v.	25 nmol per dose	SJL/J mice (5–7 weeks old) with EAE induced with PLP$_{139-151}$	BPI-Fc fusion peptides were revealed to be highly efficient in suppressing EAE. The vaccinated mice were not found to exhibit weight loss, and featured benign clinical symptoms and reduced demyelination.
PLP–cIBR Derivatives [135]	PLP$_{139-151}$	cIBR7 peptide	Studies I and II: 4, 7, and 10 days p.i. Study III: admin. on disease onset, signified by a clin. score $\geq$1, and for 3 consecutive days until the score was <1	i.v.	Study I: 100 nmol/injection/day Study II and III: 50 nmol/injection/day	SJL/J (H-2S) female mice (5–7 weeks old) with EAE induced with PLP$_{139-151}$	Vaccination with PLP–cIBR, even at low dose or less frequent i.v. injections, resulted in significant amelioration of EAE and protected CNS against demyelination.

Table 2. *Cont.*

Vaccine	Antigen	Targeting Ligand/Drug	Vaccination Type	Admin. Route	Admin. Dose	Animal Model	Vaccination Outcome
Multivalent BPI (MVB$_{MOG/PLP}$) [111]	MOG$_{38-50}$ PLP$_{139-151}$	LABL	Preclinical/Therapeutic 4, 7, and 10 days p.i.	s.c.	100 nmol/mouse	SJL/J female mice (5–7 weeks old) with EAE induced with PLP$_{139-151}$ C57BL/6 mice (4–6 weeks old) with EAE induced with MOG$_{38-50}$	MVB$_{MOG/PLP}$ was found to significantly suppress EAE in both animal models despite the evidence of epitope spreading in the C57BL/6 mice.
Antigen-drug conjugates							
PLP$_{139-151}$-DEX [61]	PLP$_{139-151}$	DEX	Preclinical/Therapeutic: 4, 7, and 10 days p.i.	s.c.		SJL/J female mice (4–6 weeks old) with EAE induced with PLP$_{139-151}$	Vaccination with PLP$_{139-151}$-DEX efficiently protected the SJL/J mice from the onset of clinical symptoms compared with DEX treatment.

MBP: myelin basic protein; b.i.: before immunization; EAE: experimental autoimmune encephalomyelitis; e.c.: epicutaneous; RR: relapsing-remitting; MOG: myelin oligodendrocyte glycoprotein; p.i.: post immunization; i.n.: intranasal; APL: altered peptide ligand; Y-MSPc: recombinant synthetic protein comprising multiple epitopes of the human myelin protein; OSP: oligodendrocyte-specific protein; MOBP: myelin associated oligodendrocyte basic protein; PLP: proteolipid protein; GMCSF: Granulocyte-macrophage colony-stimulating factor; MCSF: macrophage colony stimulating factor; DHFR: dihydrofolate reductase; i.p.: intraperitoneal; IFN: interferon; IL: interleukin; mAbs: monoclonal antibodies; scFv: single chain fragment variables; WSCH: whole spinal cord homogenate; RTL: recombinant T-cell receptor ligand; HLA: human leucocyte antigen; rhMOG: recombinant human MOG; mMOG: murine MOG; BPI: bifunctional peptide inhibitor; LABL: ICAm-I binding peptide; DEX: dexamethasone.

4.2.7. Bifunctional Peptide Inhibitors (BPIs)

Bifunctional peptide inhibitors (BPIs) are a promising novel class of peptide conjugates which are designed to selectively impede the maturation of myelin specific T cells. They comprise an immunodominant myelin protein epitope tethered to a signal-2-blocking peptide derived from lymphocyte function-associated antigen-1, LFA-1 (i.e., a T cell protein binding to intercellular adhesion molecule-1, ICAM-1) [109] (Figure 6). It is hypothesized that they bind at the same time to MHC-II and ICAM-1 on APCs thus inhibiting the immunological synapse formation during APC and T cell interactions [110]. The development of molecules that could target more than one epitope is crucial for the application of BPI technology in MS [111]. The performance of BPIs with respect to the induction antigen-specific immune tolerance has been studied in EAE animal models (Table 2).

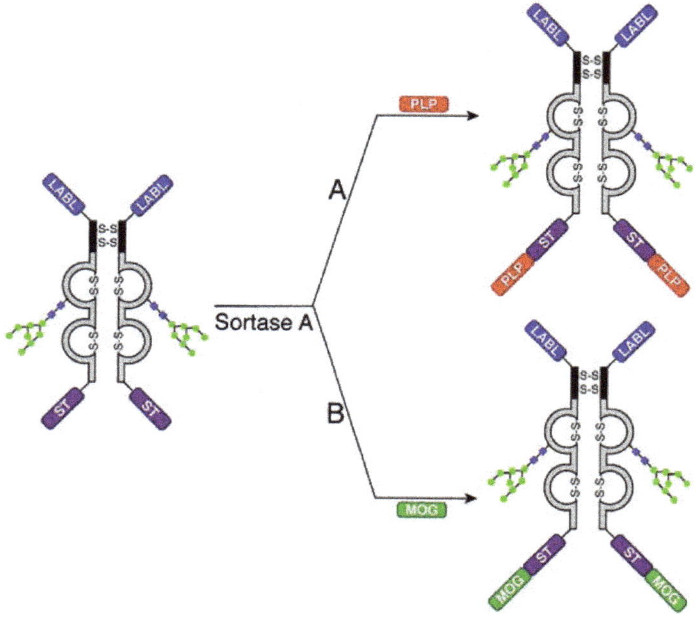

Figure 6. Sortase-mediated addition of two different antigens (A) PLP and (B) MOG to the C-terminus of LABL-Fc-ST (with permission of [109]).

4.2.8. Antigen-Drug Conjugates

Antigen drug conjugates (AgDCs) combine two therapeutic approaches (e.g., antigen-specific immunotherapies and immunomodulatory agents) to treat autoimmune diseases. Via chemical conjugation, the Ag could target the immunomodulatory agent to diseased cells thus minimizing side effects. AgDCs are assumed to exhibit increased affinity specificity through targeting cognate B cell receptors or endogenous autoantibodies. AgDCs formation entails the selection of an appropriate pair of antigen and immune modulator, and a linking scheme. An AgDC combing $PLP_{139-151}$ and dexamethasone ($PLP_{139-151}$-DEX) was administered to mice induced with EAE. It was shown that the AgDC protected the mice from developing clinical symptoms during the 25-day study [61] (Table 2).

4.3. DNA Vaccination

Deoxyribonucleic acid (DNA) vaccination is considered a promising antigen-specific approach for the treatment of MS [91,136–138]. DNA plasmid vaccines for tolerance induction in MS comprise a bacterial plasmid encoding myelin antigen(s). Expression is controlled by a mammalian promoter and a transcription terminator. They are administered either as naked DNA or with the aid of carriers (e.g., cationic lipids, cationic liposomes, polymeric particles), via the intramuscular or intradermal (e.g., "gene gun" delivering gold particles coated with pDNA vaccines) administration routes. Vaccination leads to DNA uptake and gene expression by the cells at the injection site [139,140]. Induction of immune tolerance is achieved via the following potential mechanisms (Figure 7). After intramuscular injection, myocytes are the main transfected cells, as well as few APCs. Antigens are then presented by the following mechanisms: i) myocytes process and present the antigen to T cells leading to T cell anergy ii) myocytes produce and secrete antigen that is taken up by APCs, which subsequently activate T cells. This results in loss of T cell co-stimulation through CD28, downregulation of IL-2, production of IFN-γ and reduced T cell proliferation. Intramuscular injection can also induce IFN-β via TLR9 activation due to the presence of CpG in the plasmid backbone [140], leading to downregulation of IL-12, IFN-γ, and Th17 cell responses. Following intradermal administration, DNA is delivered directly into the resident APCs (e.g., Langerhans and dermal cells). Intradermal vaccination leads to the secretion of regulatory cytokines (e.g., IL-4, IL-10, and TGF-β) thus resulting in the induction of anti-inflammatory Th2 immune responses [139,141]. Balance between tolerance induction and inflammatory immune response can be controlled by the administration route, antigen dose, and modification of the DNA-encoded antigen [141]. Numerous data from in vivo studies with the EAE animal model (Table 3), have demonstrated the efficiency of DNA plasmid vaccines at inhibiting MS via inducing T regulatory cells or anergy, clonal deletion, and immune deviation [139].

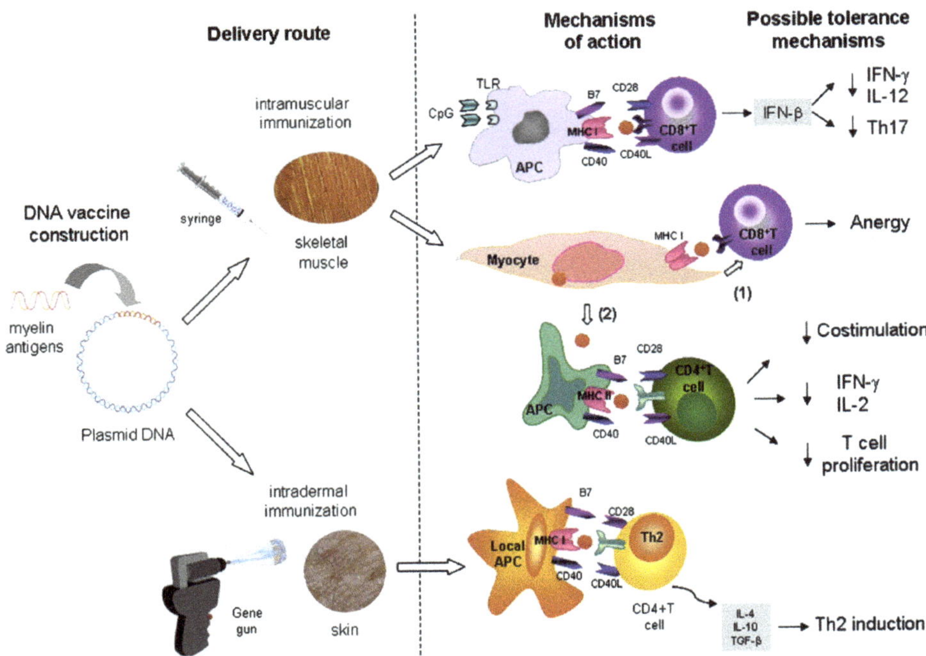

Figure 7. Mechanisms of immune tolerance induction by DNA plasmid vaccines (with permission of [139]).

Table 3. DNA vaccination.

Vaccine	Antigen/Immunosuppr.	Vaccination Type	Admin. Route	Admin. Dose	Animal Model	Vaccination Outcome
pDNA encoding IL-4 pDNA encoding PLP$_{139-151}$ pDNA encoding *MOG* [142]	PLP$_{139-151}$	Prophylactic: 17 and 10 days b.i. Therapeutic: 14 and 21 days p.i Co-vaccination with IL-4 plasmid and MOG plasmid on days 18 and 27 p.i.	i.m	100 µg of plasmid per injection	SJL/J mice with EAE induced with PLP$_{139-151}$ C57BL/6 mice with EAE induced with MOG$_{35-55}$	Co-vaccination with IL-4 and PLP$_{139-151}$ plasmids significantly protected against induction of EAE. Co-vaccination with IL-4 plasmid and MOG plasmid reversed ongoing EAE.
pMOG $_{91-108}$ pK0-MOG$_{91-108}$ (lacking CpG motifs) [143]	MOG$_{91-108}$	Prophylactic: three weeks b.i.	i.m.	200 µg DNA/injection	LEW.1AV1 (RT1av1) female rats (4–5 weeks old) with EAE induced with MOG$_{91-108}$	Vaccinated rats were protected against EAE.
pDNA encoding IL-10 pDNA encoding MBP$_{68-86}$ [144]	MBP$_{68-86}$	Admin. at the disease onset			Female Lewis rats (~6 weeks old) with EAE induced with MBP$_{68-86}$ or MBP$_{87-99}$, or with EAN induced with P2$_{57-81}$	Rats co-vaccinated with IL-10 and MBP$_{68-86}$ plasmids went into rapid remission. Co-administration of pDNA encoding IL-10 and pDNA encoding MBP$_{68-86}$ were shown to suppress EAE in rats induced either with MBP$_{68-86}$ or MBP$_{87-99}$ but not EAN.
pZZ/MOG$_{91-108}$ pMOG$_{91-108}$ pK0-MOG$_{91-108}$ pK3-MOG$_{91-108}$ [145]	MOG$_{91-108}$	Prophylactic: 3–4 weeks b.i.	i.m.	200 µg DNA/injection 100 µg of CpG DNA were added to pMOG$_{91-108}$ before the injection	Female LEW.1AV1 (RT1av1) rats (4–5 weeks old) and female DA rats with EAE induced with MOG$_{91-108}$	Vaccination with pDNA encoding MOG$_{91-108}$ (lacking the ZZ gene) reduced clinical symptoms of EAE and mortality in rats with different genetic background sharing the same MHC.
DNA encoding MBP, PLP, MOG, MAG and IL-4- [10]	MBP, PLP, MOG, MAG/GpG ODN	Therapeutic: admin. at the peak of acute EAE, when mice exhibited paralysis	i.m. i.p.	0.025 mg of each myelin peptide plasmid, 0.05 mg of IL-4 plasmid and 0.05 mg of GpG ODN	Female SJL/J and C57BL/6 (B6) mice (8–12 weeks old) with EAE induced with PLP$_{139-151}$ or MOG$_{35-55}$	Administration of myelin cocktail/IL-4 plasmids and the immunosuppressant GpG ODN resulted in dramatic improvement of the disease in mice having either chronic relapsing or chronic progressive EAE.
pMOG $_{91-108}$ pMOG-IFN-β pMOG-scr [146]	MOG$_{91-108}$	Prophylactic: three weeks b.i.	i.m.	200 µg DNA/injection	Female LEW.1AV1 (RT1av1) rats (4–5 weeks old) and female DA rats with EAE induced with MOG$_{91-108}$	The suppressive ability of DNA vaccination was found to be abrogated via silencing IFN-β.
p2MOG35 [147]	MOG$_{35-55}$/Tacrolimus (FK506)	Preclinical/Therapeutic: three and 17 days p.i.	i.m.	100 µg of p2MOG35/mouse 10 µg of FK506/mouse	Female C57BL/6 mice (6–8 weeks old) with EAE induced with MOG$_{35-55}$	Co-administration of p2MOG35 with FK506 was shown to effectively meliorate EAE in mice.

Table 3. *Cont.*

Vaccine	Antigen/Immunosuppr.	Vaccination Type	Admin. Route	Admin. Dose	Animal Model	Vaccination Outcome
pVAX-PLP, pVAX-MOG [148]	PLP, MOG	Prophylactic: four or 12 weeks b.i.	i.m.	20µg pVAX-PLP, pVAX-MOG	Female SJL/J (9H-2) mice (6 weeks old) with EAE induced with PLP$_{139-151}$ C57/B6 mice with EAE induced with MOG$_{35-55}$	EAE was found to be exacerbated in mice vaccinated with pVAX-PLP 4 weeks prior to immunization whereas both clinical and pathological symptoms were suppressed in mice vaccinated 12 weeks prior to EAE induction. In mice vaccinated with pVAX-MOG, either four or 12 weeks prior to immunization, EAE was shown to be significantly suppressed.

pDNA: plasmid DNA; IL: interleukin; MOG: myelin oligodendrocyte glycoprotein; b.i.: before immunization; p.i.: post immunization; PLP: proteolipid protein; i.m.: intramuscular; EAE: experimental autoimmune encephalomyelitis; MBP: myelin basic protein; EAN: experimental autoimmune neuritis; i.p.: intraperitoneal; GpG: GpG oligonucleotide; DA rats: dark agouti rats; IFN: interferon; pVAX: expressing vector.

4.4. Cell-Based Vaccination

4.4.1. Antigen-Specific Tolerogenic Dendritic Cells (tolDCs)

Dendritic cells (DCs) have a critical role in initiating adaptive immune responses in order to eliminate invading pathogens as well as in inducing tolerance towards innocuous components so as to maintain immune homeostasis [149]. Tolerogenic dendritic cells (TolDCs) are considered an attractive therapeutic approach for the induction of antigen-specific tolerance in autoimmune diseases [150,151]. To date various protocols have been developed for the in vitro generation of clinical-grade tolerogenic DCs ([35,152] (Figure 8) [153]) for antigen-specific immunotherapies. Autologous peripheral blood mononuclear cells (PBMCs) or bone marrow derived cells (BMDCs) are differentiated into tolDCs by numerous pharmacologic agents (e.g., immunosuppressive drugs such as rapamycin, cytotoxic T-lymphocyte-associated protein 4 (CTLA-4) Ig, corticosteroids; cyclic AMP inducers such as prostaglandin E2 and histamine; chemicals like vitamin D3, aspirin, etc.; proteins and neuropeptides like HLA-G, vasoactive intestinal peptide, etc.) and immunomodulatory cytokines (e.g., IL-10, TGF and low doses of GM-CSF) [150,153] and are further pulsed in vitro with autoantigens, encephalitogenic peptides, apoptotic cells, etc. [153]. tolDCs can display an immature or a semi-mature phenotype which is characterized by altered cytokine production and low expression of MHC and co-stimulatory molecules [150].

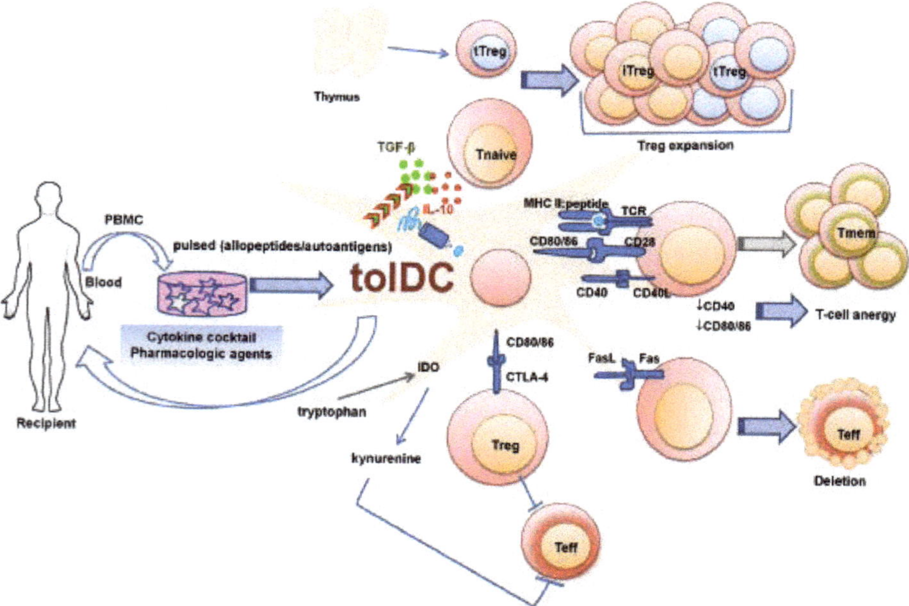

Figure 8. Strategies to generate tolDCs for clinical therapeutics [153].

Depending on the experimental protocol, the molecules used to induce tolerogenic properties, and the targeted cell population, tolDCs use different mechanisms of regulation to induce tolerance (Figure 8), including conversion to a regulatory T cell phenotype, induction of anergy, and antigen-specific deletion of T cell clones [19,35,150,152–154]. Lately, their ability to induce regulatory B cells secreting IL-10 has been also demonstrated [152]. TolDCs can be categorized into induced tolDCs (itDCs) (i.e., those acquiring their tolerogenic features in vitro or in vivo as described above and contribute to the maintenance of tolerance even under proinflammatory conditions) and natural tolDCs (ntDCs) (i.e.,

DCs present in the spleen and other lymphoid sites which inherently aid to establish tolerance in the absence of danger signals) [155].

The therapeutic potential of tolDCs has been demonstrated in the EAE model of MS (Table 4) (Figure 9). A key challenge is the translation of the in vivo results to humans. In this respect, it will be critical to correlate clinical efficiency with variation of immunological parameters and, accordingly, to define the best administration route and the effective dose of cells for this route [152]. Progress in the scientific areas of recombinant protein expression, genome editing and nanotechnology-based drug delivery systems, combined with improved immunization protocols, could further improve the promising tolDC vaccination in the furure [150].

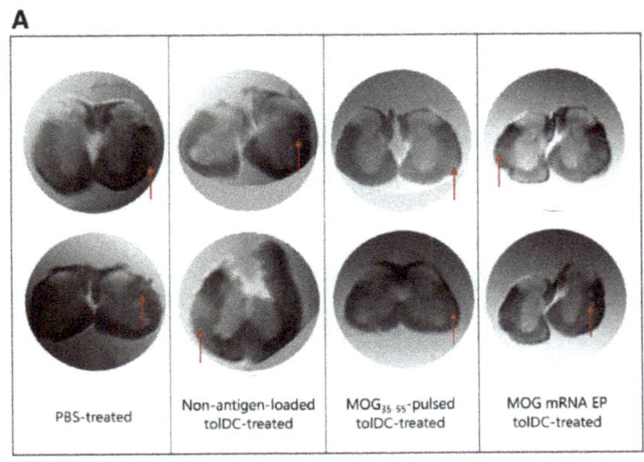

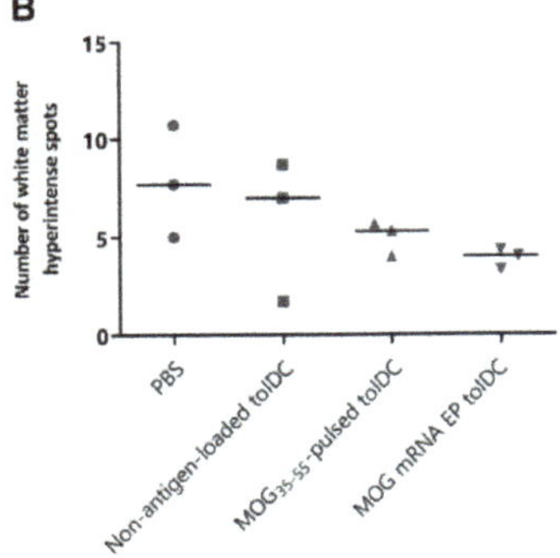

Figure 9. Evaluation of inflammatory lesion load within the spinal cord of tolDC-treated and PBS-treated mice using ex vivo MRI imaging. (**A**) Representative MRI of spinal cord with hyperintense white matter spots marked with a red arrow. Two representative axial slices are shown per treatment group. (**B**) The total number of hyperintense white matter spots along the entire spinal cord was quantified as a measure of lesion load in three mice per treatment group. Results are presented as individual scores for hyperintense spots with median [154].

4.4.2. T Cell Vaccination (TCV)

T cell vaccination involves the extraction of myelin reactive T cells from MS patients and their re-injection after irradiation in order to induce protective immunity [12,80,141,156]. To prepare T-cell vaccines, CSF mononuclear cells or blood PBMC's are stimulated with myelin antigen, and are then expanded specifically for the selected myelin peptide till an adequate population of cloned T cells is available. The latter are activated with antigen, and attenuated via exposure to radiation (6–12,000 Rads) to avoid proliferation after injection [156,157]. In clinic, the TCV protocol also involves multi-epitope TCR peptides [80]. TCV has been found to specifically suppress autoreactive T cells in MS via induction of a complicated anti-ergotypic and anti-idiotypic regulatory network or T cell deletion [80,91,156]. Various typical cytokines and lymphocyte phenotype transfer have been shown to participate in the depletion of the autoreactive T cells and the reversion of abnormal autoimmune responses [80] (Figure 10).

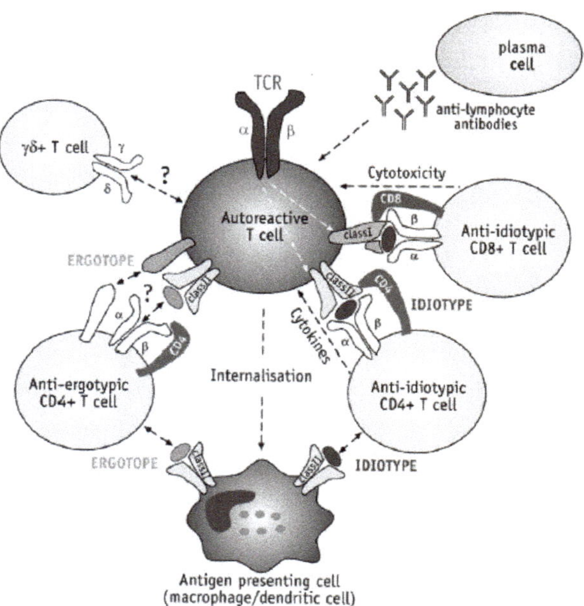

Figure 10. Complexity of anti-vaccine responses induced by TCV (with permission of [29]).

4.4.3. Antigen-Coupled Cells

Intact proteins (e.g., myelin proteins) as well as multiple peptides (e.g., MBP, PLP, and MOG derived peptides) can be coupled to a single cell (e.g., splenocyte [158,159], erythrocyte [67,160]) [86] (Table 4), thus permitting concurrent targeting of various T-cell specificities. This could be critical for antigen-specific immunotherapy in MS, where immune tolerance to multiple T-cell epitopes is considered necessary for the disease treatment due to epitope spreading. Contrary to protein/peptide-induced tolerance, vaccination with protein/peptide-coupled cells lowers the risk of anaphylaxis, since the antigen is chemically crosslinked to the cell surface. Vaccination with antigen-coupled cells has been found to prevent the active- and passive-transfer. Finally, tolerance induction with Ag-coupled cells can help define immunodominant myelin antigens, since the disease progression can be impeded by cells coupled with the spread epitope [75].

Table 4. Cell-based vaccination.

Cells	Inductive Agent/Peptide	Vaccination Type	Admin. Route	Admin. Dose	Animal Model	Vaccination Outcome
Tolerogenic Dendritic cells (tolDCs)						
BMDCs from C57BL/6 mice [161]	Atorvastatin/MOG$_{35-55}$	Preclinical/Therapeutic: days five and 13 p.i.	i.p.	1×10^6 cells per injection	Female C57BL/6 mice (8–10 weeks old) with EAE induced with MOG$_{35-55}$	MOG$_{35-55}$—specific tolDCs successfully ameliorated clinical Symptoms in mice with EAE.
BMDCs [162]	mytomycin C/MOG$_{196-204}$	Admin. of MOG196-pulsed Kb–/–Db–/– DCs to C57BL/6 (B6) mice one week b.i. and one p.i. Admin. of MOG196-pulsed B6 DCs to C57BL/6 mice three days b.i. and two and seven days p.i.	s.c.	1×10^6 cells per injection	Female C57BL/6 (B6) (8–10 weeks old) with EAE induced with MOG$_{35-55}$	Administration of MOG196-pulsed Kb–/–Db–/– DCs or MOG196-pulsed DCs ameliorated EAE in mice.
Murine BMDCs [154]	1α, 25-dihydroxy-vitamin D3/MOG-encoding mRNA or MOG$_{35-55}$	Therapeutic: 13, 17, and 21 days p.i.	i.v.	1×10^6 cells per injection	Female C57BL/6JOlaHsd mice (8–10 weeks old) with EAE induced with MOG$_{35-55}$	Vaccination with tolDCs electroporated with MOG-encoding mRNA or MOG$_{35-55}$ stabilized the clinical signs of the disease already from the first injection. MRI examination of hyperintense spots present along the spinal cord of mice was found to be in line with the clinical score (Figure 9). Administration of MOG35–55-pulsed and lentiviral transduced BMDCs led to significant decrease in the clinical symptoms of EAE in mice. The highest decrease in the clinical scores was observed with the administration of co-transduced BMDCs (BoLV-DCs).
BMDCs [163]	CD40-specific and p19-specific shRNA encoding lentiviral vectors/pyromycin/MOG$_{35-55}$	Preclinical/Thereapeutic: 3, 5, and 7 days p.i.	i.v.	2×10^6 cells per injection	C57BL/6 mice with EAE induced with MOG$_{35-55}$	
BMDCs [164]	Vitamin D3/MOG$_{40-55}$	Preclinical/Therapeutic: two and five days p.i., or five and nine days p.i. or 15, 19, 23, and 33 days p.i.	i.v.	2 or 4×10^6 cells	Female C57BL/6J mice (8–10 weeks old) with EAE induced with MOG$_{40-55}$	MOG$_{40-55}$—specific TolDCs were found to succeed in reducing EAE incidence and ameliorating its clinical signs.
BMDCs [165]	Vitamin D3/MOG$_{40-55}$/cryopreserved		i.v.	2 or 4×10^6 cells	Female C57BL/6J mice (8–10 weeks old) with EAE induced with MOG$_{40-55}$	It was shown that MOG$_{40-55}$—specific TolDCs maintain their tolerogenic properties and can efficiently ameliorate the clinical symptoms of EAE.
Murine BMDCs [166]	Tofacitinib/MOG$_{35-55}$	Therapeutic: 7, 11, and 15 days p.i.	i.v.		Twelve-week Female C57BL/6 mice (12 weeks old) with EAE induced with MOG$_{35-55}$	MOG$_{35-55}$—specific TolDCs efficiently dampened EAE severity and progression.

Table 4. *Cont.*

Cells	Inductive Agent/Peptide	Vaccination Type	Admin. Route	Admin. Dose	Animal Model	Vaccination Outcome
BMDCs [167]	1,25-dihydroxyvitamin D_3/MOG_{35-55}	Therapeutic: 10, 13, and 16 days p.i.	i.v.		Female C57BL/6 mice (6–8 weeks old) with EAE induced with MOG_{35-55}	Vitamin D3 treated MOG_{35-55}—specific. TolDCs succeeded in postponing the disease onset and reducing its clinical scores.
DCs [168]	Estriol (E3)/MOG_{35-55}	Prophylactic: one day b.i.	i.v.	$8–10 \times 10^6$ cells per mouse	Female C57BL/6 (H-2b) mice (4–6 weeks old) with EAE induced with MOG_{35-55}	Mice vaccinated with E3 MOG_{35-55}—specific TolDCs exhibited a reduced cumulative clinical score and EAE severity. They also avoided relapses and development of chronic disease.
BMDCs matured with TNF-α [169]	/MOG_{35-55}	Prophylactic: 7, 5, 3, and 1 days b.i. Preclinical: one day p.i.	i.v.	$2–2.5 \times 10^6$ cells per injection Rat anti–mouse IL-10R mAb: 0.5 mg equivalents per mouse	C57Bl/6 mice with EAE induced with MOG_{35-55}	Vaccination with MOG_{35-55}—specific TNF/DCs improved the clinical disease score. Pulsing of TNF-α/DCs with an unrelated peptide did not succeed in preventing the disease.
DCs [170]	/in vivo pulsing in Lewis rats with EAE induced with MBP_{68-86}	Prophylactic: four weeks b.i.	s.c.	1×10^6 cells per rat	Male Lewis rats with EAE induced with MBP_{68-86}	Injection of EAE DCs to rats resulted in induction of immune tolerance against the disease as demonstrated by delayed onset and marked decrease of the mean clinical score.
T cell-based vaccination						
Ob2F3 Tregs [171]	Retrovirally transduced pre-stimulated naïve CD4+ Tcells from peripheral blood mononuclear cells (PBMCs) of healthy donors using Ob2F3.	Preclinical/Therapeutic: seven days p.i.	i.v.	2×10^6 cells	Male and female HLA-DR15 transgenic mice (4.5–7.5 months old) with EAE induced with MOG_{35-55}	Ob2F3 Tregs were shown to significantly ameliorate MOG_{35-55} induced EAE via bystander suppression.
MBP-specific T-cell lines (e.g., B12 and B12-GFP) [157]		Prophylactic: admin. three times at weekly intervals, with the last injection 10 or seven days b.i.	s.c.	1×10^7 activated and irradiated T cells	Female Lewis rats (6–8 weeks old) with EAE induced via i.v. injection of antigen stimulated T cells.	Vaccination with MBP-specific T cell lines inhibited the development of EAE clinical symptoms.

Table 4. *Cont.*

Cells	Inductive Agent/Peptide	Vaccination Type	Admin. Route	Admin. Dose	Animal Model	Vaccination Outcome
Hematopoietic stem cells (HSCs)						
DC-MOG vector-transduced BM-HSC [172]	Ex vivo modification of HSCs with SIN lentivirus vectors which transcriptionally target the expression of myelin peptides to DCs.	Prophylactic: Lethally Irradiated (10.5 Gy) mice were transplanted with DC-MOG transduced BM-HSCs eight weeks b.i. BM chimeras received neomycin treatment for three weeks post transplantation.	i.v.	1–3×10^6 cells per mouse	C57BL/6 mice with EAE induced with MOG peptide.	The transplantation of DC-MOG vector-transduced BM-HSC was found to completely protect mice from developing EAE even in cases of transplantation 6 months b.i. In agreement with the clinical observations, no histological signs of the disease such as demyelination, damage of axons, etc. could be detected in the tolerized mice.
Bone marrow cells (BMC)						
BMCs expressing MOG_{40-55} [173]	liMOG	Prophylactic: mice were transplanted with BMCs transduced with liMOG 21 days b.i. Therapeutic: mice were transplanted with transduced BMCs 15–17 days p.i.	i.v.	0.7–1.6×10^6 cells per mouse	Female C57BL/6J mice (5–10 weeks old) with EAE induced with MOG_{40-55}	Transplantation of BMCs expressing MOG_{40-55} was shown to protect mice from developing EAE and reduce the disease severity in mice with established EAE.
Myeloid-derived suppressor cells (MDSCs)						
MDSCs isolated via positive selection from BMCs expressing MOG_{40-55} [174]	liMOG	Prophylactic: mice were transplanted with MDSCs transduced with liMOG seven days b.i. Therapeutic: mice were transplanted with transduced MDSCs 13–14 days p.i.	i.v.	0.5–1×10^6 cells per mouse	Female C57BL6/J mice (6–8 weeks old) with EAE induced with MOG_{40-55}	MOG_{40-55} -expressing MDSCs were found to exhibit both preventive and therapeutic effects in EAE induced with MOG_{40-55}

Table 4. *Cont.*

Cells	Inductive Agent/Peptide	Vaccination Type	Admin. Route	Admin. Dose	Animal Model	Vaccination Outcome
Antigen-cell conjugates						
Ag-SP [158]	Chemically treated Ag-coupled SPs	Administration on day −7 b.i. or at peak of disease in actively induced EAE, or two days p.i.	i.v.	50×10^6 Ag-SPs per mouse	Wild-type C57BL/6 (I-Ab), B10.S (I-As), and BALB/c (I-Ad) female mice (5–6 weeks old) with EAE induced with myelin peptide or via adoptive transfer.	It was revealed that syngeneic or allogeneic Ag-SPs can effectively protect mice against ongoing clinical EAE.
Ag-SP [159]	Chemically treated Ag-coupled SPs	Prophylactic: at indicated time points b.i.	i.v.	50×10^6 Ag-SPs or 15–20 µg Ag per mouse	SJL and C57BL/6 mice with EAE induced with myelin peptide or via adoptive transfer.	i.v. infusion of peptide antigens coupled to syngeneic splenic leukocytes (Ag-SP) was found to efficiently induce antigen-specific T cell tolerance.
Ag-RBC [160]	Genetically engineerd Kell-LPETGG RBCs, coupled with MOG $_{35-55}$ through enzymatic surface modification with sortase transpeptidase.	Prophylactic: transfusion seven days b.i. Preclinical: transfusion five days p.i. Therapeutic: Transfusion on the day of EAE onset	i.v.	200 µL RBC-MOG$_{35-55}$	C57BL/6J (CD45.2+), B6.SJL-Ptprc (CD45.1+), BALB/c Female C57BL/6 mice (10–12 weeks old) with EAE induced with MOG$_{35-55}$	The transfusion of RBC-MOG$_{35-55}$ was shown to significantly improve the clinical signs of EAE in mice.

BMDCs: Bone marrow-derived dendritic cells; p.i.: post immunization; i.p.: intraperitoneal; EAE: experimental allergicencephalomyelitis; tolDCs: tolerogenic dendritic cells; s.c.: subcutaneous; b.i.: before immunization; i.v.: intravenous; MBP: myelin basic protein; Tregs: regulatory T cells; Ob2F3: recombinant T-cell receptor (TCR) isolated from a MBP specific Tcell clone of a multiple sclerosis patient; HSCs: hematopoietic stem cells; SIN: selfinactivating; SP: splenocytes; RBCs: red blood cells; liMOG: vector encoding the murine invariant chain (Ii) containing MOG$_{40-55}$ and enhanced green fluorescent protein (EGFP).

4.5. Carrier-Aided Vaccination

In recent decades, different strategies have been pursued for the development of carriers [175–179] loaded/conjugated with myelin antigens or combinations of myelin peptides and immunomodulating agents. The developed carriers have been designed to target TCR signaling pathways, as well as cytokines and co-signaling molecules, aiming to enhance TCR-mediated tolerance [30,62,177]. Various biomaterials (e.g., polymers, lipids) have been formulated into micro- or nanoparticles, self-assembled into different structures, or formed molecular conjugates with self-antigens (e.g., conjugation of self-antigens with polymers, antibodies, small molecules). Both nanoparticles (NPs) and microparticles (MPs) can be uptaken by APCs thus enhancing the intracellular delivery of myelin antigens and imunnomodulators [180,181].

4.5.1. Polymer Particles

Polymer micro- and nanoparticles loaded with self-antigens and/or immunomodulatory molecules have recently emerged as ideal carriers for tolerogenic vaccines since their properties (e.g., particle size, composition, antigen/immunomodulator loading) can be fine-tuned to induce peripheral tolerance. Furthermore, NPs can be employed as platforms to regulate the doses and delivery times not only of the self-antigens but also of the tolerogenic adjuvants that are required to promote tolerance [70].

Poly(lactic-co-glycolic acid) (PLGA) NPs are non-toxic, biodegradable/biocompatible and have the advantage of being FDA approved for various clinical uses including drug delivery, diagnostics, etc. Additionally, surface functionalization strategies may improve their interaction with cells, thus optimizing cell targeting and vaccine performance. PLGA NPs are the most extensively assessed nanocarriers in pre-clinical models of autoimmune diseases and their effectiveness regarding antigen-specific immunotherapies (Table 5) represents a proof-of-concept of the feasibility of nanoparticle-aided tolerogenic vaccination. Furthermore, their successful application in animal models appears encouraging concerning potential translation to humans [70].

4.5.2. Soluble Antigen Arrays

Soluble antigen arrays (SAgAs) are synthesized by co-grafting the immunodominant epitope $PLP_{139-151}$ and LABL peptide (i.e., ligand of the intercellular adhesion molecule 1, ICAM-1) to hyaluronic acid (HA) via a hydrolysable oxime bond [182,183]. Their size can be fine-tuned to allow them to drain to the lymph nodes [183]. Another key factor affecting their drainage is the injection site and the molecular weight of HA. For example, following s.c. injection, HA can drain to the lymphatics and its retention time can be affected by its molecular weight [183].

The efficiency of the hydrolysable $SAgA_{PLP-LABL}$ to suppress disease in mice with EAE has been reported in various studies (Table 5) and has been attributed to the simultaneous delivery of the myelin derived antigen and the cell adhesion signal [182]. Furthermore, earlier in vitro studies indicated that SAgAs demonstrate Ag-specific binding with B lymphocytes, target the B cell receptor (BCR) and reduce BCR-mediated signaling [184]. Based on the abovementioned experimental results indicating BCR engagement as the mechanism of action of $SAgA_{PLP-LABL}$ Hartwell and coworkers developed a novel version of $SAgA_{PLP-LABL}$, the cSAgAPLP:LABL (click SAgA), employing non-hydrolysable conjugation chemistry (e.g., copper-catalyzed azide-alkyne aycloaddition) [184,185]. cSAgAPLP:LABL was found to significantly reduce or inhibit BCR-mediated signaling and to exhibit enhanced in vivo efficiency in comparison with the hydrolytically unstable $SAgA_{PLP-LABL}$ [184,185] (Figure 11).

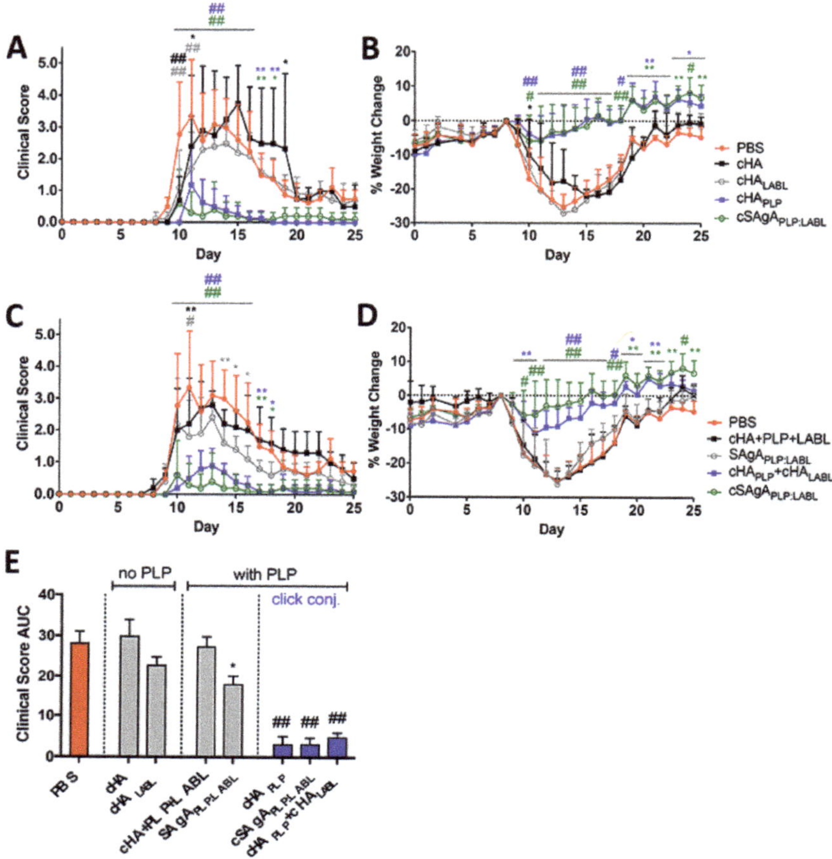

Figure 11. EAE in vivo response to click conjugates (cHA, cHALabl, cHAPLP, and cSAgAPLP:LABL) as measured by (**A**) clinical disease score and (**B**) percent weight loss. EAE in vivo response to groups containing both PLP and LABL (cHA+PLP+LABL, SAgAPLP:LABL, cHAPLP+cHALABL, and cSAgAPLP:LABL) as measured by (**C**) clinical disease score and (**D**) percent weight loss. Data represent mean ± SD ($n = 5$); statistical significance compared to PBS negative control was determined by two-way ANOVA. (**E**) Cumulative EAE in vivo response as measured by clinical disease score area under the curve (AUC) derived from subfigures A and C. Data represent mean ± SEM ($n = 5$); statistical significance compared to PBS negative control was determined by ordinary one-way ANOVA followed by Dunnett's post hoc test. (* $p < 0.05$, ** $p < 0.01$, # $p < 0.001$, ## $p < 0.0001$, color coded according to group) (with permission of [185]).

4.5.3. Immune Polyelectrolyte Multilayers (iPEMs)

It has been recently shown that excess signaling via inflammatory pathways such as toll-like receptors (TLRs) is involved in the pathogenesis of autoimmune diseases. Accordingly, the co-delivery of immunodominant myelin peptides with GpG oligonucleotide, a regulatory ligand of TLR9, could potentially limit TLR signaling during the differentiation of myelin-specific T lymphocytes, thus redirecting their differentiation towards a tolerogenic phenotype like the regulatory T cells. In this respect, immune polyelectrolyte multilayers (iPEMs) were formed using a layer-by-layer approach to co-assemble modified myelin peptides with GpG oligonucleotide. These nanostructures have key characteristics of biomaterial-based nanocarriers, such as tunable physicochemical properties and loading capacity, ability to deliver various active ingredients, etc., lacking, however, synthetic components that could exhibit inflammatory properties.

In in vitro studies, iPEMs have been shown to limit TLR9 signaling, decrease activation of DCs, and polarize myelin-specific T lymphocytes towards a tolerogenic phenotype. Additionally, they have been found to reduce inflammation and induce tolerance in mice with EAE [186,187] (Table 5).

4.5.4. pMHC-Nanoparticles (pMHC-NPs)

The "two signal theory" states that two different signals are required for the activation of naive T cells: (i) engagement of the TCR with its cognate pMHC target, and (ii) a co-stimulatory signal from molecules selectively expressed on professional APCs' surface. It is well known that engagement of the TCR on the surface of a naive T cell without co-stimulation results in the induction of apoptosis or anergy.

The development of pMHC-nanoparticles (pMHC-NPs) for the treatment of autoimmune diseases was based on the hypothesis that pMHC-coated NPs would diminish the responses of autoreactive T cells more efficiently compared with soluble pMHC complexes. This could be due to (i) their multimeric valency, (ii) their potentially superior TCR cross-linking properties compared with "artificial APCs", and (iii) the protection of the NP-bound pMHC molecules from degradation [104]. The ability of pMHC-NPs to stop the progression of EAE was assessed with in vivo experiments in mice (Table 5).

4.5.5. Mannan-Peptide Conjugates

Based on previous studies with the yeast polysaccharide, mannan, Tseveleki and coworkers, examined mannan conjugation with immunodominant myelin epitopes as an approach to divert the differentiation of myelin-specific T lymphocytes towards a regulatory phenotype, thus decreasing the mice susceptibility to EAE. It was shown that the administration of the synthesized conjugates to mice in both prophylactic and therapeutic vaccination protocols resulted in the induction of antigen-specific T cell tolerance and significant amelioration of EAE clinical and histopathological symptoms. [188] (Figure 12) (Table 5). According to these results, it was speculated that conjugation of MOG epitopes to mannan may modulate the autoimmune response in humans, thus potentially reducing the symptoms of MS [188].

4.5.6. Liposomes

Liposomes are tiny vesicles featuring an aqueous core surrounded by a lipid bilayer. They can encapsulate both hydrophilic and hydrophobic drugs and target them to specific cell surfaces via appropriate functionalization. Various types of liposomes have been already approved for clinical use (e.g., delivery of therapeutics, vaccination) and can be designed to induce or tolerate immune responses [189]. Pujol-Autonell and coworkers reported the beneficial effect of MOG peptide loaded liposomes in treating mice with EAE. Liposomes successfully delayed the onset, suppressed the severity and decreased the incidence of the disease [190]. Similarly, Belogurov and co-workers demonstrated that mannosylated liposomes containing MBP_{46-62} could significantly reduce EAE clinical signs in Dark Agouti (DA) rats [189]. Interestingly liposomes loaded with MBP_{46-62}, $MBP_{124-139}$, and $MBP_{147-170}$ and targeting CD206 were proven to be safe and well-tolerated and to normalize cytokine levels in RRMS and SPMS patients [191,192].

4.5.7. Microneedle Patches

Pires and coworkers proposed the use of minimally invasive microneedle patches for the delivery of myelin peptides, as an alternative therapeutic strategy for skin mediated antigen-specific immune tolerance in MS [178].

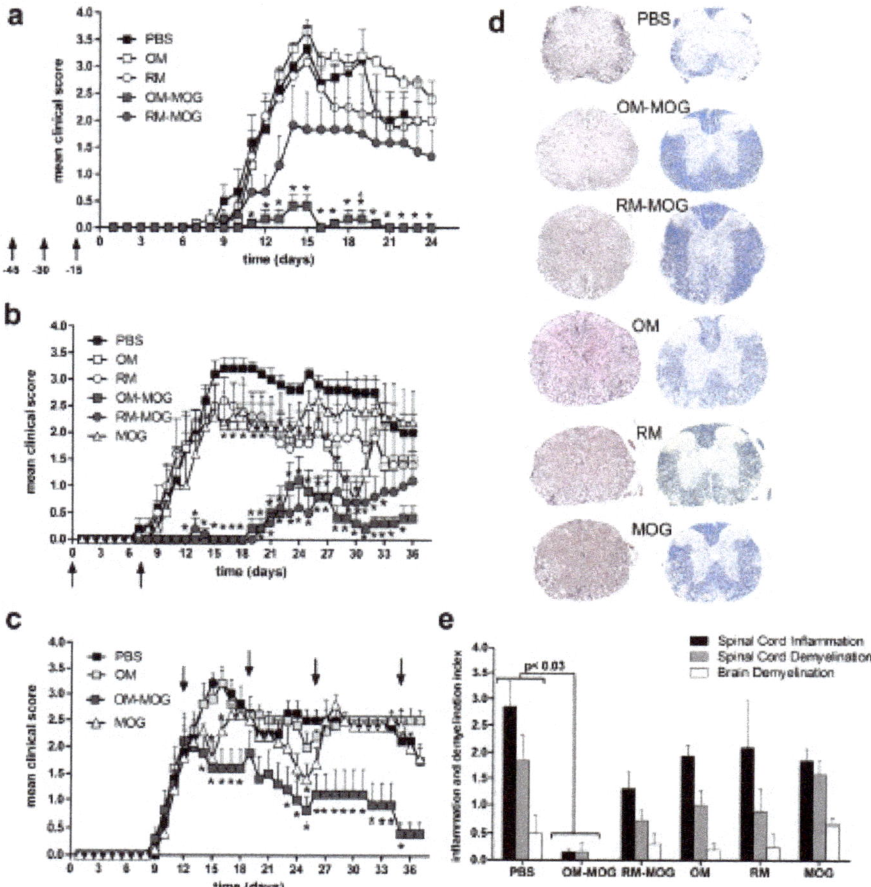

Figure 12. (**a**) Mean clinical scores of MOG-EAE in groups of mice vaccinated i.d. with OM-MOG, RM-MOG, OM, RM, or PBS at indicated time points (arrows) before immunization for EAE induction. (**b**) Mean clinical scores of MOG-EAE in groups of mice vaccinated i.d. at indicated time points (arrows) relative to immunization. (**c**) Mean clinical scores of MOG-EAE in groups of mice injected i.d. at indicated time points (arrows) after immunization. The results shown are from one representative of two (**b,c**) or three (**a**) independent experiments. (**d,e**) Vaccination with OM-MOG protects C57BL/6 mice from spinal cord inflammation and demyelination during MOG-EAE. (**d**) Inflammatory cell infiltration (left column) and demyelination (right column) were visualized on day 24 following immunization. (**e**) Quantification of spinal cord inflammation (black bars) and demyelination (grey bars) as well as brain demyelination (white bars) in all experimental groups. Representative data from five animals per group are shown. Statistical significance after comparisons between groups of mice (using the Kruskal-Wallis test) or histopathology indices (using Student's *t* test) is shown (*, $p < 0.05$). Triangles (**a**) indicate time points where pair-wise comparison between OM-MOG and RM-MOG groups also show significant differences (with permission of [188]).

Table 5. Carrier-aided vaccination.

Carrier	Particle Size (nm)	Zeta Potential (mV)	Antigen	Ag Loading (wt%)/Enc. Eff. (%)	Immunomodul. Agent	Vaccination Type	Admin. Route	Dose	Animal Model	Vaccination Outcome
Polymer particles										
PLGA NPs [193]	-	-	MOG_{35-55}	-	(r) IL-10	Prophylactic: 31 and 15 days b.i. Therapeutic: eight and 22 days p.i.	s.c.		Female C57BL/6 mice with EAE induced with MOG_{35-55}	Vaccination with mixed PLGA-MOG_{35-55} and PLGA-IL10 both in a prophylactic and therapeutic setting resulted in significant protection, decrease of EAE severity and reduction of histopathological lesions in spinal cord. i.v. vaccination with PLGA-$PLP_{139-151}$-TGF-β demonstrated improved efficiency at lower doses.
PLGA NPs [194]	-	-	$PLP_{139-151}$	8µg/mg NP	TGF-β (166ng/mg NP)	Prophylactic: seven days b.i. Therapeutic: 13 days p.i.	i.v. s.c.	2.5, 1.25, 0.0625 mg NPs	Female SJL/J mice (6–8 weeks old) with EAE induced with $PLP_{139-151}$	s.c. delivery of TGF-β-coupled to PLGA- $PLP_{139-151}$ NPs reduced the severity of relapses in EAE.
PLGA MPs [195]	800, 55,000		MOG_{35-55}	-/48.6	Vitamin D3 TGF-β1 Recombinant mouse GM-CSF	Preclinical/Therapeutic: 4, 7, and 10, days p.i.	s.c.		Female C57BL/6 mice (10–11 weeks old) with EAE induced with MOG_{35-55}	Delivery of various immunomodulators combined with MOG_{35-55} via a dual size MP platform resulted in the induction of enhanced antigen-specific autoimmune protection.
PLGA NPs [196]	151.2, 521.7	−14.1, −5.65	MOG_{35-55}	2.58, 0.96 /25.85, 9.65	-	Prophylactic: seven days b.i.	i.v. s.c.	2 mg NPs containing 20 µg MOG_{35-55}	Female C57BL/6 mice (6–8 weeks old) with EAE induced with MOG_{35-55}	The intravenous injection of PLGA-MOG_{35-55} was shown to delay EAE incidence and enhance antigen-specific immune tolerance.

Table 5. *Cont.*

Carrier	Particle Size (nm)	Zeta Potential (mV)	Antigen	Ag Loading (wt%)/Enc. Eff. (%)	Immunomodul. Agent	Vaccination Type	Admin. Route	Dose	Animal Model	Vaccination Outcome
PLGA-PEMA NPs [197]	429.9	−67.4	$PLP_{139-151}$ $PLP_{178-191}$	0.85/10.61	-	Prophylactic: 7, 25, and 50 days b.i. Preclinical/Therapeutic: 4, 14, and 18 days p.i.	i.v. i.p. s.c. oral	0.0625 0.125 0.625 1.25	Female SJL/J mice (6-8 weeks old) with EAE induced with $PLP_{178-191}$	Vaccination with PLP epitope-coupled PLGA-PEMA NPs was shown to both prevent and treat relapsing-remitting EAE. Tolerance induction was antigen-specific. The i.v. administration route was the most effective. s.c. vaccination with the tolerogenic NPs inhibited paralysis. Therapeutic s.c. treatment completely inhibited EAE relapses.
PLGA/PLA-PEG NPs [198]	-	-	$PLP_{139-151}$		rapamycin	Prophylactic: 14 and 21 days b.i. Therapeutic: 13 days p.i.	s.c. i.v.		SJL mice with EAE induced with $PLP_{139-151}$	A single therapeutic dose of tolerogenic NP sadministered i.v. near the peak of EAE resulted in complete prevention of relapses. Antigen-specific immune tolerance was successfully induced by PLP encephalitogenic epitopes, encapsulated in or conjugated with PLGA-PEMA NPs.
PLGA-PEMA NPs [199]	377.9, 621.5–834.8	−72.8, −50 to −43.7	$PLP_{139-151}$ $PLP_{178-191}$	0.58, 0.24–0.83/7.2, 4.4–16.5	-	Prophylactic: seven days b.i. Therapeutic: 18 days p.i.	i.v.		SJL/J mice with EAE induced with $PLP_{139-151}$ or $PLP_{178-191}$	Four i.v. injections of the developed NPs resulted in long-lasting amelioration of the disease by markedly reducing neuroinflammation, clinical EAE score and demyelination
PLGA NPs [181]	217	-	MOG40–54/H-2Db-Ig dimer, MOG35–55/I-Ab multimer	-	anti-Fas, PD-L1-Fc TGF-β1CD47-Fc	Therapeutic: 8, 18, 28, and 38 days p.i.	i.v.	1 mg NPs/ mouse/ injection	Female C57BL/6J mice (8–10 weeks old) with EAE induced with MOG35–55	

Table 5. *Cont.*

Carrier	Particle Size (nm)	Zeta Potential (mV)	Antigen	Ag Loading (wt%)/Enc. Eff. (%)	Immunomodul. Agent	Vaccination Type	Admin. Route	Dose	Animal Model	Vaccination Outcome
PLGA NPs PLA NPs [200]	PLGA: 351.3–436.2 PLA: 443.2	PLGA: −40.6 to −39.8 PLA: −40.2	PLP$_{139-151}$	PLGA: 0.25–0.28 PLA: 0.25	-	Preclinical/Therapeutic: seven days p.i.	i.v.	2.5, 2.0, 1.5 or 1.0 mg NPs/mouse	Female SJL/J mice (8–10 weeks old) with EAE induced with PLP$_{139-151}$	Low dose vaccination with PLA NPs resulted in long-lasting (>200 days post immunization) significant reduction of the clinical score at the chronic stage of EAE contrary to vaccination with PLGA NPs. Four injections of the multipotent particles resulted in long-lasting suppression of EAE and reduction of neuroinflammation in an antigen-specific manner. Administration of PLGA MPs resulted in slightly less efficient reduction of EAE symptoms compared with the administration of the peptide solution, but without toxicity. PLGA NPs coupled with a PLP encephalitogenic epitope were shown to efficiently induce antigen-specific tolerance in a mouse model of relapsing-remitting EAE induced either by PLP$_{139-151}$ or by both PLP$_{139-151}$ and PLP$_{178-191}$
PLGA MPs PEI-coated PLGA-MPs [201]	PLGA: 5080	PLGA: 45.3	MOG$_{35-55}$ MOG$_{40-54}$ MOG$_{40-54}$/H-2Db-Ig dimer, MOG35–55/I-Ab multimer		anti-Fas, PD-L1-Fc TGF-β1 CD47-Fc	Therapeutic: 8, 18, 28, and 38 days p.i.	i.v. i.p. s.c.		Female C57BL/6J mice with EAE induced with MOG$_{35-55}$	
PLGA MPs [202]	8000		Ac-PLP-BPI-NH2-2	24/8.2		Preclinical/Therapeutic: 4, 7, 10, and 14 days p.i.	s.c.		Female SJL/J mice (5–7 weeks old) with EAE induced with PLP139–151	
PLGA [203]	400–656	−51.3 to −38.0	PLP $_{139-151}$ PLP $_{178-191}$	0.26–0.8		Prophylactic: seven and one days b.i.	i.v.		Female SJL/J mice (6–8 weeks old) with EAE induced with PLP$_{139-151}$ or both PLP$_{139-151}$ and PLP$_{178-191}$	

Table 5. *Cont.*

Carrier	Particle Size (nm)	Zeta Potential (mV)	Antigen	Ag Loading (wt%)/Enc. Eff. (%)	Immunomodul. Agent	Vaccination Type	Admin. Route	Dose	Animal Model	Vaccination Outcome
PLGA NPs [204]	363–420	-	PLP$_{139-151}$		LABL	Preclinical/Therapeutic: 4, 7, and 10 days p.i.	s.c.	100 nmol PLP per injection	SJL/J female mice (5–7 weeks old) with EAE induced with PLP$_{139-151}$	It was shown that efficient suppression of EAE required the co-administration of PLP peptide and LABL.
PLGA [205]	538	−43	PLP$_{139-151}$	0.41–0.98	-	Preclinical/Therapeutic: seven days p.i.	i.v.	1 to 100 µg/mL NPs per injection	SJL/J mice with EAE induced with PLP$_{139-151}$	Antigen-specific, dose-dependent tolerance was successfully induced in an EAE model via the administration of PLGA NPs couple with a PLP peptide.
PLGA [206]	500	-	PLP$_{139-151}$	-	IL2	Prophylactic: secen days b.i. Therapeutic: 11 days p.i.	i.v.		SJL/J mice with EAE induced with PLP$_{139-151}$	Vaccination with PLGA NPs loaded with PLP$_{139-151}$ was found to prevent EAE onset and modulate its course.
PLGA MPs [207]	3900	-	MOG$_{35-55}$	0.73/38	Rapamycin (loading: 0.17%/enc. eff. 42.1%)	Therapeutic: 10 days p.i.	direct intra-lymph node (LN) injection	2 mg MPs per mouse or 1 mg MPs per LN	Female C57BL/6J mice (10–11 weeks old) with EAE induced with MOG$_{35-55}$	A single intra-LN injection (at the peak of EAE) of PLGA NPs containing a MOG peptide and rapamycin was revealed to permanently reverse paralysis.
Colloidal gel based on self-assembly of PLGA-CS and PLGA-Alginate NPs [208]	PLGA-CS: 400.1, PLGA-Alginate: 208.1	PLGA-CS: 23.79 PLGA-Alginate: −38.85,	Ac-PLP-BPI-NH$_2$-2		-	Prophylactic: five days b.i. Preclinical/Therapeutic: four and 30 days p.i.	s.c.	300 nmol of colloidal gel per injection	Mice (6–8 weeks old) with EAE PLP139–151	A single injection of the colloidal gel containing the Ac-PLP-BPI-NH$_2$-2 peptide led to long-term disease suppression.

Table 5. *Cont.*

Carrier	Particle Size (nm)	Zeta Potential (mV)	Antigen	Ag Loading (wt%)/Enc. Eff. (%)	Immunomodul. Agent	Vaccination Type	Admin. Route	Dose	Animal Model	Vaccination Outcome
Soluble antigen arrays (SAgAs)										
HA-peptide conjugate [209]	HA	-	PLP$_{139-151}$	-	LABL, B7AP, CD80-CAP1, sF2 (cyclized)	Preclinical/Therapeutic: 4, 7, and 10 days p.i.	s.c.	200 nmol PLP peptide	SJL/J (H-2s) female mice (4–6 weeks old) with EAE induced with PLP$_{139-151}$	SAgAs were shown to effectively reduce EAE incidence and suppress it via co-administration of an immunodominant myelin epitope and peptides targeting the B7 signaling pathway.
SAgAs [210]	HA	-	PLP$_{139-151}$	-	LABL	Preclinical/Therapeutic: 4, 7, and 10 days p.i.	s.c.	200 nmol PLP$_{139-151}$	SJL/J female mice (4–6 weeks old) with EAE induced with PLP$_{139-151}$	Co-administration via conjugation of PLP$_{139-151}$ and LABL improved the clinical scores of EAE
cSAgAs [184]	HA	-	PLP$_{139-151}$		LABL	Preclinical/Therapeutic: 4, 7, and 10 days p.i.	s.c.	50, 133, or 200 nmol PLP$_{139-151}$	SJL/J female mice (4–6 weeks old) with EAE induced with PLP$_{139-151}$	cSAgAs was found to achieve equivalent efficiency with SAgAS regarding the suppression of EAE at a quarter of the SAgAS dose.
cSAgAs (Figure 11) [185]	HA	-	PLP$_{139-151}$	-	LABL	Preclinical/Therapeutic: 4, 7, and 10 days p.i.	s.c.	50, nmol PLP$_{139-151}$	SJL/J female mice (4–6 weeks old) with EAE induced with PLP$_{139-151}$	Low dose s.c. vaccination with cSAgAS resulted in successful suppression of EAE clinical symproms and minimization of body weight loss.
SAgAs [210]	HA	-	PLP$_{139-151}$	-	LABL	Preclinical/Therapeutic: 4, 7, and 10 days p.i.	pulmonary	65.1–74.5 mg SAgAs/mouse kg	Female SJL/J mice (four weeks old) with EAE induced with PLP$_{139-151}$	The pulmonary administration of SAgAs was found to suppress the clinical score of the disease, decrease EAE incidence and improve weight gain.

Table 5. *Cont.*

Carrier	Particle Size (nm)	Zeta Potential (mV)	Antigen	Ag Loading (wt%)/Enc. Eff. (%)	Immunomodul. Agent	Vaccination Type	Admin. Route	Dose	Animal Model	Vaccination Outcome
SAgAs [183]	HA	-	PLP$_{139-151}$	-	LABL	Preclinical/Therapeutic: 4, 7, or 10 days p.i.	i.p., upper and lower i.m., upper and lower s.c., i.v. pulmonary	200 nMol PLP per 100 µL injection volume 200 nMol PLP per 50 µL injection volume	Female SJL/J mice (6–8 weeks old) with EAE induced with PLP$_{139-151}$	i.v. administration demonstrated similar efficiency with the other routes. p.i. vaccination decreased completely clinical disease scores. Single injection-based treatment resulted in decreased efficiency compared with a triple injection treatment. Decrease of SAgAs dose and/or injection volume decreased the therapeutic efficiency.
Immune polyelectrolyte multilayers (iPEMs)										
iPEMs [186]	-	-	MOG-R3	28.4–89.7%	GpG (0.7–10.3%)	Preclinical/Therapeutic: 5 and 10 p.i.	s.c.	200 µg of (MOG-R3/GpG)3 iPEMs, per injection.	C57BL/6J mice with EAE induced with a myelin antigen	s.c. delivery of iPEMs restrained inflammation and promoted autoimmune tolerance in an EAE mouse models.
iPEMs [187]	114.9–199.2	−42.5 to 33.4	MOGR1, MOGR2	0.57–9.18 µg of MOGRx	GpG 2.18 µg–4.88 µg	Preclinical/Therapeutic: seven days or 6, 12, and 18 days p.i.	s.c.	200 µg MOGR2 (85.9 µg GpG)	Female C57BL/6J mice (10 weeks old) with EAE induced with MOG$_{35-55}$	iPEMs were shown to improve the severity, progression and incidence of EAE.
Inorganic particles and pMHC-nanoparticles (pMHC-NPs)										
Quantum dots [211]	15.0–21.0	−17.6 to −4.2	MOG	Up to 55	-	Preclinical: two days p.i.	s.c.		Female C57BL/6 mice (10–12 weeks old)	Ten-fold reduction of EAE incidence. Increased numbers of QDs with lower peptide loading were more efficient regarding the induction of immune tolerance.

Table 5. *Cont.*

Carrier	Particle Size (nm)	Zeta Potential (mV)	Antigen	Ag Loading (wt%)/Enc. Eff. (%)	Immunomodul. Agent	Vaccination Type	Admin. Route	Dose	Animal Model	Vaccination Outcome
Iron oxide NPs [212]	-	-	MOG_{38-49}	-	IA^b	Therapeutic: 14 or 21 days p.i.			C57BL/6 mice with EAE induced with $pMOG_{35-55}$	By administration on day 14 the NPs were found to diminish the progression of the disease, whereas when administered on day 21 they were shown to restore the motor function of paralytic mice.
Iron oxide NPs [212]	-	-	$hPLP_{175-192}$ $hMOG_{97-108}$	-	DR4-IE	Therapeutic:			*HLA-DR4-IE-* transgenic C57BL/6 *IAb*null mice	Successful EAE suppression was observed.
Pegylated gold NPs [213]	60	-	MOG_{35-55} $PLP_{139-151}$ $PLP_{178-191}$		AhR ligand ITE	Prophylactic: admin. on the day of EAE induction Therapeutic: Admin. on day 17 post immunization. Weekly treatment of mice	parenteral	6 µg NPs per mouse	B6 mice with EAE induced with MOG_{35-55} SJL mice with EAE induced with EAE induced with $PLP_{139-151}$	Pegylated gold NPs loaded with MOG35-55 and ITE significantly suppressed the development of EAE, whereas those loaded with PLP epitopes reduced the clinical scores of the disease and the number of relapses.
Mannan-conjugated myelin peptides										
Mannan-peptide conjugates (Figure 12) [188]	-	-	MOG_{35-55}, $PLP_{139-151}$, $PLP_{178-191}$, MBP_{83-99}	-	-	Prophylactic: 45, 30, and 15 days b.i. Preclinical/Therapeutic: Admin. on day 0 and 7 p.i.	i.d.	30 µg peptide/injection 700 µg mannan/injection	C57BL/6 mice (12–14 weeks old) with EAE induced with MOG Female SJL/J mice (6–8 weeks old) with EAE induced with PLP.	Mannan-peptide conjugates were shown to generate robust antigen-specific protection of mice from the clinical disease symptoms.
Mannan-peptide conjugates [214]	-	-	Linear and cyclic MBP_{83-99} peptide analogues cyclo(83-99) [A91]MBP83-99 mutant peptide			Preclinical/Therapeutic: Admin. on day 0 and 14 p.i.	i.d.	50 µg of linear and cyclic MBP_{83-99} peptide analogues	Female SJL/J mice (6–8 weeks old) with EAE induced with linear and cyclic MBP_{83-99} peptide analogues	It was shown that the mutant peptide cyclo(83–99)[A91] MBP_{83-99} more efficiently inhibited EAE development.

Table 5. *Cont.*

Carrier	Particle Size (nm)	Zeta Potential (mV)	Antigen	Ag Loading (wt%)/Enc. Eff. (%)	Immunomodul. Agent	Vaccination Type	Admin. Route	Dose	Animal Model	Vaccination Outcome
Liposomes										
Liposomes [190]	861.3	−36.2	MOG_{40-55}	-/91.5	-	Preclinical/Therapeutic: 5 and 9 days p.i.	i.p.	1.75 mg of lipid per injection	C57BL/6 female mice (8 weeks old) with EAE induced with MOG_{40-55}	Liposomes successfully delayed the onset, suppressed the severity and decreased the incidence of the disease.
(mannosylated) SUV [189]	~85	−7.5 to −10.5	MBP_{46-62} $MBP_{124-139}$ $MBP_{147-170}$	-/90	-	Preclinical/Therapeutic: admin. on day 7 post immunization followed by five consecutive days.	s.c.		Female DA rats (8–9 weeks old) with EAE induced with a syngeneic spinal cord homogenate or with MBP_{63-81}.	It was revealed that mSUVs loaded with immunodominant epitopes of MBP could significantly suppress EAE in DA rats.
Exosomes										
mTGF-β1-EXOs [215]	50–100					Prophylactic: 8, 5, and 2 days b.i. Therapeutic: 14, 17 and 21 days p.i.	i.v.	10 μg/ mouse/ injection	Female C57BL/6 mice (6–8 weeks) with EAE induced with MOG_{35-55} Female BALB/c mice (6–8 weeks) with EAE induced with $PLP_{180-199}$	Treatment with mTGF-β1-EXOs from C57BL/6 mice successfully inhibited the development and progression of the disease in both mice strains.

Table 5. *Cont.*

Carrier	Particle Size (nm)	Zeta Potential (mV)	Antigen	Ag Loading (wt%)/Enc. Eff. (%)	Immunomodul. Agent	Vaccination Type	Admin. Route	Dose	Animal Model	Vaccination Outcome
Antigen-presenting yeast cells										
C. utilis expressing MOG$_{35-55}$ on its surface [216]	-	-	MOG$_{35-55}$ pCB13 pCB10			Prophylactic: admin. on day 7 prior to immunization and for six consecutive days	Oral	1.5×10^8 *C. utilis*	Female C57BL/6 mice (eight weeks old) with EAE induced with MOG$_{35-55}$	*C. utilis* expressing MOG$_{35-55}$ on its surface appeared to be a promising approach to protect myelin against autoimmunity by effectively inducing oral tolerance. Fungal viability was not found to affect the induction of tolerance.

PLGA: poly(lactide-co-glycolide); NPs: nanoparticles; MOG: myelin oligodendrocyte glycoprotein; (r) IL-10: recombinant interleukin; s.c.: subcutaneous; b.i.: before immunization; p.i.: post immunization; EAE: experimental autoimmune encephalomyelitis; PLP: proteolipid protein; TGF-β: transforming growth factor beta 1; i.v.: intravenous; MPs: microparticles; GM-CSF: granulocyte-macrophage colony-stimulating factor; PEMA: poly[ethylene-alt-maleic anhydride]; i.p.: intraperitoneal; PEG: polyethylene glycol; PLA: polylactide; PEI: polyethylene imine; Ac-PLP-BPI-NH2-2: (Ac-HSLGKWLGHPDKF-(AcpGAcpGAcp)2-ITDGEATDSG-NH2; Ac = acetyl, Acp = aminocaproic acid); CS: chitosan; SAgAs: soluble antigen arrays; HA: hyaluronic acid; LABL: ICAm-I binding peptide; cSAgAs: Click Soluble Antigen Arrays; i.p.: intraperitoneal; i.m.: intramuscular; iPEMs: immune polyelectrolyte multilayers; GpG: GpG oligonucleotide; MOGR3: MOG conjugated to tri-arginine; MOGR1 and MOGR2: MOG modified with either one or two cationic arginine residues; SUV: small unilamellar vesicles; mTGF-β1-EXOs: exosomes from dendritic cells expressing membrane-associated TGF-β1.

5. Clinical Trials

Various tolerance-inducing vaccination approaches (e.g., immunodominant myelin epitopes, APLs, DNA vaccination, attenuated autologous myelin reactive T cells, tolerogenic DCs, TCR peptide vaccination, nanocarriers loaded with encephalitogenic myelin peptides, etc.) with promising outcomes in experimental MS models have already reached the clinical development phase. Their safety, feasibility, and efficiency in inducing antigen-specific immune tolerance and reducing MRI-detected disease activity in patients with relapsing remitting and progressive MS have been preliminary demonstrated in phase I and II clinical trials [14,136,139] (Table 6).

6. Conclusions

Several exciting vaccination strategies targeting the induction of antigen-specific immune tolerance in MS have been developed during the last decades, based on a single epitope or cocktails of immunodominant epitopes of myelin proteins, altered peptide ligands, DNA vaccines, tolerogenic DCs pulsed with myelin peptides, attenuated autologous myelin reactive T cells, TCR peptide vaccines, conjugates of autoantigens with various types of cells, and different types of carriers (e.g., particles, vesicles, self-assembled structures, or molecular carriers) associated with myelin epitopes. Most of these approaches have demonstrated promising results in animal models of experimental autoimmune encephalomyelitis both in prophylactic and therapeutic vaccination protocols. They successfully prevented the disease or delayed the disease onset, reduced its clinical and pathological symptoms and decreased the number of relapses, or, in a therapeutic scheme, they reversed the clinical and histological signs of the disease. Accordingly, numerous of the abovementioned strategies reached the clinical development phase, and their safety, feasibility, and efficacy were assessed in both phase I and II clinical trials. However, the results from these trials have not indicated the same level of efficiency as in preclinical models. Even though different tolerance-inducing vaccination strategies were proven safe and well tolerated, and in some cases succeeded in inducing tolerogenic responses to patients, no major advances have been reported with respect to clinical efficiency. Consequently, despite the intensive research efforts, up to the present time, no FDA approved antigen-specific immunotherapy is available for treating MS patients. It appears that antigen-specific immunotherapies still face various major challenges such as the involvement of multiple autoantigens that can vary between patients, the epitope spreading, the vaccination of patients with inapparent infections, etc. These challenges need to be overcome in order to allow tolerogenic vaccines to play a major role in the treatment of MS patients. Progress in the scientific areas of recombinant protein expression, genome editing, and smartly designed carriers, combined with better understanding of MS immunopathogenesis and improved immunization protocols, could potentially improve these vaccination strategies in the future. Additionally, further clinical studies, such as phase II and III, including placebo groups, will be required in order to more realistically assess the clinical effectiveness of these interesting antigen-specific immunotherapies in both RRMS and SPMS patients.

Table 6. Clinical trials.

Objective	Phase	No. of Particip.	Antigen Immunotherapy	Admin. Route/Dose/Duration of Treatment	Results
To suppress disease activity in RRMS patients using CGP77116 [74]	II	24	CGP77116	s.c. injection/50 mg CGP77116 per week; 5 mg per week; 5 mg per month/9 months	Decrease of dose because of adverse effects. Trial termination due to treatment-related disease exacerbation.
Evaluation of NBI 5788 safety, and effect on RRMS patients [217]	II	144	NBI5788	s.c. injection/5, 20, or 50 mg NBI5788 per week/4 months	Trial suspension due to hypersensitivity reactions in some patients. No increase in relapses. Reduction of number and volume of enhancing lesions in patients who completed the trial receiving 5 mg of NBI5788 per week.
Assessment of safety, tolerability and clinical activity of AG284 in SPMS patients [218]	I	33	AG284	/0.6, 2, 6, 20, 60, 105, and 150 mg AG284/kg body weight; each dose was received daily for three alternate days/	No adverse events but also no significant therapeutic effect could be observed.
Assessment of the clinical efficiency of $MBP_{82\text{-}98}$ in patients with progressive MS [219]	II	32	$MBP_{82\text{-}98}$	i.v./500 mg $MBP_{82\text{-}98}$ per 6 months/24 months	Only patients with HLA haplotypes DR2 and/or DR4 appeared to have benefited from the treatment.
Evaluation of the safety and efficiency of $MBP_{82\text{-}98}$ in SPMS patients with HLA haplotypes DR2 and/or DR4 [220]	III	612	$MBP_{82\text{-}98}$	i.v./500 mg $MBP_{82\text{-}98}$ per 6 months/2 years	The administration of was safe and well tolerated. The treatment was not effective in SPMS patients with HLA $DR2^+$ or $DR4^+$
Evaluation of RTL1000 safety in MS patients [221]	I	34	RTL1000	i.v./2, 6, 20, 60, 200, and 100 mg of RTL/	RTL1000 was safe at doses $\leq$ 60 mg
Determination of the maximum tolerable dose and safety of RTL1000 in MS patients [222]	I	36	RTL1000	i.v./2, 6, 20, 60, 200, and 100 mg of RTL/	The maximum tolerable dose of RTL100 was 60 mg.
Examination of the effect of high dose $MBP_{82\text{-}98}$ on the number of regulatory T cells in CPMS patients [223]		10	$MBP_{82\text{-}98}$	i.v./500 mg of $MBP_{82\text{-}98}$ per 6 months/	Increase in the number of regulatory T cells in patients' PBMCs six weeks and six6 months after treatment. Renversement of the state of T cell anergy.
Assessment of safety and tolerability of autologous PBMCs coupled with 7 myelin peptides in RRMS and SPMS patients [224]	I	9	PBMCs chemically coupled with the following 7 myelin peptides: $MOG_{1\text{-}20}$, $MOG_{35\text{-}55}$, $MBP_{13\text{-}32}$, $MBP_{83\text{-}99}$, $MBP_{111\text{-}129}$, $MBP_{146\text{-}170}$, and $PLP_{139\text{-}154}$	Single infusion/1×10^3, 1×10^5, 1×10^7, 1×10^8, 1×10^9, 2.5×10^9 and 3×10^9 antigen-coupled PBMCs/3 months	The treatment was found to be safe and well-tolerated. Antigen-specific T cell responses were shown to decrease after treatment in patients who received doses $\geq 1 \times 10^9$ of antigen coupled PBMCs.
Examination of BHT-3009 safety and feasibility for immune nodulation in RRMS and SPMS patients [225]	I/II	30	BHT-3009	i.m./0.5, 1.5, and 3 mg of BHT-3009 at weeks 1, 3, 5, and 9 after patients' randomization into the clinical trial/The administration of BHT-3009 was combined or not with daily oral administration of 80 mg atorvastatin.	BHT-3009 was found to be safe and to induce antigen-specific immune tolerance in MS patients. The co-administration of atorvastatin was not considered substantially beneficial.
Assessment of the transdermal delivery of a mixture of three myelin peptides to induce immune tolerance in RRMS patients [226]		30	Mixture of the following 3 myelin peptides: $MBP_{85\text{-}99}$, $PLP_{139\text{-}151}$, and $MOG_{35\text{-}55}$	Transdermal (via an adhesive skin patch)/1 or 10 mg of each myelin peptide per week (for 4 weeks) and per month (for 11 months)/1 year	The transdermal administration of myelin peptides was proven to be tolerogenic in RRMS patients.

Table 6. *Cont.*

Objective	Phase	No. of Particip.	Antigen Immunotherapy	Admin. Route/Dose/Duration of Treatment	Results
Assessment of safety and efficiency of transdermal administration of myelin peptides in RRMS patients [227]		30	Mixture of the following three myelin peptides: MBP_{85-99}, MOG_{35-55}, and $PLP_{139-151}$	Transdermal (via an adhesive skin patch)/1 or 10 mg of each myelin peptide per week (for four weeks) and per month (for 11 months)/1 year	The transdermal delivery of myelin peptides was found to be safe, well tolerated and to reduce clinical symptoms and number of Gadolinium lesions in RRMS patients.
Evaluation of BHT-3009 regarding its safety and efficiency to induce immune tolerance in RRMS patients [228,229]	II	289	BHT-3009	i.m./0.5 and 1.5 mg of BHT-3009 at weeks 0, 2, 4, and every four weeks until week 44/The administration of BHT-3009 was combined or not with daily oral administration of 80 mg atorvastatin.	It was shown that treatment with the lower dose of BHT-3009 (e.g., 0.5 mg) succeeded in inducing antigen-specific immune tolerance in some patients in contrast with the higher dose (e.g., 1.5 mg) which was found to be ineffective.
Evaluation of ATX-MS-1467 safety in SPMS patients [117]	I	6	ATX-MS-1467	i.d/25, 50, 100, 400, and 800 µg of ATX-MS-1467/	The safety and tolerability of ATX-MS-1467 at a dose ≤ 800 µg, was successfully demonstrated in SPMS patients.
Evaluation of ATX-MS-1467 safety, tolerability and efficiency to induce tolerance in RRMS patients [230]	Ib, IIa	43, 37	ATX-MS-1467	Ib: i.d. (cohort 1) or s.c. (cohort 2)/25, 50, 100, 400 and 800 µg of ATXMS-1467 per two weeks (for eight weeks) and 800 µg per two weeks (for eight more weeks)/one year (including 32 weeks medication off study). IIa: i.d./50 µg of ATXMS-1467 (on day 1), 200 µg (on day 15), 800 µg (on day 29), and 800 µg per two weeks (for 16 more weeks)/one year (including 16 weeks medication off study).	Both treatment protocols were found to be safe. The relatively slow i.d. titration of ATX-MS-1467 followed by a longer high dose treatment period resulted in reduced GdE lesions which remained so even post treatment.
Tolerogenic DCs (tolDCs)					
Evaluation of the safety of myelin peptide loaded tolDCs and their ability of to induce immune tolerance in MS patients. [231]	I	8	Autologous tolDCs loaded with myelin peptides	i.v./50 × 10^6, 100 × 10^6, 150 × 10^6, and 300 × 10^6 tolDCs divided in three independent doses administered every two weeks/	Myelin peptide loaded tolDCs were proven to be safe and well tolerated, and to induce tolerogenic responses in MS patients.
Evaluation of the safety of intradermal and intranodal delivery myelin peptide loaded tolDCs and their efficacy regarding the induction of antigen-specific tolerization in MS patients [232]	I	9–15	Autologous peptide-mix loaded tolDCs	i.d. or intranodal/six repetitive doses of 5 × 10^6, 10 × 10^6 and 15 × 10^6 autologous peptide-mix loaded tolDCs: administration of doses 1–4 once every two weeks and of doses 5–6 once every month.	-

Table 6. *Cont.*

Objective	Phase	No. of Particip.	Antigen Immunotherapy	Admin. Route/Dose/Duration of Treatment	Results
T-cell vaccination (TCVs)					
Assessment of safety and immune efficiency of a polyclonal T cell vaccine in chronic MS patients in advanced diseases stages [233]		39	autological polyclonal TCVs	s.c./1.5–3 × 10^7 polyclonal T cells; four weekly injections followed by monthly injections.	Polyclonal TCV was proven safe and capable of inducing long-lasting, anti-inflammatory immune effects in progressive MS patients in advanced disease states.
To establish a safe and efficient dose of Tovaxin® [234]		9–15	Attenuated T cells reactive to the following myelin peptides MBP_{83-99}, $MBP_{151-170}$, PLP_{30-49}, $PLP_{180-199}$, MOG_{1-17} and MOG_{19-39}	s.c./6–9 × 10^6, 30–45 × 10^6, and 60–90 × 10^6 administered at weeks 0, 4, 12, and 20/	The study indicated the mid-dose as optimum with respect to safety, and efficiency in reducing peripheral blood myelin reactive T cells and showing a trend to improve clinical symptoms.
Evaluation of safety and efficacy of Tovaxin in RRMS patients [235]	IIb	150	T cells reactive to different immunodominant peptides from three myelin proteins	s.c./five injections at weeks 0, 4, 8, 12, and 24	s.c. administration of Tovaxin was shown to be safe. Evidence of clinical efficiency of Tovaxin® was observed during the analysis of subgroups of patients naïve to prior disease modifying therapies.
Examination of TCV safety and efficiency in progressive MS patients [236]	II	26	T-cell lines reactive to nine different peptides of MBP, MOG and PLP.	19 patients received s.c. TCV/10–30 × 10^6 T cells, on days 1, 30, 90 and 180/7 patients received sham injections.	The clinical trial demonstrated the safety of TCV in progressive MS patients and indicated its clinical efficiency.
Assessment of TCV safety and immune modulation in RRMS and CPMS patients [237]	pilot	5	CSF derived activated CD4+T cells	3 s.c. injections; 10^6 cells at months 2, 4, and 6.	TCV was safe and well tolerated. Patients were clinically stable or exhibited reduced EDSS without relapses during and post treatment. A 40% decrease in the relapses rate and a minimal decrease in EDSS was observed in RRMS patients. On the other hand, a slight increase of EDSS was detected in SPMS patients. Finally, MRI scans indicated a stabilization of the lesion activity.
Examine if the depletion of T cells reactive to MBP would have a clinical benefit for RRMS and SPMS patients [238]	Preliminary	54	Irradiated autologous T cells reactive to MBP-	3 s.c. injections at 2 month intervals, 30 × 10^6–60 × 10^6 cells per injection.	
Assess the use of T cell lines reacting with a broad range of antigens regarding targeting and depletion of specific T cells reactive to a great number of myelin antigens in SPMS patients. [239]	Pilot	4	Peripheral blood derived T cell lines reactive to bovine myelin		TCV with T cells reactive to whole bovine myelin were shown to efficiently promote depletion of circulating T cells reactive to myelin protein.
Evaluation of the TCV efficiency in patients with aggressive RRMS non-responding to DMTs [240]		20	Autologous attenuated T cell lines reactive to MBP and MOG encephalitogenic peptides.	Three s.c. injections in six- to eight-week intervals.	TCV was proven to be safe. A decrease in the relapse rate was observed. Additionally, significant decrease in the active lesions regarding number and volume as well as in T2 lesion burden was detected.
Identification of the idiotypic determinants triggering CD81 cytotoxic anti-idiotypic responses by TCV in MS patients [241]		3	Irradiated autologous T cell clones reactive to MBP_{83-99}	s.c./repetitive injections of 2 × 10^7 of each cell clone every 2 months for 8 months.	CD3-specific T cells were recognized as a representative anti-idiotypic population of T cells induced by TCV.

Table 6. *Cont.*

Objective	Phase	No. of Particip.	Antigen Immunotherapy	Admin. Route/Dose/Duration of Treatment	Results
T-cell receptor (TCR)					
To examine the therapeutic potential of a trivalent TCR vaccine in MS patients [242]		23	A trivalent TCR vaccine containing the CDR2 peptides BV5S2, BV6S5 and BV13S1	12 monthly vaccinations	The therapeutic TCR vaccine induced an extended immunoregulatory network which could control complex self-reactive responses of MS.
Liposomes					
Assessment of Xemys safety and efficiency in treating RRMS and SPMS patients non-responding to DMTs [191,192]	I	20	Xemys: Liposomes loaded with MBP_{46-62}, $MBP_{124-139}$ and $MBP_{147-170}$ And targeting CD206	s.c./six weekly injections of 50, 150, 225, 450, 900, and 900 µg Xemys	The administration of Xemys was proven to be safe and well tolerated, and to normalize cytokine levels in RRMS and SPMS patients.

RRMS: relapsing remitting multiple sclerosis; CGP77116: APL of MBP_{83-99}; APL: antigen peptide ligand; MBP: myelin basic protein; s.c.: subcutaneous; NBI 5788: APL of MBP_{83-99}; AG284: solubilized complex of HLA-DR2 with MBP_{84-102}; HLA: human leucocyte antigen; SPMS: secondary progressive multiple sclerosis; i.v.: intravenous; RTL1000: recombinant T-cell receptor ligand 1000; CPMS: chronic progressive multiple sclerosis; PBMCs: peripheral blood mononuclear cells; BHT-3009: tolerizing DNA vaccine encoding MBP; i.m.: intramuscular; ATX-MS-1467: mixture of equal quantities of synthetic peptides ATX-MS1 (MBP_{30-44}), ATX-MS4 ($MBP_{131-145}$), ATX-MS6 ($MBP_{140-154}$), and ATX-MS7 (MBP_{83-99}) in PBS; PBS: phosphate-buffered saline; i.d.: intradermal; tolDCs: tolerogenic dendritic cells; Tovaxin®: autologous T-cell immunotherapy; MOG:, PLP:; CSF:; DMTs: disease modifying therapies; CDR2: complementarity determining region 2.

Author Contributions: O.K. and C.K. contributed equally to the conceptualization, writing/preparation of the original draft, and writing—review and editing of the final paper. All authors have read and agreed to the published version of the manuscript.

Funding: This research received no external funding.

Conflicts of Interest: The authors declare no conflict of interest.

References

1. Harrington, E.P.; Bergles, D.E.; Calabresi, P.A. Immune cell modulation of oligodendrocyte lineage cells. *Neurosci. Lett.* **2020**, *715*, 134601. [CrossRef]
2. Baecher-Allan, C.; Kaskow, B.J.; Weiner, H.L. Multiple sclerosis: Mechanisms and immunotherapy. *Neuron* **2018**, *97*, 742–768. [CrossRef] [PubMed]
3. Dendrou, C.A.; Fugger, L.; Friese, M.A. Immunopathology of multiple sclerosis. *Nature Rev. Immunol.* **2015**, *15*, 545–558. [CrossRef] [PubMed]
4. Afshar, B.; Khalifehzadeh-Esfahani, Z.; Seyfizadeh, N.; Danbaran, G.R.; Hemmatzadeh, M.; Mohammadi, H. The role of immune regulatory molecules in multiple sclerosis. *J. Neuroimmunol.* **2019**, *337*, 577061. [CrossRef] [PubMed]
5. Greer, J.M.; Pender, M.P. Myelin proteolipid protein: An effective autoantigen and target of autoimmunity in multiple sclerosis. *J. Autoimmun.* **2008**, *31*, 281–287. [CrossRef] [PubMed]
6. Iwanowski, P.; Losy, J. Immunological differences between classical phenothypes of multiple sclerosis. *J. Neurol. Sci.* **2015**, *349*, 10–14. [CrossRef]
7. Lee, D.-H.; Linker, R.A. The role of myelin oligodendrocyte glycoprotein in autoimmune demyelination: A target for multiple sclerosis therapy? *Expert Opin. Ther. Targets* **2012**, *16*, 451–462. [CrossRef]
8. Rangachari, M.; Kuchroo, V.K. Using EAE to better understand principles of immune function and autoimmune pathology. *J. Autoimmun.* **2013**, *45*, 31–39. [CrossRef]
9. Lüssi, F.; Zipp, F.; Witsch, E. Dendritic cells as therapeutic targets in neuroinflammation. *Cell. Mol. Life Sci.* **2016**, *73*, 2425–2450. [CrossRef]
10. Ho, P.P.; Fontoura, P.; Platten, M.; Sobel, R.A.; DeVoss, J.J.; Lee, L.Y.; Kidd, B.A.; Tomooka, B.H.; Capers, J.; Agrawal, A.; et al. A Suppressive oligodeoxynucleotide enhances the efficacy of myelin cocktail/IL-4-tolerizing DNA vaccination and treats autoimmune disease. *J. Immunol.* **2005**, *175*, 6226–6234. [CrossRef]
11. Hemmer, B.; Nessler, S.; Zhou, D.; Kieseier, B.; Hartung, H.-P. Immunopathogenesis and immunotherapy of multiple sclerosis. *Nat. Clin. Prac. Neurol.* **2006**, *2*, 201–211. [CrossRef] [PubMed]
12. Hellings, N.; Raus, J.; Stinissen, P. T-cell based immunotherapy in multiple sclerosis: Induction of regulatory immune networks by T-cell vaccination. *Expert Rev. Clin. Immunol.* **2006**, *2*, 705–716. [CrossRef] [PubMed]
13. Zhou, Y.; Fang, L.; Peng, L.; Qiu, W. TLR9 and its signaling pathway in multiple sclerosis. *J. Neurol. Sci.* **2017**, *373*, 95–99. [CrossRef] [PubMed]
14. Willekens, B.; Cools, N. Beyond the magic bullet: Current progress of therapeutic vaccination in multiple sclerosis. *CNS Drugs* **2018**, *32*, 401–410. [CrossRef]
15. Skaper, S.D. Chapter 4—Oligodendrocyte precursor cells as a therapeutic target for demyelinating diseases, Prog. *Brain Res.* **2019**, *245*, 119–144. [CrossRef]
16. Gholamzad, M.; Ebtekar, M.; Ardestani, M.S.; Azimi, M.; Mahmodi, Z.; Mousavi, M.J.; Aslani, S. A comprehensive review on the treatment approaches of multiple sclerosis: Currently and in the future. *Inflamm. Res.* **2019**, *68*, 25–38. [CrossRef]
17. Derfuss, T. Personalized medicine in multiple sclerosis: Hope or reality? *BMC Medicine.* **2012**, *10*, 116. [CrossRef]
18. Lassmann, H. Pathogenic mechanisms associated with different clinical courses of multiple sclerosis. *Front. Immunol.* **2019**, *9*, 3116. [CrossRef]
19. Xie, Z.-X.; Zhang, H.-L.; Wu, X.-J.; Zhu, J.; Ma, D.-H.; Jin, T. Role of the immunogenic and tolerogenic subsets of dendritic cells in multiple sclerosis. *Mediat. Inflamm.* **2015**, *20*, 513295. [CrossRef]
20. Rostami, A.; Ciric, B. Role of Th17 cells in the pathogenesis of CNS inflammatory demyelination. *J. Neurol. Sci.* **2013**, *333*, 76–87. [CrossRef]
21. Baldassari, L.E.; Fox, R.J. Therapeutic advances and challenges in the treatment of progressive multiple sclerosis. *Drugs* **2018**, *78*, 1549–1566. [CrossRef] [PubMed]

22. Magliozzi, R.; Howell, O.; Vora, A.; Serafini, B.; Nicholas, R.; Puopolo, M.; Reynolds, R.; Aloisi, F. Meningeal B-cell follicles in secondary progressive multiple sclerosis associate with early onset of disease and severe cortical pathology. *Brain* **2007**, *130*, 1089–1104. [CrossRef] [PubMed]

23. Dolati, S.; Babaloo, Z.; Jadidi-Niaragh, F.; Ayromlou, H.; Sadreddini, S.; Yousefi, M. Multiple sclerosis: Therapeutic applications of advancing drug delivery systems. *Biomed. Pharmacother.* **2017**, *86*, 343–353. [CrossRef] [PubMed]

24. Wucherpfennig, K.W.; Strominger, J.L. Molecular mimicry in T cell-mediated autoimmunity: Viral peptides activate human T cell clones specific for myelin basic protein. *Cell* **1995**, *80*, 695–705. [CrossRef]

25. Fujinami, R.S.; von Herrath, M.G.; Christen, U.; Whitton, J.L. Molecular mimicry, bystander activation, or viral persistence: Infections and autoimmune disease. *Clin. Microbiol. Rev.* **2006**, *19*, 80–94. [CrossRef] [PubMed]

26. Kim, T.-S.; Shin, E.-C. The activation of bystander CD8+ T cells and their roles in viral infection. *Exp. Mol. Med.* **2019**, *51*, 154. [CrossRef]

27. Giacomini, P.S.; Bar-Or, A. Antigen-specific therapies in multiple sclerosis. *Expert Opin. Emerg. Drugs* **2009**, *14*, 551–560. [CrossRef]

28. Szczepanik, M. Mechanisms of immunological tolerance to the antigens of the central nervous system. Skin-induced tolerance as a new therapeutic concept. *J. Physiol. Pharmacol.* **2011**, *62*, 159–165.

29. Hellings, N.; Raus, J.; Stinissen, P. T-cell vaccination in multiple sclerosis: Update on clinical application and mode of action. *Autoimmun. Rev.* **2004**, *3*, 267–275. [CrossRef]

30. Irvine, D.J.; Hanson, M.C.; Rakhra, K.; Tokatlian, T. Synthetic nanoparticles for vaccines and immunotherapy. *Chem. Rev.* **2015**, *115*, 11109–11146. [CrossRef]

31. Selter, R.C.; Hemmer, B. Update on immunopathogenesis and immunotherapy in multiple sclerosis. *Immunotargets Ther.* **2013**, *2*, 21–30. [CrossRef]

32. Lim, E.T.; Giovannoni, G. Immunopathogenesis and immunotherapeutic approaches in multiple sclerosis. *Expert Rev. Neurother.* **2005**, *5*, 379–390. [CrossRef] [PubMed]

33. Grigoriadis, N.; van Pesch, V. A basic overview of multiple sclerosis immunopathology. *Eur. J. Neurol.* **2015**, *22*, 3–13. [CrossRef] [PubMed]

34. Sie, C.; Korn, T.; Mitsdoerffer, M. Th17 cells in central nervous system autoimmunity. *Exp. Neurol.* **2014**, *262*, 18–27. [CrossRef] [PubMed]

35. García-González, P.; Ubilla-Olguín, G.; Catalán, D.; Schinnerling, K.; Aguillón, J.C. Tolerogenic dendritic cells for reprogramming of lymphocyte responses in autoimmune diseases. *Autoimmun. Rev.* **2016**, *15*, 1071–1080. [CrossRef] [PubMed]

36. AGreenfield, L.; Hauser, S.L. B Cell therapy for multiple sclerosis: Entering an era. *Ann. Neurol.* **2018**, *83*, 13–26. [CrossRef] [PubMed]

37. Lu, L.-F.; Rudensky, A. Molecular orchestration of differentiation and function of regulatory T cells. *Genes Dev.* **2009**, *23*, 1270–1282. [CrossRef] [PubMed]

38. Gregori, S.; Goudy, K.S.; Roncarolo, M.G. The cellular and molecular mechanisms of immuno-suppression by human type 1 regulatory T cells. *Front. Immunol.* **2012**, *3*, 30. [CrossRef]

39. Lu, W.; Chen, S.; Lai, C.; Lai, M.; Fang, H.; Dao, H.; Kang, J.; Fan, J.; Guo, W.; Fu, L.; et al. Suppression of HIV replication by CD8(+) regulatory T cells in elite controllers. *Front. Immunol.* **2016**, *7*, 134. [CrossRef]

40. Vuddamalay, Y.; van Meerwijk, J.P.M. CD28- and CD28lowCD8+ regulatory T cells: Of mice and men. *Front. Immunol.* **2017**, *8*, 31. [CrossRef]

41. Milo, R. Therapeutic strategies targeting B-cells in multiple sclerosis. *Autoimmun. Rev.* **2016**, *15*, 714–718. [CrossRef]

42. Zhang, Y.; Salter, A.; Wallström, E.; Cutter, G.; Stüve, O. Evolution of clinical trials in multiple sclerosis. *Ther. Adv. Neurol. Disord.* **2019**, *12*, 1–14. [CrossRef]

43. Dargahi, N.; Katsara, M.; Tselios, T.; Androutsou, M.-E.; de Courten, M.; Matsoukas, J.; Apostolopoulos, V. Multiple sclerosis: Immunopathology and treatment update. *Brain Sci.* **2017**, *7*, 78. [CrossRef] [PubMed]

44. Gentile, A.; Musella, A.; de Vito, F.; Rizzo, F.R.; Fresegna, D.; Bullitta, S.; Vanni, V.; Guadalupi, L.; Bassi, M.S.; Buttari, F.; et al. Immunomodulatory effects of exercise in experimental multiple sclerosis. *Front. Immunol.* **2019**, *10*, 2197. [CrossRef] [PubMed]

45. Pasquier, R.A.D.; Pinschewer, D.D.; Merkler, D. Immunological mechanism of action and clinical profile of disease-modifying treatments in multiple sclerosis. *CNS Drugs* **2014**, *28*, 535–558. [CrossRef]

46. Tapeinos, C.; Battaglini, M.; Ciofani, G. Advances in the design of solid lipid nanoparticles and nanostructured lipid carriers for targeting brain diseases. *J. Control. Release* **2017**, *264*, 306–332. [CrossRef]

47. Cross, A.H.; Naismith, R.T. Established and novel disease-modifying treatments in multiple sclerosis. *J. Intern. Med.* **2014**, *275*, 350–363. [CrossRef]

48. Piehl, F. A changing treatment landscape for multiple sclerosis: Challenges and opportunities. *J. Intern. Med.* **2014**, *275*, 364–381. [CrossRef]

49. Wingerchuk, D.M.; Carter, J.L. Multiple Sclerosis: Current and Emerging Disease-Modifying Therapies and Treatment Strategies. *Mayo Clin. Proc.* **2014**, *89*, 225–240. [CrossRef]

50. Tramacere, I.; del Giovane, C.; Salanti, G.; D'Amico, R.; Pacchetti, I.; Filippini, G. Immunomodulators and immunosuppressants for relapsing-remitting multiple sclerosis: A network meta-analysis. *Cochrane Database Syst. Rev.* **2015**, *9*, CD011381. [CrossRef]

51. Wraith, D.C. The future of immunotherapy: A 20-year perspective. *Front. Immunol.* **2017**, *8*, 1668. [CrossRef] [PubMed]

52. D'Amico, E.; Patti, F.; Zanghì, A.; Zappia, M. A personalized approach in progressive multiple sclerosis: The current status of disease modifying therapies (DMTs) and future perspectives. *Int. J. Mol. Sci.* **2016**, *17*, 1725. [CrossRef] [PubMed]

53. Ciotti, J.R.; Cross, A.H. Disease-Modifying Treatment in Progressive Multiple Sclerosis. *Curr. Treat. Options Neurol.* **2018**, *20*, 12. [CrossRef] [PubMed]

54. Lassmann, H. Targets of therapy in progressive MS. *Mult. Scler.* **2017**, *23*, 1593–1599. [CrossRef]

55. Novartis Receives FDA Approval for Mayzent®(Siponimod), the First Oral Drug to Treat Secondary Progressive MS with Active Disease. Available online: https://novartis.gcs-web.com/Novartis-receives-FDA-approval-for-Mayzent-siponimod-the-first-oral-drug-to-treat-secondary-progressive-MS-with-active-disease?_ga=2.241998658.1110943223.1587297344-1758107691.1587297344 (accessed on 14 April 2020).

56. Zeposia (Ozanimod). Available online: https://multiplesclerosisnewstoday.com/zeposia-ozanimod-rpc1063-rrms/ (accessed on 14 April 2020).

57. Cladribine. Available online: https://en.wikipedia.org/wiki/Cladribine (accessed on 14 April 2020).

58. Wildner, P.; Selmaj, K.W. Multiple sclerosis: Skin-induced antigen-specific immune tolerance. *J. Neuroimmunol.* **2017**, *311*, 49–58. [CrossRef]

59. Sospedra, M.; Martin, R. Antigen-Specific Therapies in Multiple Sclerosis. *Int. Rev. Immunol.* **2005**, *24*, 393–413. [CrossRef]

60. Blanchfield, J.L. Antigen-specific tolerogenic vaccines inhibit autoimmune disease in a rodent model of multiple sclerosis. Ph.D. Thesis, The Faculty of the Department of Microbiology and Immunology Brody School of Medicine at East Carolina University, Greenville, USA, 2010.

61. Pickens, C.J.; Christopher, M.A.; Leon, M.A.; Pressnall, M.M.; Johnson, S.N.; Thati, S.; Sullivan, B.P.; Berkland, C. Antigen-drug conjugates as a novel therapeutic class for the treatment of antigen-specific autoimmune disorders. *Mol. Pharm.* **2019**, *16*, 2452–2461. [CrossRef]

62. Chunsong, Y.; Jingchao, X.; Meng, L.; Myunggi, A.; Haipeng, L. Bioconjugate strategies for the induction of antigen-specific tolerance in autoimmune diseases. *Bioconjug. Chem.* **2018**, *29*, 29719–29732. [CrossRef]

63. Mannie, M.D.; Curtis, A.D. II Tolerogenic vaccines for multiple sclerosis. *Hum. Vac. Immunother.* **2013**, *9*, 1032–1038. [CrossRef]

64. Yannakakis, M.P.; Tzoupis, H.; Michailidou, E.; Mantzourani, E.; Simal, C.; Tselios, T. Molecular dynamics at the receptor level of immunodominant myelin oligodendrocyte glycoprotein 35–55 epitope implicated in multiple sclerosis. *J. Mol. Graph. Model.* **2016**, *68*, 78–86. [CrossRef]

65. Lutterotti, A.; Martin, R. Antigen-specific tolerization approaches in multiple sclerosis. *Expert Opin. Investig. Drugs* **2014**, *23*, 9–20. [CrossRef] [PubMed]

66. Wraith, D. Antigen-specific immunotherapy. *Nature* **2016**, *530*, 422–423. [CrossRef] [PubMed]

67. Spence, A.; Klementowicz, J.E.; Bluestone, J.A.; Tang, Q. Targeting Treg signaling for the treatment of autoimmune diseases. *Curr. Opin. Immunol.* **2015**, *37*, 11–20. [CrossRef] [PubMed]

68. Sabatos-Peyton, C.A.; Verhagen, J.; Wraith, D.C. Antigen-specific immunotherapy of autoimmune and allergic diseases. *Curr. Opin. Immunol.* **2010**, *22*, 609–615. [CrossRef] [PubMed]

69. Steinman, L. The re-emergence of antigen-specific tolerance as a potential therapy for MS. *Mult. Scler. J.* **2015**, *21*, 1223–1238. [CrossRef] [PubMed]

70. Cappellano, G.; Comi, C.; Chiocchetti, A.; Dianzani, U. Exploiting PLGA-based biocompatible nanoparticles for next-generation tolerogenic vaccines against autoimmune disease. *Int. J. Mol. Sci.* **2019**, *20*, 204. [CrossRef]

71. Vanderlugt, C.L.; Miller, S.D. Epitope spreading inimmune-mediated diseases: Implications for immunotherapy. *Nat. Rev. Immunol.* **2002**, *2*, 85–95. [CrossRef]

72. Miller, S.D.; Turley, D.M.; Podojil, J.R. Antigen-specific tolerance strategies for the prevention and treatment of autoimmune disease. *Nat. Rev. Immunol.* **2007**, *7*, 665–677. [CrossRef]

73. Lutterotti, A.; Sospedra, M.; Martin, R. Antigen-specific therapies in MS—Current concepts and novel approaches. *J. Neurol. Sci.* **2008**, *274*, 18–22. [CrossRef]

74. Bielekova, B.; Goodwin, B.; Richert, N.; Cortese, I.; Kondo, T.; Afshar, G.; Grani, B.; Eaton, J.; Antel, J.; Frank, J.A.; et al. Encephalitogenic potential of the myelin basic protein peptide (amino acids 83–99) in multiple sclerosis: Results of a phase II clinical trial with an altered peptide ligand. *Nat. Med.* **2000**, *6*, 1167–1175. [CrossRef]

75. Turley, D.M.; Miller, S.D. Prospects for antigen-specific tolerance based therapies for the treatment of multiple sclerosis, Results Probl. *Cell Differ.* **2010**, *51*, 217–235. [CrossRef]

76. Constantinescu, C.S.; Farooqi, N.; O'Brien, K.; Gran, B. Experimental autoimmune encephalomyelitis (EAE) as a model for multiple sclerosis (MS). *Br. J. Pharmacol.* **2011**, *164*, 1079–1106. [CrossRef] [PubMed]

77. Fletcher, J.M.; Lalor, S.J.; Sweeney, C.M.; Tubridy, N.; Mills, K.H. T cells in multiple sclerosis and experimental autoimmune encephalomyelitis. *Clin. Exp. Immunol.* **2010**, *162*, 1–11. [CrossRef] [PubMed]

78. Libbey, J.E.; Fujinami, R.S. Experimental autoimmune encephalomyelitis as a testing paradigm for adjuvants and vaccines. *Vaccine* **2011**, *29*, 3356–3362. [CrossRef] [PubMed]

79. Tabansky, I.; Keskin, D.B.; Watts, D.; Petzold, C.; Funaro, M.; Sands, W.; Wright, P.; Yunis, E.J.; Najjar, S.; Diamond, B.; et al. Targeting DEC-205−DCIR2+ dendritic cells promotes immunological tolerance in proteolipid protein-induced experimental autoimmune encephalomyelitis. *Mol. Med.* **2018**, *24*, 17. [CrossRef]

80. Huang, X.; Lu, H.W.Q. The mechanisms and applications of T cell vaccination for autoimmune diseases: A comprehensive review. *Clinic. Rev. Allerg. Immunol.* **2014**, *47*, 219–233. [CrossRef] [PubMed]

81. Grau-López, L.; Raïch, D.; Ramo-Tello, C.; Naranjo-Gómez, M.; Dàvalos, A.; Pujol-Borrell, R.; Borràs, F.E.; Martínez-Cáceres, E. Myelin peptides in multiple sclerosis. *Autoimmun. Rev.* **2009**, *8*, 650–653. [CrossRef]

82. Kuchroo, V.K.; Anderson, A.C.; Waldner, H.; Munder, M.; Bettelli, E.; Nicholson, L.B. T cell response in experimental autoimmune encephalomyelitis (EAE): Role of Self and Cross-Reactive Antigens in Shaping, Tuning, and Regulating the Autopathogenic T Cell Repertoire. *Annu. Rev. Immunol.* **2002**, *20*, 101–123. [CrossRef]

83. Matsoukas, J.; Apostolopoulos, V.; Kalbacher, H.; Papini, A.-M.; Tselios, T.; Chatzantoni, K.; Biagioli, T.; Lolli, F.; Deraos, S.; Papathanassopoulos, P.; et al. Design and synthesis of a novel potent myelin basic protein epitope 87-99 cyclic analogue: Enhanced stability and biological properties of mimics render them a potentially new class of immunomodulators. *J. Med. Chem.* **2005**, *48*, 1470–1480. [CrossRef]

84. Reindl, M.; Waters, P. Myelin oligodendrocyte glycoprotein antibodies in neurological disease. *Nat. Rev. Neurol.* **2019**, *15*, 89–102. [CrossRef]

85. Kaushansky, N.; Eisenstein, M.; Zilkha-Falb, R.; Ben-Nun, A. The myelin-associated oligodendrocytic basic protein (MOBP) as a relevant primary target autoantigen in multiple sclerosis. *Autoimmun. Rev.* **2010**, *9*, 233–236. [CrossRef] [PubMed]

86. Androutsou, M.E.; Tapeinou, A.; Vlamis-Gardikas, A.; Tselios, T. Myelin oligodendrocyte glycoprotein and multiple sclerosis. *Med. Chem.* **2018**, *14*, 120–128. [CrossRef] [PubMed]

87. Tselios, T.; Aggelidakis, M.; Tapeinou, A.; Tseveleki, V.; Kanistras, I.; Gatos, D.; Matsoukas, J. Rational design and synthesis of altered peptide ligands based on human myelin oligodendrocyte glycoprotein 35–55 epitope: Inhibition of chronic experimental autoimmune encephalomyelitis in mice. *Molecules* **2014**, *19*, 17968–17984. [CrossRef] [PubMed]

88. Deraos, G.; Kritsi, E.; Matsoukas, M.-T.; Christopoulou, K.; Kalbacher, H.; Zoumpoulakis, P.; Apostolopoulos, V.; Matsoukas, J. Design of linear and cyclic mutant analogues of dirucotide peptide (MBP82–98) against multiple sclerosis: Conformational and binding studies to MHC Class II. *Brain Sci.* **2018**, *8*, 213. [CrossRef] [PubMed]

89. Tapeinou, A.; Giannopoulou, E.; Hansen, C.S.B.E.; Kalofonos, H.; Apostolopoulosd, V.; Vlamis-Gardikas, A.; Tselios, T. Design, synthesis and evaluation of an anthraquinone derivative conjugated to myelin basic protein immunodominant (MBP85-99) epitope: Towards selective immunosuppression. *Eur. J. Med. Chem.* **2018**, *143*, 621–631. [CrossRef]

90. Yannakakis, M.-P.; Simal, C.; Tzoupis, H.; Rodi, M.; Dargahi, N.; Prakash, M.; Mouzaki, A.; Platts, J.A.; Apostolopoulos, V.; Tselios, T.V. Design and synthesis of non-peptide mimetics mapping the immunodominant myelin basic protein (MBP83–96) epitope to function as T-cell receptor antagonists. *Int. J. Mol. Sci.* **2017**, *18*, 1215. [CrossRef]

91. Correale, J.; Farez, M.; Gilmore, W. Vaccines for multiple sclerosis: Progress to date. *CNS Drugs* **2008**, *22*, 175–198. [CrossRef]

92. Mantzourani, E.D.; Platts, J.A.; Brancale, A.; Mavromoustakos, T.M.; Tselios, T.V. Molecular dynamics at the receptor level of immunodominant myelin basic protein epitope 87–99 implicated in multiple sclerosis and its antagonists altered peptide ligands: Triggering of immune response. *J. Mol. Graph. Model.* **2007**, *26*, 471–481. [CrossRef]

93. Kaushansky, N.; de Rosbo, N.K.; Zilkha-Falb, R.; Yosef-Hemo, R.; Cohen, L.; Ben-Nun, A. 'Multi-epitope-targeted' immune-specific therapy for a multiple sclerosis-like disease via engineered multi-epitope protein is superior to peptides. *PLoS ONE* **2011**, *6*, e27860. [CrossRef]

94. Kaushansky, N.; Kaminitz, A.; Allouche-Arnon, H.; Ben-Nun, A. Modulation of MS-like disease by a multi epitope protein is mediated by induction of CD11c+ CD11b+ Gr1+ myeloid-derived dendritic cells. *J. Neuroimmunol.* **2019**, *333*, 476953. [CrossRef]

95. Moorman, C.D.; Curtis, A.D., 2nd; Bastian, A.G.; Elliott, S.E.; Mannie, M.D. A GMCSF-Neuroantigen Tolerogenic Vaccine Elicits Systemic Lymphocytosis of CD4+ CD25high FOXP3+ Regulatory T Cells in Myelin-Specific TCR Transgenic Mice Contingent Upon Low-Efficiency T Cell Antigen Receptor Recognition. *Front. Immunol.* **2019**, *9*, 3119. [CrossRef] [PubMed]

96. Mannie, M.D.; Blanchfield, J.L.; Islam, S.M.T.; Abbott, D.J. Cytokine-neuroantigen fusion proteins as a new class of tolerogenic, therapeutic vaccines for treatment of inflammatory demyelinating disease in rodent models of multiple sclerosis. *Front. Immunol.* **2012**, *3*, 255. [CrossRef] [PubMed]

97. Abbott, D.J.; Blanchfield, J.L.; Martinson, D.A.; Russell, S.C.; Taslim, N.; Curtis, A.D.; Mannie, M.D. Neuroantigen-specific, tolerogenic vaccines: GMCSF is a fusion partner that facilitates tolerance rather than immunity to dominant self-epitopes of myelin in murine models of experimental autoimmune encephalomyelitis (EAE). *BMC Immunol.* **2011**, *12*, 72. [CrossRef] [PubMed]

98. Blanchfield, J.L.; Mannie, M.D. A GMCSF-neuroantigen fusion protein is a potent tolerogen in experimental autoimmune encephalomyelitis (EAE) that is associated with efficient targeting of neuroantigen to APC. *J. Leukoc. Biol.* **2010**, *87*, 509–521. [CrossRef]

99. Petzold, C.; Schallenberg, S.; Stern, J.N.H.; Kretschmer, K. Targeted antigen delivery to DEC-205+ dendritic cells for tolerogenic vaccination. *Rev. Diabet. Stud.* **2012**, *9*, 305–318. [CrossRef]

100. Idoyaga, J.; Fiorese, C.; Zbytnuik, L.; Lubkin, A.; Miller, J.; Malissen, B.; Mucida, D.; Merad, M.; Steinman, R.M. Specialized role of migratory dendritic cells in peripheral tolerance induction. *J. Clin. Investig.* **2013**, *123*, 844–854. [CrossRef]

101. Stern, J.N.H.; Keskin, D.B.; Kato, Z.; Waldner, H.; Schallenberg, S.; Anderson, A.; von Boehmer, H.; Kretschmer, K.; Strominger, J.L. Promoting tolerance to proteolipid protein-induced experimental autoimmune encephalomyelitis through targeting dendritic cells. *Proc. Natl. Acad. Sci. USA* **2010**, *107*, 17280–17285. [CrossRef]

102. Ring, S.; Maas, M.; Nettelbeck, D.M.; Enk, A.H.; Mahnke, K. Targeting of autoantigens to DEC205+ dendritic cells in vivo suppresses experimental allergic encephalomyelitis in mice. *J. Immunol.* **2013**, *191*, 2938–2947. [CrossRef]

103. Kasagi, S.; Wang, D.; Zhang, P.; Zanvit, P.; Chen, H.; Zhang, D.; Li, J.; Che, L.; Maruyama, T.; Nakatsukasa, H.; et al. Combination of apoptotic T cell induction and self-peptide administration for therapy of experimental autoimmune encephalomyelitis. *EBioMedicine* **2019**, *44*, 50–59. [CrossRef]

104. Clemente-Casares, X.; Tsai, S.; Yang, Y.; Santamaria, P. Peptide-MHC-based nanovaccines for the treatment of autoimmunity: A "one size fits all" approach? *J. Mol. Med.* **2011**, *89*, 733–742. [CrossRef]

105. Offner, H.; Sinha, S.; Wang, C.; Burrows, G.G.; Vandenbark, A.A. Recombinant T-cell receptor ligands: Immunomodulatory, neuroprotective and neuroregenerative effects suggest application as therapy for multiple sclerosis. *Rev. Neurosci.* **2008**, *19*, 327–339. [CrossRef] [PubMed]

106. Sinha, S.; Subramanian, S.; Emerson-Webber, A.; Lindner, M.; Burrows, G.G.; Grafe, M.; Linington, C.; Vandenbark, A.A.; Bernard, C.C.A.; Offner, H. Recombinant TCR ligand reverses clinical signs and CNS damage of EAE induced by recombinant human MOG. *J. Neuroimmune Pharmacol.* **2010**, *5*, 231–239. [CrossRef] [PubMed]

107. Gong, Y.; Wang, Z.; Liang, Z.; Duan, H.; Ouyang, L.; Yu, Q.; Xu, Z.; Shen, G.; Weng, X.; Wu, X. Soluble MOG35-55/I-Ab Dimers Ameliorate Experimental Autoimmune Encephalomyelitis by Reducing Encephalitogenic T Cells. *PLoS ONE* **2012**, *7*, e47435. [CrossRef] [PubMed]

108. Vandenbark, A.A.; Rich, C.; Mooney, J.; Zamora, A.; Wang, C.; Huan, J.; Fugger, L.; Offner, H.; Jones, R.; Burrows, G.G. Recombinant TCR ligand induces tolerance to myelin oligodendrocyte glycoprotein 35-55 peptide and reverses clinical and histological signs of chronic experimental autoimmune encephalomyelitis in HLA-DR2 transgenic mice. *J. Immunol.* **2003**, *171*, 127–133. [CrossRef]

109. White, D.R.; Khedri, Z.; Kiptoo, P.; Siahaan, T.J.; Tolbert, T.J. Synthesis of a bifunctional peptide inhibitor–IgG1 Fc fusion that suppresses experimental autoimmune encephalomyelitis. *Bioconjug. Chem.* **2017**, *28*, 1867–1877. [CrossRef]

110. Ridwan, R.; Kiptoo, P.; Kobayashi, N.; Weir, S.; Hughes, M.; Williams, T.; Soegianto, R.; Siahaan, T.J. Antigen-specific suppression of experimental autoimmune encephalomyelitis by a novel bifunctional peptide inhibitor: Structure optimization and pharmacokinetics. *JPET* **2010**, *332*, 1136–1145. [CrossRef]

111. Badawi, A.H.; Siahaan, T.J. Suppression of MOG- and PLP-induced experimental autoimmune encephalomyelitis using a novel multivalent bifunctional peptide inhibitor. *J. Neuroimmunol.* **2013**, *263*, 20–27. [CrossRef]

112. Majewska, M.; Zając, K.; Srebro, Z.; Sura, P.; Książek, L.; Zemelka, M.; Szczepanik, M. Epicutaneous immunization with myelin basic protein protects from the experimental autoimmune encephalomyelitis. *Pharmacol. Rep.* **2007**, *59*, 74–79.

113. Szczepanik, M.; Tutaj, M.; Bryniarski, K.; Dittel, B.N. Epicutaneously induced TGF-h-dependent tolerance inhibits experimental autoimmune encephalomyelitis. *J. Neuroimmunol.* **2005**, *164*, 105–114. [CrossRef]

114. Tutaj, M.; Szczepanik, M. Epicutaneous (EC) immunization with myelin basic protein (MBP) induces TCRabþ CD4þ CD8þ double positive suppressor cells that protect from experimental autoimmune encephalomyelitis (EAE). *J. Autoimmun.* **2007**, *28*, 208–215. [CrossRef]

115. Li, H.; Zhang, G.-X.; Chen, Y.; Xu, H.; Fitzgerald, D.C.; Zhao, Z.; Rostami, A. CD11c+CD11b+ Dendritic cells play an important role in intravenous tolerance and the suppression of experimental autoimmune encephalomyelitis. *J. Immunol.* **2008**, *181*, 2483–2493. [CrossRef] [PubMed]

116. Lourbopoulos, A.; Deraos, G.; Matsoukas, M.-T.; Touloumi, O.; Giannakopoulou, A.; Kalbacher, H.; Grigoriadis, N.; Apostolopoulos, V.; Matsoukas, J. Cyclic MOG35–55 ameliorates clinical and neuropathological features of experimental autoimmune encephalomyelitis, Bioorgan. *Med. Chem.* **2017**, *25*, 4163–4174. [CrossRef] [PubMed]

117. Streeter, H.B.; Rigden, R.; Martin, K.F.; Scolding, N.J.; Wraith, D.C. Preclinical development and first-in-human study of ATX-MS-1467 for immunotherapy of MS. *Neurol. Neuroimmunol. Neuroinflamm.* **2015**, *2*, e93. [CrossRef] [PubMed]

118. Billetta, R.; Ghahramani, N.; Morrow, O.; Prakken, B.; de Jong, H.; Meschter, C.; Lanza, P.; Albani, S. Epitope-specific immune tolerization ameliorates experimental autoimmune encephalomyelitis. *Clin. Immunol.* **2012**, *145*, 94–101. [CrossRef]

119. Peron, J.P.S.; Yang, K.; Chen, M.-L.; Brandao, W.N.; Basso, A.S.; Commodaro, A.G.; Weiner, H.L.; Rizzo, L.V. Oral tolerance reduces Th17 cells as well as the overall inflammation in the central nervous system of EAE mice. *J. Neuroimmunol.* **2010**, *227*, 10–17. [CrossRef]

120. Song, F.; Guan, Z.; Gienapp, I.E.; Shawler, T.; Benson, J.; Whitacre, C.C. The thymus plays a role in oral tolerance in experimental autoimmune encephalomyelitis. *J. Immunol.* **2006**, *177*, 1500–1509. [CrossRef]

121. Deraos, G.; Rodi, M.; Kalbacher, H.; Chatzantoni, K.; Karagiannis, F.; Synodinos, L.; Plotas, P.; Papalois, A.; Dimisianos, N.; Papathanasopoulos, P.; et al. Properties of myelin altered peptide ligand cyclo(8–99)(Ala91,Ala96) MBP87-99 render it a promising drug lead for immunotherapy of multiple sclerosis. *Eur. J. Med.Chem.* **2015**, *101*, 13–23. [CrossRef]

122. Islam, S.M.T.; Curtis, A.D., 2nd; Taslim, N.; Wilkinson, D.S.; Mannie, M.D. GM-CSF-neuroantigen fusion proteins reverse experimental autoimmune encephalomyelitis and mediate tolerogenic activity in adjuvant-primed environments: Association with inflammation-dependent, inhibitory antigen presentation. *J. Immunol.* **2014**, *193*, 2317–2329. [CrossRef]

123. Mannie, M.D.; Clayson, B.A.; Buskirk, E.J.; DeVine, J.L.; Hernandez, J.J.; Abbott, D.J. IL-2/Neuroantigen fusion proteins as antigen-specific tolerogens in experimental autoimmune encephalomyelitis (EAE): Correlation of T cell-mediated antigen presentation and tolerance induction. *J. Immunol.* **2007**, *178*, 2835–2843. [CrossRef]

124. Link, J.M.; Rich, C.M.; Korat, M.; Burrows, G.G.; Offner, H.; Vandenbark, A.A. Monomeric DR2/MOG-35–55 recombinant TCR ligand treats relapses of experimental encephalomyelitis in DR2 transgenic mice. *Clin. Immunol.* **2007**, *123*, 95–104. [CrossRef]

125. Huan, J.; Subramanian, S.; Jones, R.; Rich, C.; Link, J.; Mooney, J.; Bourdette, D.N.; Vandenbark, A.A.; Burrows, G.G.; Offner, H. Monomeric recombinant TCR ligand reduces relapse rate and severity of experimental autoimmune encephalomyelitis in SJL/J mice through cytokine switch. *J. Immunol.* **2004**, *172*, 4556–4566. [CrossRef] [PubMed]

126. Offner, H.; Subramanian, S.; Wang, C.; Afentoulis, M.; Vandenbark, A.A.; Huan, J.; Burrows, G.G. Treatment of passive experimental autoimmune encephalomyelitis in SJL mice with a recombinant TCR ligand induces IL-13 and prevents axonal injury. *J. Immunol.* **2005**, *175*, 4103–4111. [CrossRef] [PubMed]

127. Sinha, S.; Subramanian, S.; Proctor, T.M.; Kaler, L.J.; Grafe, M.; Dahan, R.; Huan, J.; Vandenbark, A.A.; Burrows, G.G.; Offner, H. A promising therapeutic approach for multiple sclerosis: Recombinant T-cell receptor ligands modulate experimental autoimmune encephalomyelitis by reducing interleukin-17 production and inhibiting migration of encephalitogenic cells into the CNS. *J. Neurosci.* **2007**, *27*, 12531–12539. [CrossRef] [PubMed]

128. Sinha, S.; Subramanian, S.; Miller, L.; Proctor, T.M.; Roberts, C.; Burrows, G.G.; Vandenbark, A.A.; Offner, H. Cytokine switch and bystander suppression of autoimmune responses to multiple antigens in experimental autoimmune encephalomyelitis by a single recombinant T-Cell receptor ligand. *J. Neurosci.* **2009**, *29*, 3816–3823. [CrossRef] [PubMed]

129. Sinha, S.; Miller, L.; Subramanian, S.; McCarty, O.; Proctor, T.; Meza-Romero, R.; Burrows, G.G.; Vandenbark, A.A.; Offner, H. Binding of recombinant T cell receptor ligands (RTL) to antigen presenting cells prevents upregulation of CD11b and inhibits T cell activation and transfer of experimental autoimmune encephalomyelitis. *J. Neuroimmunol.* **2010**, *225*, 52–61. [CrossRef] [PubMed]

130. Wang, C.; Gold, B.G.; Kaler, L.J.; Yu, X.; Afentoulis, M.E.; Burrows, G.G.; Vandenbark, A.A.; Bourdette, D.N.; Offner, H. Antigen-specific therapy promotes repair of myelin and axonal damage in established EAE. *J. Neurochem.* **2006**, *98*, 1817–1827. [CrossRef] [PubMed]

131. Badawi, A.H.; Kiptoo, P.; Siahaan, T.J. Immune tolerance induction against experimental Autoimmune Encephalomyelitis (EAE) Using A New PLP-B7AP Conjugate that simultaneously targets B7/CD28 costimulatory signal and TCR/MHC-II signal. *J. Mult. Scler. (Foster City)* **2015**, *2*, 1000131.

132. Badawi, A.H.; Kiptoo, P.; Wang, W.-T.; Choi, I.-Y.; Lee, P.; Vines, C.M.; Siahaan, T.J. Suppression of EAE and prevention of blood-brain barrier breakdown after vaccination with novel bifunctional peptide inhibitor. *Neuropharmacology* **2012**, *62*, 1874–1881. [CrossRef]

133. Kobayashi, N.; Kobayashi, H.; Gu, L.; Malefyt, T.; Siahaan, T.J. Antigen-specific suppression of experimental autoimmune encephalomyelitis by a novel bifunctional peptide inhibitor. *JPET* **2007**, *322*, 879–886. [CrossRef]

134. Kobayashi, N.; Kiptoo, P.; Kobayashi, H.; Ridwan, R.; Brocke, S.; Siahaan, T.J. Prophylactic and therapeutic suppression of experimental autoimmune encephalomyelitis by a novel bifunctional peptide inhibitor. *Clin. Immunol.* **2008**, *129*, 69–79. [CrossRef]

135. Kiptoo, P.; Büyüktimkin, B.; Badawi, A.H.; Stewart, J.; Ridwan, R.; Siahaan, T.J. Controlling immune response and demyelination using highly potent bifunctional peptide inhibitors in the suppression of experimental autoimmune encephalomyelitis. *Clin. Exp. Immunol.* **2012**, *172*, 23–36. [CrossRef] [PubMed]

136. Fissolo, N.; Montalban, X.; Comabella, M. DNA vaccination techniques. *Methods Mol. Biol.* **2016**, *1304*, 39–50. [CrossRef] [PubMed]

137. Garren, H. DNA vaccines for autoimmune diseases. *Expert Rev. Vaccines* **2009**, *8*, 1195–1203. [CrossRef] [PubMed]

138. Fontoura, P.; Garren, H.; Steinman, L. Antigen-specific therapies in multiple sclerosis: Going beyond proteins and peptides. *Int. Rev. Immunol.* **2005**, *24*, 415–446. [CrossRef] [PubMed]

139. Fissolo, N.; Montalban, X.; Comabella, M. DNA-based vaccines for multiple sclerosis: Current status and future directions. *Clin. Immunol.* **2012**, *142*, 76–83. [CrossRef]

140. Stuve, O.; Cravens, P.; Eagar, T.N. DNA-based vaccines: The future of multiple sclerosis therapy? *Expert Rev. Neurother.* **2008**, *8*, 351–360. [CrossRef] [PubMed]

141. Jakimovski, D.; Weinstock-Guttman, B.; Ramanathan, M.; Dwyer, M.G.; Zivadinov, R. Infections, vaccines and autoimmunity: A multiple sclerosis perspective. *Vaccines* **2020**, *8*, 50. [CrossRef]

142. Garren, H.; Ruiz, P.J.; Watkins, T.A.; Fontoura, P.; Nguyen, L.-V.T.; Estline, E.R.; Hirschberg, D.L.; Steinman, L. Combination of gene delivery and DNA vaccination to protect from and reverse Th1 autoimmune disease via deviation to the Th2 pathway. *Immunity* **2001**, *15*, 15–22. [CrossRef]

143. Wefer, J.; Harris, R.A.; Lobell, A. Protective DNA vaccination against experimental autoimmune encephalomyelitis is associated with induction of IFNh. *J. Neuroimmunol.* **2004**, *149*, 66–76. [CrossRef]

144. Schif-Zuck, S.; Wildbaum, G.; Karin, N. Coadministration of plasmid DNA constructs encoding an encephalitogenic determinant and IL-10 elicits regulatory T cell-mediated protective immunity in the central nervous system. *J. Immunol.* **2006**, *177*, 8241–8247. [CrossRef]

145. Lobell, A.; Weissert, R.; Eltayeb, S.; de Graaf, K.L.; Wefer, J.; Storch, M.K.; Lassmann, H.; Wigzell, H.; Olsson, T. Suppressive DNA vaccination in myelin oligodendrocyte glycoprotein peptide-induced experimental autoimmune encephalomyelitis involves a T1-biased immune response. *J. Immunol.* **2003**, *170*, 1806–1813. [CrossRef] [PubMed]

146. Andersson, A.; Isaksson, M.; Wefer, J.; Norling, A.; Flores-Morales, A.; Rorsman, F.; Kämpe, O.; Harris, R.A.; Lobell, A. Impaired autoimmune T helper 17 cell responses following DNA vaccination against rat experimental autoimmune encephalomyelitis. *PLoS ONE* **2008**, *3*, e3682. [CrossRef] [PubMed]

147. Kang, Y.; Sun, Y.; Zhang, J.; Gao, W.; Kang, J.; Wang, Y.; Wang, B.; Xia, G. Treg cell resistance to apoptosis in DNA vaccination for experimental autoimmune encephalomyelitis treatment. *PLoS ONE* **2012**, *7*, e49994. [CrossRef] [PubMed]

148. Walczak, A.; Szymanska, B.; Selmaj, K. Differential prevention of experimental autoimmune encephalomyelitis with antigen-specific DNA vaccination. *Clin. Neurol. Neurosurg.* **2004**, *106*, 241–245. [CrossRef] [PubMed]

149. Liu, J.; Cao, X. Regulatory dendritic cells in autoimmunity: A comprehensive review. *J. Autoimmun.* **2015**, *63*, 1–12. [CrossRef] [PubMed]

150. Domogalla, M.P.; Rostan, P.V.; Raker, V.K.; Steinbrink, K. Tolerance through education: How tolerogenic dendritic cells shape immunity. *Front. Immunol.* **2017**, *8*, 1764. [CrossRef]

151. Van Brussel, I.; Lee, W.P.; Rombouts, M.; Nuyts, A.H.; Heylen, M.; DeWinter, B.Y.; Cools, N.; Schrijvers, D.M. Tolerogenic dendritic cell vaccines to treat autoimmune diseases: Can the unattainable dream turn into reality? *Autoimmun. Rev.* **2014**, *13*, 138–150. [CrossRef]

152. Flórez-Grau, G.; Zubizarreta, I.; Cabezón, R.; Villoslada, P.; Benitez-Ribas, D. Tolerogenic dendritic cells as a promising antigen-specific therapy in the treatment of multiple sclerosis and neuromyelitis optica from preclinical to clinical trials. *Front. Immunol.* **2018**, *9*, 1169. [CrossRef]

153. Obregon, C.; Kumar, R.; Pascual, M.A.; Vassalli, G.; Golshayan, D. Update on dendritic cell-induced immunological and clinical tolerance. *Front. Immunol.* **2017**, *8*, 1514. [CrossRef]

154. Derdelinckx, J.; Mansilla, M.J.; de Laere, M.; Lee, W.-P.; Navarro-Barriuso, J.; Wens, I.; Nkansah, I.; Daans, J.; de Reu, H.; Keliris, A.J.; et al. Clinical and immunological control of experimental autoimmune encephalomyelitis by tolerogenic dendritic cells loaded with MOG-encoding RNA. *J. Neuroinflam.* **2019**, *16*, 167. [CrossRef]

155. Iberg, C.A.; Hawiger, D. Natural and induced tolerogenic dendritic cells. *J. Immunol.* **2020**, *204*, 733–744. [CrossRef] [PubMed]

156. Vandenbark, A.A.; Abulafia-Lapid, R. Autologous T-cell vaccination for multiple sclerosis: A perspective on progress. *BioDrugs* **2008**, *22*, 265–273. [CrossRef] [PubMed]

157. Volovitz, I.; Marmora, Y.; Mor, F.; Flügel, A.; Odoardi, F.; Eisenbach, L.; Cohen, I.R. T cell vaccination induces the elimination of EAE effector T cells: Analysis using GFP-transduced, encephalitogenic T cells. *J. Autoimmun.* **2010**, *35*, 135–144. [CrossRef] [PubMed]

158. Turley, D.M.; Miller, S.D. Peripheral tolerance induction using ethylenecarbodiimide-fixed APCs uses both direct and indirect mechanisms of antigen presentation for prevention of experimental autoimmune encephalomyelitis. *J. Immunol.* **2007**, *178*, 2212–2220. [CrossRef] [PubMed]

159. Getts, D.R.; Turley, D.M.; Smith, C.E.; Harp, C.T.; McCarthy, D.; Feeney, E.M.; Getts, M.T.; Martin, A.J.; Luo, X.; Terry, R.R.; et al. Tolerance Induced by Apoptotic Antigen-Coupled Leukocytes is Induced by PD-L1+, IL-10-Producing Splenic Macrophages and Maintained by Tregs. *J. Immunol.* **2011**, *187*, 2405–2417. [CrossRef] [PubMed]

160. Pishesha, N.; Bilate, A.M.; Wibowo, M.C.; Huang, N.-J.; Lia, Z.; Deshycka, R.; Bousbaine, D.; Lia, H.; Pattersona, H.C.; Dougana, S.K.; et al. Engineered erythrocytes covalently linked to antigenic peptides can protect against autoimmune disease. *Proc. Natl. Acad. Sci. USA* **2017**, *114*, 3157–3162. [CrossRef]

161. Chen, Z.; Yang, D.; Peng, X.; Lin, J.; Suc, Z.; Lia, J.; Zhang, X.; Weng, Y. Beneficial effect of atorvastatin-modified dendritic cells pulsed with myelin oligodendrocyte glycoprotein autoantigen on experimental autoimmune encephalomyelitis. *Neuroreport* **2018**, *29*, 317–327. [CrossRef]

162. Wang, X.; Zhang, J.; Baylink, D.J.; Li1, C.-H.; Watts, D.M.; Xu, Y.; Qin, X.; Walter, M.H.; Tang, X. Targeting non-classical myelin epitopes to treat experimental autoimmune encephalomyelitis. *Sci. Rep.* **2016**, *6*, 36064. [CrossRef]

163. Kalantari, T.; Karimi, M.H.; Ciric, B.; Yan, Y.; Rostami, A.; Kamali-Sarvestani, E. Tolerogenic dendritic cells produced by lentiviral-mediated CD40- and interleukin-23p19-specific shRNA can ameliorate experimental autoimmune encephalomyelitis by suppressing T helper type 17 cells. *Clin. Exp. Immunol.* **2014**, *176*, 180–189. [CrossRef]

164. Mansilla, M.J.; Sellès-Moreno, C.; Fàbregas-Puig, S.; Amoedo, J.; Navarro-Barriuso, J.; Teniente-Serra, A.; Grau-López, L.; Ramo-Tello, C.; Martínez-Cáceres, E.M. Beneficial effect of tolerogenic dendritic cells pulsed with MOG autoantigen in experimental autoimmune encephalomyelitis. *CNS Neurosci. Ther.* **2015**, *21*, 222–230. [CrossRef]

165. Mansilla, M.J.; Contreras-Cardone, R.; Navarro-Barriuso, J.; Cools, N.; Berneman, Z.; Ramo-Tello, C.; Martínez-Cáceres, E.M. Cryopreserved vitamin D3-tolerogenic dendritic cells pulsed with autoantigens as a potential therapy for multiple sclerosis patients. *J. Neuroinflamm.* **2016**, *13*, 113. [CrossRef] [PubMed]

166. Zhou, Y.; Leng, X.; Luo, S.; Su, Z.; Luo, X.; Guo, H.; Mo, C.; Zou, Q.; Liu, Y.; Wang, Y. Tolerogenic dendritic cells generated with tofacitinib ameliorate experimental autoimmune encephalomyelitis through modulation of Th17/Treg balance. *J. Immunol. Res.* **2016**, *2016*, 5021537. [CrossRef] [PubMed]

167. Xie, Z.; Chen, J.; Zheng, C.; Wu, J.; Cheng, Y.; Zhu, S.; Lin, C.; Cao, Q.; Zhu, J.; Jin, T. 1,25-dihydroxyvitamin D3-induced dendritic cells suppress experimental autoimmune encephalomyelitis by increasing proportions of the regulatory lymphocytes and reducing T helper type 1 and type 17 cells. *Immunology* **2017**, *152*, 414–424. [CrossRef] [PubMed]

168. Papenfuss, T.L.; Powell, N.D.; McClain, M.A.; Bedarf, A.; Singh, A.; Gienapp, I.E.; Shawler, T.; Whitacre, C.C. Estriol generates tolerogenic dendritic cells in vivo that protect against autoimmunity. *J. Immunol.* **2011**, *15*, 186–3346. [CrossRef] [PubMed]

169. Menges, M.; Rößner, S.; Voigtländer, C.; Schindler, H.; Kukutsch, N.A.; Bogdan, C.; Erb, K.; Schuler, G.; Lutz, M.B. Repetitive injections of dendritic cells matured with tumor necrosis factor α induce antigen-specific protection of mice from autoimmunity. *J. Exp. Med.* **2002**, *195*, 15–21. [CrossRef] [PubMed]

170. Xiao, B.-G.; Huang, Y.-M.; Yang, J.-S.; Xu, L.-Y.; Link, H. Bone marrow-derived dendritic cells from experimental allergic encephalomyelitis induce immune tolerance to EAE in Lewis rats. *Clin. Exp. Immunol.* **2001**, *125*, 300–309. [CrossRef] [PubMed]

171. Kim, Y.C.; Zhang, A.-H.; Yoon, J.; Culp, W.E.; Leesa, J.R.; Wucherpfennig, K.W.; Scott, D.W. Engineered MBP-specific human Tregs ameliorate MOG-induced EAE through IL-2-triggered inhibition of effector T cells. *J. Autoimmun.* **2018**, *92*, 77–86. [CrossRef]

172. Pereira, B.d.; Fraefel, C.; Hilbe, M.; Ackermann, M.; Dresch, C. Transcriptional targeting of DCs with lentiviral vectors induces antigen-specific tolerance in a mouse model of multiple sclerosis. *Gene Ther.* **2013**, *20*, 556–566. [CrossRef]

173. Eixarch, H.; Espejo, C.; Gómez, A.; Mansilla, M.J.; Castillo, M.; Mildner, A.; Vidal, F.; Gimeno, R.; Prinz, M.; Montalban, X.; et al. Tolerance induction in experimental autoimmune encephalomyelitis using non-myeloablative hematopoietic gene therapy with autoantigen. *Mol. Ther.* **2009**, *17*, 897–905. [CrossRef]

174. Casacuberta-Serra, S.; Costa, C.; Eixarch, H.; Mansilla, M.J.; López-Estévez, S.; Martorell, L.; Parés, M.; Montalban, X.; Espejo, C.; Barquinero, J. Myeloid-derived suppressor cells expressing a self-antigen ameliorate experimental autoimmune encephalomyelitis. *Exp. Neurol.* **2016**, *286*, 50–60. [CrossRef]

175. Tabansky, I.; Messina, M.D.; Bangeranye, C.; Goldstein, J.; Blitz-Shabbir, K.M.; Machado, S.; Jeganathan, V.; Wright, P.; Najjar, S.; Cao, Y.; et al. Advancing drug delivery systems for the treatment of multiple sclerosis. *Immunol. Res.* **2015**, *63*, 58–69. [CrossRef] [PubMed]

176. Ballerini, C.; Baldi, G.; Aldinucci, A.; Maggi, P. Nanomaterial applications in multiple sclerosis inflamed Brain. *J. Neuroimmune Pharmacol.* **2015**, *10*, 1–13. [CrossRef] [PubMed]

177. Gharagozloo, M.; Majewski, S.; Foldvari, M. Therapeutic applications of nanomedicine in autoimmune diseases: From immunosuppression to tolerance induction. *Nanomed. Nanotechnol.* **2015**, *11*, 1003–1018. [CrossRef] [PubMed]

178. Pires, L.R.; Marques, F.; Sousa, J.C.; Cerqueira, J.; Pinto, I.M. Nano- and micro-based systems for immunotolerance induction in multiple sclerosis. *Hum. Vaccines Immunother.* **2016**, *12*, 1886–1890. [CrossRef] [PubMed]

179. Veld, R.H.I.; da Silva, C.G.; Kaijzel, E.L.; Chan, A.B.; Cruz, L.J. The Potential of nano-vehicle mediated therapy in vasculitis and multiple sclerosis. *Curr. Pharm. Des.* **2017**, *23*, 1985–1992. [CrossRef]

180. Gammon, J.M.; Jewell, C.M. Engineering immune tolerance with biomaterials. *Adv. Healthc. Mater.* **2019**, *8*, e1801419. [CrossRef]

181. Pei, W.; Wan, X.; Shahzad, K.A.; Zhang, L.; Song, S.; Jin, X.; Wang, L.; Zhao, C.; Shen, C. Direct modulation of myelin-autoreactive CD4+ and CD8+ T cells in EAE mice by a tolerogenic nanoparticle co-carrying myelin peptide-loaded major histocompatibility complexes, CD47 and multiple regulatory molecules. *Int. J. Nanomed.* **2018**, *13*, 3731–3750. [CrossRef]

182. Sestak, J.O.; Sullivan, B.P.; Thati, S.; Northrup, L.; Hartwell, B.; Antunez, L.; Forrest, M.L.; Vines, C.M.; Siahaan, T.J.; Berkland, C. Codelivery of antigen and an immune cell adhesion inhibitor is necessary for efficacy of soluble antigen arrays in experimental autoimmune encephalomyelitis. *Mol. Ther. Methods Clin. Dev.* **2014**, *1*, 14008. [CrossRef]

183. Thati, S.; Kuehl, C.; Hartwell, B.; Sestak, J.; Siahaan, T.; Forrest, M.L.; Berkland, C. Routes of Administration and dose optimization of soluble antigen arrays in mice with experimental autoimmune encephalomyelitis. *J. Pharm. Sci.* **2015**, *104*, 714–721. [CrossRef]

184. Hartwell, B.L.; Pickens, C.J.; Leon, M.; Berkland, C. Multivalent antigen arrays exhibit high avidity binding and modulation of B cell receptor-mediated signaling to drive efficacy against experimental autoimmune encephalomyelitis. *Biomacromolecules* **2017**, *18*, 1893–1907. [CrossRef]

185. Hartwell, B.L.; Pickens, C.J.; Leon, M.; Northrup, L.; Christopher, M.; Griffin, J.D.; Martinez-Becerra, F.; Berkland, C. Soluble antigen arrays disarm antigen-specific B cells to promote lasting immune tolerance in experimental autoimmune encephalomyelitis. *J. Autoimmun.* **2018**, *93*, 76–88. [CrossRef] [PubMed]

186. Tostanoski, L.H.; Chiu, Y.-C.; Andorko, J.I.; Guo, M.; Zeng, X.; Zhang, P.; Royal, W., III; Jewell, C.M. Design of polyelectrolyte multilayers to promote immunological tolerance. *ACS Nano* **2016**, *10*, 9334–9345. [CrossRef] [PubMed]

187. Hess, K.L.; Andorko, J.I.; Tostanoski, L.H.; Jewell, C.M. Polyplexes assembled from self-peptides and regulatory nucleic acids blunt toll-like receptor signaling to combat autoimmunity. *Biomaterials* **2017**, *118*, 51–62. [CrossRef] [PubMed]

188. Tseveleki, V.; Tselios, T.; Kanistras, I.; Koutsoni, O.; Karamita, M.; Vamvakas, S.-S.; Apostolopoulos, V.; Dotsika, E.; Matsoukas, J.; Lassmann, H.; et al. Mannan-conjugated myelin peptides prime non-pathogenic Th1 and Th17 cells and ameliorate experimental autoimmune encephalomyelitis. *Exp. Neurol.* **2015**, *267*, 254–267. [CrossRef] [PubMed]

189. Belogurov, A.A., Jr.; Stepanov, A.V.; Smirnov, I.V.; Melamed, D.; Bacon, A.; Mamedov, A.E.; Boitsov, V.M.; Sashchenko, L.P.; Ponomarenko, N.A.; Sharanova, S.N.; et al. Liposome-encapsulated peptides protect against experimental allergic encephalitis. *FASEB J.* **2013**, *27*, 222–231. [CrossRef] [PubMed]

190. Pujol-Autonell, I.; Mansilla, M.-J.; Rodriguez-Fernandez, S.; Cano-Sarabia, M.; Navarro-Barriuso1, J.; Ampudia, R.-M.; Rius, A.; Garcia-Jimeno, S.; Perna-Barrull, D.; Martinez-Caceres, E.; et al. Liposome-based immunotherapy against autoimmune diseases: Therapeutic effect on multiple sclerosis. *Nanomedicine* **2017**, *12*, 1231–1242. [CrossRef]

191. Belogurov, A.; Zakharov, K.; Lomakin, Y.; Surkov, K.; Avtushenko, S.; Kruglyakov, P.; Smirnov, I.; Makshakov, G.; Lockshin, C.; Gregoriadis, G.; et al. CD206-targeted liposomal myelin basic protein peptides in patients with multiple sclerosis resistant to first-line disease-modifying therapies: A first-in-human, proof-of-concept dose-escalation study. *Neurotherapeutics* **2016**, *13*, 895–904. [CrossRef]

192. Lomakin, Y.; Belogurov, A., Jr.; Glagoleva, I.; Stepanov, A.; Zakharov, K.; Okunola, J.; Smirnov, I.; Genkin, D.; Gabibov, A. Administration of myelin basic protein peptides encapsulated in mannosylated liposomes normalizes level of serum TNF-α and IL-2 and chemoattractants CCL2 and CCL4 in multiple sclerosis patients. *Mediat. Inflamm.* **2016**, *2016*, 2847232. [CrossRef]

193. Cappellano, G.; Woldetsadik, A.D.; Orilieri, E.; Shivakumar, Y.; Rizzi, M.; Carniato, F.; Gigliotti, C.L.; Boggio, E.; Clemente, N.; Comi, C.; et al. Subcutaneous inverse vaccination with PLGA particles loaded with aMOG peptide and IL-10 decreases the severity of experimental autoimmune encephalomyelitis. *Vaccine* **2014**, *32*, 5681–5689. [CrossRef]

194. Casey, L.M.; Pearson, R.M.; Hughes, K.R.; Liu, J.M.H.; Rose, J.A.; North, M.G.; Wang, L.Z.; Lei, M.; Miller, S.D.; Shea, L.D. Conjugation of transforming growth factor Beta to antigen-loaded Poly(lactide-co-glycolide) nanoparticles enhances efficiency of antigen-specific tolerance. *Bioconjug. Chem.* **2018**, *29*, 813–823. [CrossRef]

195. Cho, J.J.; Stewart, J.M.; Drashansky, T.T.; Brusko, M.A.; Zuniga, A.N.; Lorentsen, K.J.; Keselowsky, B.G.; Avram, D. An antigen-specific semi-therapeutic treatment with local delivery of tolerogenic factors through a dual-sized microparticle system blocks experimental autoimmune encephalomyelitis. *Biomaterials* **2017**, *143*, 79–92. [CrossRef] [PubMed]

196. Gholamzad, M.; Ebtekar, M.; Ardestani, M.S. Intravenous injection of myelin oligodendrocyte glycoprotein-coated PLGA microparticles have tolerogenic effects in experimental autoimmune encephalomyelitis. *Iran J. Allergy Asthma Immunol.* **2017**, *16*, 27–281.

197. Hunter, Z.; McCarthy, D.P.; Yap, W.T.; Harp, C.T.; Getts, D.R.; Shea, L.D.; Miller, S.D. A biodegradable nanoparticle platform for the induction of antigen-specific immune tolerance for treatment of autoimmune disease. *ACS Nano* **2014**, *8*, 2148–2160. [CrossRef] [PubMed]

198. Maldonado, R.A.; LaMothe, R.A.; Ferrari, J.D.; Zhang, A.-H.; Rossi, R.J.; Kolte, P.N.; Griset, A.P.; O'Neil, C.; Altreuter, D.H.; Browning, E.; et al. Polymeric synthetic nanoparticles for the induction of antigen-specific immunological tolerance. *Proc. Natl. Acad. Sci. USA* **2014**, *112*, E156–E165. [CrossRef]

199. McCarthy, D.P.; Yap, J.W.-T.; Harp, C.T.; Song, W.K.; Chen, J.; Pearson, R.M.; Miller, S.D.; Shea, L.D. An antigen-encapsulating nanoparticle platform for TH1/17 immune tolerance therapy. *Nanomedicine* **2017**, *13*, 191–200. [CrossRef]

200. Saito, E.; Kuo, R.; Kramer, K.R.; Gohel, N.; Giles, D.A.; Moore, B.B.; Miller, S.D.; Shea, L.D. Design of biodegradable nanoparticles to modulate phenotypes of antigen presenting cells for antigen-specific treatment of autoimmune disease. *Biomaterials* **2019**, *222*, 119432. [CrossRef]

201. Wan, X.; Pei, W.; Shahzad, K.A.; Zhang, L.; Song, S.; Jin, X.; Wang, L.; Zhao, C.; Shen, C. A Tolerogenic artificial APC durably ameliorates experimental autoimmune encephalomyelitis by directly and selectively modulating myelin peptide–autoreactive CD4+ and CD8+ T cell. *J. Immunol.* **2018**, *201*, 1194–1210. [CrossRef]

202. Zhao, H.; Kiptoo, P.; Williams, T.D.; Siahaan, T.J.; Topp, E.M. Immune response to controlled release of immunomodulating peptides in a murine experimental autoimmune encephalomyelitis (EAE) model. *J. Control. Release* **2010**, *141*, 145–152. [CrossRef]

203. Pearson, R.M.; Casey, L.M.; Hughes, K.R.; Wang, L.Z.; North, M.G.; Getts, D.R.; Miller, S.D.; Shea, L.D. Controlled delivery of single or multiple antigens in tolerogenic nanoparticles using peptide-polymer bioconjugates. *Mol. Ther.* **2017**, *25*, 1655–1664. [CrossRef]

204. Sestak, J.O.; Fakhari, A.; Badawi, A.H.; Siahaan, T.J.; Berkland, C. Structure, size, and solubility of antigen arrays determines efficacy in experimental autoimmune encephalomyelitis. *AAPS J.* **2014**, *16*, 1185–1193. [CrossRef]

205. Kuo, R.; Saito, E.; Miller, S.D.; Shea, L.D. Peptide-conjugated nanoparticles reduce positive co-stimulatory expression and T cell activity to induce tolerance. *Mol. Ther.* **2017**, *25*, 1676–1685. [CrossRef] [PubMed]

206. Getts, D.R.; Martin, A.J.; McCarthy, D.P.; Terry, R.L.; Hunter, Z.N.; Yap, W.T.; Getts, M.T.; Pleiss, M.; Luo, X.; King, N.J.C.; et al. Microparticles bearing encephalitogenic peptides induce T-cell tolerance and ameliorate experimental autoimmune encephalomyelitis. *Nat. Biotechnol.* **2012**, *30*, 1217–1224. [CrossRef] [PubMed]

207. Tostanoski, L.H.; Chiu, Y.-C.; Gammon, J.M.; Simon, T.; Andorko, J.I.; Bromberg, J.S.; Jewell, C.M. Reprogramming the local lymph node microenvironment promotes tolerance that is systemic and antigen-specific. *Cell Rep.* **2016**, *16*, 2940–2952. [CrossRef] [PubMed]

208. Büyüktimkin, B.; Wang, Q.; Kiptoo, P.; Stewart, J.M.; Berkland, C.; Siahaan, T.J. Vaccine-like controlled-release delivery of an immunomodulating peptide to treat experimental autoimmune encephalomyelitis. *Mol. Pharm.* **2012**, *9*, 979–985. [CrossRef]

209. Northrup, L.; Sestak, J.O.; Sullivan, B.P.; Thati, S.; Hartwell, B.L.; Siahaan, T.J.; Vines, C.M.; Berkland, C. Co-delivery of autoantigen and B7 pathway modulators suppresses experimental autoimmune encephalomyelitis. *AAPS J.* **2014**, *16*, 1204–1213. [CrossRef]

210. Kuehl, C.; Thati, S.; Sullivan, B.; Sestak, J.; Thompson, M.; Siahaan, T.; Berkland, C. Pulmonary administration of soluble antigen arrays is superior to antigen in treatment of experimental autoimmune encephalomyelitis. *J. Pharm. Sci.* **2017**, *106*, 3293–3302. [CrossRef]

211. Hess, K.L.; Oh, E.; Tostanoski, L.H.; Andorko, J.I.; Susumu, K.; Deschamps, J.R.; Medintz, I.L.; Jewell, C.M. Engineering immunological tolerance using quantum dots to tune the density of self-antigen display. *Adv. Funct. Mater.* **2017**, *27*. [CrossRef]

212. Clemente-Casares, X.; Blanco, J.; Ambalavanan, P.; Yamanouchi, J.; Singha, S.; Fandos, C.; Tsai, S.; Wang, J.; Garabatos, N.; Izquierdo, C.; et al. Expanding antigen-specific regulatory networks to treat autoimmunity. *Nature* **2016**, *530*, 434–440. [CrossRef]

213. Yeste, A.; Nadeau, M.; Burns, E.J.; Weiner, H.L.; Qu, F.J. Nanoparticle-mediated codelivery of myelin antigen and a tolerogenic small molecule suppresses experimental autoimmune encephalomyelitis. *Proc. Natl. Acad. Sci. USA* **2012**, *109*, 11270–11275. [CrossRef]

214. Katsara, M.; Deraos, G.; Tselios, T.; Matsoukas, J.; Apostolopoulos, V. Design of novel cyclic altered peptide ligands of myelin basic protein MBP83–99 that modulate immune responses in SJL/J mice. *J. Med. Chem.* **2008**, *51*, 3971–3978. [CrossRef]

215. Yu, L.; Yang, F.; Jiang, L.; Chen, Y.; Wang, K.; Xu, F.; Wei, Y.; Cao, X.; Wang, J.; Cai, Z. Exosomes with membrane-associated TGF-β1 from gene-modified dendritic cells inhibit murine EAE independently of MHC restriction. *Eur. J. Immunol.* **2013**, *43*, 2461–2472. [CrossRef] [PubMed]

216. Buerth, C.; Mausberg, A.K.; Heininger, M.K.; Hartung, H.-P.; Kieseier, B.C.; Ernst, J.F. Oral tolerance induction in experimental autoimmune encephalomyelitis with Candida utilis expressing the immunogenic MOG35-55 peptide. *PLoS ONE* **2016**, *11*, e0155082. [CrossRef] [PubMed]

217. Kappos, L.; Comi, G.; Panitch, H.; Oger, J.; Antel, J.; Conlon, P.; Steinman, L. The Altered peptide ligand in relapsing MS study group, Induction of a non-encephalitogenic type 2 T helper-cell autoimmune response in multiple sclerosis after administration of an altered peptide ligand in a placebo-controlled, randomized phase II trial. *Nat. Med.* **2000**, *6*, 1176–1182. [CrossRef] [PubMed]

218. Goodkin, D.E.; Shulman, M.; Winkelhake, J.; Waubant, E.; Andersson, P.; Stewart, T.; Nelson, S.; Fischbein, N.; Coyle, P.K.; Frohman, E.; et al. A phase I trial of solubilized DR2:MBP84-102 (AG284) in multiple sclerosis. *Neurology* **2000**, *54*, 1414–1420. [CrossRef] [PubMed]

219. Warren, K.G.; Catz, I.; Ferenczi, L.Z.; Krantz, M.J. Intravenous synthetic peptide MBP8298 delayed disease progression in an HLA Class II-defined cohort of patients with progressive multiple sclerosis: Results of a 24-month double-blind placebo-controlled clinical trial and 5 years of follow-up treatment. *Eur. J. Neurol.* **2006**, *13*, 887–895. [CrossRef]

220. Freedman, M.S.; Bar-Or, A.; Oger, J.; Traboulsee, A.; Patry, D.; Young, C.; Olsson, T.; Li, D.; Hartung, H.-P.; Krantz, M.; et al. A phase III study evaluating the efficacy and safety of MBP8298 in secondary progressive MS. *Neurology* **2011**, *77*, 1551–1560. [CrossRef]

221. Offner, H.; Burrows, G.G.; Ferro, A.J.; Vandenbark, A.A. RTL therapy for multiple sclerosis: A Phase I clinical study. *J. Neuroimmunol.* **2011**, *231*, 7–14. [CrossRef]

222. Yadav, V.; Bourdette, D.N.; Bowen, J.D.; Lynch, S.G.; Mattson, D.; Preiningerova, J.; Bever, C.T.; Simon, J.; Goldstein, A.; Burrows, G.G.; et al. Recombinant T-cell receptor ligand (RTL) for treatment of multiple sclerosis: A double-blind, placebo-controlled, Phase 1, dose-escalation study. *Autoimmune Dis.* **2012**, *2012*, 954739. [CrossRef]

223. Loo, E.W.; Krantz, M.J.; Agrawal, B. High dose antigen treatment with a peptide epitope of myelin basic protein modulates T cells in multiple sclerosis patients. *Cell. Immunol.* **2012**, *280*, 10–15. [CrossRef]

224. Lutterotti, A.; Yousef, S.; Sputtek, A.; Stürner, K.H.; Stellmann, J.-P.; Breiden, P.; Reinhardt, S.; Schulze, C.; Bester, M.; Heesen, C.; et al. Antigen-specific tolerance by autologous myelin peptide–coupled cells: A Phase 1 trial in multiple sclerosis. *Sci. Transl. Med.* **2013**, *5*, 188ra75. [CrossRef]

225. Bar-Or, A.; Vollmer, T.; Antel, J.; Arnold, D.L.; Bodner, C.A.; Campagnolo, D.; Gianettoni, J.; Jalili, F.; Kachuck, N.; Lapierre, Y.; et al. Induction of antigen-specific tolerance in multiple sclerosis after immunization with a DNA encoding myelin basic protein in a randomized, placebo-controlled Phase I/II trial. *Arch. Neurol.* **2007**, *64*, 1407–1415. [CrossRef] [PubMed]

226. Juryńczyk, M.; Walczak, A.; Jurewicz, A.; Jesionek-Kupnicka, D.; Szczepanik, M.; Selmaj, K. Immune regulation of multiple sclerosis by transdermally applied myelin peptides. *Ann. Neurol.* **2010**, *68*, 593–601. [CrossRef] [PubMed]

227. Walczak, A.; Siger, M.; Ciach, A.; Szczepanik, M.; Selmaj, K. Transdermal application of myelin peptides in multiple sclerosis treatment. *JAMA Neurol.* **2013**, *70*, 1105–1109. [CrossRef] [PubMed]

228. Garren, H.; Robinson, W.H.; Krasulova´, E.; Havrdova´, E.; Nadj, C.; Selmaj, K.; Losy, J.; Nadj, I.; Radue, E.-W.; Kidd, B.A.; et al. Steinman, and the BHT-3009 Study Group, Phase 2 trial of a DNA vaccine encoding myelin basic protein for multiple sclerosis. *Ann. Neurol.* **2008**, *63*, 611–620. [CrossRef]

229. Papadopoulou, A.; von Felten, S.; Traud, S.; Rahman, A.; Quan, J.; King, R.; Garren, H.; Steinman, L.; Cutter, G.; Kappos, L.; et al. Evolution of MS lesions to black holes under DNA vaccine treatment. *J. Neurol.* **2012**, *259*, 1375–1382. [CrossRef]

230. Chataway, J.; Martin, K.; Barrell, K.; Sharrack, B.; Stolt, P.; Wraith, D.C. Effects of ATX-MS-1467 immunotherapy over 16 weeks in relapsing multiple sclerosis. *Neurology* **2018**, *90*, e955–e962. [CrossRef]

231. Zubizarreta, I.; Flórez-Graub, G.; Vila, G.; Cabezón, R.; España, C.; Andorra, M.; Saiza, A.; Llufriu, S.; Sepulveda, M.; Sola-Valls, N.; et al. Immune tolerance in multiple sclerosis and neuromyelitis optica with peptide-loaded tolerogenic dendritic cells in a phase 1b trial. *Proc. Natl. Acad. Sci. USA* **2019**, *116*, 8463–8470. [CrossRef]

232. Willekens, B.; Presas-Rodríguez, S.; Mansilla, M.J.; Derdelinckx, J.; Lee, W.-P.; Nijs, G.; de Laere, M.; Wens, I.; Cras, P.; Parizel, P.; et al. On behalf of the RESTORE consortium, Tolerogenic dendritic cell-based treatment for multiple sclerosis (MS): A harmonized study protocol for two phase I clinical trials comparing intradermal and intranodal cell administration. *BMJ Open* **2019**, *9*, e030309. [CrossRef]

233. Seledtsova, G.V.; Ivanova, I.P.; Shishkov, A.A.; Seledtsov, V.I. Immune responses to polyclonal T-cell vaccination in patients with progressive multiple sclerosis. *J. Immunotoxicol.* **2016**, *13*, 879–884. [CrossRef]

234. Loftus, B.; Newsom, B.; Montgomery, M.; von Gynz-Rekowski, K.; Riser, M.; Inman, S.; Garces, P.; Rill, D.; Zhang, J.; Williams, J.C. Autologous attenuated T-cell vaccine (Tovaxin®) dose escalation in multiple sclerosis relapsing–remitting and secondary progressive patients nonresponsive to approved immunomodulatory therapies. *Clin. Immunol.* **2009**, *131*, 202–215. [CrossRef]

235. Fox, E.; Wynn, D.; Cohan, S.; Rill, D.; McGuire, D.; Markowitz, C. A randomized clinical trial of autologous T-cell therapy in multiple sclerosis: Subset analysis and implications for trial design. *Mult. Scler. J.* **2012**, *18*, 843–852. [CrossRef] [PubMed]

236. Karussis, D.; Shor, H.; Yachnin, J.; Lanxner, N.; Amiel, M.; Baruch, K.; Keren-Zur, Y.; Haviv, O.; Filippi, M.; Petrou, P.; et al. T Cell vaccination benefits relapsing progressive multiple sclerosis patients: A randomized, double-blind clinical trial. *PLoS ONE* **2012**, *7*, e50478. [CrossRef] [PubMed]

237. Van der, A.A.; Hellings, N.; Medaer, R.; Gelin, G.; Palmers, Y.; Rauss, J.; Stinissen, P. T cell vaccination in multiple sclerosis patients with autologous CSF-derived activated T cells: Results from a pilot study. *Clin. Exp. Immunol.* **2003**, *131*, 155–168. [CrossRef] [PubMed]

238. Zhang, J.Z.; Rivera, V.M.; Tejada-Simon, M.V.; Yang, D.; Hong, J.; Li, S.; Haykal, H.; Killian, J.; Zang, Y.C.Q. T cell vaccination in multiple sclerosis: Results of a preliminary study. *J. Neurol.* **2002**, *249*, 212–218. [CrossRef] [PubMed]

239. Correale, J.; Lunda, B.; McMillan, M.; Koa, D.Y.; McCarthy, K.; Weiner, L.P. T cell vaccination in secondary progressive multiple sclerosis. *J. Neuroimmunol.* **2000**, *107*, 130–139. [CrossRef]

240. Achiron, A.; Lavie, G.; Kishner, I.; Stern, Y.; Sarova-Pinhas, I.; Ben-Aharon, T.; Barak, Y.; Raz, H.; Lavie, M.; Barliya, T.; et al. T cell vaccination in multiple sclerosis relapsing–remitting nonresponders patients. *Clin. Immunol.* **2004**, *113*, 155–160. [CrossRef]

241. Zang, Y.C.Q.; Hong, J.; Rivera, V.M.; Killian, J.; Zhang, J.Z. Preferential recognition of TCR hypervariable regions by human anti-idiotypic T cells induced by T cell vaccination. *J. Immunol.* **2000**, *164*, 4011–4017. [CrossRef]

242. Vandenbark, A.A.; Culbertson, N.E.; Bartholomew, R.M.; Huan, J.; Agotsch, M.; LaTocha, D.; Yadav, V.; Mass, M.; Whitham, R.; Lovera, J.; et al. Therapeutic vaccination with a trivalent T-cell receptor (TCR) peptide vaccine restores deficient FoxP3 expression and TCR recognition in subjects with multiple sclerosis. *Immunology* **2007**, *123*, 66–78. [CrossRef]

Review

Promising Nanotechnology Approaches in Treatment of Autoimmune Diseases of Central Nervous System

Maria Chountoulesi and Costas Demetzos *

Section of Pharmaceutical Technology, Department of Pharmacy, School of Health Sciences, National and Kapodistrian University of Athens, Panepistimioupolis Zografou, 15771 Athens, Greece; mchountoules@pharm.uoa.gr
* Correspondence: demetzos@pharm.uoa.gr; Tel.: +30-210-727-4596

Received: 4 May 2020; Accepted: 30 May 2020; Published: 2 June 2020

Abstract: Multiple sclerosis (MS) is a chronic, autoimmune, neurodegenerative disease of the central nervous system (CNS) that yields to neuronal axon damage, demyelization, and paralysis. Although several drugs were designed for the treatment of MS, with some of them being approved in the last few decades, the complete remission and the treatment of progressive forms still remain a matter of debate and a medical challenge. Nanotechnology provides a variety of promising therapeutic tools that can be applied for the treatment of MS, overcoming the barriers and the limitations of the already existing immunosuppressive and biological therapies. In the present review, we explore literature case studies on the development of drug delivery nanosystems for the targeted delivery of MS drugs in the pathological tissues of the CNS, providing high bioavailability and enhanced therapeutic efficiency, as well as nanosystems for the delivery of agents to facilitate efficient remyelination. Moreover, we present examples of tolerance-inducing nanocarriers, being used as promising vaccines for antigen-specific immunotherapy of MS. We emphasize on liposomes, as well as lipid- and polymer-based nanoparticles. Finally, we highlight the future perspectives given by the nanotechnology field toward the improvement of the current treatment of MS and its animal model, experimental autoimmune encephalomyelitis (EAE).

Keywords: multiple sclerosis; nanotechnology; drug delivery nanosystems; lipids; polymers; vaccines; nanoparticles; antigen-specific immunotherapy; experimental autoimmune encephalomyelitis; neurodegeneration

---

1. Introduction

Multiple sclerosis (MS) is a chronic, autoimmune, demyelinating disease of the central nervous system (CNS), accompanied by a relapsing/remitting (RR) or a progressive course that is followed by axon damage and paralysis, including symptoms of muscle weakness, weak reflexes, muscle spasm, difficulty in movement, miscoordination, unbalance, vertigo, fatigue, and pain. Other symptoms that are usually referred are optic nerve dysfunction, loss of vision, diplopia, pyramidal tract dysfunction, ataxia, tremor, bladder and bowel dysfunction, sexual dysfunction, depression, anxiety, swallowing dysfunction, memory loss, sleep disturbance, and obstructive sleep apnea [1–5]. Unfortunately, the exact etiology of MS remains unknown, while many different risk factors were referred, characterizing MS as a heterogeneous, multifactorial disease. The occurrence is 2–3 times higher in females than males. MS is the most common neurologically disabling disease in young adults, while older people and children can also acquire MS [4,6]. Our understanding of the immune processes that contributes to MS led to the approval or clinical development of some disease-modifying therapies (DMTs) that are effective in relapsing forms of MS. However, few treatments are effective for the progressive forms of the disease [7,8].

Nanotechnology provides a variety of promising therapeutic tools that can be applied for the treatment of CNS-related disorders, such as MS, overcoming the barriers and the restrictions of the already existing conventional therapies. Extensive research is being carried out for the development of drug delivery nanosystems for the targeted delivery of MS drugs in the pathological tissues of CNS, providing high bioavailability and enhanced therapeutic efficiency. In addition, remyelination is an attractive, innovative strategy toward MS therapy [9], where nanoparticles can also contribute, via the targeted delivery of remyelinating agents to specific cells, leading to the improvement of their therapeutic performance. Moreover, tolerance-inducing vaccines, based on tolerance-inducing nanocarriers for antigen-specific immunotherapies, are considered to be another promising strategy toward the treatment of MS [10,11].

In the present review study, literature examples of the aforementioned nanocarriers that were designed for MS treatment are presented, highlighting the future perspectives given by the nanotechnology field toward the improvement of the current treatment of MS. We focus on liposomes, as well as lipid- and polymer- based nanocarriers.

2. Multiple Sclerosis (MS)

MS is an autoimmune, chronic, neurodegenerative disorder, targeting the myelin sheaths (a protective layer surrounding the nerve fibers) of the CNS. The caused damage of myelin sheaths provokes nerve demyelination, followed by axon damage and, thus, interruption of signal transmission to and from the CNS. As with many other neurodegenerative diseases, the real and exact origin of MS is still unidentified, although the literature describes many different potential triggering factors that may stimulate the autoimmune responses, which harm the brain tissues and spinal cord. More particularly, genetic predisposition and environmental factors, as well as microbial and viral infections, smoking, toxins, low concentrations of vitamin D, and circadian rhythm disruption, can contribute to the onset of this disorder [12–16]. Regarding genetic predisposition, the major histocompatibility complex (MHC) class II phenotype, the human leukocyte antigen (HLA)-DR2, and HLA-DR4 are reported as the most commonly affected, while the incidence of MS is also increased 10-fold in monozygotic twins, as compared to siblings of patients with MS [17,18].

MS is categorized into three distinct types, primarily based on its clinical course, which are characterized by increasing severity. Relapsing/remitting MS (RRMS) is the most common form, which involves relapses followed by silent remission with any MS symptoms. RRMS generally switches to a chronic progressive course several years after onset, while a minor number of patients display primary chronic progressive course without the RR phase. Subsequently, we have secondary progressive MS (SPMS), which develops over time following diagnosis of RRMS, and primary progressive MS (PPMS), noted as gradual continuous neurologic deterioration, which provides continuous disease progression [19–21].

The symptoms may vary person-to-person and produce temporary, long-lasting, or even permanent losses due to the disrupted signal transmission, including mood swings, memory-related issues, tingling, fatigue, numbness, partial or complete blindness, pain, and partial or even whole-body paralysis, depending upon the severity of disease. The disrupted signal transmission is responsible for various complex and erratic indications. MS broadly presents three stages, starting from a pre-clinical stage, followed by a relapsing/remitting stage and a progressive clinical stage, in which, after typically 10–20 years, neurologic dysfunction progressively worsens, eventually leading to impaired mobility, cognition, and a progressive loss of nerve functions [22–25].

2.1. Immunopathophysiology of MS

Myelin sheath damage and inflammation contribute to the formation of lesions [26]. MS lesions appear in the white matter inside the visual neuron, brain stem, and spinal cord [26]. In MS, the immune system starts to recognize myelin components as foreign, leading to its destruction [12]. Plaque formation and disease symptoms are widely accepted as the result of immune cell infiltration, with the release of cytokines and inflammatory mediators, leading to inflammation, myelin destruction, oligodendrocyte loss, and loss of neuronal function and eventual axonal degeneration. On the other hand, demyelination further increases the activation of inflammatory processes, causing damage of the blood–brain barrier (BBB), stimulation of oxidative stress pathways, and macrophage activation [27]. Microglial cells upregulate MHC class I and II molecules, as well as cell surface co-stimulatory molecules, while they also secrete cytokines and chemokines. Several types of immune infiltrates can be found in the white matter lesions, where myelin is damaged, including monocytes, B cells, T cells (for example, TH cells and myelin-reactive auto-T cells), and dendritic cells [25,28–32]. Substantial fractions of cluster of differentiation 4 (CD4)+ and CD8+ T cells being isolated from MS lesions and cerebrospinal fluid (CSF) indicate that antigen-specific T-cell responses contribute to the disease process [33]. It seems that T helper 1 (TH1) and TH17 cells are the main pathogenic populations in the immunopathogenesis of MS, along with regulatory T cells (Treg) and natural killer T (NKT) cells [34–37]. The pro-inflammatory state of macrophages is correlated with MS because macrophages interact with T and B cells, instructing demyelination, axonal loss, and degeneration, and they are presented as of the most predominant cell type in patient lesions [38,39]. In Figure 1, there is a detailed presentation of the cells that are involved in the immune process of MS [25].

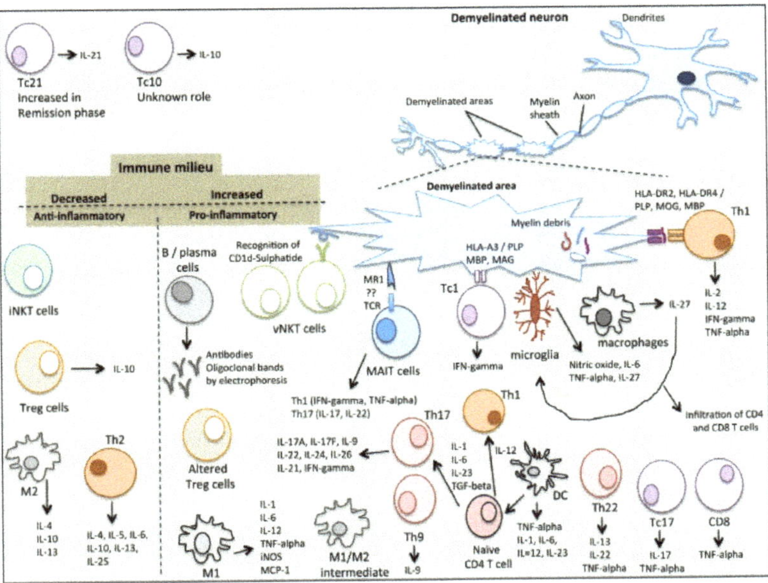

Figure 1. The immunological complexity of the immune/cytokine network in multiple sclerosis (MS). Adapted from Dargahi et al. [25].

Traditionally, the etiology of MS was based on an "outside-in" autoimmune hypothesis, whereby dysregulated auto-reactive T cells in the periphery cross into the CNS parenchyma and, together with macrophages and B cells, proceed to attack myelin. Contrariwise, the "inside-out" hypothesis argues that MS is a primary degenerative disease, where the initial malfunction occurs within the CNS and is accompanied by varying degrees of inflammation, as a secondary response, leading to the release of various antigenic cell components. According to the "inside-out" hypothesis, the primary degeneration is present from the start (probably years before the first overt clinical symptoms) and continues throughout the entire course of the disease. Whether early neurodegeneration drives autoimmune injury, or whether ongoing inflammation reaches a threshold to trigger neurodegeneration is still unclear, while another question is whether neurodegeneration is independent or not of chronic inflammation. The lack of understanding with regard to mechanisms of progression phase, when the most irreversible disability takes place, is responsible for the extremely limited treatment options that are currently available to patients with progressive MS. Until now, many different treatment approaches for progressive MS were proposed including mitochondrion-protective strategies, anti-inflammatory strategies, strategies targeting microglia and astrocytes, inhibitors of microglial activity, remyelination therapies, and strategies targeting lymphocytes [20,23,40].

2.2. Current Therapeutics of MS

Until now, the Food and Drug Administration (FDA) approved over a dozen therapeutic agents to reduce the number of attacks and delay MS progression in terms of available DMTs. Disease-modifying agents are commonly shown to reduce the rate of relapses, reduce magnetic resonance imaging (MRI) lesions, and stabilize or delay MS disability. However, MS remains incurable. Interferon beta (IFN-β) and glatiramer acetate (GA), being the first two introduced drugs, are able to alter T-cell responses, are injectable, and still remain the "first line" therapies for MS, owing to their relative safety and proven efficacy. Oral DMTs available for RRMS include immunosuppressives, which inhibit lymphocyte trafficking, namely, fingolimod and the teriflunomide (TFM) that inhibit activated T and B cells, as well as the immunomodulatory/immunosuppressive dimethyl fumarate (DMF) that alters T-cell responses. The above are characterized as long-term treatments. In addition, a number of humanized monoclonal antibody-based therapeutics for RRMS were developed, including natalizumab, alemtuzumab, and ocrelizumab. Unfortunately, their disadvantage is the high risk of side effects, including progressive multifocal leukoencephalopathy (PML) and the development of secondary autoimmune diseases [4,12,25,26,39,41–45].

2.3. Experimental Autoimmune Encephalomyelitis (EAE) as an Animal Model for MS

Experimental autoimmune encephalomyelitis (EAE) is a widely used animal model of MS because it shares several features with the human disease, including neurological dysfunction and perivascular inflammation in the CNS. It is considered to be the most frequently used model to study the demyelinating and immune pathology of MS. EAE can be induced in several mammalian species by directly immunizing animals with CNS homogenate or myelin proteins, such as myelin oligodendrocyte glycoprotein (MOG), myelin basic protein (MBP), and proteolipid protein (PLP), or using small peptides derived from these proteins; alternatively, the transfer of isolated activated CD4+ T cells or less commonly CD8+ T cells, may be introduced to a naïve animal. The fact that demyelination and lesion formation occur predominantly in the spinal cord rather than the CNS is a disadvantage indicating that EAE does not fully recapitulate human MS pathology [46–50]. However, there is no other model representing better the pathophysiology of MS. EAE extensively contributed toward the knowledge and understanding of the pathophysiology of the MS, as well as contributed to the discovery and the understanding of the mechanism of action of some of the current approved therapies [39,51–53].

3. Nanotechnology and MS

Drug delivery nanosystems are considered as innovative technological platforms that are able to transport bioactive molecules to target tissues, modifying their solubility and improving their bioavailability, by altering their pharmacokinetic profile [54]. The existence of physiological barriers, such as the BBB and the blood–cerebrospinal fluid barrier (BCSFB), limits the access of several therapeutic agents to the CNS and downgrades their therapeutic activity. In this regard, nanotechnology is considered to be a promising strategy to improve drug targeting to the brain, as well as increase bioavailability. There is great research attention toward the employment of nanotechnology, as well as the development of new therapies and improvement of the therapeutic efficacy for MS [55]. Taking advantage of their size, nanoparticles are easily internalized by the cells, being suitable carriers for drugs, immunomodulatory molecules, or antigens. Nanosystems are able to improve the drug solubility, provide targeted delivery, diminish potential side effects from high doses, and establish controlled drug release. Moreover, nanoparticles can be administrated through various routes apart from systemic administration, such as intranasally, for example, in cases where nose-to-brain delivery is desired.

In the following paragraphs, different literature cases on nanoparticles and MS are examined. There are referred nanoparticles that are used as drug delivery systems, in order to improve the pharmacokinetics and bioavailability, while enhancing the therapeutic efficacy when compared to the free administered drugs; furthermore, nanoparticles can be used as vectors for antigen-specific immunomodulation. Through antigen-specific immunomodulation, the immune system is repeatedly exposed to a specific antigen, which results in immunomodulation from the disease to the tolerance state. Antigen-loaded nanoparticles provide several advantages for antigen-specific immunomodulation, such as sustained antigen release, the co-delivery of antigens and adjuvants, the formation of antigen depots at the injection site, the effective presentation of B-cell epitopes, and an increase in the uptake and stimulation of cell-mediated immune responses against acellular antigens. These nanoparticles are coupled with specific antigens related to the autoimmune response in MS and their epitopes. These nanoparticles are able to regulate T-cell function, as well as induce Treg cell and dendritic cell differentiation, restoring immunological tolerance [56]. More details are analytically described throughout the paragraphs below.

Another subject of great scientific interest regarding MS, which is also correlated with the development of new therapeutics, is remyelination. While some patients suffer from progressive neurological deficits, other patients occasionally display improved neurological function. The mechanism via which some patients experience neurological improvement remains unclear. However, increasing evidence shows that remyelination occurs in the MS/EAE [57–60]. In addition, recent studies showed that activated neural stem/progenitor cells around region sites contribute to remyelination [61,62]. Promoting remyelination, therefore, provides an additional line of defense against the axonal damage that follows the loss of myelin. Several potential drugs that target CNS inflammation and the different aspects and stages of remyelination (for example, by inducing oligodendrocyte precursor cell (OPC) differentiation) are being identified [9,63]. Nanoparticles themselves can also act as remyelinating agents. Most recently, Robinson et al. [64] reported a new nanocatalytic therapeutic candidate for remyelination, consisting of a suspension of clean-surfaced, faceted nanocrystals of gold, which was found to demonstrate robust remyelinating activity in response to demyelinating agents, in both chronic cuprizone and acute lysolecithin rodent animal models. More analytically, the administrated nanoparticles induced differentiation of OPCs and enhanced activities of neurons and oligodendrocytes through enhancement of bioenergetic processes of key indicators of aerobic glycolysis. According to the in vivo results, the oral delivery of gold nanocrystals improved the motor functions of cuprizone-treated mice in both open-field and kinematic gait studies. Additional in vitro data indicated an upregulation of myelin synthesis-related genes, collectively resulting in functional myelin generation.

However, there is still no available treatment to regenerate myelin, and several strategies are being scrutinized, while there are difficulties in translating these potential drug targets into practical therapies for patients, as these must cross the BBB to reach OPCs in the CNS, and ideally should be delivered directly to OPCs, to avoid off-target effects. Once again, nanotechnology may solve these problems, with nanoparticles crossing the BBB and facilitating targeting to specific cells, as highlighted in the below-described literature examples.

3.1. Lipid-Based Nanosystems

Lipid nanocarriers such as liposomes, solid lipid nanoparticles (SLNs), nanostructured lipid carriers (NLCs), and nanoemulsions (Figure 2) are considered an ideal strategy for effective delivery of the therapeutic agents against MS at the CNS, enhancing brain transport. They possess the ability to cross the BBB by naturally entering the brain capillary endothelial cells, reducing the peripheral side effects. Furthermore, through suitable surface decoration, lipid nanocarriers can be engineered to interact with particular types of molecules or cell receptors presented in the BBB, even delivering drugs which normally cannot cross the BBB [65,66].

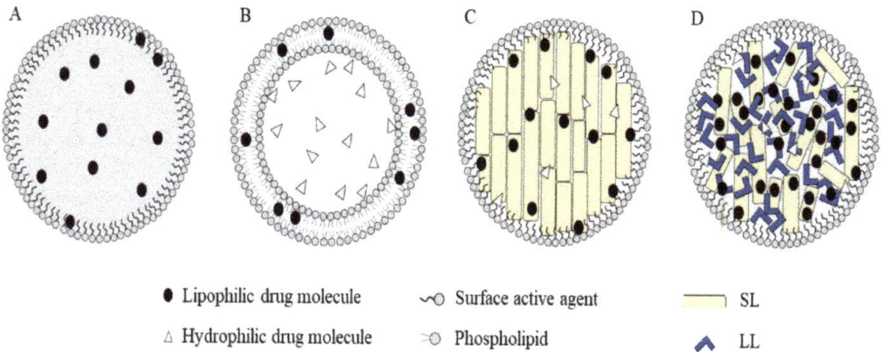

Figure 2. Different types of lipid-based nanoparticles: (**A**) nanoemulsions; (**B**) liposomes; (**C**) solid lipid nanoparticles (SLNs); (**D**) nanostructured lipid carriers (NLCs). Adapted from Haider et al. [67].

3.1.1. Nanolipid Carriers (NLCs) and Solid Lipid Nanoparticles (SLNs)

The lipid-based nanoparticulate system with a solid matrix primarily originated from an oil-in-water type emulsion, by replacing the oil phase or liquid lipids with solid lipids to make it solid at body temperature. Solid lipid nanoparticles (SLNs) are usually considered as the first-generation lipid nanoparticle developed from solid lipid, while nanolipid carriers (NLCs) are known as second-generation lipid nanoparticles, comprising a solid and liquid lipid blend, as well as a surfactant (Figure 2C,D) [68,69]. Although the formulation contains a liquid lipid, NLCs remain in the solid state at body and room temperatures, by adjusting the levels of the liquid lipid. SLNs are colloidal particles with increased physical stability derived from oil-in-water (O/W) emulsions, by replacing liquid lipids with a lipid matrix that is solid at both room and body temperature. The lipid core of SLNs typically consists of fatty acids, monoglycerides, diglycerides, triglycerides, waxes, or steroids and is stabilized by surfactants [70,71]. SLNs and NLCs differ in the composition and organization of the lipids of the matrix, which causes different morphological structures. Commonly used solid lipids for NLC formulation are stearic acid, stearyl alcohol, glycerol monostearate, mono-stearin, etc., while common examples of liquid lipids involve the use of olive oil, sesame oil, almond oil, peanut oil, soyabean oil, oleic acid, corn oil, soy lecithin, phosphatidyl choline, vitamin E, etc. [69].

Recently, Gadhave et al. [72] formulated intranasal nanolipid carriers loaded with TFM, an inhibitor of dihydroorotate dehydrogenase, exhibiting anti-inflammatory activity. The TFM-loaded NLCs were prepared via the melt emulsification ultrasonication method, while the Box–Behnken statistical design was applied to optimize the formulation. The lipid nanocarriers were composed of Compritol® 888 ATO (solid lipid), maisine 35–1 (liquid lipid), gelucire 44/14 (stabilizer) (all by Gattefosse, Mumbai, India), and its aqueous phase from water and Tween-20 (surfactant) (S.D. Fine Chemicals, Pune, India). They were evaluated regarding their particle size, entrapment efficiency (%), in vitro and ex vivo permeation, and their pharmacological and toxicological properties. According to the results, the optimized formulation exhibited particle size, surface charge, and entrapment efficiency of 99.82 nm, −22.29 mV, and 83.39%, respectively. The formulation was enriched with mucoadhesive and gelling agents. The ex vivo drug permeation study and the permeation flux of the prepared mucoadhesive nanosystem was higher than the plain one. From the therapeutic point of view, the intranasal administration of the prepared nanostructures facilitates the rapid remyelination of damaged neurons in the cuprizone-treated rat model, without any significant change in hepatic biomarkers and sub-acute toxicity of the drug; thus, it can be considered as effective and safe delivery for brain disorders.

Kumar et al. [73] developed nanolipid carriers for the delivery of DMF and tocopherol acetate, in an effort to enhance the brain permeability of DMF, improve its gastric tolerance, and reduce its side effects. They used stearic acid (M/s Central Drug House, New Delhi, India) as the solid-phase lipid, tocopherol acetate as the liquid-phase lipid, and Tween 80 (M/s Fisher Scientific India Pvt. Limited, Mumbai, India) as an emulsifier, while the hot microemulsion technique was utilized as the preparation method. The prepared formulation was then evaluated by physicochemical, Caco-2 cellular permeability, in vitro drug release, in vivo pharmacokinetics, and biodistribution studies. The physicochemical study revealed characteristics suitable for brain drug delivery, namely, an average size of 69.70 nm, polydispersity index (PDI) of 0.317, and zeta potential of −9.71 mV. At the same time, the loading efficiency and entrapment efficiency of the formulation were observed as 20.13% and 90.12%, respectively. The obtained controlled release profile, in both gastric and intestinal pH (up to 68% of drug in 24 h), was attributed to the better encapsulation of DMF within the NLC. The drug release was found to best fit the Higuchi equation of release kinetics. Cellular uptake studies on Caco-2 and SH-SY5Y monolayers confirmed better intestinal absorption, due to its biodegradable, lipidic formulation and small size, as well as higher neuronal uptake of the developed system. Furthermore, the in vivo pharmacokinetic data showed a significant improvement in C_{max} and $t_{1/2}$, three times higher drug absorption, and a reduction in the drug clearance, volume of distribution, T_{max}, and drug elimination. The findings are promising and offer preclinical evidence for better brain bioavailability of DMF, which can improve the clinical therapy of MS through reduction of the dosing frequency.

Ghasemian et al. [74] developed baclofen-loaded nanolipid carriers for effective brain drug delivery, in order to reach the site of baclofen action in the CNS. Although baclofen is not an MS-specific therapy, it is used in the treatment of multiple sclerosis, in order to eliminate the spasticity that usually appears in MS. However, the effect of oral baclofen is limited due to its hydrophilic nature, which creates insufficient concentrations in the brain and CSF. The proposed lipid nanocarrier was prepared via the double emulsification solvent evaporation technique and was composed of glyceryl monostearate, glyceryl distearate, glyceryl trioleate (all by Gattefossè, France), and an aqueous phase containing Tween 80 (Sigma-Aldrich, Darmstadt, Germany), as a surface acting agent. Taking into account the obtained physicochemical results, the formulation gave suitable ranges of particle size, size distribution, and zeta potential, while encapsulation efficiencies ranged between 39% and 42%. Glyceryl trioleate lipid gave the minimum particle size. The obtained sustained release profile (up to 74.6% of drug in 28 h) was found to best fit the Higuchi equation of release kinetics and to take place via Fickian diffusion. The formulation of baclofen in nanolipid carriers increased the half-life of drug in plasma and brain by up to 10 and 1.5 times, respectively, and provided a prolonged effect compared to the solution formulation.

Most recently, Kumar et al. [75] loaded methylthioadenosine in SLNs for oral delivery to the brain for the management of MS. The SLNs were prepared via the well-reported microencapsulation technique and were composed of stearic acid (M/s Central Drug House, New Delhi, India), phospholipid 90 G (IPCA Laboratories, Mumbai, India), and Tween-80. The obtained SLNs exhibited sizes below 100 nm, being well within the range to offer a promise of enhanced BBB permeability and bypassing the reticuloendothelial system, with almost neutral zeta potential; they also offered higher drug entrapment and drug loading. Cuprizone-induced demyelination model in mice was employed to mimic the MS-like conditions of demyelination. We should mention that the cuprizone model does not accurately capture the MS disease state, but it is useful for pro-remyelination investigations. It is mainly neurodegenerative-based, exhibiting toxin-based demyelination, rather than immune-mediated demyelination [9]. The symptoms were monitored to some level by plain methylthioadenosine and to a major extent by the SLN version of this nucleoside. According to the pharmacokinetic studies, the methylthioadenosine was better absorbed from SLNs vis-à-vis plain methylthioadenosine, and the biological residence was substantially enhanced so as to reach the target site, thus improving bioavailability and enhancing bioresidence. Methylthioadenosine-loaded SLNs were able to maintain the normal metabolism, locomotor activity, motor coordination, balancing, and grip strength of the rodents compared to plain MTA. The pharmacokinetics corroborated the pharmacodynamic findings, indicating that the orally administered SLNs can be substantially delivered to the brain and can effectively remyelinate the neurons.

In another case of SLNs, Gandomi et al. [76] developed polyethylene glycol (PEG)ylated SLNs in order to efficiently deliver the glycocorticosteroid methylprednisolone to the brain. The corticosteroids, including the glucocorticosteroids, which are mentioned in many literature cases of drug delivery nanosystems described in the present review article, are used in pulse curses, in order to clinically treat significant relapses in MS patients, in an attempt to hasten recovery. Although they are one of the most common clinically prescribed drugs for reducing MS symptoms, due to their anti-inflammatory and immunosuppressive effects, we should note that they are not a specific treatment of MS. The SLNs were surface-modified by using two targeting moieties and, more specifically, glycoprotein antigens, either anti-Contactin2 or anti-Neurofascin, which are two axo-glial-glycoprotein antigens located in the node of Ranvier and are considered to be the main targets of autoimmune reaction in MS. Myelin-based SLNs were prepared via the solvent evaporation method and were composed of myelin lipids including cholesterol, sphingosine, phosphatidylethanolamine, phosphatidylcholine, sphingomyelin, and phosphatidylserine (all by Sigma Aldrich, St. Louis, MO, USA). In order to obtain targeted SLNs, the antibody, either anti-Contactin2 or anti-Neurofascin, was conjugated to the drug-loaded polyethylene glycol (PEG)-covered SLNs (PEGylated SLNs). The prepared SLNs were physicochemically characterized, while their in vitro release profile, cell viability, and cell uptake were studied. Their brain uptakes were also probed following injections into MS-induced mice. The formulation differentiated the particle size; for example, smaller particle sizes following the antibody bindings were observed, due to the lesser hydration of polymer after the antibody conjugation. The SLNs presented good release profiles, where variations in the lipid type slightly affected the drug release profile, while the PEG and/or antibody coating provided a sink boundary condition for the drug diffusion within the lipoid matrix. It was found that the targeted PEGylated SLNs had no significant cytotoxicity on U87MG-cells, although their cellular uptake was increased four- and eight-fold when surface-modified with anti-Contactin2 or anti-Neurofascin, respectively, compared to control. However, the non-surface-modified SLNs exhibited better penetration ability of the BBB compared to the PEGylated and antibody-functionalized SLNs. The authors suggested that the antibodies may facilitate the adsorption of the SLNs to myelin due to binding to the related antigens, i.e., Contactin2 or Neurofascin, on the CNS axon, and the above information would help toward the development of more efficient nanocarriers for the treatment of MS.

3.1.2. Liposomes

Liposomes (Figure 2B) are considered to be one of the most well-investigated drug delivery nanosystems, presenting major advantages, such as biocompatibility and biodegradability, while also exhibiting great versatility, because they can be easily surface-modified with functional biomaterials, in order to acquire advanced properties, such as escaping rapid clearance in circulation and presenting increased targeting to pathological tissues [54,77].

In addition to lipid nanoparticles, liposomes and especially the PEGylated ones were also reported as potential drug carriers for MS and specially for the delivery of glycosteroids. Schmidt et al. [78] developed a novel formulation of PEG-coated long-circulating liposomes, prepared from dipalmitoyl phosphatidylcholine (Lipoid GmbH, Ludwigshafen, Germany), encapsulating prednisolone that was administrated to the CNS of rats exhibiting EAE. The authors tried to achieve ultra-high tissue concentrations of glucocorticosteroids in the inflamed target organ as compared to an equivalent dose given as free drug, along with a much lower systemic concentration with a reduction of unwanted side effects. Radioactive labeling showed the accumulation of liposomes in the inflamed target organ. More specifically, ^{3}H-labeled prednisolone liposomes showed selective targeting to the inflamed CNS, where up to 4.5-fold higher radioactivity was achieved compared to healthy control animals. Moreover, much higher and more persistent levels of prednisolone in the spinal cord were detected by liposomal administration than in the case of free administrated drug. Gold-labeled liposomes were used, while they could be detected in the spinal cord within the vascular endothelium, as well as in inflammatory macrophages, microglial cells, and astrocytes. The BBB disruption, the T-cell and macrophage infiltration, and the percentage of tumor necrosis factor-α (TNF-α) in these cells were monitored, indicating superior performance by the liposomal administration of prednisolone. It was also reported that a single injection of the liposomes clearly ameliorated the course of adoptive transfer EAE and EAE induced by immunization. The authors finally stated that the liposomal prednisolone was found to be highly effective in the treatment of EAE, being superior to a five-fold higher dose of free methylprednisolone, possibly due to liposomal targeting.

Later, Gailard et al. [79] used PEGylated liposomes conjugated to the brain-targeting ligand glutathione (GSH-PEG), in order to optimally improve the therapeutic window of methylprednisolone, composed of hydro soy phosphatidylcholine lipid (Lipoid, Cham, Switzerland). The authors chose the GSH-PEG liposomes because they include the safety of the liposomal constituents, the ability to encapsulate compounds without modification, the prolonged plasma exposure, and the possibility to enhance drug delivery to the brain. The prepared liposomes were administrated to rats with acute EAE. Apart from the prolonged plasma circulation and increased brain uptake as revealed by the pharmacokinetic analysis, the treatment with GSH-PEG liposomes was found to be significantly more effective in EAE as compared to PEG liposomes, while the same dose level of free methylprednisolone even worsened disease outcome. The rats received intravenous treatment, before disease onset, at disease onset, or at the peak of disease. Free methylprednisolone and non-targeted pegylated (PEG) liposomal methylprednisolone served as control treatments. It was reported that, when the treatment was initiated at disease onset, free methylprednisolone showed no effect, while GSH-PEG liposomal methylprednisolone significantly reduced the clinical signs to 42% ± 6.4% of the saline control, confirming that GSH-PEG liposomes improve the therapeutic availability of methylprednisolone and, therefore, a lower dose and lower dosing frequency can be used to obtain an effective brain concentration.

Lee et al. [80] further exploited the efficacy of the GSH-PEG liposomes carrying methylprednisolone, by using mice exhibiting murine myelin oligodendrocyte-induced EAE (MOG-EAE). This animal model mimics many neurodegenerative features of MS, including axonal damage. The experimental protocol was as follows: after the disease onset, mice were randomized to receive saline, three injections of free drug (high dose methylprednisolone), two injections of free drug (low dose methylprednisolone), or two injections of liposomes. The infiltration of T cells and macrophage/microglia, the amount of astrocyte activation, the extent of axonal loss, and the demyelination in spinal cord lesions were also monitored, indicating good performance of liposomes compared to a low dose of the free drug. Treatment with a low dose of liposomes significantly reduced the severity of EAE, similar to treatment with high-dose free drug but at one-tenth of the dosage, while a low dose of free methylprednisolone was not effective. The proposed liposomes were clinically and histologically effective as a high dose of free drug, thus allowing treatment by liposomal administration at a lower application frequency.

More recently, sterically stabilized liposomes were designed by Turjeman et al. [81] in order to carry glucocorticosteroids and treat the neuroinflammation presented in MS. More specifically, the authors investigated the remote loading of the "water-soluble", amphipathic weak acid glucocorticosteroid prodrug methylprednisolone hemisuccinate (MPS) or the amphipathic weak base nitroxide tempamine (TMN) and compared the effect of passive targeting alone and of active targeting based on short peptide fragments of ApoE or of β-amyloid. The stealthiness of the liposomes was achieved by incorporating the PEG-DSPE-2000 lipid (Genzyme Pharmaceuticals, Liestal, Switzerland). The peptide-conjugated sterically stabilized liposomes (actively targeted) were prepared via the covalent attachment of either ApoE or β-amyloid to dioleoyl (DO)-succinate, in order to form the lipidated peptides. These two peptides can be transported through the BBB. In addition to the physicochemical and thermotropic characterization of the fabricated liposomes, the EAE mice model was used, in order to monitor their therapeutic efficacy, while its mechanism of action in both the acute and the adoptive transfer EAE models was investigated. According to the results, for the liposomes carrying

MPS, active targeting is not superior to passive targeting. For both groups of liposomes, carrying MPS or TMN, it was demonstrated that these nano-drugs ameliorated the clinical signs and the pathology of EAE. The authors concluded that the highly efficacious anti-inflammatory therapeutic feature of these two nano-drugs meets the criteria of disease-modifying drugs and supports further development and evaluation of these nano-drugs as potential therapeutic agents for diseases with an inflammatory component.

The prevalence of autoimmunity is on the rise, and there is no cure for any autoimmune disease, caused by the loss of tolerance to self. Apart from delivering drugs, liposomes were applied as immunotherapy, due to their ability of apoptosis to induce immunological tolerance. One of the mechanisms to maintain self-tolerance is the efficient removal of apoptotic cells. For example, liposomes were generated mimicking apoptotic β-cells, where they arrest autoimmunity in type 1 diabetes, through specific and definitive re-establishment of tolerance. This type of liposome was also investigated to be applied in other autoimmune diseases, such as MS. More analytically, phosphatidylserine-rich liposomes were prepared and loaded with MS-specific autoantigen and, more specifically, the myelin-oligodendrocyte glycoprotein peptide 40–55 (MOG_{40-55}), in order to co-deliver a double signal of tolerance and specificity to arrest autoimmunity in a synergistic, effective, and safe manner. Phosphatidylserine is considered to be the main "eat me" and "tolerate me" signal of the apoptotic cell membrane, which allows recognition and phagocytosis by antigen-presenting cells, such as dendritic cells. According to the results, the prepared liposomes induced a tolerogenic phenotype in dendritic cells and arrested autoimmunity, while they were efficiently phagocytosed by dendritic cells and induced tolerogenic features in dendritic cells. Moreover, after immunization administration, the MOG-loaded phosphatidylserine liposomes reduced the incidence and severity of EAE, as well as delayed the onset of the disease, correlating with an increase in the regulatory CD25+ FoxP3– CD4+ T-cell subset. The authors concluded that the proposed nanosystems exhibit high potential to operate as a platform for MS and autoimmune diseases [82].

3.1.3. Other Lipid-Based Nanocarriers

Binyamin et al. [83] developed a nanodroplet formulation of pomegranate seed oil (PSO), denominated as nano-PSO, which was an oil-in-water (O/W) nanoemulsion (Figure 2A) administrated in an EAE model. PSO comprises high levels of punicic acid, a unique poly-unsaturated fatty acid considered as one of the strongest natural antioxidants. Nano-PSO was supposed to enhance the bioavailability and activity of PSO. According to the results, the beneficial effect of PSO was increased significantly when EAE mice were treated with nano-PSO of specific size nanodroplets (200 nm) size and much lower concentrations of the oil. Nano-PSO was also beneficial to the EAE mice when treatment commenced close to disease manifestation (day 7) and not only when administered concomitant with disease induction, highlighting its potential to abrogate the disease progression and not only prevent it. Pathological examinations revealed that nano-PSO administration dramatically reduced demyelination and oxidation of lipids in the brains of the affected animals, even in the existence of immune infiltrates in the CNS. Nano-PSO may be a good choice for individuals at the initial stages of MS, as well as at later stages, where it may be used in combination with advanced MS treatments such as natalizumab or antioxidant formulations. Last but not least, nano-PSO was also beneficial in the prevention and treatment of genetic prion disease model, indicating that reagents that can prevent lipid oxidation may be beneficial for an array of neurodegenerative diseases.

Most recently, Lu et al. [84] tried to take advantage of the significant role of the monocytes in the process of MS and, thus, developed a targeting immunomodulatory carrier from high-density lipoprotein-mimicking peptide–phospholipid scaffold (HPPS), which can target monocytes, in order to improve the bioavailability of curcumin. Monocytes are considered to be mediators and immunomodulation targets because they are considered to be similar to most of the amplified inflammatory monocytes crossing the BBB, to promote neuron injury and recruit more immune cells to infiltrate the CNS. More specifically, peripheral monocytes were chosen as the immunomodulation targets and curcumin as the anti-inflammatory agent, which was delivered by HPPS nanoparticles. The nanoparticles were taken up efficiently, specifically by monocytes through the scavenger receptor class B type I (SR-B1) receptor, which is a receptor of high-density lipoprotein (HDL) that is highly expressed in the peripheral monocytes; thus, the proposed nanoparticles could be used for early detection of CNS inflammation in EAE. The nanoparticle distribution in monocytes was confirmed in vivo by using optical imaging, while the nanoparticles were loaded with a fluorescent dye. According to the results, the delivery of curcumin by HPPS nanoparticles hindered inflammatory monocytes across the BBB in EAE mice, owing to the downregulation of intercellular adhesion molecules 1 (ICAM-1) and macrophage-1 antigen (MAC-1) expression in the monocytes by inhibiting the activation of the nuclear factor-κB (NF-κB). Moreover, the proposed nanoparticles inhibited the proliferation of microglia and restricted the infiltration of other effect or immune cells, such as TH1, TH17, and myeloid cells, due to the blockade of inflammatory monocyte infiltration, resulting in the reduction of EAE morbidity from 100% (10 of 10 mice exhibited EAE pathology in the phosphate-buffered saline (PBS) group) to 30% (three of 10 mice showed mild clinical signs). The authors concluded that the targeted modulation of monocytes with such HPPS nanoparticles, carrying therapeutic and/or imaging agents, offers a novel strategy for MS diagnosis and treatment.

MicroRNAs were postulated as a promising tool to induce OPC differentiation and, therefore, remyelination, albeit exhibiting significant limitations in terms of the administration of microRNAs to the CNS. Osorio-Querejeta et al. [85] utilized three different categories of nanosystems for miR-219a-5p encapsulation, release, and remyelination promotion, namely, liposomes from 1,2-dioctadecanoyl-sn-glycero-3-phosphocholine (DSPC) lipid (Avanti Polar Lipids, Alabaster, AL, USA), poly(lactic-*co*-glycolic) acid polymeric nanoparticles, and finally biologically engineered extracellular vesicles overexpressing miR-219a-5p, which are also lipid-based, due to their biological origin. Extracellular vesicles are biological delivery systems that contain other proteins, lipids, and genetic material, apart from RNA, which can be integrated into the cell in several ways, and which contain microRNA-processing molecules. The three nanosystem categories were compared by assessing their ability to induce OPC differentiation in a primary oligodendrocyte precursor cell culture and cross the BBB. According to the results, on the one hand, the liposomes and the polymeric nanoparticles were able to entrap higher amounts of miR-219a-5p and showed higher uptake levels than extracellular vesicles. On the other hand, the extracellular vesicles, due to their biological complexity, were surprisingly the only delivery system that was able to induce a significant OPC differentiation, while they also showed the highest BBB permeability levels. Finally, the EAE animal model was used to study the remyelination potential of the extracellular vesicles. The intranasally administered miR-219a-5p-enriched extracellular vesicles successfully decreased clinical scores in the EAE model, without affecting the tested anti-inflammatory pathways. As the authors concluded, the significant differences in clinical score observed after the disease peak indicate that extracellular vesicles might be increasing the myelin production.

3.2. Polymer-Based Nanosystems

Several types of polymers and their respective nanoparticles were studied for efficient BBB crossing, the delivery of therapeutic agents to the CNS, and their contribution to the treatment of neurodegenaration diseases, such as MS. Polymer nanoparticles can be prepared using various synthetic and natural monomers/polymers and via different preparation methods. Their surface can also be functionalized for specific brain targeting. It is crucial for the used polymers in manufacturing these nanocarriers to be biocompatible and biodegradable. The mechanisms for brain uptake and drug release to the CNS of the polymeric nanoparticles involve endocytosis or transcytosis through the endothelial cells, as well as accumulation in the brain capillaries, resulting in transfer to the brain parenchyma, owing to the high concentration gradient, and membrane fluidization through lipid solubilization, due to the surfactant effect and tight junctions opening. One of their most significant advantage is their potential to be easily surface-functionalized by the conjugation of targeting peptides or cell-penetrating ligands, in order to improve their targeting ability and crossing through the BBB [86].

3.2.1. Poly(Lactic-*co*-Glycolic Acid) (PLGA) Polymeric Nanoparticles

Polymeric poly(lactic-*co*-glycolic acid) (PLGA) or poly(lactide-*co*-glycolide) (PLG) nanoparticles are attractive carriers due to their biodegradability, bio-compatibility, and approval by the FDA. One of the key advantages of using PGLA nanoparticles is that they can be easily loaded with a wide variety of molecules. The degradation kinetics of nanoparticles and the release rate of the encapsulated molecules can be easily monitored by controlling the PGLA physicochemical properties, such as the lactide-to-glycolide ratio, molecular weight, crystal profile, storage temperature, and surface coating materials [87–91].

"Inverse vaccination" includes antigen-specific tolerogenic immunization treatments that are capable of inhibiting autoimmune responses, exhibiting great therapeutic potential in MS. Antigen-specific treatments are highly desirable for autoimmune diseases in contrast to treatments which induce systemic immunosuppression. Several myelin proteins, such as MBP, PLP, and MOG, were implicated as targets of autoreactive T cells in MS and EAE [92]. The use of protein-based inverse vaccines in the EAE model, loaded in polymeric biodegradable PLGA nanoparticles, in order to obtain the sustained release of antigens and regulatory adjuvants, can be an alternative strategy to overcome the main obstacles of the administration of free myelin antigens, which exhibit rapid clearance, or of the DNA-based vaccines encoding for myelin autoantigens, whose potential risks limit their use in humans. These data suggest that subcutaneous PLGA nanoparticle-based inverse vaccination maybe an effective tool to treat autoimmune diseases (Figure 3), such as MS, and reduce the significant toxicity, as well as the high costs, caused by the chronic therapies [93,94].

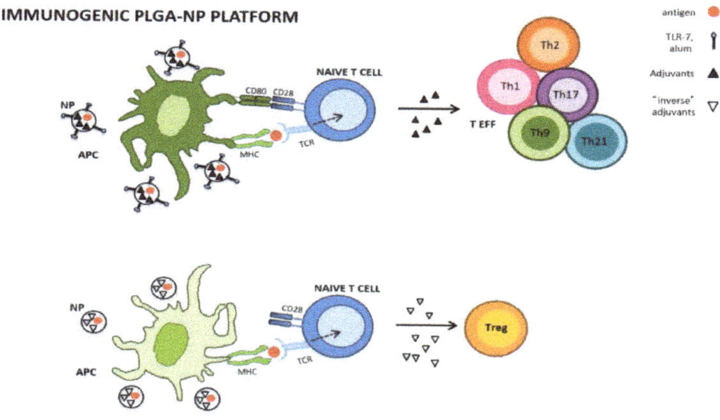

Figure 3. Immunogenic and tolerogenic poly(lactic-*co*-glycolic acid) (PLGA) NP (nanoparticle) platforms. Immunogenic PLGA-NPs may include standard adjuvants such as alum or Toll-like receptor 7 (TLR-7) agonists, capable of activating APCs (antigen-presenting cells) and promoting naïve T-cell differentiation into effector T cells (TEFFs); on the contrary, in the absence of costimulatory signals, tolerogenic PLGA-NPs by employing "inverse adjuvants" induce tolerogenic APCs that lead to the expansion of antigen-specific regulatory T cells (Tregs). The dotted arrow indicates activation of naïve T cells, and the solid arrow indicates its differentiation into TEFFs or Tregs. Adapted from Cappellano et al. [94].

Cappellano et al. [93] developed PLGA nanoparticles loaded with either the immunodominant 35–55 epitope of MOG (MOG$_{35-55}$) in C57BL/6 mice or the recombinant interleukin-10 (IL-10), used as an inverse adjuvant, for prophylactic and therapeutic treatment of a chronic progressive model of EAE.

The authors prepared 65:35 PLGA nanoparticles because they slowly release the loaded molecule for several weeks, which was also confirmed in vitro, and they display minimal cell toxicity along with low intrinsic adjuvant activity. More specifically, the prepared nanoparticles did not display cytotoxic or proinflammatory activity and were partially endocytosed by phagocytes. The prepared nanoparticles loaded with IL-10 completely lost their ability to induce secretion of TNF-α in vitro in peripheral blood mononuclear cells, while the in vivo subcutaneous injection of the prepared nanoparticles in C57BL/6 mice, loaded with either MOG$_{35-55}$ or IL-10, being both simultaneously injected, ameliorated the course of EAE in both prophylactic and therapeutic vaccination. The EAE onset was earlier in the prophylactic vaccination than in the therapeutic vaccination, which suggests that pre-treatment with the PLGA nanoparticles may precondition the induction of EAE by stimulating the innate immunity. Contrariwise, immunization with only one type of nanoparticle, loaded with either MOG$_{35-55}$ or IL-10, did not have any effect. Moreover, they decreased the histopathological lesions in the central nervous tissue, the inflammation, and the T-cell infiltration in the CNS, as well as the secretion of proinflammatory cytokines IL-17 and interferon (IFN)-γ induced by MOG$_{35-55}$ in splenic T cells in vitro.

Maldonaldo et al. [95] developed synthetic, biodegradable PLGA nanoparticles loaded with the immunodominant 139–151 epitope of myelin proteolipid protein (PLP$_{139-151}$) together with the tolerogenic immunomodulator rapamycin, used as an inverse adjuvant, in order to induce durable and antigen-specific immune tolerance, even in the presence of potent Toll-like receptor agonists and control both cellular and humoral immune responses. The nanoparticles were administrated in a relapse remitting model of EAE in mice. According to the results, the treatment, by either subcutaneous or intravenous administration, with tolerogenic nanoparticles results in the inhibition of the activation of the antigen-specific CD4+ and CD8+ T cell, an increase in regulatory cells, durable B-cell tolerance with resistance to multiple immunogenic challenges, and the inhibition of antigen-specific hypersensitivity reactions at relapsing EAE. More specifically, when Swiss Jack Lambert (SJL) mice were immunized with the PLP$_{139-151}$ peptide in complete Freud's adjuvant (PLP$_{139-151}$/CFA) and treated therapeutically with a single dose of tolerogenic nanoparticles at the peak of disease, the mice were completely protected from developing relapsing paralysis. Prophylactic treatment using these nanoparticles inhibited the onset of EAE, whereas the therapeutic treatment inhibited relapse. Moreover, an inhibition of the neutralizing antibody responses against coagulation factor VIII in hemophilia A mice, was observed, even in animals previously sensitized to antigen. Only the encapsulated rapamycin and not the free form in solution could induce immunological tolerance, preventing both cellular and humoral immunity. The authors concluded that therapy with tolerogenic nanoparticles can be applied against allergies and autoimmune diseases, such as MS, as well as for the prevention of antidrug antibodies against biologic therapies.

Cho et al. [96] developed a formulation of dual-sized, polymeric microparticles from PLGA, referred as dMPs, which were loaded with multiple immunomodulatory factors, namely, specific antigen and tolerizing factors, targeting both intra- and extracellular tolerogenic receptors, in order to block the autoimmunity. The epitope MOG$_{35-55}$, along with vitamin D, was encapsulated in the phagocytosable small particles, while transforming growth factor beta 1 (TGF-β1) and granulocyte-macrophage colony-stimulating factor (GM-CSF) were encapsulated in the large non-phagocytosable particles. According to the results, the administration of the dMPs resulted in a reduction of infiltrating CD4+ T cells, inflammatory cytokine-producing pathogenic CD4+ T cells, activated macrophages, and microglia in the CNS, as well as reduced frequency of CD86^{hi}MHCII^{hi} dendritic cells in draining lymph nodes of EAE mice. Thus, the formulation was successfully recruited and modulated the dendritic cells toward a tolerogenic phenotype, while also exhibiting a local controlled release and achieving robust durable antigen-specific autoimmune protection.

Tolerogenic PLGA nanoparticles can also function as on-target and direct modulators of myelin-autoreactive T cells without eliciting the intervention of tolerogenic dendritic cells. Pei et al. [97] developed PLGA nanoparticles for the encapsulation of multiple regulatory molecules. More specifically, the TGF-β was encapsulated in nanoparticles that were surface-decorated by multimers of MHC class I and II molecules loaded with myelin peptides to target autoreactive T cells (MOG_{40-54}/H-2D^b-Ig dimer, MOG_{35-55}/I-A^b multimer), by the regulatory molecules (anti-Fas, PD-L1-Fc), being capable of inducing apoptosis or dysfunction of the autoreactive T cells bound to the MHC multimers, or by the "self-marker" CD47-Fc, being able to inhibit nanoparticle phagocytosis. The suggested nanoparticles exhibited a size of 217 nm and were administrated by intravenous infusion, where they were capable of durably ameliorating EAE, with a marked reduction of clinical score, neuroinflammation, and demyelination. Their mechanism of action was based on the inhibition of the myelin-autoreactive T-cell surface presentation of multiple ligands and the paracrine release of cytokine. According to the results, the MOG_{35-55}-reactive Th1 and Th17 cells, as well as the MOG_{40-55}-reactive Tc1 and Tc17 cells, were decreased, and the regulatory T cells were increased, while inhibited T-cell proliferation and elevated T-cell apoptosis in the spleen took place.

Another report on inducing peripheral, antigen-specific T-cell tolerance for treatment of MS describes the use of biodegradable PLG tolerogenic nanoparticles, being loaded with myelin antigens ($PLP_{139-151}$ and $OVA_{323-339}$ (ovalbumin)) and administrated in mice with relapsing/remitting EAE. PLG particles were fabricated on-site via an emulsion process with modifications using poly(ethylene-*co*-maleic acid) as a surfactant (PLG-PEMA), in order to increase their ability to couple with peptides and to prevent disease induction. In addition to the characterization of the physical properties of the particles, their safety and their therapeutic efficacy for EAE were evaluated by clinical score, histology, and flow cytometric assessment of CNS inflammatory cell infiltration. Their intravenous infusion yielded a significant improvement in the ongoing disease and subsequent relapses when administered at onset or at peak of acute disease, as well as a minimization of epitope spreading when administered during disease remission, resulting in complete long-term protection from disease with superior effects to the already existing commercial PLG nanoparticles. The tolerance was induced by the combined effects of T-cell anergy and the activation of Treg. According to the results, there were reduced CNS infiltration of encephalitogenic Th1 (IFN-γ) and Th17 (IL-17a) cells, as well as inflammatory monocytes/macrophages and demyelination in the CNS of the treated animals [98].

Casey et al. [99] compared the differences between the intravenous and subcutaneous administration of PLG nanoparticles containing disease-relevant antigens (denoted as silver nanoparticles (Ag-NPs)), being surface decorated with TGF-β. The purpose of Ag-NP therapies is to treat autoimmunity. Although the Ag-NPs being intravenously administrated exhibit antigen (Ag)-specific immune tolerance in models of autoimmunity, there is a lower efficacy with subcutaneous administration. Thus, the authors tried to investigate whether the co-delivery of the immunomodulatory cytokine TGF-β using the Ag-NPs would modulate the immune response to Ag-NPs and improve the efficiency of tolerance induction. The selected antigen was $PLP_{139-151}$. An in vitro co-culture system of bone marrow-derived dendritic cells (BMDCs) and naive T cells was used to evaluate the bioactivity and the immunomodulatory effects of the Ag-NPs with TGF-β, by monitoring the surface costimulatory markers and inflammatory cytokine production. According to the results, the Ag-NPs with TGF-β provided Ag-specific T-cell stimulation, decreased co-stimulatory molecule presentation, and suppressed inflammatory cytokine secretions. The in vivo Treg frequency and number were measured following the injection of particles into OT-II mice that expressed $OVA_{323-339}$-restricted T-cell receptors, resulting in the surface binding of TGF-β to PLGA-NPs loaded with OVA being required to efficiently induce tolerance to OVA in an antigen-specific manner. Finally, the ability of surface-bound TGF-β to enhance the tolerogenicity of Ag-NPs was evaluated using the EAE mouse model in the context of intravenous and subcutaneous administration routes, revealing improved efficacy at lower doses by intravenous administration and significantly reduced disease severity by subcutaneous administration. Thus, subcutaneous inverse vaccination needs inverse adjuvants in order to be effective.

Getts et al. [100] used either polystyrene or biodegradable PLG microparticles bearing encephalitogenic peptides, being administrated by intravenous infusion, in order to prevent the onset and modify the course of the disease. These antigen-decorated microparticles, exhibiting an approximate 500-nm diameter induced a long-term T-cell tolerance in mice with relapsing EAE. The chosen antigen for the nanoparticle decoration was MOG_{35-55}, being covalently linked to the surface of the nanoparticle. This treatment reduced the inflammatory cell infiltration and the damage of the CNS. The authors stated that the beneficial effect of these antigen-linked particles requires the scavenger receptor MARCO, being expressed at the marginal zone macrophages. Moreover, the results of the treatment were also monitored by the activity of regulatory T cells, the abortive T-cell activation, and the T-cell anergy. The obtained results highlighted the potential for using microparticles in order to target natural apoptotic clearance pathways, to inactivate pathogenic T cells, to stop the disease process in autoimmunity, and to treat T-cell-based autoimmune disorders, by inducing T-cell tolerance.

Kuo et al. [101] investigated the effect of the amount of antigen conjugated to PLG nanoparticles, as well as of the nanoparticle dose for the induction of immune tolerance, during the treatment of MS with disease-relevant antigens. More specifically, different amounts of the $PLP_{139-151}$ antigen were loaded in PLG nanoparticles that were intravenously administrated in vivo, in different doses to mice with EAE. As a result, the amounts of antigen conjugation and nanoparticle dose were correlated with the severity of EAE. More specifically, the high dose of PLG nanoparticles carrying high amounts of $PLP_{139-151}$ significantly decreased the severity of the EAE with a more durable immune tolerance, also preventing relapses, while a low dose of the nanoparticles carrying high amounts of the antigen or a high dose of the nanoparticles carrying low antigen amounts were not so efficient, as indicated by the observed relapses. The increase of nanoparticle dose and antigen amount was also correlated with the suppression of inflammatory signaling pathways in vitro. Through the analysis of the cells expressing MHC-restricted antigen, significant decreases in positive co-stimulatory molecules (CD86, CD80, and CD40) and a high expression of a negative co-stimulatory molecule (PD-L1) were found in high doses of both nanoparticles and with high antigen conjugation. Tolerance induction was evaluated by co-culturing cells administrated with the nanoparticles and autoreactive T cells isolated from mice immunized against $PLP_{139-151}$. According to the results, reduced T-cell proliferation, increased T-cell apoptosis, and a stronger anti-inflammatory response were observed.

More recently, Saito et al. [102] investigated whether the use of multiple nanoparticle formulations and the extent of antigen loading at the carrier could impact the immune cell internalization and polarization of the immune cells that get associated with the particles and the subsequent disease progression. The nanoparticles were administrated for antigen-specific immunotherapy for the efficient induction of tolerance in the EAE disease model. More specifically, the formulation was composed of three polymeric carriers, 50:50 poly (DL-lactide-*co*-glycolide) with inherent viscosity (IV) = 0.17 dL/g and 0.66 dL/g (termed agPLG-L and agPLG-H, respectively), and poly(DL-lactide) (PLA) particles with IV = 0.21 dL/g (termed agPLA) loaded with the disease-specific antigen myelin protein ($PLP_{139-151}$), being administrated in a single injection for mice with EAE. According to the results, at a low particle dose, mice treated with PLA-based particles had significantly lower clinical scores at the chronic stage, while neither PLG-based particles nor OVA control particles reduced the clinical scores. Moreover, the higher antigen loading PLA-based particles helped to reduce inflammation during the acute stage of the disease, as well as completely ameliorate EAE over 200 days, along with the inhibition of Th1 and Th17 polarization, allowing a smaller particle dosage and, thus, fewer potential adverse effects. PLA, which was correlated with a more tolerogenic polarization of the antigen-presenting cells, was found to be a more efficacious platform relative to PLG. Fluorescently labeled particles were employed to examine the biodistributions among the organs, the interaction with specific antigen-presenting cells within the organs, and the resulting phenotypes. Compared to PLG-based particles, PLA-based particles, interacting with antigen-presenting cells, were largely associated with Kupffer cells and liver sinusoidal endothelial cells, reducing the CD4+ T-cell populations that were activated locally and not trafficked in the CNS.

In addition to antigen-specific tolerogenic immunization, the PLGA nanoparticles were also employed in the remyelination strategy. Rittchen et al. [103] developed a PLGA nanoparticle-based strategy for the targeted delivery of leukemia inhibitory factor (LIF) to OPCs, in order to promote their differentiation into mature oligodendrocytes that are able to repair myelin. Among the potential therapeutics that promote remyelination is the LIF, a cytokine known to play a key regulatory role in self-tolerant immunity and recently identified as a promyelination factor. However, LIF is rapidly degraded in vivo, and high doses may have unwanted off-target effects; therefore, a nanotechnology strategy could help toward its targeted delivery to OPCs. In detail, PLGA-based nanoparticles of ~120 nm diameter were prepared, loaded with LIF (LIF-NP), and functionalized with surface antibodies against NG-2 chondroitin sulfate proteoglycan, expressed on OPCs. The nanoparticle platform used in this study previously featured in clinical trials (e.g., Clinicaltrials.gov NCT01812746, NCT01792479). According to the in vitro results, the NG2-targeted LIF-NPs bound to OPCs activated pSTAT-3 signaling and induced OPC differentiation into mature oligodendrocytes. As per the in vivo results, where a model of focal CNS demyelination was used, the NG2-targeted LIF-NP increased myelin repair, at the level of both increased number of myelinated axons and increased thickness (maturity of sheath) of myelin per axon. The authors also noted that the potency was high, because even a single dose of nanoparticles delivering picomolar quantities of LIF was sufficient to increase remyelination. The nanoparticles were added intralesionally, as a proof of concept, confirming the powerful potential of this delivery system as a targeted approach to deliver bioactive molecules for in situ myelin repair. The authors concluded that, for translation into clinical use, it is notable that the intravenously delivered nanoparticles should be able to cross the BBB, at least in rodents, or they can be delivered intranasally directly to the CNS, bypassing the BBB.

3.2.2. Other Polymeric Nanoparticles

Führmann et al. [104] tried to exploit the physiopathology of MS, regarding the presence of leaky permeable blood vessels, where fibrinogen and nidogen are progressively upregulated after disease onset. Peptide-modified polymeric nanoparticles that are able to target the blood clots and the extracellular matrix (ECM) molecules, such as nidogen, can be used for the targeted drug delivery and the reduction of "off-target" effects. More specifically, poly(ethylene glycol)-*block*-poly(caprolactone) (PEG-*b*-PCL) nanoparticles, functionalized with fibrin peptides, were developed. The prepared nanoparticles were administrated in rats that exhibited EAE along with upregulation of fibrin and nidogen/entactin-1. The administrated nanoparticles showed enhanced binding to these targets and to lesion sites ex vivo and in vivo, compared to non-functionalized or scrambled-peptide control nanoparticles. By using a minimally invasive technique, the active targeting of leaky blood vessels in the diseased spinal cord of animals with EAE can be achieved after the systemic injection of fibrin-targeting nanoparticles. Thus, increased drug concentrations at the site of injury, reduction of undesirable side effects of systemic delivery, and enhanced efficacy of hydrophobic drugs can be achieved in MS conditions.

In another study [105] employing PCL nanoparticles, the recombinant human myelin basic protein (rhMBP) was purified from the milk of transgenic cows, by using a vacuum-driven cation exchanger, and it was formulated into PCL nanoparticles in order to achieve an rhMBP controlled release kinetic, fabricating a therapeutic vaccine for protection against EAE symptoms in mice. RhMBP-loaded PCL nanoparticles were prepared and characterized in terms of entrapment efficiency, size, morphology, charge, and release pattern. The nanoparticles were optimized, until discrete spherical, rough-surfaced rhMBP nanoparticles with small particle size, high surface charge, lower concentration of stabilizer, and intermediate concentration of polymer were obtained, in order to achieve high entrapment efficiency and a controlled release pattern. According to the in vivo results, after the subcutaneous administration of free or rhMBP nanoparticles before EAE induction, the average behavioral score in EAE mice was reduced, and only mild histological alterations and preservation of myelin sheath were observed, indicating increased protection and enhanced efficacy of rhMBP as a therapeutic vaccine due to the nanoformulation. Being hydrophobic, the PCL nanoparticles provided an improved drug delivery system for rhMBP, enabling it to cross the BBB and deliver rhMBP into the mice brains, while the differences in activity between the immediate release and sustained release of the nanoparticle formulations were highlighted. Moreover, the analysis of inflammatory cytokines (IFN-γ and IL-10) in mice brains revealed that the pre-treatment with free or rhMBP nanoparticles significantly protected against EAE-induced behavioral, histopathological, and inflammatory changes.

Lunin et al. [106] developed poly(butylcyanoacrylate) (PBCA) nanoparticles, in order to deliver the thymic peptide thymulin and prolong its presence in the blood of mice with relapsing/remitting EAE (rEAE). The increase in the blood content of thymulin, which is depleted with age, and the prolongation of its blood half-life, alone or in combination with other treatments, may be a prospective strategy for treatment of chronic inflammatory conditions, such as MS. More analytically, as revealed by the results, thymulin significantly decreased symptoms of rEAE and lowered plasma cytokine levels, both in early and in later stages of rEAE, as well as decreased the NF-κB and stress-activated protein kinase/Jun amino-terminal kinase (SAPK/JNK) cascade activation, as confirmed by ELISA measurements of cytokine levels in the blood. In terms of the cytokine response in rEAE, it was multi-staged, where an early phase was accompanied by an increase in plasma IFN-γ, while the IL-17 response was markedly increased at a later stage. According to the results, the nanoparticle-bound thymulin had enhanced efficacy in comparison to the effect of free thymulin.

Kondiah et al. [107] synthesized a pH-sensitive copolymer (TMC-PEGDMA-MAA) from trimethyl-chitosan (TMC), poly(ethylene glycol)dimethacrylate (PEGDMA), and methacrylic acid (MAA), via free radical suspension polymerization, in order to produce microparticles for the oral delivery of IFN-β. The polymer was designed to possess pH-sensitive, mucoadhesive, and hydrophilic properties. A Box–Behnken experimental design was used for optimization of the formulation, where varying concentrations of TMC and percentage crosslinker (polyethylene glycol diacrylate) were investigated. The prepared copolymeric microparticulate system was characterized for its morphological, porositometric, and mucoadhesive properties. The optimized microparticles with 0.5 g/100 mL TMC and 3% crosslinker had an IFN-β loading efficiency of 53.25%. The microparticles were subsequently compressed into a suitable oral tablet formulation. The in vitro release of IFN-β had a pH-sensitive pattern, being increased in intestinal (pH 6.8) and decreased in gastric (pH 1.2) environments. Regarding the in vivo results, the tablets were orally administrated in the New Zealand White rabbit, and the plasma concentration of IFN-β was compared to a known subcutaneous formulation during a 24-h blood sampling procedure. The IFN-β-loaded particulate system demonstrated a remarkable drug release in vivo, greater than the subcutaneous commercial formulation (Rebif®) over a 24-h duration, with a larger bioavailable IFN-β concentration after 2 h.

Youssef et al. [108] investigated the effect of intranasal administration of LINGO-1-directed small interfering RNA (siRNA)-loaded chitosan nanoparticles on demyelination and remyelination processes in a rat model of demyelination. Chitosan nanoparticles are one of the most studied polymers in non-viral siRNA delivery, due to their polycationic nature and biocompatibility. Suppression of LINGO-1 by different strategies, such as LINGO-1 gene knockout or infusion of LINGO-1 antagonists, was associated with enhancement of remyelination in different animal models of CNS demyelination. The RNA interference is a new strategy to block LINGO-1 expression in the target lesions with low doses and fewer systemic side effects, provided that there are efficient gene delivery carriers. In detail, the authors studied whether the nasal administration of LINGO-1-directed siRNA-loaded chitosan nanoparticles in rats could inhibit pontine LINGO-1 expression and enhance the remyelination in a model of demyelination induced by ethidium bromide (EB). The chitosan nanoparticle dose was given alone after the induction of EB in rats that were categorized into demyelination and remyelination groups. According to the results, following the LINGO-1-directed siRNA–chitosan nanoparticle treatment, the animals performed better than controls. Specifically, the remyelination-treated group showed better motor performance than the demyelination group. LINGO-1 downregulation was associated with signs of repair in histopathological sections, higher expression of pontine myelin basic protein (MBP) messenger RNA (mRNA) and protein, and lower levels of caspase-3 activity, indicating neuroprotection and remyelination enhancement.

4. Conclusions

Drug delivery into the CNS is one of the most inhibitory points in the treatment of MS. Nanotechnology opened a new window for the treatment of MS and created promising opportunities by providing drug delivery nanosystems, as well as tolerance-inducing nanocarriers. On the one hand, several novel promising drug delivery systems are being developed for the targeted delivery of MS therapeutics and remyelinating agents into the CNS, in order to increase their therapeutic efficiency and decrease the unwanted side effects that are caused by the high doses. On the other hand, tolerance-inducing nanocarriers can be used as promising vaccines for antigen-specific immunotherapy. Currently, numerous studies describe the successful administration and the promising results of various nanoparticles, both lipidic and polymeric, in the EAE animal model at the preclinical stage, although there is a lack of clinical data in MS patients. Finally, the great potential in the improvement of nanotechnological formulations via the proper monitoring of their physical characteristics or the right choice and functionalization of their biomaterials continually creates future perspectives for the upgrade of current MS therapies.

Author Contributions: Writing—original draft preparation, M.C.; writing—review and editing, C.D.; visualization, M.C.; supervision, C.D. All authors read and agreed to the published version of the manuscript.

Funding: This research received no external funding.

Conflicts of Interest: The authors declare no conflicts of interest.

References

1. Compston, A.; Coles, A. Multiple sclerosis. *Lancet* **2002**, *359*, 1221–1231. [CrossRef]
2. Passos, G.R.D.; Sato, D.K.; Becker, J.; Fujihara, K. Th17Cells pathways in multiple sclerosis and neuromyelitis optica spectrum disorders: Pathophysiological and therapeutic implications. *Mediat. Inflamm.* **2016**, *2016*, 5314541. [CrossRef]
3. Koch, M.W.; Metz, L.M.; Kovalchuk, O. Epigenetic changes in patients with multiple sclerosis. *Nat. Rev. Neurol.* **2013**, *9*, 35–43. [CrossRef] [PubMed]
4. Tillery, E.E.; Clements, J.N.; Howard, Z. What's new in multiple sclerosis? *Ment. Health Clin.* **2018**, *7*, 213–220. [CrossRef] [PubMed]
5. Boissy, A.R.; Cohen, J.A. Multiple sclerosis symptom management. *Expert Rev. Neurother.* **2007**, *7*, 1213–1222. [CrossRef] [PubMed]
6. Pugliatti, M.; Rosati, G.; Carton, H.; Riise, T.; Drulovic, J.; Vecsei, L.; Milanov, I. The Epidemiology of Multiple Sclerosis in Europe. *Eur. J. Neurol.* **2006**, *13*, 700–722. [CrossRef]
7. Dendrou, C.A.; Fugger, L.; Friese, M.A. Immunopathology of multiple sclerosis. *Nat. Rev. Immunol.* **2015**, *15*, 545–558. [CrossRef]
8. Diebold, M. Immunological treatment of multiple sclerosis. *Semin Hematol.* **2016**, *53*, 54–57. [CrossRef]
9. Melchor, G.S.; Khan, T.; Reger, J.F.; Huang, J.K. Remyelination Pharmacotherapy Investigations Highlight Diverse Mechanisms Underlying Multiple Sclerosis Progression. *ACS Pharmacol. Transl. Sci.* **2019**, *2*, 372–386. [CrossRef]
10. Ghalamfarsa, G.; Hojjat-Farsangi, M.; Mohammadnia-Afrouzi, M.; Anvari, E.; Farhadi, S.; Yousefi, M.; Jadidi-Niaragh, F. Application of nanomedicine for crossing the blood–brain barrier: Theranostic opportunities in multiple sclerosis. *J. Immunotoxicol.* **2016**, *13*, 603–619. [CrossRef]
11. Ballerini, C.; Baldi, G.; Aldinucci, A.; Maggi, P. Nanomaterial applications in multiple sclerosis inflamed brain. *J. Neuroimmune Pharmacol.* **2015**, *10*, 1–13. [CrossRef] [PubMed]
12. Gironi, M.; Arno, C.; Comi, G.; Penton-Rol, G.; Furlan, R. Chapter 4—Multiple Sclerosis and Neurodegenerative Diseases. In *Immune Rebalancing: The Future of Immunosuppression*; Elsevier Inc.: Amsterdam, The Netherlands, 2016; pp. 63–84.
13. Leray, E.; Moreau, T.; Fromont, A.; Edan, G. Epidemiology of multiple sclerosis. *Rev. Neurol.* **2016**, *172*, 3–13. [CrossRef] [PubMed]

14. Beecham, A.H.; Patsopoulos, N.A.; Xifara, D.K.; Davis, M.F.; Kemppinen, A.; Cotsapas, C.; Shah, T.S.; Spencer, C.; Booth, D.; Goris, A.; et al. Analysis of Immune-Related Loci Identifies 48 New Susceptibility Variants for Multiple Sclerosis. *Nat. Genet.* **2013**, *45*, 1353–1360.

15. Belbasis, L.; Bellou, V.; Evangelou, E.; Ioannidis, J.P.A.; Tzoulaki, I. Environmental Risk Factors and Multiple Sclerosis: An Umbrella Review of Systematic Reviews and Meta-Analyses. *Lancet Neurol.* **2015**, *14*, 263–273. [CrossRef]

16. Hedström, A.K.; Åkerstedt, T.; Hillert, J.; Olsson, T.; Alfredsson, L. Shift Work at Young Age Is Associated with Increased Risk for Multiple Sclerosis. *Ann. Neurol.* **2011**, *70*, 733–741. [CrossRef]

17. Sadovnick, A.D.; Ebers, G.C.; Dyment, D.A.; Risch, N.J. Evidence for genetic basis of multiple sclerosis. *Lancet* **1996**, *347*, 1728–1730. [CrossRef]

18. Hemmer, B.; Nessler, S.; Zhou, D.; Kieseier, B.; Hartung, H.P. Immunopathogenesis and immunotherapy of multiple sclerosis. *Nat. Clin. Pract. Neurol.* **2006**, *2*, 201–211. [CrossRef]

19. Lublin, F.D.; Reingold, S.C.; Cohen, J.A.; Cutter, G.R.; Sørensen, P.S.; Thompson, A.J.; Wolinsky, J.S.; Balcer, L.J.; Banwell, B.; Barkhof, F.; et al. Defining the clinical course of multiple sclerosis: The 2013 revisions. *Neurology* **2014**, *83*, 278–286. [CrossRef]

20. Faissner, S.; Plemel, J.R.; Gold, R.; Yong, V.W. Progressive multiple sclerosis: From pathophysiology to therapeutic strategies. *Nat. Rev. Drug Discov.* **2019**, *18*, 905–922. [CrossRef]

21. Eckstein, C.; Bhatti, M.T. Currently approved and emerging oral therapies in multiple sclerosis: An update for the ophthalmologist. *Surv. Ophthalmol.* **2016**, *61*, 318–332. [CrossRef]

22. Ian Douglas, J.K.C.; Rompani, P.; Singhal, B.S.; Thompson, A. Multiple sclerosis. In *Neurological Disorders: Public Health Challenges*; Saraceno, B., Ed.; WHO (World Health Organization): Geneva, Switzerland, 2005; pp. 85–94.

23. Correale, J.; Gaitan, M.I.; Ysrraelit, M.C.; Fiol, M.P. Progressive multiple sclerosis: From pathogenic mechanisms to treatment. *Brain J. Neurol.* **2017**, *140*, 527–546. [CrossRef]

24. Baecher-Allan, C.; Kaskow, B.J.; Weiner, H.L. Multiple Sclerosis: Mechanisms and Immunotherapy. *Neuron* **2018**, *97*, 742–768. [CrossRef]

25. Dargahi, N.; Katsara, M.; Tselios, T.; Androutsou, M.E.; de Courten, M.; Matsoukas, J.; Apostolopoulos, V. Multiple sclerosis: Immunopathology and treatment update. *Brain Sci.* **2017**, *7*, 78. [CrossRef] [PubMed]

26. Reich, D.S.; Lucchinetti, C.F.; Calabresi, P.A. Multiple sclerosis. *N. Engl. J. Med.* **2018**, *378*, 169–180. [CrossRef] [PubMed]

27. Kallaur, A.P.; Lopes, J.; Oliveira, S.R.; Simão, A.N.; Reiche, E.M.; de Almeida, E.R.; Morimoto, H.K.; de Pereira, W.L. Immune-Inflammatory and oxidative and nitrosative stress biomarkers of depression symptoms in subjects with multiple sclerosis: Increased peripheral inflammation but less acute neuroinflammation. *Mol. Neurobiol.* **2015**, *53*, 5191–5202. [CrossRef] [PubMed]

28. McFarland, H.F.; Martin, R. Multiple Sclerosis: A Complicated Picture of Autoimmunity. *Nat. Immunol.* **2007**, *8*, 913–919. [CrossRef] [PubMed]

29. Fletcher, J.M.; Lalor, S.J.; Sweeney, C.M.; Tubridy, N.; Mills, K.H.G. T Cells in Multiple Sclerosis and Experimental Autoimmune Encephalomyelitis. *Clin. Exp. Immunol.* **2010**, *162*, 1–11. [CrossRef] [PubMed]

30. Kaskow, B.J.; Baecher-Allan, C. Effector T Cells in Multiple Sclerosis. *Cold Spring Harb. Perspect. Med.* **2018**, *8*, 029025. [CrossRef]

31. Farjam, M.; Zhang, G.X.; Ciric, B.; Rostami, A. Emerging immunopharmacological targets in multiple sclerosis. *J. Neurol. Sci.* **2015**, *358*, 22–30. [CrossRef]

32. Xie, Z.X.; Zhang, H.L.; Wu, X.J.; Zhu, J.; Ma, D.H.; Jin, T. Role of the immunogenic and tolerogenic subsets of dendritic cells in multiple sclerosis. *Mediators Inflamm.* **2015**, *2015*, 513295. [CrossRef]

33. Hemmer, B.; Kerschensteiner, M.; Korn, T. Role of the innate and adaptive immune responses in the course of multiple sclerosis. *Lancet Neurol.* **2015**, *14*, 406–419. [CrossRef]

34. Jadidi-Niaragh, F.; Shegarfi, H.; Naddafi, F.; Mirshafiey, A. The role of natural killer cells in Alzheimer's disease. *Scand. J. Immunol.* **2012**, *76*, 451–456. [CrossRef]

35. Jadidi-Niaragh, F.; Mirshafiey, A. Regulatory T-cell as orchestra leader in immunosuppression process of multiple sclerosis. *Immunopharmacol. Immunotoxicol.* **2011**, *33*, 545–567. [CrossRef] [PubMed]

36. Jadidi-Niaragh, F.; Mirshafiey, A. Th17 cell, the new player of neuroinflammatory process in multiple sclerosis. *Scand. J. Immunol.* **2011**, *74*, 1–13. [CrossRef] [PubMed]

37. Ghalamfarsa, G.; Hadinia, A.; Yousefi, M.; Jadidi-Niaragh, F. The role of natural killer T cells in B cell malignancies. *Tumor Biol.* **2013**, *34*, 1349–1360. [CrossRef]

38. Mishra, M.K.; Yong, V.W. Myeloid Cells—Targets of Medication in Multiple Sclerosis. *Nat. Rev. Neurol.* **2016**, *12*, 539–551. [CrossRef]

39. Nally, F.K.; De Santi, C.; McCoy, C.E. Nanomodulation of Macrophages in Multiple Sclerosis. *Cells* **2019**, *8*, 543. [CrossRef] [PubMed]

40. Stys, P.K.; Tsutsui, S. Recent advances in understanding multiple sclerosis. *F1000Res* **2019**, *8*, F1000 Faculty Rev-2100. [CrossRef] [PubMed]

41. Wingerchuk, D.M.; Carter, J.L. Multiple Sclerosis: Current and Emerging Disease-Modifying Therapies and Treatment Strategies. *Mayo Clin. Proc.* **2014**, *89*, 225–240. [CrossRef]

42. Cross, A.H.; Naismith, R.T. Established and Novel Disease-Modifying Treatments in Multiple Sclerosis. *J. Intern. Med.* **2014**, *275*, 350–363. [CrossRef]

43. Polman, C.H.; O'Connor, P.W.; Havrdova, E.; Hutchinson, M.; Kappos, L.; Miller, D.H.; Phillips, J.T.; Lublin, F.D.; Giovannoni, G.; Wajgt, A.; et al. A Randomized, Placebo-Controlled Trial of Natalizumab for Relapsing Multiple Sclerosis. *N. Engl. J. Med.* **2006**, *354*, 899–910. [CrossRef]

44. Coles, A.J.; Twyman, C.L.; Arnold, D.L.; Cohen, J.A.; Confavreux, C.; Fox, E.J.; Hartung, H.-P.; Havrdova, E.; Selmaj, K.W.; Weiner, H.L.; et al. Alemtuzumab for Patients with Relapsing Multiple Sclerosis after Disease-Modifying Therapy: A Randomised Controlled Phase 3 Trial. *Lancet* **2012**, *380*, 1829–1839. [CrossRef]

45. Katsara, M.; Matsoukas, J.; Deraos, G.; Apostolopoulos, V. Towards immunotherapeutic drugs and vaccines against multiple sclerosis. *Acta Biochim. Biophys. Sin.* **2008**, *40*, 636–642. [CrossRef] [PubMed]

46. Bettelli, E. Building different mouse models for human MS. *Ann. N. Y. Acad. Sci.* **2007**, *1103*, 11–18. [CrossRef]

47. Burt, R.K.; Testori, A.; Craig, R.; Cohen, B.; Suffit, R.; Barr, W. Hematopoietic stem cell transplantation for autoimmune diseases: What have we learned? *J. Autoimmun.* **2008**, *30*, 116–120. [CrossRef]

48. Lassmann, H.; Bradl, M. Multiple Sclerosis: Experimental Models and Reality. *Acta Neuropathol.* **2017**, *133*, 223–244. [CrossRef] [PubMed]

49. Terry, R.L.; Ifergan, I.; Miller, S.D. Experimental Autoimmune Encephalomyelitis in Mice. *Methods Mol. Biol.* **2016**, *1304*, 145–160. [PubMed]

50. Sun, D.; Whitaker, J.N.; Huang, Z.; Liu, D.; Coleclough, C.; Wekerle, H.; Raine, C.S. Myelin Antigen-Specific CD8+ T Cells Are Encephalitogenic and Produce Severe Disease in C57BL/6 Mice. *J. Immunol.* **2001**, *166*, 7579–7587. [CrossRef] [PubMed]

51. Bjelobaba, I.; Begovic-Kupresanin, V.; Pekovic, S.; Lavrnja, I. Animal Models of Multiple Sclerosis: Focus on Experimental Autoimmune Encephalomyelitis. *J. Neurosci. Res.* **2018**, *96*, 1021–1042. [CrossRef]

52. Robinson, A.P.; Harp, C.T.; Noronha, A.; Miller, S.D. The Experimental Autoimmune Encephalomyelitis (EAE) Model of MS. In *Handbook of Clinical Neurology*; Elsevier: Amsterdam, The Netherlands, 2014; Volume 122, pp. 173–189.

53. Kipp, M.; Nyamoya, S.; Hochstrasser, T.; Amor, S. Multiple Sclerosis Animal Models: A Clinical and Histopathological Perspective. *Brain Pathol.* **2017**, *27*, 123–137. [CrossRef]

54. Demetzos, C.; Pippa, N. Advanced drug delivery nanosystems (aDDnSs): A mini-review. *Drug Deliv.* **2014**, *21*, 250–257. [CrossRef] [PubMed]

55. Ortiz, G.G.; Pacheco-Moisés, F.P.; Macías-Islas, M.Á.; Flores-Alvarado, L.J.; Mireles-Ramírez, M.A.; González-Renovato, E.D.; Hernández-Navarro, V.E.; Sánchez-López, A.L.; Alatorre-Jiménez, M.A. Role of the blood-brain barrier in multiple sclerosis. *Arch. Med. Res.* **2014**, *45*, 687–697. [CrossRef]

56. Shakya, A.K.; Nandakumar, K.S. Antigen-Specific Tolerization and Targeted Delivery as Therapeutic Strategies for Autoimmune Diseases. *Trends Biotechnol.* **2018**, *36*, 686–699. [CrossRef]

57. Prinz, J.; Karacivi, A.; Stormanns, E.R.; Recks, M.S.; Kuerten, S. Time-Dependent Progression of Demyelination and Axonal Pathology in MP4-Induced Experimental Autoimmune Encephalomyelitis. *PLoS ONE* **2015**, *10*, e0144847. [CrossRef] [PubMed]

58. Keough, M.B.; Yong, V.W. Remyelination therapy for multiple sclerosis. *Neurotherapeutics* **2013**, *10*, 44–54. [CrossRef] [PubMed]

59. Aharoni, R.; Herschkovitz, A.; Eilam, R.; Blumberg-Hazan, M.; Sela, M.; Bruck, W.; Arnon, R. Demyelination arrest and remyelination induced by glatiramer acetate treatment of experimental autoimmune encephalomyelitis. *Proc. Natl. Acad. Sci. USA* **2008**, *105*, 11358–11363. [CrossRef] [PubMed]

60. Deshmukh, V.A.; Tardif, V.; Lyssiotis, C.A.; Green, C.C.; Kerman, B.; Kim, H.J.; Padmanabhan, K.; Swoboda, J.G.; Ahmad, I.; Kondo, T.; et al. A regenerative approach to the treatment of multiple sclerosis. *Nature* **2013**, *502*, 327–332. [CrossRef]

61. Maeda, Y.; Nakagomi, N.; Nakano-Doi, A.; Ishikawa, H.; Tatsumi, Y.; Bando, Y.; Yoshikawa, H.; Matsuyama, T.; Gomi, F.; Nakagomi, T. Potential of Adult Endogenous Neural Stem/Progenitor Cells in the Spinal Cord to Contribute to Remyelination in Experimental Autoimmune Encephalomyelitis. *Cells* **2019**, *8*, 1025. [CrossRef]

62. Arvidsson, L.; Covacu, R.; Estrada, C.P.; Sankavaram, S.R.; Svensson, M.; Brundin, L. Long-distance effects of inflammation on differentiation of adult spinal cord neural stem/progenitor cells. *J. Neuroimmunol.* **2015**, *288*, 47–55. [CrossRef]

63. Münzel, E.J.; Williams, A. Promoting remyelination in multiple sclerosis—Recent advances. *Drugs* **2013**, *73*, 2017–2029. [CrossRef] [PubMed]

64. Robinson, A.P.; Zhang, J.Z.; Titus, H.E.; Karl, M.; Merzliakov, M.; Dorfman, A.R.; Karlik, S.; Stewart, M.G.; Watt, R.K.; Facer, B.D.; et al. Nanocatalytic activity of cleansurfaced, faceted nanocrystalline gold enhances remyelination in animal models of multiple sclerosis. *Sci. Rep.* **2020**, *10*, 1936. [CrossRef] [PubMed]

65. Zhang, T.T.; Li, W.; Meng, G.; Wang, P.; Liao, W. Strategies for transporting nanoparticles across the blood-brain barrier. *Biomater. Sci.* **2016**, *4*, 219–229. [CrossRef] [PubMed]

66. Niu, X.; Chen, J.; Gao, J. Nanocarriers as a powerful vehicle to overcome bloodbrain barrier in treating neurodegenerative diseases: Focus on recent advances. *Asian J. Pharm. Sci.* **2019**, *14*, 480–496. [CrossRef]

67. Haider, M.; Abdin, S.M.; Kamal, L.; Orive, G. Nanostructured Lipid Carriers for Delivery of Chemotherapeutics: A Review. *Pharmaceutics* **2020**, *12*, 288. [CrossRef]

68. Puri, A.; Loomis, K.; Smith, B.; Lee, J.H.; Yavlovich, A.; Heldman, E.; Blumenthal, R. Lipid-based nanoparticles as pharmaceutical drug carriers: From concepts to clinic. *Crit. Rev. Ther. Drug Carrier Syst.* **2009**, *26*, 523–580. [CrossRef]

69. Agrawal, M.; Saraf, S.; Saraf, S.; Dubey, S.K.; Puri, A.; Patel, R.J.; Ajazuddin; Ravichandiran, V.; Murty, U.S.; Amit, A. Recent strategies and advances in the fabrication of nano lipid carriers and their application towards brain targeting. *J. Control. Release* **2020**, *321*, 372–415. [CrossRef] [PubMed]

70. Shah, R.; Eldridge, D.; Palombo, E.; Harding, I. Composition and structure. In *Lipid Nanoparticles: Production, Characterization and Stability*; Shah, R., Eldridge, D., Palombo, E., Harding, I., Eds.; Springer International Publishing: Cham, Switzerland, 2014; pp. 11–22.

71. Ganesan, P.; Narayanasamy, D. Lipid nanoparticles: Different preparation techniques, characterization, hurdles, and strategies for the production of solid lipid nanoparticles and nanostructured lipid carriers for oral drug delivery. *Sustain. Chem. Pharm.* **2017**, *6*, 37–56. [CrossRef]

72. Gadhave, D.G.; Kokare, C.R. Nanostructured lipid carriers engineered for intranasal delivery of teriflunomide in multiple sclerosis: Optimization and in vivo studies. *Drug Dev. Ind. Pharm.* **2019**, *45*, 839–851. [CrossRef]

73. Kumar, P.; Sharma, G.; Kumar, R.; Malik, R.; Singh, B.; Katare, O.P.; Raza, K. Enhanced brain delivery of dimethyl fumarate employing tocopherol-acetate-based nanolipidic carriers: Evidence from pharmacokinetic, biodistribution, and cellular uptake studies. *ACS Chem. Neurosci.* **2017**, *8*, 860–865. [CrossRef]

74. Ghasemian, E.; Vatanara, A.; Navidi, N.; Rouini, M.R. Brain delivery of baclofen as a hydrophilic drug by nanolipid carriers: Characteristics and pharmacokinetics evaluation. *J. Drug Deliv. Sci. Technol.* **2017**, *37*, 67–73. [CrossRef]

75. Kumar, P.; Sharma, G.; Gupta, V.; Kaur, R.; Thakur, K.; Malik, R.; Kumar, A.; Kaushal, N.; Katare, O.P.; Raza, K. Oral Delivery of Methylthioadenosine to the Brain Employing Solid Lipid Nanoparticles: Pharmacokinetic, Behavioral, and Histopathological Evidences. *AAPS PharmSciTech.* **2019**, *20*, 74. [CrossRef]

76. Gandomi, N.; Varshochian, R.; Atyabi, F.; Ghahremani, M.H.; Sharifzadeh, M.; Amini, M.; Dinarvand, R. Solid lipid nanoparticles surface modified with anti-Contactin-2 or anti-Neurofascin for brain-targeted delivery of medicines. *Pharm. Dev. Technol.* **2017**, *22*, 426–435. [CrossRef]

77. Chountoulesi, M.; Naziris, N.; Pippa, N.; Pispas, S.; Demetzos, C. Differential scanning calorimetry (DSC): An invaluable tool for the thermal evaluation of advanced chimeric liposomal drug delivery nanosystems. In *Thermodynamics and Biophysics of Biomedical Nanosystems. Applications and Practical Considerations. Series: Series in BioEngineering*; Demetzos, C., Pippa, N., Eds.; Springer: Berlin/Heidelberg, Germany, 2019; pp. 297–337.

78. Schmidt, J.; Metselaar, J.M.; Wauben, M.H.M.; Toyka, K.V.; Storm, G.; Gold, R. Drug targeting by long-circulating liposomal glucocorticosteroids increases therapeutic efficacy in a model of multiple sclerosis. *Brain* **2003**, *126*, 1895–1904. [CrossRef]

79. Gaillard, P.J.; Appeldoorn, C.C.; Rip, J.; Dorland, R.; van der Pol, S.M.; Kooij, G.; de Vries, H.E.; Reijerkerk, A. Enhanced brain delivery of liposomal methylprednisolone improved therapeutic efficacy in a model of neuroinflammation. *J. Control. Release* **2012**, *164*, 364–369. [CrossRef]

80. Lee, D.H.; Rötger, C.; Appeldoorn, C.C.; Reijerkerk, A.; Gladdines, W.; Gaillard, P.J.; Linker, R.A. Glutathione PEGylated liposomal methylprednisolone (2B3-201) attenuates CNS inflammation and degeneration in murine myelin oligodendrocyte glycoprotein induced experimental autoimmune encephalomyelitis. *J. Neuroimmunol.* **2014**, *274*, 96–101. [CrossRef]

81. Turjeman, K.; Bavli, Y.; Kizelsztein, P.; Schilt, Y.; Allon, N.; Katzir, T.B.; Sasson, E.; Raviv, U.; Ovadia, H.; Barenholz, Y. Nano-Drugs Based on Nano Sterically Stabilized Liposomes for the Treatment of Inflammatory Neurodegenerative Diseases. *PLoS ONE* **2015**, *10*, e0130442. [CrossRef]

82. Pujol-Autonell, I.; Mansilla, M.J.; Rodriguez-Fernandez, S.; Cano-Sarabia, M.; Navarro-Barriuso, J.; Ampudia, R.M.; Rius, A.; Garcia-Jimeno, S.; Perna-Barrull, D.; Martinez-Caceres, E.; et al. Liposome-based immunotherapy against autoimmune diseases: Therapeutic effect on multiple sclerosis. *Nanomedicine* **2017**, *12*, 1231–1242. [CrossRef]

83. Binyamin, O.; Larush, L.; Frid, K.; Keller, G.; Friedman-Levi, Y.; Ovadia, H.; Abramsky, O.; Magdassi, S.; Gabizon, R. Treatment of a multiple sclerosis animal model by a novel nanodrop formulation of a natural antioxidant. *Int. J. Nanomed.* **2015**, *10*, 7165–7174. [CrossRef]

84. Lu, L.; Qi, S.; Chen, Y.; Luo, H.; Huang, S.; Yu, X.; Luo, Q.; Zhang, Z. Targeted immunomodulation of inflammatory monocytes across the blood-brain barrier by curcumin-loaded nanoparticles delays the progression of experimental autoimmune encephalomyelitis. *Biomaterials* **2020**, *245*, 119987. [CrossRef]

85. Osorio-Querejeta, I.; Carregal-Romero, S.; Ayerdi-Izquierdo, A.; Mäger, I.; Nash, L.A.; Wood, M.; Egimendia, A.; Betanzos, M.; Alberro, A.; Iparraguirre, L.; et al. MiR-219a-5p Enriched Extracellular Vesicles Induce OPC Differentiation and EAE Improvement More Efficiently Than Liposomes and Polymeric Nanoparticles. *Pharmaceutics* **2020**, *12*, 186. [CrossRef]

86. El-Say, K.M.; El-Sawy, H.S. Polymeric nanoparticles: Promising platform for drug delivery. *Int. J. Pharm.* **2017**, *528*, 675–691. [CrossRef]

87. Houchin, M.L.; Topp, E.M. Chemical degradation of peptides and proteins in PLGA:a review of reactions and mechanisms. *J. Pharm. Sci.* **2008**, *97*, 2395–2404. [CrossRef]

88. Makadia, H.K.; Siegel, S.J. Poly Lactic-co-Glycolic Acid (PLGA) as Biodegradable Controlled Drug Delivery Carrier. *Polymers* **2011**, *3*, 1377–1397. [CrossRef]

89. Gentile, P.; Chiono, V.; Carmagnola, I.; Hatton, P.V. An overview of poly(lactic-co-glycolic) acid (PLGA)-based biomaterials for bone tissue engineering. *Int. J. Mol. Sci.* **2014**, *15*, 3640–3659. [CrossRef]

90. Samavedi, S.; Poindexter, L.K.; Van Dyke, M.; Goldstein, A.S. Synthetic biomaterials for regenerative medicine applications. *Regen. Med. Appl. Organ. Transpl.* **2014**, 81–99. [CrossRef]

91. Houchin, M.L.; Topp, E.M. Physical properties of PLGA films during polymer degradation. *J. Appl. Polym. Sci.* **2009**, *114*, 2848–2854. [CrossRef]

92. Johnson, D.; Hafler, D.A.; Fallis, R.J.; Lees, M.B.; Brady, R.O.; Quarles, R.H.; Weiner, H.L. Cell-mediated immunity to myelin-associated glycoprotein, proteolipid pro-tein, and myelin basic protein in multiple sclerosis. *J. Neuroimmunol.* **1986**, *13*, 99–108. [CrossRef]

93. Cappellano, G.; Woldetsadik, A.D.; Orilieri, E.; Shivakumar, Y.; Rizzi, M.; Carniato, F.; Gigliotti, C.L.; Boggio, E.; Clemente, N.; Comi, C.; et al. Subcutaneous inverse vaccination with PLGA particles loaded with a MOG peptide and IL-10 decreases the severity of experimental autoimmune encephalomyelitis. *Vaccine* **2014**, *32*, 5681–5689. [CrossRef]

94. Cappellano, G.; Comi, C.; Chiocchetti, A.; Dianzani, U. Exploiting PLGA-Based Biocompatible Nanoparticles for Next-Generation Tolerogenic Vaccines against Autoimmune Disease. *Int. J. Mol. Sci.* **2019**, *20*, 204. [CrossRef]

95. Maldonado, R.A.; LaMothe, R.A.; Ferrari, J.D.; Zhang, A.H.; Rossi, R.J.; Kolte, P.N.; Griset, A.P.; O'Neil, C.; Altreuter, D.H.; Browning, E.; et al. Polymeric synthetic nanoparticles for the induction of antigen-specific immunological tolerance. *Proc. Natl. Acad. Sci. USA* **2015**, *112*, 156–165. [CrossRef]

96. Cho, J.J.; Stewart, J.M.; Drashansky, T.T.; Brusko, M.A.; Zuniga, A.N.; Lorentsen, K.J.; Keselowsky, B.G.; Avram, D. An antigen-specific semi-therapeutic treatment with local delivery of tolerogenic factors through a dual-sized microparticle system blocks experimental autoimmune encephalomyelitis. *Biomaterials* **2017**, *143*, 79–92. [CrossRef]

97. Pei, W.; Wan, X.; Shahzad, K.A.; Zhang, L.; Song, S.; Jin, X.; Wang, L.; Zhao, C.; Shen, C. Direct modulation of myelin-autoreactive CD4(+) and CD8(+) T cells in EAE mice by a tolerogenic nanoparticle co-carrying myelin peptide-loaded major histocompatibility complexes, CD47 and multiple regulatory molecules. *Int. J. Nanomed.* **2018**, *13*, 3731–3750. [CrossRef]

98. Hunter, Z.; McCarthy, D.P.; Yap, W.T.; Harp, C.T.; Getts, D.R.; Shea, L.D.; Miller, S.D. A biodegradable nanoparticle platform for the induction of antigen-specific immune tolerance for treatment of autoimmune disease. *ACS Nano* **2014**, *8*, 2148–2160. [CrossRef]

99. Casey, L.M.; Pearson, R.M.; Hughes, K.R.; Liu, J.M.H.; Rose, J.A.; North, M.G.; Wang, L.Z.; Lei, M.; Miller, S.D.; Shea, L.D. Conjugation of Transforming Growth Factor Beta to Antigen-Loaded Poly(lactideco-glycolide) Nanoparticles Enhances Efficiency of Antigen-Specific Tolerance. *Bioconjug. Chem.* **2018**, *29*, 813–823. [CrossRef]

100. Getts, D.R.; Martin, A.J.; McCarthy, D.P.; Terry, R.L.; Hunter, Z.N.; Yap, W.T.; Getts, M.T.; Pleiss, M.; Luo, X.; King, N.J.; et al. Microparticles bearing encephalitogenic peptides induce T-cell toleranceand ameliorate experimental autoimmune encephalomyelitis. *Nat. Biotechnol.* **2012**, *30*, 1217–1224. [CrossRef]

101. Kuo, R.; Saito, E.; Miller, S.D.; Shea, L.D. Peptide-Conjugated Nanoparticles Reduce Positive Co-stimulatory Expression and T Cell Activity to Induce Tolerance. *Mol. Ther.* **2017**, *25*, 1676–1685. [CrossRef]

102. Saito, E.; Kuo, R.; Kramer, K.R.; Gohel, N.; Giles, D.A.; Moore, B.B.; Miller, S.D.; Shea, L.D. Design of biodegradable nanoparticles to modulate phenotypes of antigen-presenting cells for antigen-specific treatment of autoimmune disease. *Biomaterials* **2019**, *222*, 119432. [CrossRef]

103. Rittchen, S.; Boyd, A.; Burns, A.; Park, J.; Fahmy, T.M.; Metcalfe, S.; Williams, A. Myelin repair in vivo is increased by targeting oligodendrocyte precursor cells with nanoparticles encapsulating leukaemia inhibitory factor (LIF). *Biomaterials* **2015**, *56*, 78–85. [CrossRef]

104. Führmann, T.; Ghosh, M.; Otero, A.; Goss, B.; Dargaville, T.R.; Pearse, D.D.; Dalton, P.D. Peptide-functionalized polymeric nanoparticles for active targeting of damaged tissue in animals with experimental autoimmune encephalomyelitis. *Neurosci. Lett.* **2015**, *602*, 126–132. [CrossRef]

105. Al-Ghobashy, M.A.; El Meshad, A.N.; Abdelsalam, R.M.; Nooh, M.M.; Al-Shorbagy, M.; Laible, G. Development and Pre-Clinical Evaluation of Recombinant Human Myelin Basic Protein Nano Therapeutic Vaccine in Experimental Autoimmune Encephalomyelitis Mice Animal Model. *Sci. Rep.* **2017**, *7*, 46468. [CrossRef]

106. Lunin, S.M.; Khrenov, M.O.; Glushkova, O.V.; Parfenyuk, S.B.; Novoselova, T.V.; Novoselova, E.G. Protective Effect of PBCA Nanoparticles Loaded with Thymulin Against the Relapsing-Remitting Form of Experimental Autoimmune Encephalomyelitis in Mice. *Int. J. Mol. Sci.* **2019**, *20*, 5374. [CrossRef]

107. Kondiah, P.P.; Tomar, L.K.; Tyagi, C.; Choonara, Y.E.; Modi, G.; du Toit, L.C.; Kumar, P.; Pillay, V. A novel pH-sensitive interferon-β (INF-β) oral delivery system for application in multiple sclerosis. *Int. J. Pharm.* **2013**, *456*, 459–472. [CrossRef]

108. Youssef, A.E.H.; Dief, A.E.; El Azhary, N.M.; Abdelmonsif, D.A.; El-fetiany, O.S. LINGO-1 siRNA nanoparticles promote central remyelination in ethidium bromide-induced demyelination in rats. *J. Physiol. Biochem.* **2019**, *75*, 89–99. [CrossRef]

Article

Comprehensive Analysis of the Immune and Stromal Compartments of the CNS in EAE Mice Reveal Pathways by Which Chloroquine Suppresses Neuroinflammation

Rodolfo Thome, Alexandra Boehm, Larissa Lumi Watanabe Ishikawa, Giacomo Casella, Jaqueline Munhoz, Bogoljub Ciric, Guang-Xian Zhang and Abdolmohamad Rostami *

Department of Neurology, Thomas Jefferson University, Philadelphia, PA 19107, USA;
rodolfo.thome@jefferson.edu (R.T.); alexandra.boehm@jefferson.edu (A.B.);
larissa.ishikawa@jefferson.edu (L.L.W.I.); giacomo.casella@jefferson.edu (G.C.); jlima.munhoz@gmail.com (J.M.);
bogoljub.ciric@jefferson.edu (B.C.); guang-xian.zhang@jefferson.edu (G.-X.Z.)
* Correspondence: a.m.rostami@jefferson.edu

Received: 27 April 2020; Accepted: 3 June 2020; Published: 5 June 2020

Abstract: Multiple sclerosis (MS) and experimental autoimmune encephalomyelitis (EAE) are neuroinflammatory diseases of the central nervous system (CNS), where leukocytes and CNS resident cells play important roles in disease development and pathogenesis. The antimalarial drug chloroquine (CQ) has been shown to suppress EAE by modulating dendritic cells (DCs) and Th17 cells. However, the mechanism of action by which CQ modulates EAE is far from being elucidated. Here, we comprehensively analyzed the CNS of CQ and PBS-treated EAE mice to identify and characterize the cells that are affected by CQ. Our results show that leukocytes are largely modulated by CQ and have a reduction in the expression of inflammatory markers. Intriguingly, CQ vastly modulated the CNS resident cells astrocytes, oligodendrocytes (OLs) and microglia (MG), with the latter producing IL-10 and IL-12p70. Overall, our results show a panoramic view of the cellular components that are affect by CQ and provide further evidence that drug repurposing of CQ will be beneficial to MS patients.

Keywords: chloroquine; EAE; dendritic cells; microglia; astrocytes; oligodendrocytes

---

1. Introduction

MS and EAE are inflammatory diseases of the CNS. Although the mechanisms that trigger MS are not fully elucidated, studies in EAE have uncovered the major role played by T cells in disease development, severity, and recovery [1]. Inflammatory CD4$^+$ T cells differentiate into IFN-γ^+ Th1 cells and IL-17$^+$ Th17 cells and migrate to the CNS where they induce inflammation and the chemoattraction of myeloid cells [2–5]. We and others have shown that the cytokine GM-CSF is an essential mediator of pathogenic Th17 cells [6–8]. Moreover, GM-CSF is induced in Th17 cells by IL-23R stimulation [7,9]. Conversely, Foxp3-expressing regulatory T (Treg) cells suppress Th1/Th17 cells and overall inflammation through contact-dependent and independent mechanisms [10]. Additionally, Treg cells were shown to promote remyelination [11,12]. In this context, antigen-presenting cells, such as dendritic cells (DCs), may direct T cell differentiation to Th17/Th1 and Treg cell phenotypes [13–15].

CNS resident cells are affected by local inflammation and may contribute to neurodegeneration in EAE. At early stages of EAE, microglia (MG) acquire an inflammatory M1 profile with the production of IL-1β, IL-6 and nitric oxide (NO), which shifts towards a tissue repair-associated M2 phenotype at later stages [16–18]. MG-derived IL-1β promote neuroinflammation and MG proliferation in an autocrine manner [19]. Moreover, MG-astrocyte crosstalk greatly influences the outcome of neuroinflammation in EAE [20,21]. Inflammatory A1 astrocytes actively induce damage to oligodendrocytes (OLs) and neurons, promote leukocyte infiltration and aid neuroinflammatory processes [22,23]. These observations illustrate the complex network and interactions among peripherally-derived leukocytes and CNS resident cells in EAE and MS. Thus, drugs that target multiple cell types while inducing tissue repair are of great importance in MS therapy.

Chloroquine (CQ) is a known antimalarial drug with anti-inflammatory properties [24]. CQ modulates monocyte activation and TNF-α production [25,26]. CQ also inhibits lysosomal maturation and antigen-processing [27–29]. By modulating DCs and Th17 cells, CQ has been shown to suppress EAE [27,30,31]. However, the mechanisms by which CQ suppresses EAE are far from being elucidated. CQ induces tolerogenic DCs in a STAT1-dependent manner [32]. Interestingly, CQ inhibited Th17 cell differentiation in a STAT-1-independent and T-bet-dependent fashion [31]. These observations suggest that CQ interferes with multiple cell types and influences diverse signaling pathways to suppress inflammation. The identification of such cell types and pathways may provide a better understanding of the mechanisms by which CQ suppresses EAE and may shed light on new therapeutic targets in MS.

In this study, we comprehensively characterized the phenotype and activity of peripherally derived leukocytes and CNS resident cells in CQ-treated EAE mice. We observed that CQ suppressed inflammatory leukocytes while also modulating MG and astrocytes. Further analyses revealed that CQ reduced IL-23R expression in CD4$^+$ T cells and IL-23 production by MG. Moreover, CQ stimulated MG to produce IL-10 and IL-12p70, which stimulated IL-10 production in T cells and refrained astrocyte activation. Together, our results show that CQ has a broad action by modulating leukocytes and CNS resident cells to suppress EAE and provide further evidence that drug repurposing of CQ is beneficial to patients with MS.

2. Methods

2.1. Animals

We used 8–12 week old male and female C57BL/6 and B6SJLF1 mice from the Jackson Laboratory (Bar Harbor, ME, USA). Mice were acclimated in clean cages in a controlled environment with food and water ad libitum. Experiments detailed in this study were approved by Thomas Jefferson University's IACUC under protocol numbers 01970 and 02034.

2.2. EAE Induction and Evaluation

To induce and evaluate EAE, we followed previously described protocols [33,34]. Mice were subcutaneously immunized with 200 µg of MOG$_{35-55}$ peptide (Genscript) and equal volume of Complete Freund's adjuvant supplemented with 10 mg/mL of heat-killed *Mycobacterium tuberculosis* H37Ra. Additionally, mice were intraperitoneally injected with 240 ng of Pertussis toxin at 0 and 2 days after immunization. EAE development was analyzed daily and scored on a 0–5 scale, where: 0—no clinical sign, 1—limp tail, 2—hind paw weakness, 3—hind paw paralysis, 4—hind paw paralysis and front paw weakness, 5—full paralysis/death.

2.3. CQ Treatment

The dosage for CQ treatment has been assessed before [35]. Mice were treated with CQ (chloroquine diphosphate salt, Sigma-Aldrich) at a 5 mg/kg concentration via i.p. injections. The pH in CQ solution was 7.2. Control mice were injected with diluent solution (phosphate-buffered saline 0.02 M pH 7.2).

2.4. Isolation of Mononuclear Cells in the CNS of Mice with EAE

Mononuclear cells from the CNS of EAE mice were isolated by Percol gradient centrifugation following published reports [32–34]. In brief, euthanized mice were perfused with ice-cold PBS and the CNS tissue was collected and incubated with 700 µg/mL Liberase TL (Sigma-Aldrich, St. Louis, MO,

USA) at 37 °C for 30 min. To remove myelin debris, the digested tissue was centrifuged in a 30% Percol solution. MNCs were recovered from the bottom of the tube and used for flow cytometry analyses.

2.5. Flow Cytometry

For detection of intracellular cytokines by flow cytometry, cells were stimulated with PMA (50 ng/mL), ionomycin (500 ng/mL) and GolgiPlug (1 μg/mL) in IMDM complete medium for 3 h at 37 °C. Cells were washed in FACS buffer (PBS/2% FBS) and stained with fluorochrome labeled Abs to surface molecules for 20 min at 4 °C. Cells were then fixed and permeabilized (Invitrogen/ThermoFisher, Waltham, MA, USA) and incubated with antibodies against intracellular antigens for 18 h at 4 °C. Immediately before acquisition, cells were washed and resuspended in PBS. We utilized a FACSAria Fusion (BD Biosciences) flow cytometer for acquisition and FlowJo VX (Tristar Inc., Ashland, OR, USA) for analyses. Antibodies used in this study were anti-mouse: CD45 (30-F11), TCR-β (H57-597), CD4 (GK1.5), CD8 (53-6.7), GFAP (2E1.E9), CD11b (M1/70), Ly6C (HK1.4), CD11c (N418), MHC-II (M5/114.15.2), CD80 (16-10A1), CD86 (GL-1), pSTAT1 (A15158B), pSTAT3 (13A3-1), mTOR (O21-404, from BD Biosciences), IL-1β (NJTEN3, from eBioscience/ThermoFisher), IL-6 (MP5-20F3), IL-10 (JES5-16E3), IL-12p70 (C15.6, from BD Biosciences), IL-17A (TC11-18H10.1), IL-23 (N71-1183, from BD Biosciences), GM-CSF (MP1-22E9), Foxp3 (FJK-16s, from eBioscience/ThermoFisher, Waltham, MA, USA), IL-23R (12B2B64), IL-10R (1B1.3a), Granzyme B (GB11), and IRF8 (V3GYWCH, from eBioscience/ThermoFisher). All antibodies used in this study were purchased from Biolegend, San Diego, CA, USA, except where mentioned otherwise.

2.6. Isolation of Primary MG and CQ Treatment

CD11b$^+$ MG were isolated from MNCs obtained from the CNS of P0–P3 pups using magnetic beads (Miltenyi Biotec., Auburn, CA, USA). This isolation procedure yielded a consistent purity of 95% of CD11b$^+$ cells assessed by flow cytometry. MG were activated with LPS (100 ng/mL) with or without CQ (50 μM) for 18 h at 37 °C. The optimal CQ concentration for in vitro treatment of myeloid cells has been determined before [27]. At the end of culture time, MG cells were processed for flow cytometry, RNA extraction and co-culture.

2.7. PCRArray and Gene Ontology Analysis

RNA was extracted and reverse-transcribed from primary MG utilizing commercially available kits (RNAeasy extraction kit and high capacity RNA-to-cDNA kit, respectively, both from ThermoFisher). The cDNA was tested for quality and purity in a nanodrop equipment before being subjected to PCRArray (ThermoFisher). Gene ontology analysis was performed with CytoScape v3.8 (CytoScape.org).

2.8. Co-Culture of MG and T Cells

Primary MG were treated as above and extensively washed with Iscove's Modified Dulbecco Medium (IMDM) to remove LPS and CQ from the cells. In total, 50,000 MG were seeded into each well of a 96 U-bottom well plate. Then, CD4$^+$ T cells were isolated from spleens of naïve mice using magnetic beads (Miltenyi Biotec) and 100,000 cells were seeded on each well on top of MG. As controls, T cells were cultured without MG. As stimulus, cells were incubated with agonistic anti-CD3 antibody at a concentration of 0.5 μg/mL. For these analyses, co-stimulation was provided by MG without the need for anti-CD28 antibodies. Cells were incubated for 72 h at 37 °C as described [32], and then, analyzed by flow cytometry.

2.9. Co-Culture of MG and Astrocytes

Primary MG were treated as above and extensively washed with Iscove's Modified Dulbecco Medium (IMDM) to remove LPS and CQ from the cells. In total, 50,000 MG were seeded into each well of a 96 U-bottom well plate. Then, 50,000 C8D30 astrocytes (ATCC-CRL-2534) were seeded on each well on top of MG. As controls, astrocytes were cultured without MG. Cells were incubated for 48 h at 37 °C before being collected and analyzed by flow cytometry.

2.10. Statistical Analyses

Daily clinical scores among experimental groups in EAE were compared by two-way ANOVA and post-tested with Sidak. Comparisons between two groups were carried out with unpaired Student's *t* test and Welch's correction. Values of $p < 0.05$ were defined as significant.

3. Results

3.1. CQ Reduces Ongoing Inflammation in Relapsing EAE

CQ has been shown to prevent EAE development when given prophylactic and to reduced EAE severity when given at disease onset [31,35]. However, if CQ is able to suppress ongoing relapsing EAE is unknown. B6SJLF1 mice, the offspring of C57BL/6 and SJL mice, develop relapsing EAE when immunized with MOG_{35-55} [36,37]. We immunized mice to develop EAE and CQ treatment started at the onset of clinical signs of EAE. Mice treated with PBS fully developed disease clinical signs and the characteristic relapsing feature of this model (Figure 1A). However, mice treated with CQ developed a significantly less severe EAE compared with PBS-treated mice (Figure 1A). These results are on par with those observed in chronic EAE in C57BL/6 mice (Figure 1B). B6SJLF1 mice treated with CQ had a significant decrease in the infiltration of leukocytes to the CNS compared with those treated with PBS (Figure 1C). A similar trend was observed in treated C57BL/6 mice (Figure 1C). These results show that CQ is a potent agent in suppressing the clinical development of two models of EAE.

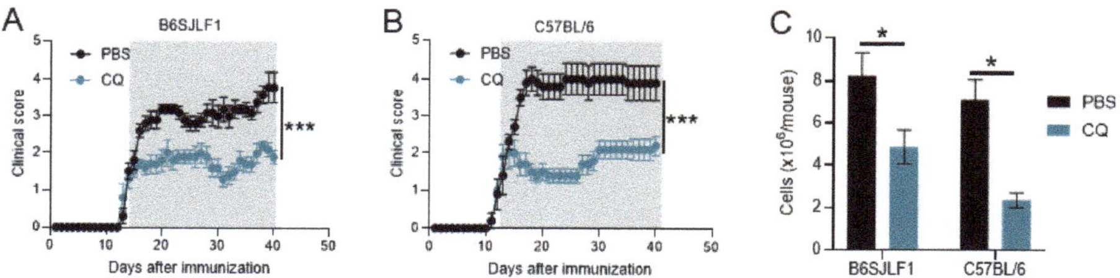

Figure 1. Chloroquie (CQ) suppresses neuroinflammation in two models of experimental autoimmune encephalomyelitis (EAE). C57BL/6 and B6SJLF1 mice (n = 5/group) were immunized to induce EAE. At the onset and until the end of experimentation, mice were treated with CQ (5 mg/kg) via i.p. everyday. (**A**) Disease development in B6SJLF1 mice and in (**B**) C57BL/6 mice. (**C**) CNS cells were analyzed at day 21 post-immunization. Values of $p < 0.05$ (*) and < 0.001 (***) were considered statistically significative. Results are from three independent experiments.

3.2. Peripherally-Derived Leukocytes Are Modulated in CQ-Treated Mice

We analyzed the phenotype of leukocytes in the CNS of CQ-treated EAE B6SJLF1 mice. We observed a significant reduction in numbers of $CD45^{high}$ leukocyte in mice treated with CQ compared with those treated with PBS (Figure 1C). Furthermore, DCs and macrophages from CQ-treated mice presented a reduced expression of the molecules involved in antigen-presentation and activation: class II MHC, CD80, CD86 in comparison with controls (Figure 2A). Analyses of the signaling mediators pSTAT1, pSTAT3, and mTOR in peripherally derived myeloid cells from CQ-treated mice revealed a decrease in the expression of pSTAT3 and mTOR in DCs and macrophages, but not in neutrophils in comparison with those from PBS-treated mice (Figure 2C). pSTAT1 was significantly upregulated in DCs and macrophages while pSTAT3 was upregulated in neutrophils (Figure 2C). There was no significant difference in mTOR expression in neutrophils between the two groups of mice (Figure 2C).

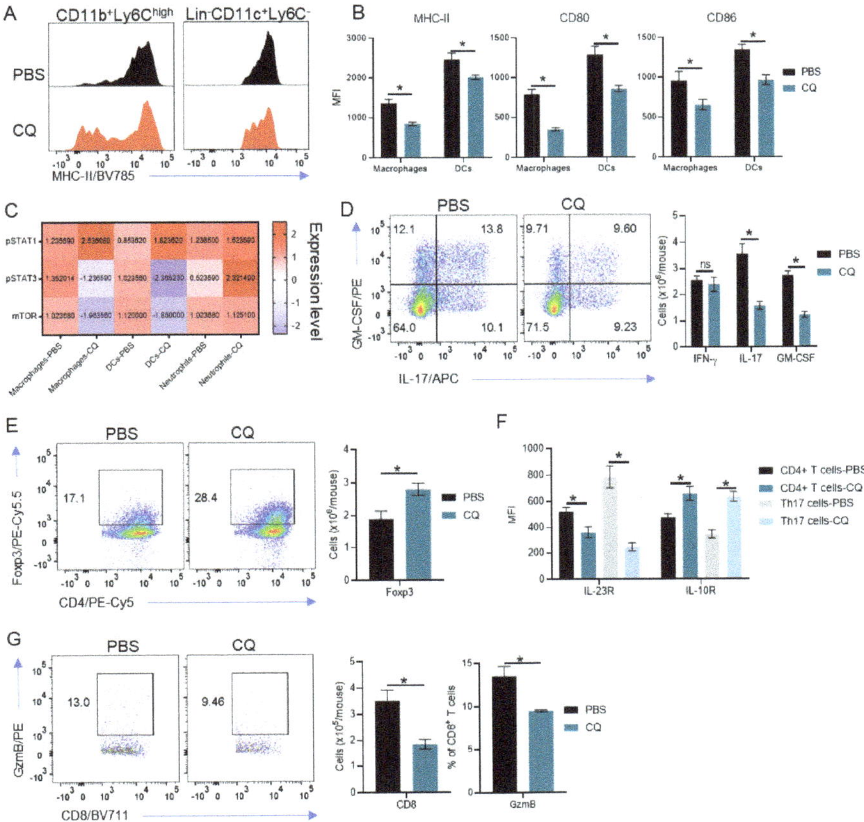

Figure 2. Comprehensive analysis of the immune compartment in the central nervous system (CNS) of CQ-treated mice. CNS cells from EAE B6SJLF1 mice (n = 5/group) at day 21 post-immunization were analyzed by flow cytometry. (**A**) Representative histogram for MHC-II expression in CD11b⁺Ly6C^high macrophages and in Lin⁻CD11c⁺Ly6C⁻ DCs. (**B**) MHC-II, CD80 and CD86 expression in macrophages and dendritic cells (DCs). (**C**) Mean fluorescent values for pSTAT1, pSTAT3 and mTOR in macrophages, DCs and Ly6G⁺ neutrophils were measured by flow cytometry. (**D**) Infiltrating GM-CSF-, IFN-γ and IL-17-producing CD4⁺ T cells in the CNS of EAE mice. (**E**) Numbers of Foxp3⁺ Treg cells. (**F**) Analysis of IL-23R and IL-10R in total CD4⁺ T cells and in Th17 cells in the CNS of EAE mice. (**G**) Analysis of total CD8⁺ T cell GzmB-producing CD8⁺ T cell infiltration in the CNS of EAE mice. Values of $p < 0.05$ (*) were considered statistically significative. Ns: not significative. Results are from three independent experiments.

We also observed a significant decrease in IL-17⁺ and GM-CSF⁺CD4⁺ T cells in the CNS of CQ-treated mice compared with PBS-treated ones (Figure 2D). No significant differences were observed in Th1 cells (Figure 2D). Foxp3⁺ Treg cells were upregulated in CQ-treated mice compared with controls (Figure 2E). Interestingly, we observed a significant decrease in the expression of IL-23R and an increase in the expression of IL-10R in total CD4⁺ T cells and more drastically in Th17 cells (Figure 2F). CD8⁺ T cells numbers were reduced and had lower expression of Granzyme B in CQ-treated mice compared with PBS-treated mice (Figure 2G). B cells did not show significant differences in numbers, MHC-II expression and in cytokine production (not shown). Combined, these results show that CQ modulates both myeloid and lymphoid leukocyte populations in the CNS of EAE mice.

3.3. MG and Astrocytes Are Modulated by CQ

We then analyzed the effect of CQ treatment on CNS resident cells of EAE mice. We observed that GFAP$^+$ astrocytes presented as two populations GFAP$^+$ and GFAP^{high} (Figure 3A), which reflect their activation profile. GFAP^{high} are reactive astrocytes that are involved in scar formation and oligodendrocyte death [38,39]. We observed a significant decrease in GFAP^{high} astrocytes in CQ-treated mice compared with PBS-treated controls (Figure 3A). Moreover, astrocytes from CQ-treated mice showed an increase in IL-10 production and a decrease in IL-6 and IL-1β production in comparison with cells from PBS-treated mice (Figure 3B).

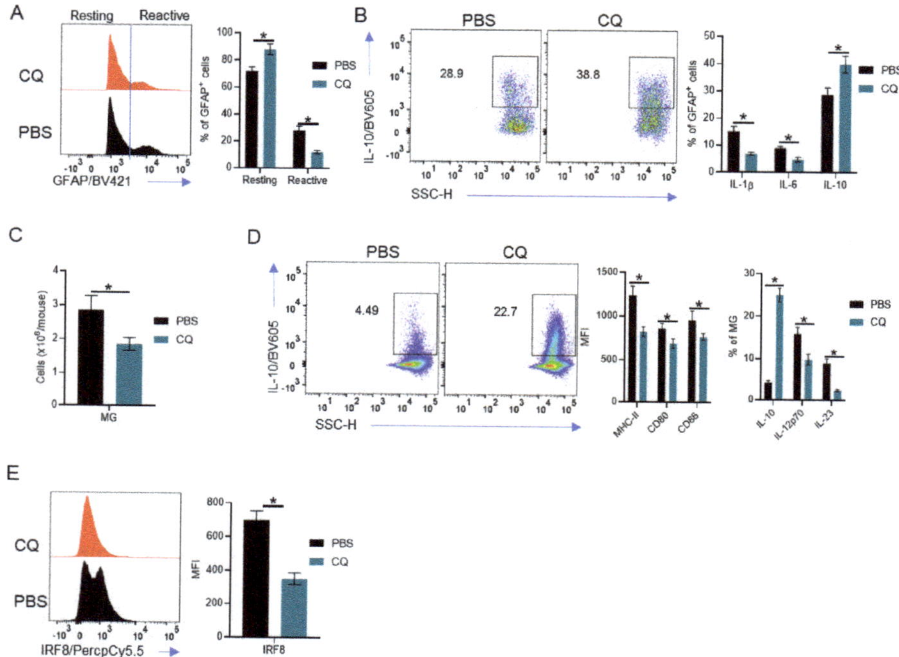

Figure 3. Comprehensive analysis of the stromal compartment in the CNS of CQ-treated mice. CNS cells from EAE B6SJLF1 mice (*n* = 5/group) at day 21 post-immunization were analyzed by flow cytometry. (**A**) Analysis of resting (GFAP$^{+/low}$) and reactive (GFAP^{high}) astrocytes. (**B**) Analysis of IL-1β, IL-6 and IL-10 production by astrocytes. (**C**) Absolute numbers of MG (CD45$^{+/low}$CD11b$^+$Ly6C$^-$) in the CNS of CQ- and PBS-treated mice. (**D**) Analysis of the expression of the antigen-presenting molecules MHC-II, CD80 and CD86 and of the cytokine profile of MG. (**E**) IRF8 expression in MG from CQ- and PBS-treated mice. Values of *p* < 0.05 (*) were considered statistically significative. Results are from three independent experiments.

Furthermore, MG greatly influences astrocyte activation [40,41] and provides additional antigen stimulation to T cells in the CNS [42–44]. Thus, we investigated the phenotype of MG in mice treated with CQ. Our results showed that the numbers of MG were reduced in the CNS of CQ-treated mice compared with those from PBS-treated mice (Figure 3C). Interestingly, MG from CQ-treated mice had an increase in IL-10 and IL-12p70 cytokine production while MHC-II, CD80 and CD86 levels were decreased in comparison with MG from PBS-treated mice (Figure 3D). Moreover, we observed a significant decrease in IL-23 production in MG from CQ-treated mice compared with controls (Figure 3D). The transcriptional factor IRF8, which promotes MG differentiation and aids in IL-1β production by MG [45–48], was significantly reduced in MG from CQ-treated mice (Figure 3E). Together, these results reveal a portrait of CNS resident cells in EAE mice treated with CQ, where astrocytes and MG acquire an immunomodulatory phenotype.

3.4. CQ Induces IL-10 and IL-12p70 in MG Which Enhances IL-10 Production by T Cells and Reduces Astrocyte Activation

Finally, we investigated whether CQ modulates MG directly. We isolated primary MG and activated the cells with LPS for 18 h in the presence or absence of CQ. Then, the RNA from MG was analyzed by PCRArray. We observed that CQ reduced the expression of genes associated with inflammation (Cd40, H2-eb, Il1b, Il6, Tnfa, Il23a), while increasing those related to immune modulation (Il10, Lag3, Entpd1, Cd274) (Figure 4A). Moreover, CQ-treated MG produced significantly more IL-10 and IL-12p70 and less IL-23 than PBS-treated MG (Figure 4B).

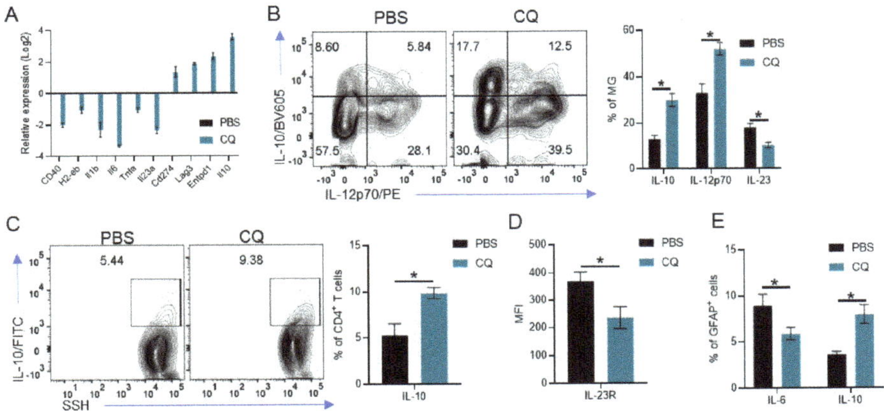

Figure 4. CQ-treated MG modulates T cells and astrocytes. Primary MG were activated with LPS (100 ng/mL) in the presence or absence of CQ (50 µM) for 18 h. (**A**) Gene expression analysis in CQ-treated MG compared with PBS-treated ones. (**B**) Analysis of IL-10, IL-12p70 and IL-23 production by MG. (**C**) MG were co-cultured with CD4$^+$ T cells and the production of IL-10 was analyzed by flow cytometry. (**D**) IL-23R in CD4$^+$ T cells cultured with MG. (**E**) MG were cultured with C8D30 astrocytic cells and the production of IL-6 and IL-10 was analyzed by flow cytometry. Cultures were carried out in triplicate. Values of $p < 0.05$ (*) were considered statistically significative. Results from three independent experiments.

To test their effect on CD4$^+$ T cell activation, we co-cultured CQ-treated MG with CD4$^+$ T cells isolated from naïve WT mice. We observed that CQ-treated MG induced significantly more IL-10 production in T cells than PBS-treated MG (Figure 4C). Moreover, IL-23R in T cells cultured with CQ-treated MG was downregulated when compared with T cells cultured with PBS-treated MG (Figure 4D).

To test the effect of CQ-treated MG on astrocyte activation, we co-cultured MG with the astrocytic cell line C8D30. We observed that CQ-treated MG induced a significant increase in IL-10 production by astrocytes, while PBS-treated MG induced higher IL-6 production (Figure 4E). Collectively, our results show a pivotal role of CNS resident cells, especially MG, on CQ-induced immunosuppression in EAE.

4. Discussion

In this study, we show that CQ modulates a broad array of cell subtypes to reduce EAE severity. Although the therapeutic effect of CQ in EAE was shown before, its underlying mechanism of action remains to be fully elucidated. Here, our results have uncovered a portrait of the CNS resident and infiltrating cells after CQ treatment. We observed that CQ reduced the numbers and phenotype of inflammatory leukocytes and also upregulated the expression of modulatory mediators in leukocytes and CNS resident cells.

EAE is a T cell-dependent model of MS, where pathogenic Th1 and Th17 cells play a major role in disease severity and development [1]. In addition, GM-CSF production grants pathogenicity of Th17 cells [7]. Thus, we investigated the profile of CD4[+] T cells in CQ-treated mice and compared them with those found in PBS-treated mice. Our results showed that CQ inhibited differentiation of Th17 and GM-CSF[+] CD4[+] T cells without disturbing the frequencies of Th1 cells. This finding is in line with a recent publication from our group, where we show that Th17 cells are more sensitive to CQ treatment [31]. Moreover, CQ reduces the differentiation of Th17 cells by inducing the expression of T-bet [31]. GM-CSF production in Th17 cells is induced through stimulation of the IL-23 receptor [7,9]. Interestingly, CQ reduced the expression of IL-23R on CD4[+] T cells. These results show that CQ modulates T cells through different mechanisms: either by suppressing their differentiation into inflammatory Th17 cells or by reducing their pathogenicity through downregulation of IL-23R.

We also observed a reduction in CD8[+] T cells numbers in the CNS of CQ-treated mice. Although EAE is mainly a Th cell model, CD8[+] T cells were shown to play an important role shaping the phenotype of CD4[+] T cells during EAE through the Qa-1 receptor [49,50]. In fact, pathogenic MOG-reactive CD8[+] T cells cooperate with Th cells to promote sustained CNS inflammation in EAE [51], whereas dual modulation of CD8[+] and CD4[+] T cells proved to be more efficient in suppressing EAE [52]. Our results showed that CQ induced suppression of both CD8[+] and CD4[+] T cells, which may account for its efficiency in reducing EAE severity.

We have showed previously that CQ induces Foxp3[+] Treg cells indirectly by stimulating tolerogenic DCs [35]. Additionally, CQ directly modulates DCs by inducing iNOS and pSTAT1 expression, and the lack of these molecules abrogated the tolerogenic phenotype of CQ-treated DCs [30,32]. A similar increase in Foxp3[+] Treg cells in models of inflammatory bowel disease and lupus was observed [53,54]. CQ directly induced Foxp3 expression in T cells in a Nurr1-dependent manner [53]. Thus, the literature data show that CQ induces STAT1 in DCs, T-bet in Th17 cells and Nurr1 in naïve T cells [32,53,54]. In line with this observation, CQ modulates tumor infiltrating macrophages through stimulation of the NF-kB signaling pathway [29]. In the present study, we show that STAT3 and mTOR signaling pathways in myeloid cells were also affected by CQ. Specifically, pSTAT1 was upregulated in DCs and macrophages, while pSTAT3 was upregulated in neutrophils. mTOR was decreased in DCs and macrophages, and unaffected in neutrophils. These results show that CQ induces different signaling cascades in various leukocytes.

MG are CNS resident cells that closely resemble peripheral macrophages [55,56]. MG act as a first line defense in the CNS against infections [16,21,46]. In EAE, MG cooperate with local inflammation and provide antigen-stimulation to infiltrating T cells [40,43]. Thus, we analyzed MG phenotype in CQ-treated mice, and we found that CQ inhibited MG activation. Moreover, we observed a reduced production of IL-23 and an increase in IL-10 and IL-12p70 cytokines in MG from CQ-treated mice. CQ modulated MG directly in in vitro experiments and conferred an anti-inflammatory profile in them. In co-culture experiments with total CD4[+] T cells, CQ-treated MG reduced GM-CSF production and increased IL-10 production. These results are in line with our in vivo observations, where CD4[+] T cells had a reduced GM-CSF and increased IL-10 production while IL-23R expression was also reduced.

In prophylactical treatment with CQ, it is possible that EAE amelioration relied on the modulation of antigen-presenting cells, such as DCs and macrophages [27,35]. However, during ongoing EAE, there is a possibility that CQ acts directly on Th17 cells in the CNS and on MG to refrain further T cell activation and promote an immunosuppressive microenvironment [31,57,58]. In line with this, MG modulated T cells and astrocytes. Astrocytes cultured with CQ-treated MG showed a reduced production of IL-6 and IL-1β and an increase in IL-10 production. Ultimately, paralysis and weakness are a result of the loss of neuronal transmission, and OL play an important role in maintaining neuronal growth and providing the myelin sheaths [59–61]. We observed that CQ-treated mice had more MBP[+] OLs than PBS-treated mice (not shown). We did not investigate whether CQ induces myelination directly or if this finding is a by-stander result of reduced inflammation in the CNS, which warrant further investigations.

Overall, results presented in this study clearly show that CQ suppresses ongoing neuroinflammation by acting in both leukocytes and CNS resident cells. This finding places CQ as a promising therapeutic agent due to its modulation of both inflammation and neurodegeneration, and strengthen the basis for the use of CQ in patients with MS. The majority of FDA-approved drugs to treat MS target leukocytes and inflammation with limited effect on CNS resident cells [62–64]. Moreover, CQ has high bioavailability when taken orally [65,66], which increases acceptability among patients. Thus, we believe that CQ would be as efficient as, if not better than, current FDA-approved drugs in the treatment of MS.

5. Conclusions

Our results show that CQ is a potent modulator of the immune system and CNS resident cells. Moreover, CQ suppressed EAE progression and interfered with the activation of leukocytes and MG. CQ-treated MG-derived cytokines modulated T cells and astrocytes towards an anti-inflammatory phenotype that likely influenced the inflammation in the CNS. Overall, our data reveal a novel pathway by which CQ suppresses neuroinflammation that can be harnessed to develop novel strategies to treat MS.

Author Contributions: R.T., A.B., L.L.W.I., and J.M. performed the experiments. R.T., G.C., B.C. and G.-X.Z. analyzed the data. R.T. and A.R. wrote the manuscript. A.R. supervised the study. All authors have read and agreed to the published version of the manuscript.

Funding: This study was supported with funds from the National MS Society (RG-1607-25366 to A.R. and FG-1608-25579 to R.T.) and the National Institutes of Health (R01AI106026 to A.R.).

Acknowledgments: We would like to thank Katherine Regan for help editing this manuscript.

Conflicts of Interest: The authors declare that no conflict of interests exists.

References

1. Glatigny, S.; Bettelli, E. Experimental Autoimmune Encephalomyelitis (EAE) as Animal Models of Multiple Sclerosis (MS). *Cold Spring Harb. Perspect. Med.* **2018**, *8*, a028977. [CrossRef]
2. Baron, J.L.; Madri, J.A.; Ruddle, N.H.; Hashim, G.; Janeway, C.A. Surface expression of alpha 4 integrin by CD4 T cells is required for their entry into brain parenchyma. *J. Exp. Med.* **1993**, *177*, 57–68. [CrossRef] [PubMed]
3. Voskuhl, R.R.; Martin, R.; Bergman, C.; Dalal, M.; Ruddle, N.H.; McFarland, H.F. T helper 1 (Th1) functional phenotype of human myelin basic protein-specific T lymphocytes. *Autoimmunity* **1993**, *15*, 137–143. [CrossRef] [PubMed]
4. Veldhoen, M.; Hocking, R.J.; Flavell, R.A.; Stockinger, B. Signals mediated by transforming growth factor-beta initiate autoimmune encephalomyelitis, but chronic inflammation is needed to sustain disease. *Nat. Immunol.* **2006**, *7*, 1151–1156. [CrossRef] [PubMed]
5. Zhang, Z.; Zhenyi, X.; Ying, L.; Hongkun, L.; Xiangdong, G.; Yan, L.; Hongwei, Y.; Lijuan, Z.; Yurong, D.; Zhi, Y. MicroRNA-181c promotes Th17 cell differentiation and mediates experimental autoimmune encephalomyelitis. *Brain Behav. Immun.* **2018**, *70*, 305–314. [CrossRef]
6. Pare, A.; Mailhot, B.; Lévesque, S.A.; Juzwik, C.; Doss, P.M.I.A.; Lécuyer, M.; Prat, A.; Rangachari, M.; Fournier, A.; Lacroix, S. IL-1beta enables CNS access to CCR2(hi) monocytes and the generation of pathogenic cells through GM-CSF released by CNS endothelial cells. *Proc. Natl. Acad. Sci. USA* **2018**, *115*, E1194–E1203. [CrossRef]
7. El-Behi, M.; Ciric, B.; Hong, D.; Yaping, Y.; Cullimore, M.; Safavi, F.; Guang-Xian, Z.; Dittel, B.N.; Rostami, A. The encephalitogenicity of T(H)17 cells is dependent on IL-1- and IL-23-induced production of the cytokine GM-CSF. *Nat. Immunol.* **2011**, *12*, 568–575. [CrossRef]

8. Codarri, L.; Gyülvészi, G.; Tosevski, V.; Hesske, L.; Fontana, A.; Magnenat, L.; Suter, T.; Bechere, B. RORgammat drives production of the cytokine GM-CSF in helper T cells, which is essential for the effector phase of autoimmune neuroinflammation. *Nat. Immunol.* **2011**, *12*, 560–567. [CrossRef]

9. Komuczki, J.; Tuzlak, S.; Friebel, E.; Hartwig, T.; Spath, S.; Rosenstiel, P.; Waisman, A.; Opitz, L.; Oukka, M.; Schreiner, B.; et al. Fate-Mapping of GM-CSF Expression Identifies a Discrete Subset of Inflammation-Driving T Helper Cells Regulated by Cytokines IL-23 and IL-1beta. *Immunity* **2019**, *50*, 1289–1304. [CrossRef]

10. Dominguez-Villar, M.; Hafler, D.A. Regulatory T cells in autoimmune disease. *Nat. Immunol.* **2018**, *19*, 665–673. [CrossRef]

11. McIntyre, L.L.; Greilach, S.A.; Othy, S.; Sears-Kraxberger, I.; Wi, B.; Ayala-Angulo, J.; Vu, E.; Pham, Q.; Silva, J.; Dang, K.; et al. Regulatory T cells promote remyelination in the murine experimental autoimmune encephalomyelitis model of multiple sclerosis following human neural stem cell transplant. *Neurobiol. Dis.* **2020**, *140*, 104868. [CrossRef] [PubMed]

12. Dombrowski, Y.; O'Hagan, T.; Dittmer, M.; Penalva, R.; Mayoral, S.R.; Bankhead, P.; Fleville, S.; Eleftheriadis, G.; Chao, Z.; Naughton, M.; et al. Regulatory T cells promote myelin regeneration in the central nervous system. *Nat. Neurosci.* **2017**, *20*, 674–680. [CrossRef] [PubMed]

13. Bailey, S.L.; Schreiner, B.; McMahon, E.J.; Miller, S.D. CNS myeloid DCs presenting endogenous myelin peptides 'preferentially' polarize CD4+ T(H)-17 cells in relapsing EAE. *Nat. Immunol.* **2007**, *8*, 172–180. [CrossRef] [PubMed]

14. Louveau, A.; Herz, J.; Alme, M.N.; Salvador, A.F.; Dong, M.Q.; Viar, K.E.; Herod, S.G.; Knopp, J.; Setliff, J.C.; Lupi, A.L.; et al. CNS lymphatic drainage and neuroinflammation are regulated by meningeal lymphatic vasculature. *Nat. Neurosci.* **2018**, *21*, 1380–1391. [CrossRef] [PubMed]

15. Mundt, S.; Mrdjen, D.; Utz, S.G.; Greter, M.; Schreiner, B.; Becher, B. Conventional DCs sample and present myelin antigens in the healthy CNS and allow parenchymal T cell entry to initiate neuroinflammation. *Sci. Immunol.* **2019**, *4*, eaau8380. [CrossRef]

16. Chu, F.; Shi, M.; Zheng, C.; Shen, D.; Zhu, J.; Zheng, X.; Cui, L. The roles of macrophages and microglia in multiple sclerosis and experimental autoimmune encephalomyelitis. *J. Neuroimmunol.* **2018**, *318*, 1–7. [CrossRef]

17. Tanabe, S.; Saitoh, S.; Miyajima, H.; Itokazu, T.; Yamashita, T. Microglia suppress the secondary progression of autoimmune encephalomyelitis. *Glia* **2019**, *67*, 1694–1704. [CrossRef]

18. Zabala, A.; Vazquez-Villoldo, N.; Rissiek, B.; Gejo, J.; Martin, A.; Palomino, A.; Perez-Samartín, A.; Pulagam, K.R.; Lukowiak, M.; Capetillo-Zarate, E.; et al. P2X4 receptor controls microglia activation and favors remyelination in autoimmune encephalitis. *EMBO Mol. Med.* **2018**, *10*, e8743. [CrossRef]

19. Zhang, C.J.; Jiang, M.; Zhou, H.; Liu, W.; Wang, C.; Kang, Z.; Han, B.; Zhang, Q.; Chen, X.; Xiao, J.; et al. TLR-stimulated IRAKM activates caspase-8 inflammasome in microglia and promotes neuroinflammation. *J. Clin. Investig.* **2018**, *128*, 5399–5412. [CrossRef]

20. Rothhammer, V.; Borucki, D.M.; Tjon, E.M.; Takenaka, M.C.; Chao, C.C.; Ardura-Fabregat, A.; Alves de Lima, K.; Gutiérrez-Vázquez, C.; Hewson, P.; Staszewski, O.; et al. Microglial control of astrocytes in response to microbial metabolites. *Nature* **2018**, *557*, 724–728. [CrossRef]

21. Domingues, H.S.; Portugal, C.C.; Socodato, R.; Relvas, J.B. Oligodendrocyte, Astrocyte, and Microglia Crosstalk in Myelin Development, Damage, and Repair. *Front. Cell Dev. Biol.* **2016**, *4*, 71. [PubMed]

22. Chrzanowski, U.; Bhattarai, S.; Scheld, M.; Clarner, T.; Fallier-Becker, P.; Beyer, C.; Rohr, S.O.; Schmitz, C.; Hochstrasser, T.; Schweiger, F.; et al. Oligodendrocyte degeneration and concomitant microglia activation directs peripheral immune cells into the forebrain. *Neurochem. Int.* **2019**, *126*, 139–153. [CrossRef] [PubMed]

23. Liddelow, S.A.; Guttenplan, K.A.; Clarke, L.E.; Bennett, F.C.; Bohlen, C.J.; Schirmer, L.; Bennett, M.L.; Münch, A.E.; Chung, W.; Peterson, T.C.; et al. Neurotoxic reactive astrocytes are induced by activated microglia. *Nature* **2017**, *541*, 481–487. [CrossRef] [PubMed]

24. Thome, R.; Lopes, S.C.P.; Costa, F.T.M.; Verinaud, L. Chloroquine: Modes of action of an undervalued drug. *Immunol. Lett.* **2013**, *153*, 50–57. [CrossRef] [PubMed]

25. Jang, C.H.; Choi, J.H.; Byun, M.S.; Jue, D.M. Chloroquine inhibits production of TNF-alpha, IL-1beta and IL-6 from lipopolysaccharide-stimulated human monocytes/macrophages by different modes. *Rheumatology* **2006**, *45*, 703–710. [CrossRef] [PubMed]

26. Long, Y.; Liu, X.; Wang, N.; Zhou, H.; Zheng, J. Chloroquine attenuates LPS-mediated macrophage activation through miR-669n-regulated SENP6 protein translation. *Am. J. Transl. Res.* **2015**, *7*, 2335–2345.

27. Thome, R.; Issayama, L.K.; DiGangi, R.; Bombeiro, A.L.; da Costa, T.A.; Ferreira, I.T.; de Oliveira, A.L.R.; Verinaud, L. Dendritic cells treated with chloroquine modulate experimental autoimmune encephalomyelitis. *Immunol. Cell Biol.* **2014**, *92*, 124–132. [CrossRef]

28. Mauthe, M.; Orhon, I.; Rocchi, C.; Zhou, X.R.; Luhr, M.; Hijlkema, K.-J.; Coppes, R.P.; Engedal, N.; Mari, M.; Reggiori, F. Chloroquine inhibits autophagic flux by decreasing autophagosome-lysosome fusion. *Autophagy* **2018**, *14*, 1435–1455. [CrossRef]

29. Chen, D.; Xie, J.; Fiskesund, R.; Dong, W.Q.; Liang, X.Y.; Lv, J.D.; Jin, X.; Liu, J.Y.; Mo, S.Q.; Zhang, T.Z.; et al. Chloroquine modulates antitumor immune response by resetting tumor-associated macrophages toward M1 phenotype. *Nat. Commun.* **2018**, *9*, 873. [CrossRef]

30. Verinaud, L.; Issayama, L.K.; Zanucoli, F.; de Carvalho, A.C.; da Costa, T.A.; Di Gangi, R.; Bonfanti, A.P.; Ferreira, I.T.; de Oliveira, A.L.R.; Machado, D.R.S.; et al. Nitric oxide plays a key role in the suppressive activity of tolerogenic dendritic cells. *Cell Mol. Immunol.* **2015**, *12*, 384–386. [CrossRef]

31. Thome, R.; Boehm, A.; Ishikawa, L.L.W.; Casella, G.; Munhoz, J.; Ciric, B.; Zhang, G.X.; Rostami, A. Chloroquine reduces Th17 cell differentiation by stimulating T-bet expression in T cells. *Cell Mol. Immunol.* **2020**.

32. Thome, R.; Bonfanti, A.P.; Rasouli, J.; Mari, E.R.; Zhang, G.X.; Rostami, A.; Verinaud, L. Chloroquine-treated dendritic cells require STAT1 signaling for their tolerogenic activity. *Eur. J. Immunol.* **2018**, *48*, 1228–1234. [CrossRef] [PubMed]

33. Thome, R.; Moore, J.N.; Mari, E.R.; Rasouli, J.; Hwang, D.; Yoshimura, S.; Ciric, B.; Zhang, G.X.; Rostami, A.M. Induction of Peripheral Tolerance in Ongoing Autoimmune Inflammation Requires Interleukin 27 Signaling in Dendritic Cells. *Front. Immunol.* **2017**, *8*, 1392. [CrossRef] [PubMed]

34. Yoshimura, S.; Thome, R.; Konno, S.; Mari, E.R.; Rasouli, J.; Hwang, D.; Boehm, A.; Li, Y.; Zhang, G.X.; Ciric, B.; et al. IL-9 Controls Central Nervous System Autoimmunity by Suppressing GM-CSF Production. *J. Immunol.* **2020**, *204*, 531–539. [CrossRef] [PubMed]

35. Thome, R.; Moraes, A.S.; Bombeiro, A.L.; Farias Ados, S.; Francelin, C.; da Costa, T.A.; Di Gangi, R.; dos Santos, L.M.; de Oliveira, A.L.; Verinaud, L. Chloroquine treatment enhances regulatory T cells and reduces the severity of experimental autoimmune encephalomyelitis. *PLoS ONE* **2013**, *8*, e65913. [CrossRef]

36. Skundric, D.S.; Dai, R.; Zakarian, V.L.; Zhou, W. Autoimmune-induced preferential depletion of myelin-associated glycoprotein (MAG) is genetically regulated in relapsing EAE (B6 x SJL) F1 mice. *Mol. Neurodegener.* **2008**, *3*, 7. [CrossRef]

37. Zhang, G.X.; Yu, S.; Gran, B.; Li, J.; Calida, D.; Ventura, E.; Chen, X.; Rostami, A. T cell and antibody responses in remitting-relapsing experimental autoimmune encephalomyelitis in (C57BL/6 x SJL) F1 mice. *J. Neuroimmunol.* **2004**, *148*, 1–10. [CrossRef]

38. Frik, J.; Merl-Pham, J.; Plesnila, N.; Mattugini, N.; Kjell, J.; Kraska, J.; Gómez, R.M.; Hauck, S.M.; Sirko, S.; Götz, M. Cross-talk between monocyte invasion and astrocyte proliferation regulates scarring in brain injury. *EMBO Rep.* **2018**, *19*, e45294. [CrossRef]

39. Anderson, M.A.; Ao, Y.; Sofroniew, M.V. Heterogeneity of reactive astrocytes. *Neurosci. Lett.* **2014**, *565*, 23–29. [CrossRef]

40. Norden, D.M.; Trojanowski, P.J.; Villanueva, E.; Navarro, E.; Godbout, J.P. Sequential activation of microglia and astrocyte cytokine expression precedes increased Iba-1 or GFAP immunoreactivity following systemic immune challenge. *Glia* **2016**, *64*, 300–316. [CrossRef]

41. Shinozaki, Y.; Shibata, K.; Yoshida, K.; Shigetomi, E.; Gachet, C.; Ikenaka, K.; Tanaka, K.F.; Koizumi, S. Transformation of Astrocytes to a Neuroprotective Phenotype by Microglia via P2Y1 Receptor Downregulation. *Cell Rep.* **2017**, *19*, 1151–1164. [CrossRef] [PubMed]

42. Wlodarczyk, A.; Løbner, M.; Cédile, O.; Owens, T. Comparison of microglia and infiltrating CD11c(+) cells as antigen presenting cells for T cell proliferation and cytokine response. *J. Neuroinflamm.* **2014**, *11*, 57. [CrossRef] [PubMed]

43. Matsumoto, Y.; Ohmori, K.; Fujiwara, M. Immune regulation by brain cells in the central nervous system: Microglia but not astrocytes present myelin basic protein to encephalitogenic T cells under in vivo-mimicking conditions. *Immunology* **1992**, *76*, 209–216. [PubMed]

44. Hu, J.; He, H.; Yang, Z.; Zhu, G.; Kang, L.; Jing, X.; Lu, H.; Song, W.; Bai, B.; Tang, H. Programmed Death Ligand-1 on Microglia Regulates Th1 Differentiation via Nitric Oxide in Experimental Autoimmune Encephalomyelitis. *Neurosci. Bull.* **2016**, *32*, 70–82. [CrossRef]

45. Zhou, N.; Liu, K.; Sun, Y.; Cao, Y.; Yang, J. Transcriptional mechanism of IRF8 and PU.1 governs microglial activation in neurodegenerative condition. *Protein Cell* **2019**, *10*, 87–103. [CrossRef]

46. Kierdorf, K.; Erny, D.; Goldmann, T.; Sander, V.; Schulz, C.; Perdiguero, E.G.; Wieghofer, P.; Heinrich, A.; Riemke, P.; Hölscher, C.; et al. Microglia emerge from erythromyeloid precursors via Pu.1- and Irf8-dependent pathways. *Nat. Neurosci.* **2013**, *16*, 273–280. [CrossRef]

47. Masuda, T.; Iwamoto, S.; Mikuriya, S.; Tozaki-Saitoh, H.; Tamura, T.; Tsuda, M.; Inoue, K. Transcription factor IRF1 is responsible for IRF8-mediated IL-1beta expression in reactive microglia. *J. Pharmacol. Sci.* **2015**, *128*, 216–220. [CrossRef]

48. Masuda, T.; Tsuda, M.; Yoshinaga, R.; Tozaki-Saitoh, H.; Ozato, K.; Tamura, T.; Inoue, K. IRF8 is a critical transcription factor for transforming microglia into a reactive phenotype. *Cell Rep.* **2012**, *1*, 334–340. [CrossRef]

49. Jiang, H.; Braunstein, N.S.; Yu, B.; Winchester, R.; Chess, L. CD8+ T cells control the TH phenotype of MBP-reactive CD4+ T cells in EAE mice. *Proc. Natl. Acad. Sci. USA* **2001**, *98*, 6301–6306. [CrossRef]

50. Leuenberger, T.; Paterka, M.; Reuter, E.; Herz, J.; Niesner, R.A.; Radbruch, H.; Bopp, T.; Zipp, F.; Siffrin, V. The role of CD8+ T cells and their local interaction with CD4+ T cells in myelin oligodendrocyte glycoprotein35-55-induced experimental autoimmune encephalomyelitis. *J. Immunol.* **2013**, *191*, 4960–4968. [CrossRef]

51. Bettini, M.; Rosenthal, K.; Evavold, B.D. Pathogenic MOG-reactive CD8+ T cells require MOG-reactive CD4+ T cells for sustained CNS inflammation during chronic EAE. *J. Neuroimmunol.* **2009**, *213*, 60–68. [CrossRef] [PubMed]

52. Pei, W.; Wan, X.; Shahzad, K.A.; Zhang, L.; Song, S.; Jin, X.; Wang, L.; Zhao, C.; Shen, C. Direct modulation of myelin-autoreactive CD4(+) and CD8(+) T cells in EAE mice by a tolerogenic nanoparticle co-carrying myelin peptide-loaded major histocompatibility complexes, CD47 and multiple regulatory molecules. *Int. J. Nanomed.* **2018**, *13*, 3731–3750. [CrossRef] [PubMed]

53. Park, T.Y.; Jang, Y.; Kim, W.; Shin, J.; Toh, H.T.; Kim, C.H.; Yoon, H.S.; Leblanc, P.; Kim, K.S. Chloroquine modulates inflammatory autoimmune responses through Nurr1 in autoimmune diseases. *Sci. Rep.* **2019**, *9*, 15559. [CrossRef] [PubMed]

54. An, N.; Chen, Y.; Wang, C.; Yang, C.; Wu, Z.H.; Xue, J.; Ye, L.; Wang, S.; Liu, H.F.; Pan, Q. Chloroquine Autophagic Inhibition Rebalances Th17/Treg-Mediated Immunity and Ameliorates Systemic Lupus Erythematosus. *Cell Physiol. Biochem.* **2017**, *44*, 412–422. [CrossRef] [PubMed]

55. Prinz, M.; Erny, D.; Hagemeyer, N. Ontogeny and homeostasis of CNS myeloid cells. *Nat. Immunol.* **2017**, *18*, 385–392. [CrossRef]

56. Bennett, M.L.; Bennett, F.C.; Liddelow, S.A.; Ajami, B.; Zamanian, J.L.; Fernhoff, N.B.; Mulinyawe, S.B.; Bohlen, C.J.; Adil, A.; Tucker, A.; et al. New tools for studying microglia in the mouse and human CNS. *Proc. Natl. Acad. Sci. USA* **2016**, *113*, E1738–E1746. [CrossRef]

57. Chu, T.; Tran, T.; Yang, F.; Beech, W.; Cole, G.M.; Frautschy, S.A. Effect of chloroquine and leupeptin on intracellular accumulation of amyloid-beta (A beta) 1-42 peptide in a murine N9 microglial cell line. *FEBS Lett.* **1998**, *436*, 439–444. [CrossRef]

58. Koch, M.W.; Zabad, R.; Giuliani, F.; Hader, W., Jr.; Lewkonia, R.; Metz, L.; Wee Yong, V. Hydroxychloroquine reduces microglial activity and attenuates experimental autoimmune encephalomyelitis. *J. Neurol. Sci.* **2015**, *358*, 131–137. [CrossRef]

59. Barateiro, A.; Brites, D.; Fernandes, A. Oligodendrocyte Development and Myelination in Neurodevelopment: Molecular Mechanisms in Health and Disease. *Curr. Pharm. Des.* **2016**, *22*, 656–679. [CrossRef]

60. Domingues, H.S.; Cruz, A.; Chan, J.R.; Relvas, J.B.; Rubinstein, B.; Pinto, I.M. Mechanical plasticity during oligodendrocyte differentiation and myelination. *Glia* **2018**, *66*, 5–14. [CrossRef]

61. Foster, A.Y.; Bujalka, H.; Emery, B. Axoglial interactions in myelin plasticity: Evaluating the relationship between neuronal activity and oligodendrocyte dynamics. *Glia* **2019**, *67*, 2038–2049. [CrossRef]

62. Aharoni, R.; Eilam, R.; Domev, H.; Labunskay, G.; Sela, M.; Arnon, R. The immunomodulator glatiramer acetate augments the expression of neurotrophic factors in brains of experimental autoimmune encephalomyelitis mice. *Proc. Natl. Acad. Sci. USA* **2005**, *102*, 19045–19050. [CrossRef] [PubMed]

63. Maarouf, A.; Boutière, C.; Rico, A.; Audoin, B.; Pelletier, J. How much progress has there been in the second-line treatment of multiple sclerosis: A 2017 update. *Rev. Neurol.* **2018**, *174*, 429–440. [CrossRef] [PubMed]

64. Vargas, D.L.; Tyor, W.R. Update on disease-modifying therapies for multiple sclerosis. *J. Investig. Med.* **2017**, *65*, 883–891. [CrossRef] [PubMed]

65. Krishna, S.; White, N.J. Pharmacokinetics of quinine, chloroquine and amodiaquine. *Clin. Pharmacokinet.* **1996**, *30*, 263–299. [CrossRef]

66. Gustafsson, L.L.; Walker, O.; Alván, G.; Beermann, B.; Estevez, F.; Gleisner, L.; Lindström, B.; Sjöqvist, F. Disposition of chloroquine in man after single intravenous and oral doses. *Br. J. Clin. Pharmacol.* **1983**, *15*, 471–479. [CrossRef]

Review

A Journey to the Conformational Analysis of T-Cell Epitope Peptides Involved in Multiple Sclerosis

Catherine Koukoulitsa [1], Eleni Chontzopoulou [1], Sofia Kiriakidi [1], Andreas G. Tzakos [2] and Thomas Mavromoustakos [1,*]

1   Department of Chemistry, National and Kapodistrian University of Athens, Zographou, 15784 Athens, Greece; ckoukoulitsa@chem.uoa.gr (C.K.); elenichontzo@chem.uoa.gr (E.C.); sofki@chem.uoa.gr (S.K.)
2   Department of Chemistry, Section of Organic Chemistry and Biochemistry, University of Ioannina, 45110e Ioannina, Greece; atzakos@uoi.gr
*   Correspondence: tmavrom@chem.uoa.gr; Tel.: +30-210-7274475

Received: 13 May 2020; Accepted: 2 June 2020; Published: 8 June 2020

Abstract: Multiple sclerosis (MS) is a serious central nervous system (CNS) disease responsible for disability problems and deterioration of the quality of life. Several approaches have been applied to medications entering the market to treat this disease. However, no effective therapy currently exists, and the available drugs simply ameliorate the destructive disability effects of the disease. In this review article, we report on the efforts that have been conducted towards establishing the conformational properties of wild-type myelin basic protein (MBP), myelin proteolipid protein (PLP), myelin oligodendrocyte glycoprotein (MOG) epitopes or altered peptide ligands (ALPs). These efforts have led to the aim of discovering some non-peptide mimetics possessing considerable activity against the disease. These efforts have contributed also to unveiling the molecular basis of the molecular interactions implicated in the trimolecular complex, T-cell receptor (TCR)–peptide–major histocompatibility complex (MHC) or human leucocyte antigen (HLA).

Keywords: conformational analysis; peptides; altered peptide ligands; multiple sclerosis; MS; NMR spectroscopy; NOE-constraints; molecular dynamic; trimolecular complex; experimental autoimmune encephalomyelitis

---

1. Introduction

Multiple sclerosis (MS) is a serious disease of the central nervous system (CNS). MS affects almost 3.3 million people worldwide [1]. It affects more females than males between the ages of 20 and 40 [2]. MS-related disability significantly affects the quality of life (e.g., restraints on daily life activities) [3]. As the number of patients continuously increases, negative effects on social and economic aspects have been observed [4,5]. Factors such as genetic, environment, metabolism and viral infections considerably progress the disease [6,7].

MS is classified into four subclasses according to the increase of the neurologic deterioration of the disease:

1.   Relapsing-remitting MS (RRMS): This is the most frequently occurring and affects ca. 85% of all MS patients. The patients with RRMS suffer from relapses and remissions of their neurological symptoms.
2.   Secondary progressive MS (SPMS): This follows the development of RRMS and causes further worsening of the disease.
3.   Primary progressive MS (PPMS): This affects 8–10% of patients and is characterized by the gradual further worsening of the disease.

4. Progressive-relapsing MS (PRMS): This is the least often occurring class, affecting less than 5% of patients and progressing from onset [8–10].

MS takes place in brain and spinal cord regions containing myelin. As shown in Figure 1, MS lesions involve demyelination and inflammation of B-cells, T-cells, macrophages and activated microglia. Then follows tissue damage, which includes loss of neurons and oligodendrocytes, astrogliosis and remyelination [11,12].

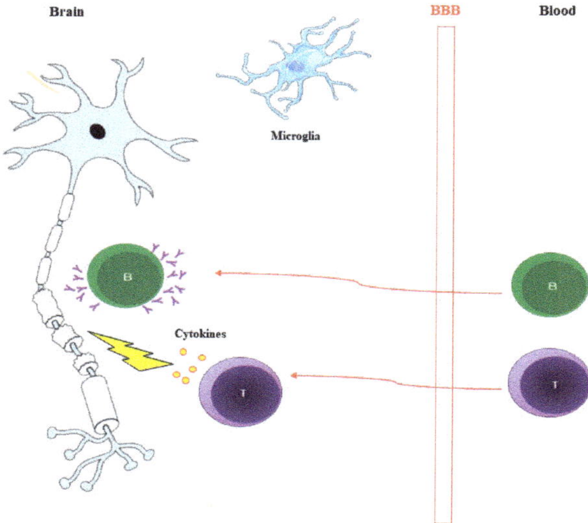

Figure 1. T-cells enter the blood brain barrier (BBB) and release cytokines which degrade the myelin. The cytokines can also recruit some other cells as B-cells. These cells enter the BBB and produce antibodies which target the myelin for further degradation. Activated microglia are also involved in myelin degradation.

The cause of autoimmune disease MS is still mostly unknown. It is hypothesized that environment induces MS in individuals prone to the disease. The molecular mimicry theory has been used to explain the pathogenesis of MS. The gathered evidence proposes that viral peptidic epitopes bearing sequence homology to protein regions of normal human tissue are responsible for the initiation of the disease. The immune response of T-cells targets mainly the viral epitopes. However, cross-reaction with the normal human tissue leads to the autoimmune disease [13,14].

The myelin basic protein (MBP), the proteolipid protein (PLP), the myelinoligodendrocyte glycoprotein (MOG), and the myelin associated oligodendrocytic basic protein (MOBP), have been associated as T-cell epitopes in MS. These peptides have been utilized to trigger experimental autoimmune encephalomyelitis (EAE). EAE is the most frequently and broadly used animal model that simulates MS [15–21].

Although advances in MS treatment have proceeded impressively, the currently available medications are not fully in line to respond to the future and emerging needs raised by the complicated nature of MS [22].

One of the major approaches for the treatment of MS is the peptidic or peptidomimetic therapeutic approach [23,24]. There are different steps involved in the development of peptidomimetic drugs in a rational design strategy. In the first step the minimal peptide amino acid sequence that exerts the activity (epitope) and serves as a lead compound is identified. In the second step the information derived from nuclear magnetic resonance (NMR) spectroscopy, and/or molecular modeling and/or x-ray crystallography is utilized in order to define a putative bioactive conformation of the minimal peptide sequence [25]. In the third step the resultant 3D architecture is used for the development of non-peptide mimetics that are prone to metabolic clearance.

Activated encephalitogenic T-cells, triggered by the formation of a trimolecular complex between the T-cell receptor (TCR), the peptide (antigen)—with identical residue sequence to a fragment of a protein of the myelin sheath—and the major histocompatibility complex (MHC) or human leukocyte antigen (HLA), initiate the onset of MS. The potential of the peptide–HLA complex to activate T-cells parallels the strength of its binding affinity with TCR [26–28]. It follows the stimulation, or not, of T-cells that cause MS [29–33].

The dimer HLA class II receptors contain two polypeptide chains named as α and β [34,35]. Their joined polypeptide chains form a single receptor suitable to form a complex with the antigen binders. This complex is recognized by the T-cell receptors on the cell surface. The formed trimolecular complex leads to the activation of T-cells through a series of biochemical alterations and the triggering of the immune response to the antigen [36].

This review summarizes the conformational analysis of peptides involved in multiple sclerosis. In addition the impact of these conformational changes on rational drug design is described.

2. Results and Discussion

Mouzaki et al. [37] pointed out that peptides constitute a class of administered molecules as immunomodulatory drugs due to their rapid and cost-effective synthesis. The peptides that can cause EAE in animals are called agonists and those that can compete the action of the agonists and treat EAE are called antagonists.

In the discussed studies peptides are used that either map to wild-type MBP, PLP [38] or MOG epitopes or are mutants (altered peptide ligands, APLs), which are linear or cyclized variants that are more resistant to in vivo enzymatic degradation [39]. APLs differ from their parent encephalitogenic peptides by single amino acid substitutions and can inhibit autoimmune mediated disease through several mechanisms.

For many years we have made an effort to explore the conformational properties that govern various epitopes related to EAE with their agonist and antagonists both in solution and in trimolecular complexes (drug:TCR:HLA). In this review we will outline the most significant results obtained from these studies.

The first step in these studies is to extract favored averaged conformations of the epitopes in solution using NMR spectroscopy. These conformations after energy minimization serve as initial conformations for applying molecular dynamics (MD) simulations in the generation of the trimolecular complex. The results will lead to the synthesis of antagonist peptides which could potentially provide useful mechanistic information to combat MS (Figure 2).

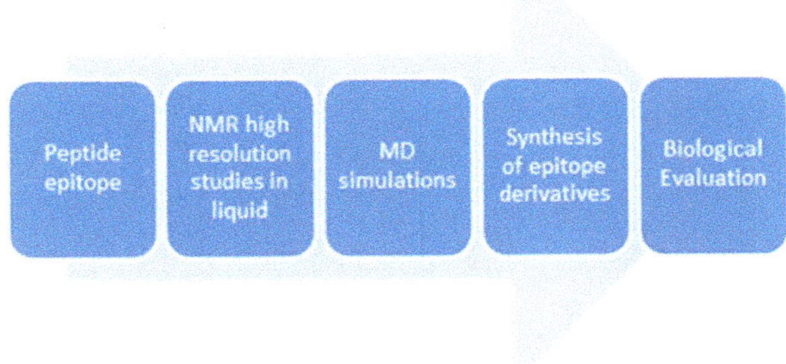

Figure 2. Steps of a rational design process aiming to aid the development of peptide mimics against multiple sclerosis (MS).

The conformational analysis of hMOG$_{35-55}$ epitope (Met$_{35}$-Glu-Val-Gly-Trp-Tyr-Arg-Pro^{42}-Pro-Phe-Ser-Arg-Val-Val-His-Leu-Tyr-Arg-Asn-Gly-Lys$_{55}$) and its mutants (hMOG$_{35-55}$(Ala^{41}) and hMOG$_{35-55}$(Ala^{41,46})) alone and in the trimolecular complex containing HLA and TCR have been studied using MD simulations [36]. The results showed that the hMOG$_{35-55}$ epitope in the MD trajectory does not retain the linear conformation. Its dominant conformation shows two bends in the polypeptide backbone between residues Trp^{39}, Tyr^{40} and Arg^{41} and Val^{48} and Arg^{52}.

This conformation is similar to that published for the rat/mouse MOG$_{35-55}$ peptide by Ntountaniotis et al. [40] in DMSO and D$_2$O solvents (Figure 3).

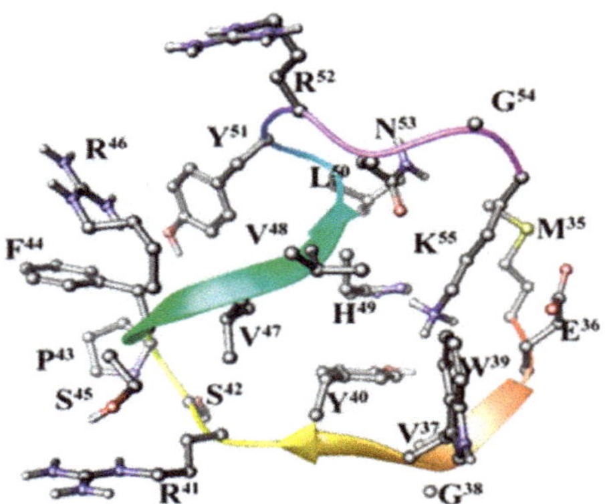

Figure 3. Low energy conformer of hMOG$_{35-55}$ (myelinoligodendrocyte glycoprotein) derived from in silico molecular synamics (MD) calculations restricted with nuclear Overhauser effect (NOE)-constraints.

During the formation of the trimolecular complex the amino acids Arg^{41} and Arg^{46} of hMOG$_{35-55}$ anchor at TCR and Tyr^{40} interacts with HLA. The amino acids Arg^{41} and Arg^{46} form an extensive hydrogen bonding (HB) network with both receptors. Substitution of Arg^{41} or Arg^{41} and Arg^{46} with Ala leads to the two mutants hMOG$_{35-55}$ (Ala^{41}) and hMOG$_{35-55}$ (Ala^{41}, Ala^{46}). These mutations lead to the elimination of key interactions with TCR but leave intact the binding affinity towards the HLA receptor. These two mutants function as EAE inhibitors. This finding is significant as it provides basic mechanistic aspects of the action of agonist versus antagonist peptides (Figure 4).

The conformational analysis of MBP$_{77-89}$ and the antagonist altered ligands (Arg^{91}, Ala^{96}) MBP$_{87-99}$ and (Ala^{91,96}) MBP$_{87-99}$ have been studied. All the three molecules showed an extended conformation in DMSO environment with no long-range nuclear Overhauser effects (NOEs) [41] in disagreement with the observations recorded in other chemical environments [29].

Interestingly, X-ray results existed for a peptide analogue of MBP$_{87-99}$ that formed a trimolecular complex with a human TCR and HLA-DR2b [42]. A bioactive conformation of APL that resembled that of the crystallized peptide was derived from the molecular dynamics trajectories (Root-Mean Square Deviation (RMSD) value of 0.95 Å). The two peptides were oriented similarly to the two TCR anchor residues, His^{88} and Phe^{89}, and the HLA anchor residue Phe^{90}.

These two amino acids orient variably in the trimolecular complex for (Arg^{91}, Ala^{96}) MBP$_{87-99}$ and (Ala^{91,96}) MBP$_{87-99}$, and remain buried in HLA grooves and cannot interact with the TCR. This finding may explain the antagonism of the two altered ligands (Figure 5).

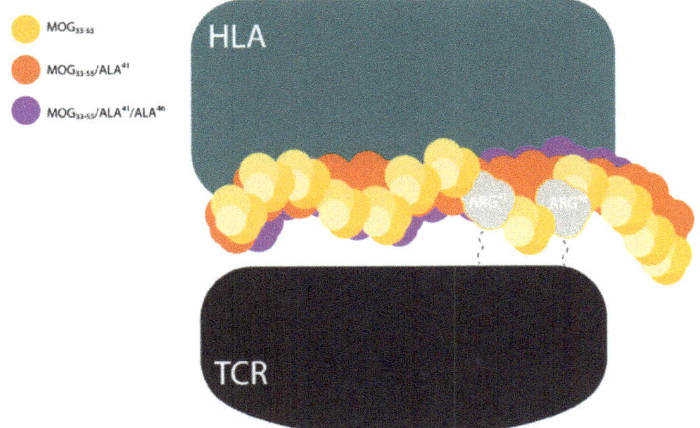

Figure 4. The replacement of Arg[41] or Arg[41] and Arg[46] of hMOG$_{35-55}$ with Ala interrupts the hydrogen bonding (HB) with the amino acids Asp[98], Ser[101], and Asn[104] of T-cell receptors (TCR). This may be due to the decrease of polarity of Ala vs. Arg (disruption of the interaction network) and may lead to a reduced bending of Ala in the low energy conformation of hMOG$_{35-55}$. HLA—Human Leukocyte Antigen.

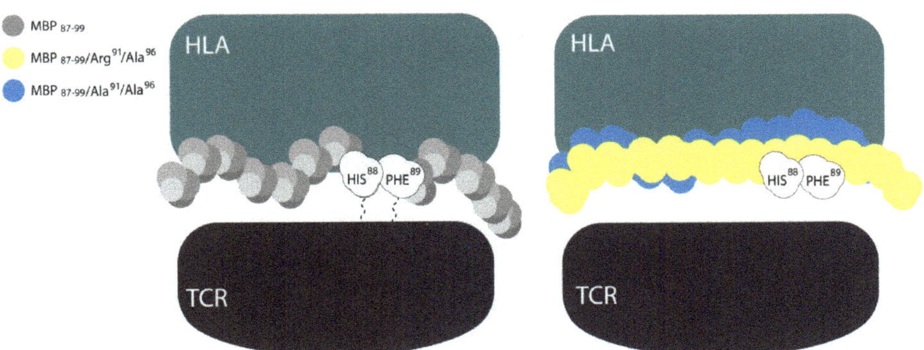

Figure 5. (**left**) His[88] and Phe[89] of hMBP$_{87-99}$ (myelin basic protein) interact with the TCR receptor. (**right**) In the two antagonists (Arg[91], Ala[96]) MBP$_{87-99}$ and (Ala[91,96]) MBP$_{87-99}$ this interaction is lost as the two amino acids are buried in HLA grooves.

The cyclo (91–99)(Ala[96])MBP$_{87-99}$, cyclo(87–99)(Ala[91,96])MBP$_{87-99}$ and cyclo(87–99)(Arg[91], Ala[96])MBP$_{87-99}$ (Figure 6), except the wild-type linear MBP$_{87-99}$, were found to strongly inhibit MBP$_{72-85}$- induced EAE in Lewis rats. Cyclo(87–99)(Arg[91], Ala[96])MBP$_{87-99}$ provided long protection for the EAE induction [39,43,44].

Conformational analysis was achieved for the three cyclo(87–99) MBP$_{87-99}$, cyclo(87–99) (Ala[91,96]) MBP$_{87-99}$, and cyclo(87–99) (Arg[91], Ala[96]) MBP$_{87-99}$ analogs using 2D NMR spectroscopy and computational analysis. The conformational analysis of the three synthetic analogues showed that their bioactivity, or its absence, may be attributed to the distinct local conformation, overall topology and exposed area after binding with MHC II. An overall larger solvent accessible area may occlude the approach and binding of the TCR on the APL-MHC complex. In contrast, more compact structures do not block weak interactions as TCR approaches and can induce EAE antagonism. These results led to the generation of the pharmacophore model described in Figure 7 [45].

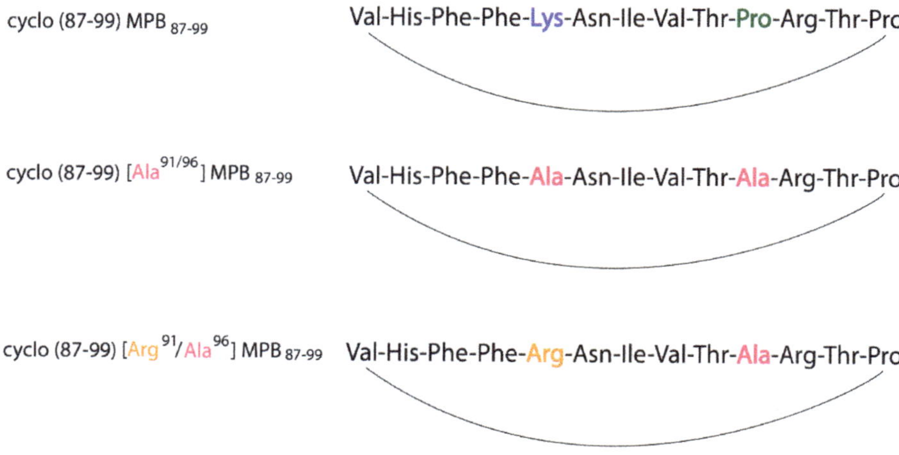

Figure 6. Structures of cyclo(91–99)(Ala^{96})MBP$_{87-99}$, cyclo(87–99)(Ala^{91},^{96})MBP$_{87-99}$ and cyclo(87–99)(Arg^{91}, Ala^{96})MBP$_{87-99}$.

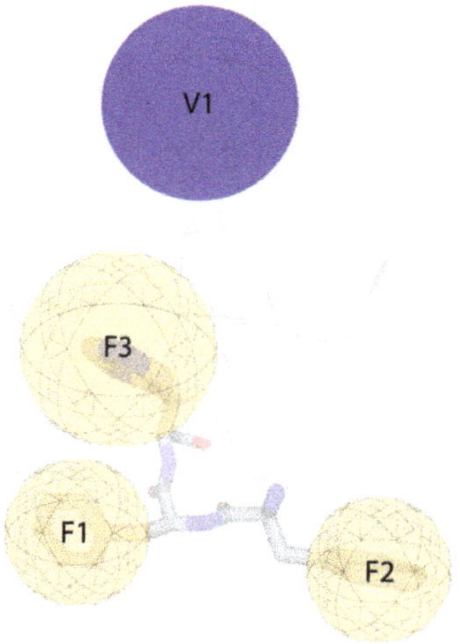

Figure 7. Pharmacophore model depicted using the conformational properties obtained from the conformational analysis for the cyclic altered peptide ligands (APLs). Exclusion volume V1 is presented with a blue sphere, feature F1 (Phe^{90}), F2 (Phe^{89}), F3 (Phe^{88}) with a yellow sphere.

Two citrullinated peptides, the linear (Cit^{91}, Ala^{96}, Cit^{97})MBP$_{87-99}$ and cyclo(87–99)(Cit^{91}, Ala^{96}, Cit^{97})MBP$_{87-99}$ have been synthesized by citrullinating the Arg residues 91 and 97 in the antagonists, linear (Arg91, Ala96)MBP$_{87-99}$ and cyclo(87–99)(Arg^{91}, Ala^{96})MBP$_{87-99}$ peptides. In contrast to the antagonists, these citrullinated molecules induced EAE. Molecular modeling results pointed out that both Cit^{91} and Cit^{97} residues are oriented toward the TCR and possibly are interacting with the complementarity-determining region (CDR3) loops of the TCR, thus triggering an altered cytokine response [46].

Another epitope which is shown to induce EAE in guinea pigs is the linear peptide MBP$_{74-85}$ (Gln^1-Lys^2-Ser^3-Gln^4-Arg^5-Ser^6-Gln^7-Asp^8-Glu^9-Asn^{10}-Pro^{11}-Val^{12}-NH$_2$). A Rotating frame Overhauser Effect Spectroscopy (ROESY) connectivity was observed for the molecule in DMSO between αVal^{12}-αGln^1, suggesting a cyclic conformation. This intriguing result prompted the synthesis of the cyclic analogue by tethering the εNH_2 of Lys and $\gamma COOH$ of Glu at positions 2 and 9, respectively. Cyclic peptides are well known to be more stable and less susceptible to enzymatic degradation than linear peptides. Moreover, cyclic peptides are an important intermediate step in the rational design and development of non-peptide mimetics [47].

This cyclic analogue illustrated comparable bioactivity with the linear one, confirming that the possible bioactive conformation of MBP_{74-85} resembles that of the cyclic variant or the cyclic variant resembles more, from the available ensemble, the structure of the linear peptide that is of biological significance. The structures of the linear and cyclic analogues are shown in Figure 8. The same relationship was observed with the linear Ala^{81} MBP_{74-85} and its cyclic analogue [25].

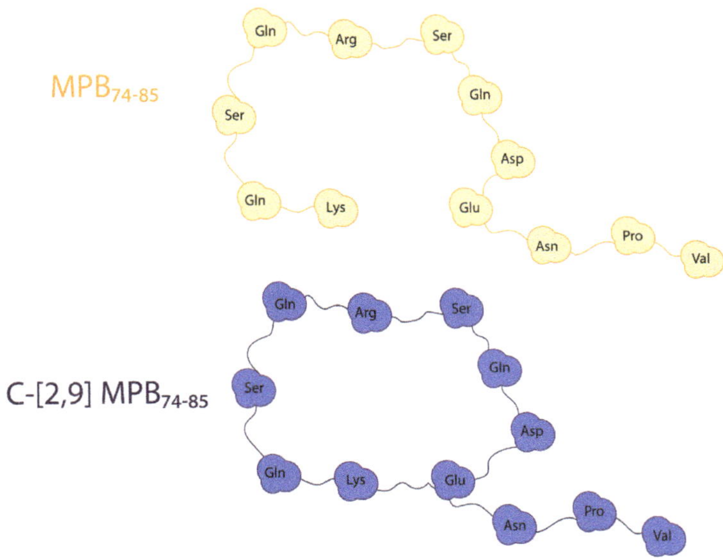

Figure 8. 2D and 3D models of linear MBP_{74-85} (**top**) and its cyclic analogue (**bottom**).

Tzakos et al. [48] applied NMR and molecular dynamic simulations to study the conformational properties of agonist MBP, Gln^{74}-Lys^{75}-Ser^{76}-Gln^{77}-Arg^{78}-Ser^{79}-Gln^{80}-Asp^{81}-Glu^{82}-Asn^{83}-Pro^{84}-Val^{85} ($MBP_{(74-85)}$), and its antagonist analogue $Ala^{81}MBP_{(74-85)}$. The agonist $MBP_{(74-85)}$ adopted a compact conformation attributed to electrostatic interactions of Arg^{78} with the side chains of Asp^{81} and Glu^{82}. Arg^{78} adopted a well-defined conformation, which did not depend on the solvent. Such electrostatic interactions were not observed in the antagonist Ala^{81} $MBP_{(74-85)}$, and a high flexibility of the side chain of Arg^{78} was observed. The positively charged residue Arg^{78} is suggested to stabilize the local microdomains (epitopes) of the integral protein. Flexible docking calculations point out that Gln^{74}, Ser^{76} and Ser^{79} are MHC II anchor residues. Lys^{75}, Arg^{78} and Asp^{81} are the mainly solvent-exposed residues and this may signify their participation in the formation of the trimolecular T-cell receptor–$MBP_{(74-85)}$–MHC II complex.

In another study the conformational analysis of the immunodominant epitope of acetylated myelin basic protein residues 1–11 (Ac-MBP1–11) and its ALPs, mutated at position 4 to an alanine (Ac-MBP1–11(4A)) or a tyrosine residue (Ac-MBP1–11(4Y)), was achieved. The amino acids constituting the MBP^{1-11} are Ala-Ser-Gln-Lys-Arg-Pro-Ser-Gln-Arg-His-Gly (Ac-MBP1–11). The Ac-MBP1–11(4A) analogue inhibited EAE symptoms induced by encephalitogenic Ac-MBP1–11 epitope when co-injected in (PL/J × SJL)F1 mice. These results are interpreted to suggest that Ac-MBP1–11(4A) induced immunomodulation that inhibits EAE in vivo [49]. Studies indicated that the residue at position 4 in MBP1–11 peptide plays a major role in binding of the peptide to MHC class II, I–Au [50,51].

The mutated analogue Ac-MBP1–11(4A) binds to I–Au with a minimum of 50-fold higher affinity in comparison to the native Ac-MBP1–11 [52]. In addition, the mutation at position 4 of Lys to Tyr (Ac-MBP1–11(4Y)) increases the stability of the I–Au-peptide complex by enhancing 1500-fold the affinity, which triggers Ac-MBP1–11 T-cells more effectively in relation to Ac-MBP1–11(4A) [53].

The conformational analysis of the three analogues showed that they adopt an extended conformation in deuterated DMSO solvent due to the absence of long-distance NOEs. Furthermore, they adopt a similar conformation when bound to the active site of the MHC II. Gln^3 residue is a TCR contact site and has a different orientation in the mutated analogues. Specifically, its side chain is not solvent exposed, and it is not available for interaction with the TCR. The main MHC contact residues (Ser^2, Pro^6 and Ser^7) stand in the same position for all peptides [54].

The conformational properties of MBP$_{83-99}$ have been studied using NMR spectroscopy in DMSO to simulate the biological environment. The results showed that the peptide exists in a rather extended conformation and forms a helix between Val^{87} and Phe^{90} [55].

Two analogues of the MBP$_{83-99}$ epitope substituted at Lys^{91} (primary TCR contact) with Phe (MBP$_{83-99}$ (Phe^{91})) or Tyr (MBP$_{83-99}$ (Tyr^{91})) were synthesized (Figure 9).

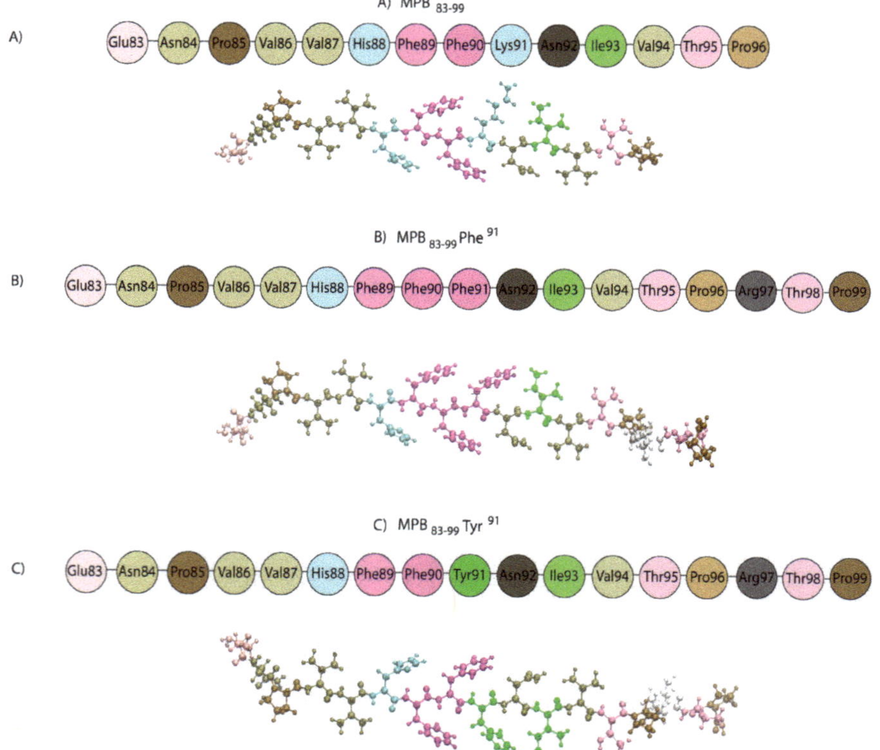

Figure 9. Sequences and 3D low energy structures of (**A**) MBP$_{83-96}$ (numbering is according to the human MBP$_{83-99}$ epitope) and the two synthetic analogs (**B**) MBP$_{83-99}$ (Phe^{91}) and (**C**) MBP$_{83-99}$ (Tyr^{91}).

The two analogues showed distinct antagonistic activity versus the agonistic activity of the MBP$_{83-99}$ epitope. The conformational analysis of the two APLs was performed using NMR spectroscopy and MD. Both synthetic analogues show an extended conformation in agreement with the structural features of the peptides that interact with the HLA-DR2 and TCR receptors. MD simulations of the two analogues in complex with HLA-DR2 (DRA, DRB1*1501) and TCR revealed their modes of interactions. MBP_{83-99} (Phe^{91}) analogue adopts more interactions during the formation of the trimolecular complex relatively to MBP_{83-99} (Tyr^{91}), as their trajectory profiles confirmed. This may explain the improved biological profile of the latter. The two analogues differ in the way of binding relatively to the wild epitope MBP_{83-96}. This is attributed to the fact that mutation of Lys^{91} by either Tyr or Phe alters their stereoelectronic properties.

This alteration of the stereoelectronic properties affects the binding mode of the regional amino acids and explains their antagonistic or agonistic activity. Such binding mode differences have been observed and outlined above with the MBP_{87-99} epitope [45,56–60].

It is important to note that although the two peptides mentioned above differ only in a small segment, they possess distinct biological profiles. The tyrosine^{91} in MBP_{83-99} (Tyr^{91}) possesses a phenolic hydroxyl group that induces differential biological activity. This is in agreement with a plethora of literature data pointing out the key role of the phenolic group in drug bioactivity [61–72] (Figure 10).

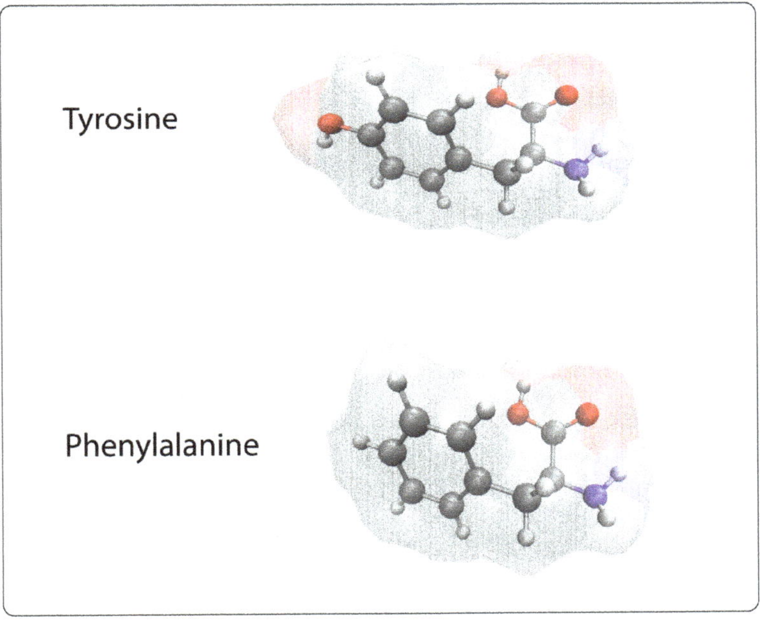

Figure 10. The absence of the phenolic hydroxyl group in Phe is responsible for the different biological properties between the two synthetic analogues MBP_{83-99} (Tyr^{91}) and MBP_{83-99} (Phe^{91}).

The superimposition of the two peptides at the binding site of the trimolecular complex shows that Phe^{91} and Tyr^{91} occupy almost identical areas. However, they induce different conformations to other vicinal amino acids Asn^{92} and Ile^{93}, as the phenolic hydroxyl group lies in a relatively hydrophobic environment. Their apparently small structural difference induces a sequence of distinct interactions that determine their fingerprint of biological action. $MBP_{(85-99)}$ is an immuno-dominant epitope of MBP which binds to the MHC haplotype HLA-DR2 and is associated with the pathogenesis of MS. The synthetic 15-mer peptide J5n (Figure 11), was designed and was found to antagonize $MBP_{(85-99)}$ through the binding of $MBP_{(85-99)}$ to soluble HLA-DR2b [73]. The therapeutic efficacy of J5 is limited, probably due to its low biological half-life or bioavailability. The structural features of J5 in relation to its parent (i.e., $MBP_{(85-99)}$) are shown in Figure 11. Phe at position P4 has been replaced with Tyr, Val at position P1 has been retained and His, Phe, and Lys at P2, P3 and P5 have been replaced with Glu, Ala and Lys, respectively.

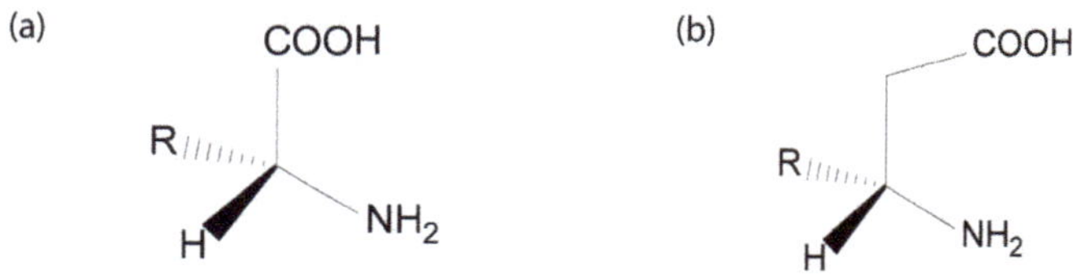

Figure 11. (a) α-amino acid, (b) Homo-β-amino acid, (c) MBP85-99, J5 and S18 analogue.

In another study J5 was derivatized into analogs possessing superior biological half-lives and antagonistic activities. This is achieved by substitution of some of its residues with homo-β-amino acids. S18 (Figure 11), the most active analog, ameliorated symptoms of EAE at least twice more effectively than glatiramer acetate or J5. S18 showed high resistance to proteolysis, which contributed to a delayed clinical onset of disease and prolonged therapeutic benefits [74].

The conformational analysis studies of MBP$_{83-96}$ epitope led the group of Professor T. Tselios to search for the mining and synthesis of non-peptide mimetic molecules. In particular, they sought molecules that inhibit the trimolecular complex formation and consequently the proliferation of activated T-cells. They generated a structure-based pharmacophore and used ZINC as a chemical database to extract candidates (Figure 12). Semi-empirical and density functional theory (DFT) methods were performed to predict the binding energy between the proposed non-peptide mimetics and the TCR. From the six synthesized molecules the following 15 and 16 were the most promising as they inhibited the stimulation of T-cells by the immunodominant MBP$_{83-99}$ from immunized mice [75].

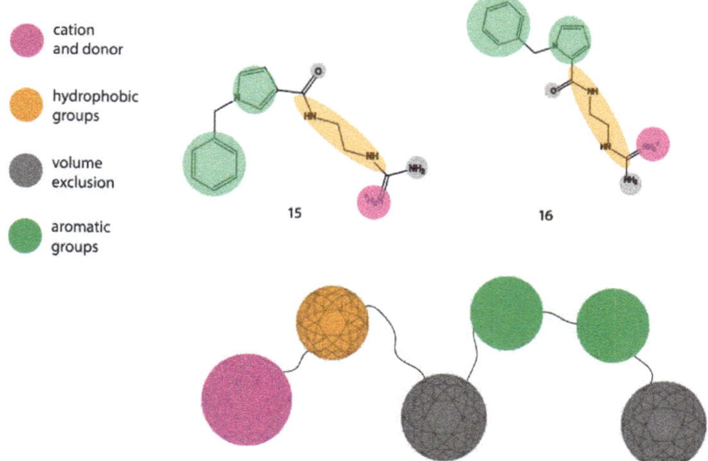

Figure 12. Structure-based pharmacophore derived from ZINC database data.

3. Conclusions

An extensive effort has been made the last years to explore the conformational properties of key peptides involved in MS. The conformational analysis of the different epitopes, consisting of in silico MD and pharmacophore studies, along with NMR spectroscopy, has led to the rational design of some bioactive non-peptide mimetics and provided some mechanistic input of the agonistic and antagonistic action of ALPs. However, there is still a long way towards the generation of more potent compounds. Interestingly, in a study it was illustrated that the extent of MHC or TCR competition does not successfully predict the EAE treatment [76]. Other routes to treat MS had also limited success [21,77].

Such an example is the immunomodulatory co-polymer 1 (Copaxone, glatiramer acetate) drug. This contains synthetic peptides composed of nonspecific sequences of four amino acids: L-alanine, L-lysine, L-glutamic acid and L-tyrosine (Figure 13). As its composition is based on the amino acid structure of MBP it exerts an antagonistic action to the 82–100 epitope of MBP [78].

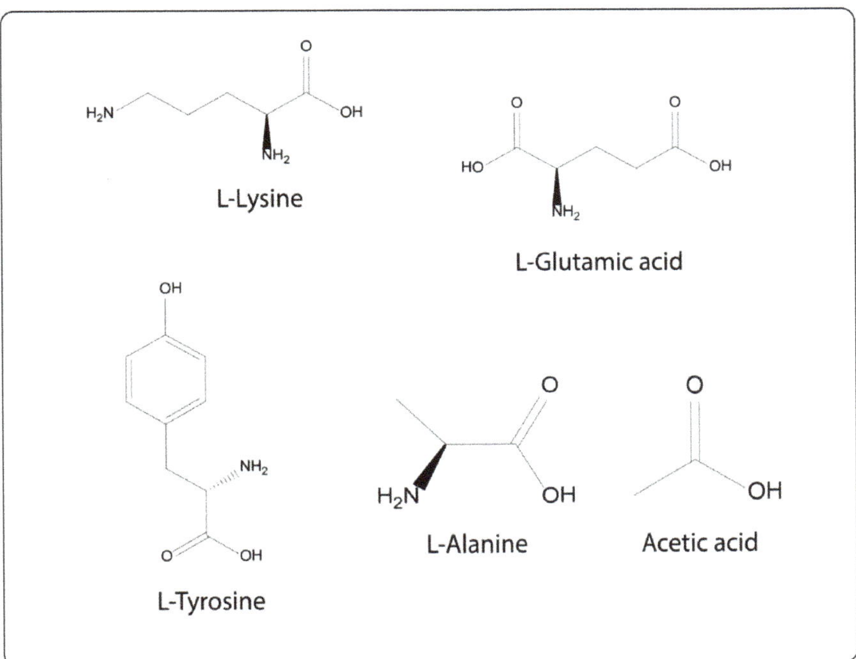

Figure 13. Synthetic peptides of Cop1 structure.

Recently, semi-empirical calculations have been applied to detect peptides associated with MS. It was found that the A_31:01 allele may be associated with the MS disease and the peptide Leu-Ile-Ile-Cys-Tyr-Asn-Trp-Leu-His-Arg may serve as a potential epitope to this allele. This finding must be confirmed by experimental evidence [79].

The multifactorial aspects of MS, especially in its severe state, makes the task of finding a drug against MS tremendously difficult. This must reinforce the efforts in order to advance the progress of understanding and treating the disease.

Author Contributions: Conceptualization, C.K., E.C., S.K., A.G.T.; writing—original draft preparation, T.M.; writing—review and editing, C.K., E.C., S.K., visualization, C.K., E.C., S.K., A.G.T.; supervision, T.M.; project administration, T.M. All authors have read and agreed to the published version of the manuscript.

Funding: This research received no external funding.

Conflicts of Interest: The authors declare no conflict of interest.

References

1. Wallin, M.T.; Culpepper, W.J.; Campbell, J.D.; Nelson, L.M.; Langer-Gould, A.; Marrie, R.A.; Cutter, G.R.; Kaye, W.E.; Wagner, L.; Tremlett, H.; et al. The prevalence of MS in the United States. *Neurology* **2019**, *92*, 1029–1040. [CrossRef] [PubMed]
2. Howard, J.; Trevick, S.; Younger, D.S. Epidemiology of Multiple Sclerosis. *Neurol. Clin.* **2016**, *34*, 919–939. [CrossRef]
3. Dziedzic, A.; Miller, E.; Saluk-Bijak, J.; Bijak, M. The gpr17 receptor—A promising goal for therapy and a potential marker of the neurodegenerative process in multiple sclerosis. *Int. J. Mol. Sci.* **2020**, *21*, 1852. [CrossRef] [PubMed]
4. Jennum, P.; Wanscher, B.; Frederiksen, J.; Kjellberg, J. The socioeconomic consequences of multiple sclerosis: A controlled national study. *Eur. Neuropsychopharmacol.* **2012**, *22*, 36–43. [CrossRef]
5. Raj, R.; Kaprio, J.; Korja, M.; Mikkonen, E.D.; Jousilahti, P.; Siironen, J. Risk of hospitalization with neurodegenerative disease after moderate-to-severe traumatic brain injury in the working-age population: A retrospective cohort study using the Finnish national health registries. *PLoS Med.* **2017**, *14*, e1002316. [CrossRef] [PubMed]
6. Melchor, G.S.; Khan, T.; Reger, J.F.; Huang, J.K. Remyelination Pharmacotherapy Investigations Highlight Diverse Mechanisms Underlying Multiple Sclerosis Progression. *ACS Pharmacol. Transl. Sci.* **2019**, *2*, 372–386. [CrossRef] [PubMed]
7. Khan, F.; Amatya, B. Rehabilitation in Multiple Sclerosis: A Systematic Review of Systematic Reviews. *Arch. Phys. Med. Rehabil.* **2017**, *98*, 353–367. [CrossRef]
8. Eckstein, C.; Bhatti, M.T. Currently approved and emerging oral therapies in multiple sclerosis: An update for the ophthalmologist. *Surv. Ophthalmol.* **2016**, *61*, 318–332. [CrossRef]
9. Lublin, F.D.; Reingold, S.C. Defining the clinical course of multiple sclerosis: Results of an international survey. *Neurology* **1996**, *46*, 907–911. [CrossRef]
10. Lublin, F.D.; Baier, M.; Cutter, G. Effect of relapses on development of residual deficit in multiple sclerosis. *Neurology* **2003**, *61*, 1528–1532. [CrossRef]
11. Hemmer, B.; Archelos, J.J.; Hartung, H.P. New concepts in the immunopathogenesis of multiple sclerosis. *Nat. Rev. Neurosci.* **2002**, *3*, 291–301. [CrossRef] [PubMed]
12. Lassmann, H.; Raine, C.S.; Antel, J.; Prineas, J.W. Immunopathology of multiple sclerosis: Report on an international meeting held at the Institute of Neurology of the University of Vienna. *J. Neuroimmunol.* **1998**, 213–217. [CrossRef]
13. Tzakos, A.; Kursula, P.; Troganis, A.; Theodorou, V.; Tselios, T.; Svarnas, C.; Matsoukas, J.; Apostolopoulos, V.; Gerothanassis, I. Structure and Function of the Myelin Proteins: Current Status and Perspectives in Relation to Multiple Sclerosis. *Curr. Med. Chem.* **2005**, *12*, 1569–1587. [CrossRef] [PubMed]
14. Kouskoff, V.; Lacaud, G.; Nemazee, D. T cell-independent rescue of B lymphocytes from peripheral immune tolerance. *Science* **2000**, *287*, 2501–2513. [CrossRef]
15. Tuusa, J.; Raasakka, A.; Ruskamo, S.; Kursula, P. Myelin-derived and putative molecular mimic peptides share structural properties in aqueous and membrane-like environments. *Mult. Scler. Demyelinating Disord.* **2017**, *2*. [CrossRef]
16. Bielekova, B.; Sung, M.-H.; Kadom, N.; Simon, R.; McFarland, H.; Martin, R. Expansion and Functional Relevance of High-Avidity Myelin-Specific CD4 + T Cells in Multiple Sclerosis. *J. Immunol.* **2004**, *172*, 3893–3904. [CrossRef]
17. Greer, J.M.; Csurhes, P.A.; Cameron, K.D.; McCombe, P.A.; Good, M.F.; Pender, M.P. Increased immunoreactivity to two overlapping peptides of myelin proteolipid protein in multiple sclerosis. *Brain* **1997**, *120*, 1447–1460. [CrossRef]
18. Mendel, I.; De Rosbo, N.K.; Ben-Nun, A. Delineation of the minimal encephalitogenic epitope within the immunodominant region of myelin oligodendrocyte glycoprotein: Diverse Vβ gene usage by T cells recognizing the core epitope encephalitogenic for T cell receptor Vβb and T cell receptor Vβa H-2. *Eur. J. Immunol.* **1996**, *26*, 2470–2479. [CrossRef]

19. De Rosbo, N.K.; Kaye, J.F.; Eisenstein, M.; Mendel, I.; Hoeftberger, R.; Lassmann, H.; Milo, R.; Ben-Nun, A. The Myelin-Associated Oligodendrocytic Basic Protein Region MOBP15–36 Encompasses the Immunodominant Major Encephalitogenic Epitope(s) for SJL/J Mice and Predicted Epitope(s) for Multiple Sclerosis-Associated HLA-DRB1*1501. *J. Immunol.* **2004**, *173*, 1426–1435. [CrossRef]

20. Holz, A.; Bielekova, B.; Martin, R.; Oldstone, M.B.A. Myelin-Associated Oligodendrocytic Basic Protein: Identification of an Encephalitogenic Epitope and Association with Multiple Sclerosis. *J. Immunol.* **2000**, *164*, 1103–1109. [CrossRef]

21. Denic, A.; Johnson, A.J.; Bieber, A.J.; Warrington, A.E.; Rodriguez, M.; Pirko, I. The relevance of animal models in multiple sclerosis research. *Pathophysiology* **2011**, *18*, 21–29. [CrossRef] [PubMed]

22. Gholamzad, M.; Ebtekar, M.; Ardestani, M.S.; Azimi, M.; Mahmodi, Z.; Mousavi, M.J.; Aslani, S. A comprehensive review on the treatment approaches of multiple sclerosis: Currently and in the future. *Inflamm. Res.* **2019**, *68*, 25–38. [CrossRef]

23. Shahrizaila, N.; Yuki, N. Guillain-Barré syndrome animal model: The first proof of molecular mimicry in human autoimmune disorder. *J. Biomed. Biotechnol.* **2011**, *2011*, 829129. [CrossRef]

24. Moise, L.; Beseme, S.; Tassone, R.; Liu, R.; Kibria, F.; Terry, F.; Martin, W.; De Groot, A.S. T cell epitope redundancy: Cross-conservation of the TCR face between pathogens and self and its implications for vaccines and autoimmunity. *Expert Rev. Vaccines* **2016**, *15*, 607–617. [CrossRef] [PubMed]

25. Matsoukas, J.; Apostolopoulos, V.; Mavromoustakos, T. Designing Peptide Mimetics for the Treatment of Multiple Sclerosis. *Mini Rev. Med. Chem.* **2005**, *1*, 273–282. [CrossRef] [PubMed]

26. Compston, A.; Coles, A. Multiple sclerosis. *Lancet* **2008**, *372*, 1502–1517. [CrossRef]

27. Davis, M.M.; Boniface, J.J.; Reich, Z.; Lyons, D.; Hampl, J.; Arden, B.; Chien, Y. Ligand Recognition By alpha beta T Cell Receptors. *Annu. Rev. Immunol.* **1998**, *16*, 523–544. [CrossRef] [PubMed]

28. Wootla, B.; Eriguchi, M.; Rodriguez, M. Is multiple sclerosis an autoimmune disease? *Autoimmune Dis.* **2012**, *2012*, 969657. [CrossRef] [PubMed]

29. Hare, B.J.; Wyss, D.F.; Osburne, M.S.; Kern, P.S.; Reinherz, E.L.; Wagner, G. Structure, specificity and CDR mobility of a class II restricted single- chain T-cell receptor. *Nat. Struct. Biol.* **1999**, *6*, 574–581. [CrossRef]

30. Rudolph, M.G.; Wilson, I.A. The specificity of TCR/pMHC interaction. *Curr. Opin. Immunol.* **2002**, *14*, 52–65. [CrossRef]

31. Rudolph, M.G.; Stanfield, R.L.; Wilson, I.A. How TCRS Bind MHCS, Peptides, And Coreceptors. *Annu. Rev. Immunol.* **2006**, *24*, 419–466. [CrossRef] [PubMed]

32. Lassmann, H. Axonal and neuronal pathology in multiple sclerosis: What have we learnt from animal models. *Exp. Neurol.* **2010**, *225*, 2–8. [CrossRef] [PubMed]

33. Reuter, E.; Gollan, R.; Grohmann, N.; Paterka, M.; Salmon, H.; Birkenstock, J.; Richers, S.; Leuenberger, T.; Brandt, A.U.; Kuhlmann, T.; et al. Cross-Recognition of a myelin peptide by CD8+ T cells in the CNS is not sufficient to promote neuronal damage. *J. Neurosci.* **2015**, *35*, 4837–4850. [CrossRef] [PubMed]

34. Madden, D.R. The Three-Dimensional Structure of Peptide-MHC Complexes. *Annu. Rev. Immunol.* **1995**, *13*, 587–622. [CrossRef]

35. Adams, E.J.; Luoma, A.M. The Adaptable Major Histocompatibility Complex (MHC) Fold: Structure and Function of Nonclassical and MHC Class I–Like Molecules. *Annu. Rev. Immunol.* **2013**, *31*, 529–561. [CrossRef]

36. Yannakakis, M.P.; Tzoupis, H.; Michailidou, E.; Mantzourani, E.; Simal, C.; Tselios, T. Molecular dynamics at the receptor level of immunodominant myelin oligodendrocyte glycoprotein 35–55 epitope implicated in multiple sclerosis. *J. Mol. Graph. Model.* **2016**, *68*, 78–86. [CrossRef]

37. Mouzaki, A.; Tselios, T.; Papathanassopoulos, P.; Matsoukas, I.; Chatzantoni, K. Immunotherapy for Multiple Sclerosis: Basic Insights for New Clinical Strategies. *Curr. Neurovasc. Res.* **2005**, *1*, 325–340. [CrossRef]

38. Tselios, T.; Daliani, I.; Deraos, S.; Thymianou, S.; Matsoukas, E.; Troganis, A.; Gerothanassis, I.; Mouzaki, A.; Mavroustakos, T.; Probert, L.; et al. Treatment of experimental allergic encephalomyelitis (EAE) by a rationally designed cyclic analogue of myelin basic protein (MBP) epitope 72-85. *Bioorg. Med. Chem. Lett.* **2000**, *10*, 2713–2717. [CrossRef]

39. Tselios, T.; Apostolopoulos, V.; Daliani, I.; Deraos, S.; Grdadolnik, S.; Mavroustakos, T.; Melachrinou, M.; Thymianou, S.; Probert, L.; Mouzaki, A.; et al. Antagonistic effects of human cyclic MBP87-99 altered peptide ligands in experimental allergic encephalomyelitis and human T-cell proliferation. *J. Med. Chem.* **2002**, *45*, 275–283. [CrossRef]

40. Ntountaniotis, D.; Vanioti, M.; Kordopati, G.G.; Kellici, T.F.; Marousis, K.D.; Mavromoustakos, T.; Spyroulias, G.A.; Golic Grdadolnik, S.; Tselios, T.V. A combined NMR and molecular dynamics simulation study to determine the conformational properties of rat/mouse 35-55 myelin oligodendrocyte glycoprotein epitope implicated in the induction of experimental autoimmune encephalomyelitis. *J. Biomol. Struct. Dyn.* **2017**, *35*, 1559–1567. [CrossRef]

41. Mantzourani, E.D.; Platts, J.A.; Brancale, A.; Mavromoustakos, T.M.; Tselios, T.V. Molecular dynamics at the receptor level of immunodominant myelin basic protein epitope 87-99 implicated in multiple sclerosis and its antagonists altered peptide ligands: Triggering of immune response. *J. Mol. Graph. Model.* **2007**, *26*, 471–481. [CrossRef] [PubMed]

42. Willcox, B.E.; Gao, G.F.; Wyer, J.R.; Ladbury, J.E.; Bell, J.I.; Jakobsen, B.K.; Van der Merwe, P.A. TCR binding to peptide-MHC stabilizes a flexible recognition interface. *Immunity* **1999**, *10*, 357–365. [CrossRef]

43. Matsoukas, J.; Apostolopoulos, V.; Kalbacher, H.; Papini, A.M.; Tselios, T.; Chatzantoni, K.; Biagioli, T.; Lolli, F.; Deraos, S.; Papathanassopoulos, P.; et al. Design and synthesis of a novel potent myelin basic protein epitope 87-99 cyclic analogue: Enhanced stability and biological properties of mimics render them a potentially new class of immunomodulators. *J. Med. Chem.* **2005**, *48*, 1470–1480. [CrossRef] [PubMed]

44. Emmanouil, M.; Tseveleki, V.; Triantafyllakou, I.; Nteli, A.; Tselios, T.; Probert, L. A cyclic altered peptide analogue based on myelin basic protein 87-99 provides lasting prophylactic and therapeutic protection against acute experimental autoimmune encephalomyelitis. *Molecules* **2018**, *23*, 304. [CrossRef] [PubMed]

45. Spyranti, Z.; Dalkas, G.A.; Spyroulias, G.A.; Mantzourani, E.D.; Mavromoustakos, T.; Friligou, I.; Matsoukas, J.M.; Tselios, T.V. Putative bioactive conformations of amide linked cyclic myelin basic protein peptide analogues associated with experimental autoimmune encephalomyelitis. *J. Med. Chem.* **2007**, *50*, 6039–6047. [CrossRef] [PubMed]

46. Deraos, G.; Chatzantoni, K.; Matsoukas, M.T.; Tselios, T.; Deraos, S.; Katsara, M.; Papathanasopoulos, P.; Vynios, D.; Apostolopoulos, V.; Mouzaki, A.; et al. Citrullination of linear and cyclic altered peptide ligands from myelin basic protein (MBP87-99) epitope elicits a Th1 polarized response by T cells isolated from multiple sclerosis patients: Implications in triggering disease. *J. Med. Chem.* **2008**, *51*, 7834–7842. [CrossRef] [PubMed]

47. Matsoukas, J.; Apostolopoulos, V.; Lazoura, E.; Deraos, G.; Matsoukas, M.-T.; Katsara, M.; Tselios, T.; Deraos, S. Round and Round we Go: Cyclic Peptides in Disease. *Curr. Med. Chem.* **2006**. [CrossRef]

48. Tzakos, A.G.; Fuchs, P.; Van Nuland, N.A.J.; Troganis, A.; Tselios, T.; Deraos, S.; Matsoukas, J.; Gerothanassis, I.P.; Bonvin, A.M.J.J. NMR and molecular dynamics studies of an autoimmune myelin basic protein peptide and its antagonist: Structural implications for the MHC II (I-A u)-peptide complex from docking calculations. *Eur. J. Biochem.* **2004**, *271*, 3399–3413. [CrossRef]

49. Wraith, D.C.; Smilek, D.E.; Mitchell, D.J.; Steinman, L.; McDevitt, H.O. Antigen recognition in autoimmune encephalomyelitis and the potential for peptide-mediated immunotherapy. *Cell* **1989**, *59*, 247–255. [CrossRef]

50. Lee, C.; Liang, M.N.; Tate, K.M.; Rabinowitz, J.D.; Beeson, C.; Jones, P.P.; McConnell, H.M. Evidence that the autoimmune antigen myelin basic protein (MBP) Ac1-9 binds towards one end of the major histocompatibility complex (MHC) cleft. *J. Exp. Med.* **1998**, *187*, 1505–1516. [CrossRef]

51. He, X.L.; Radu, C.; Sidney, J.; Sette, A.; Ward, E.S.; Garcia, K.C. Structural snapshot of aberrant antigen presentation linked to autoimmunity: The immunodominant epitope of MBP complexed with I-Au. *Immunity* **2002**, *17*, 83–94. [CrossRef]

52. Fugger, L.; Liang, J.; Gautam, A.; Rothbard, J.B.; McDevitt, H.O. Quantitative analysis of peptides from myelin basic protein binding to the MHC class II protein, I-A(u), which confers susceptibility to experimental allergic encephalomyelitis. *Mol. Med.* **1996**, *2*, 181–188. [CrossRef] [PubMed]

53. Metzler, B.; Wraith, D.C. Inhibition of experimental autoimmune encephalomyelitis by inhalation but not oral administration of the encephalitogenic peptide: Influence of MHC binding affinity. *Int. Immunol.* **1993**, *5*, 1159–1165. [CrossRef] [PubMed]

54. Laimou, D.; Lazoura, E.; Troganis, A.N.; Matsoukas, M.T.; Deraos, S.N.; Katsara, M.; Matsoukas, J.; Apostolopoulos, V.; Tselios, T.V. Conformational studies of immunodominant myelin basic protein 1-11 analogues using NMR and molecular modeling. *J. Comput. Aided. Mol. Des.* **2011**, *25*, 1019–1032. [CrossRef]

55. Spyranti, Z.; Tselios, T.; Deraos, G.; Matsoukas, J.; Spyroulias, G.A. NMR structural elucidation of myelin basic protein epitope 83-99 implicated in multiple sclerosis. *Amino Acids* **2010**, *38*, 929–936. [CrossRef]

56. Mantzourani, E.; Mavromoustakos, T.; Platts, J.; Matsoukas, J.; Tselios, T. Structural Requirements for Binding of Myelin Basic Protein (MBP) Peptides to MHC II: Effects on Immune Regulation. *Curr. Med. Chem.* **2005**, *12*, 1521–1535. [CrossRef]

57. Mantzourani, E.D.; Tselios, T.V.; Grdadolnik, S.G.; Platts, J.A.; Brancale, A.; Deraos, G.N.; Matsoukas, J.M.; Mavromoustakos, T.M. Comparison of proposed putative active conformations of myelin basic protein epitope 87-99 linear altered peptide ligands by spectroscopic and modelling studies: The role of positions 91 and 96 in T-cell receptor activation. *J. Med. Chem.* **2006**, *49*, 6683–6691. [CrossRef]

58. Mantzourani, E.D.; Tselios, T.V.; Grdadolnik, S.G.; Brancale, A.; Platts, J.A.; Matsoukas, J.M.; Mavromoustakos, T.M. A putative bioactive conformation for the altered peptide ligand of myelin basic protein and inhibitor of experimental autoimmune encephalomyelitis [Arg91, Ala96] MBP87-99. *J. Mol. Graph. Model.* **2006**, *25*, 17–29. [CrossRef]

59. Mantzourani, E.; Laimou, D.; Matsoukas, M.; Tselios, T. Peptides as Therapeutic Agents or Drug Leads for Autoimmune, Hormone Dependent and Cardiovascular Diseases. *Antiinflamm. Antiallergy Agents Med. Chem.* **2008**, *7*, 294–306. [CrossRef]

60. Mantzourani, E.D.; Blokar, K.; Tselios, T.V.; Matsoukas, J.M.; Platts, J.A.; Mavromoustakos, T.M.; Grdadolnik, S.G. A combined NMR and molecular dynamics simulation study to determine the conformational properties of agonists and antagonists against experimental autoimmune encephalomyelitis. *Bioorg. Med. Chem.* **2008**, *16*, 2171–2182. [CrossRef]

61. Martel, P.; Makriyannis, A.; Mavromoustakos, T.; Kelly, K.; Jeffrey, K.R. Topography of tetrahydrocannabinol in model membranes using neutron diffraction. *BBA Biomembr.* **1993**, *1151*, 51–58. [CrossRef]

62. Mavromoustakos, T.; Daliani, I. Effects of cannabinoids in membrane bilayers containing cholesterol. *Biochim. Biophys. Acta Biomembr.* **1999**, *1420*, 252–265. [CrossRef]

63. Mavromoustakos, T.; Theodoropoulou, E. A combined use of 13C-cross polarization/magic angle spinning, 13C-magic angle spinning and 31P-nuclear magnetic resonance spectroscopy with differential scanning calorimetry to study cannabinoid-membrane interactions. *Chem. Phys. Lipids* **1998**, *92*, 37–52. [CrossRef]

64. Mavromoustakos, T.; De-Ping, Y.; Broderick, W.; Fournier, D.; Makriyannis, A. Small angle X-ray diffraction studies on the topography of cannabinoids in synaptic plasma membranes. *Pharmacol. Biochem. Behav.* **1991**, *40*, 547–552. [CrossRef]

65. Mavromoustakos, T.; Yang, D.P.; Makriyannis, A. Topography of alphaxalone and Δ16-alphaxalone in membrane bilayers containing cholesterol. *BBA Biomembr.* **1994**, *1194*, 69–74. [CrossRef]

66. Mavromoustakos, T.; Yang, D.P.; Makriyannis, A. Small angle X-ray diffraction and differential scanning calorimetric studies on O-methyl-(-)-Δ8-tetrahydrocannabinol and its 5′ iodinated derivative in membrane bilayers. *BBA Biomembr.* **1995**, *1237*, 183–188. [CrossRef]

67. Mavromoustakos, T.; Yang, D.P.; Makriyannis, A. Effects of the anesthetic steroid alphaxalone and its inactive Δ16-analog on the thermotropic properties of membrane bilayers. A model for membrane perturbation. *BBA Biomembr.* **1995**, *1239*, 257–264. [CrossRef]

68. Mavromoustakos, T.; Yang, D.P.; Makriyannis, A. Topography and thermotropic properties of cannabinoids in brain sphingomyelin bilayers. *Life Sci.* **1996**, *59*, 1969–1979. [CrossRef]

69. Mavromoustakos, T.; Theodoropoulou, E.; Papahatjis, D.; Kourouli, T.; Yang, D.P.; Trumbore, M.; Makriyannis, A. Studies on the thermotropic effects of cannabinoids on phosphatidylcholine bilayers using differential scanning calorimetry and small angle X-ray diffraction. *Biochim. Biophys. Acta Biomembr.* **1996**, *1281*, 235–244. [CrossRef]

70. Mavromoustakos, T.; Theodoropoulou, E.; Yang, D.P. The use of high-resolution solid-state NMR spectroscopy and differential scanning calorimetry to study interactions of anaesthetic steroids with membrane. *Biochim. Biophys. Acta Biomembr.* **1997**, *1328*, 65–73. [CrossRef]

71. Mavromoustakos, T.; Papahatjis, D.; Laggner, P. Differential membrane fluidization by active and inactive cannabinoid analogues. *Biochim. Biophys. Acta Biomembr.* **2001**, *1512*, 183–190. [CrossRef]

72. Koukoulitsa, C.; Durdagi, S.; Siapi, E.; Villalonga-Barber, C.; Alexi, X.; Steele, B.R.; Micha-Screttas, M.; Alexis, M.N.; Tsantili-Kakoulidou, A.; Mavromoustakos, T. Comparison of thermal effects of stilbenoid analogs in lipid bilayers using differential scanning calorimetry and molecular dynamics: Correlation of thermal effects and topographical position with antioxidant activity. *Eur. Biophys. J.* **2011**, *40*, 865–875. [CrossRef] [PubMed]

73. Stern, J.N.H.; Illés, Z.; Reddy, J.; Keskin, D.B.; Fridkis-Hareli, M.; Kuchroo, V.K.; Strominger, J.L. Peptide 15-mers of defined sequence that substitute for random amino acid copolymers in amelioration of experimental autoimmune encephalomyelitis. *Proc. Natl. Acad. Sci. USA* **2005**, *102*, 1620–1625. [CrossRef] [PubMed]

74. Kant, R.; Pasi, S.; Surolia, A. Homo-β-amino acid containing MBP(85-99) analogs alleviate experimental autoimmune encephalomyelitis. *Sci. Rep.* **2015**, *3*, 8205. [CrossRef]

75. Yannakakis, M.P.; Simal, C.; Tzoupis, H.; Rodi, M.; Dargahi, N.; Prakash, M.; Mouzaki, A.; Platts, J.A.; Apostolopoulos, V.; Tselios, T.V. Design and synthesis of non-peptide mimetics mapping the immunodominant myelin basic protein (MBP83-96) epitope to function as T-cell receptor antagonists. *Int. J. Mol. Sci.* **2017**, *18*, 1215. [CrossRef]

76. Karin, N.; Mitchell, D.J.; Brocke, S.; Ling, N.; Steinman, L. Reversal of experimental autoimmune encephalomyelitis by a soluble peptide variant of a myelin basic protein epitope: T cell receptor antagonism and reduction of interferon γ and tumor necrosis factor α production. *J. Exp. Med.* **1994**, *180*, 2227–2237. [CrossRef]

77. Katsara, M.; Matsoukas, J.; Deraos, G.; Apostolopoulos, V. Towards immunotherapeutic drugs and vaccines against multiple sclerosis. *Acta Biochim. Biophys. Sin.* **2008**, *40*, 636–642. [CrossRef]

78. Aharoni, R.; Teitelbaum, D.; Arnon, R.; Sela, M. Copolymer 1 acts against the immunodominant epitope 82-100 of myelin basic protein by T cell receptor antagonism in addition to major histocompatibility complex blocking. *Proc. Natl. Acad. Sci. USA* **1999**, *96*, 634–639. [CrossRef]

79. Mohammadi-Milasi, F.; Mahnam, K.; Shakhsi-Niaei, M. In silico study of the association of the HLA-A*31:01 allele (human leucocyte antigen allele 31:01) with neuroantigenic epitopes of PLP (proteolipid protein), MBP (myelin basic protein) and MOG proteins (Myelin oligodendrocyte glycoprotein) for studying t. *J. Biomol. Struct. Dyn.* **2020**, 1–17. [CrossRef]

Article

*HLA-DPB1*03* as Risk Allele and *HLA-DPB1*04* as Protective Allele for Both Early- and Adult-Onset Multiple Sclerosis in a Hellenic Cohort

Maria Anagnostouli [1,2,*], **Artemios Artemiadis** [2,3], **Maria Gontika** [2], **Charalampos Skarlis** [2], **Nikolaos Markoglou** [2], **Serafeim Katsavos** [2], **Konstantinos Kilindireas** [4], **Ilias Doxiadis** [5] and **Leonidas Stefanis** [6]

[1] Faculty of Neurology in Demyelinating Disease Unit & Director of Immunogenetics Laboratory, 1st Department of Neurology, Medical School, National and Kapodistrian University of Athens, NKUA, Aeginition Hospital, 115 28 Athens, Greece

[2] Immunogenetics Laboratory, 1st Department of Neurology, Medical School, National and Kapodistrian University of Athens, NKUA, Aeginition Hospital, 115 28 Athens, Greece; kmwartem@yahoo.com (A.A.); mary212009@windowslive.com (M.G.); charskarlis@med.uoa.gr (C.S.); nmarkoglou@yahoo.gr (N.M.); serafkatsavos@gmail.com (S.K.)

[3] Medical School, University of Cyprus, Nicosia 1678, Cyprus

[4] Demyelinating Diseases Unit, 1st Department of Neurology, Medical School, National and Kapodistrian University of Athens, NKUA, Aeginition Hospital, 115 28 Athens, Greece; kildrcost@med.uoa.gr

[5] Transplantation Immunological Laboratory, Institute of Transfusion Medicine, University Hospital Leipzig, 04103 Leipzig, Germany; ilidox@planet.nl

[6] 1st Department of Neurology, Medical School, National and Kapodistrian University of Athens, NKUA, Aeginition Hospital, 115 28 Athens, Greece; lstefanis@med.uoa.gr

* Correspondence: managnost@med.uoa.gr

Received: 11 May 2020; Accepted: 14 June 2020; Published: 16 June 2020

Abstract: Background: Human Leucocyte Antigens (HLA) represent the genetic loci most strongly linked to Multiple Sclerosis (MS). Apart from *HLA-DR* and *HLA–DQ*, *HLA-DP* alleles have been previously studied regarding their role in MS pathogenesis, but to a much lesser extent. Our objective was to investigate the risk/resistance influence of *HLA-DPB1* alleles in Hellenic patients with early- and adult-onset MS (EOMS/AOMS), and possible associations with the *HLA-DRB1*15:01* risk allele. **Methods**: One hundred MS-patients (28 EOMS, 72 AOMS) fulfilling the McDonald-2010 criteria were enrolled. HLA genotyping was performed with standard low-resolution Sequence-Specific Oligonucleotide techniques. Demographics, clinical and laboratory data were statistically processed using well-defined parametric and nonparametric methods and the SPSSv22.0 software. **Results**: No significant *HLA-DPB1* differences were found between EOMS and AOMS patients for 23 distinct *HLA-DPB1* and 12 *HLA-DRB1* alleles. The *HLA-DPB1*03* allele frequency was found to be significantly increased, and the *HLA-DPB1*02* allele frequency significantly decreased, in AOMS patients compared to controls. The *HLA-DPB1*04* allele was to be found significantly decreased in AOMS and EOMS patients compared to controls. **Conclusions**: Our study supports the previously reported risk susceptibility role of the *HLA-DPB1*03* allele in AOMS among Caucasians. Additionally, we report for the first time a protective role of the *HLA-DPB1*04* allele among Hellenic patients with both EOMS and AOMS.

Keywords: Multiple Sclerosis; early-onset; adult-onset; Human Leucocyte Antigens; immunogenetics; clinical phenotype; clinical outcome; therapeutics

---

1. Introduction

Multiple sclerosis (MS) is considered a complex, multifactorial disease entity, as both environmental and genetic factors have been implicated in its pathogenesis [1]. The Major Histocompatibility Complex (MHC) represents a cluster of highly polymorphic genes, including mainly the Human Leukocyte Antigens (HLA) system, namely Class I (A, B, C) and II (DR, DQ, DP) genes, and genes encoding for some other immune factors, like complement components, Bf, C2, C4 and TNF, in Class III and IV loci [2]. HLA molecules mediate antigen presentation to T-lymphocytes, playing a crucial role in immune response and affecting all clinical and neuroimaging characteristics and response to treatment in MS [3,4]. Linkage studies in various populations have consistently demonstrated that the MHC and its polymorphisms represent the genetic locus most strongly linked to MS [3–5], and that the MHC class II (*HLA-DR, HLA-DQ, HLA-DP*) region is the susceptibility complex that accounts for the majority of familial clustering in MS [6].

The MHC class II linkage to MS differs in various populations, with the highest association conferred by the *HLA-DRB1*15:01/HLA-DQB1*06:02* haplotype, present in Caucasians [5]. In 2011, in a collaborative European study, the *HLA-DRB1*15:01* allele exhibited the strongest association with MS, along with the *HLA-DRB1* 03:01* and *HLA-DRB1*13:01* alleles [7], although *DRB1*15:01* was recently found to be hypomethylated and predominantly expressed in monocytes among carriers of *DRB1*15:01*, suggesting putative therapeutic strategies targeting methylation-mediated regulation of this major risk gene [8].

Recent studies have further established the role of *HLA-DRB1*15:01* in early-onset (pediatric and adolescent) MS (EOMS), which accounts for 3–5% of all MS cases, while the role of *HLA-DRB1*04* and *HLA-DRB1*03* remains to be clarified [9–11].

Apart from the well examined *HLA-DR* and *HLA-DQ* genes, other class II genes and their products, *HLA-DP* alleles, have been previously studied regarding their role in MS pathogenesis. One of the earliest studies regarding *HLA-DP* genotyping was performed three decades ago using a small sample of 45 Swedish patients with MS in comparison with 166 Danish controls [12]. Since then, few studies have been published on the role of the *HLA- DPB1* locus concerning genetic risk in adult-onset MS (AOMS), either in Asian [13–16] or European populations [17–21], and no such studies have been performed on EOMS. In 2013, Patsopoulos et al. used single nucleotide polymorphisms (SNP) data from genome-wide studies and tested classical alleles and polymorphisms in eight classical HLA genes in 5091 AOMS cases and 9595 controls [22]. Among a total of 11 identified statistically independent effects, they confirmed a possible association of *HLA-DPB1*03:01*, and also highlighted a more statistically significant effect at amino acid position 65 in the peptide binding groove of *HLA-DPB1** [22]. So far, *HLA-DPB1** alleles have been mainly correlated with neuromyelitis optica spectrum disorders (NMOSD) in Asian but not Caucasian populations [23], while a series of studies suggest a possible role in other autoimmune disorders as well, including juvenile idiopathic arthritis [24], type I diabetes [25] and atopic myelitis in Japanese [26].

The present study attempts to expand the existing data on HLA and MS by investigating the influence of *HLA-DPB1** alleles on disease risk and resistance in a Hellenic sample of 100 patients of both EOMS and AOMS, using healthy controls (HC) for comparisons, given the pre-existing difference in *HLA-DRB1* allele frequencies in EOMS and AOMS in our ethnic group [11] and the total absence of information on *HLA-DPB1* genotyping in the Hellenic MS population.

Additionally, we examined, the putative positive or negative association between the well-defined *HLA-DRB1*15:01* allele and the various *HLA-DPB1** alleles, given the extensive epistatic mechanisms that exist in HLA loci, as clearly illustrated in previous reports [12,19,27,28].

2. Materials and Methods

2.1. Patients

One hundred patients with MS (62 females, 38 males, mean age 36.9 ± 11.4 years old) were selected, fulfilling the McDonald criteria for MS diagnosis [29]. These patients were enrolled from the outpatient clinic at the Neurology Department of the Aeginition University Hospital (Athens, Greece) after providing written informed consent. The study received ethical approval by the Hospital's Ethics Committee (ethic approval code number: 117/2-4-13), as it was found consistent with the Declaration of Helsinki. At the time this study, 42 patients had the relapsing-remitting type of the disease (RRMS), and 10 patients were identified with primary progressive MS (PPMS), while the rest had the secondary progressive type (SPMS). For all patients, the mean age of disease onset was 27.8 ± 10.8 years old, the mean disease duration was 100.9 ± 80.4 months and the median Expanded Disability Status Scale (EDSS) was 3.0 (range:1.0–8.0) [30]. There were two MS onset groups; 28 in the ≤19 years old or early onset MS (EOMS) group and 72 in the >19 years old or adult onset MS (AOMS) group. Valid Magnetic Resonance Imaging (MRI) and cerebrospinal fluid (CSF) (i.e., presence of oligoclonal bands and IgG index calculation) assessments were available 60 (60%) of the patients. Missing data for MRIs were attributed to the lack of recent MRI scans. With regards to CSF, some but not all patients had been subjected to CSF analysis, since this was not a prerequisite for the MS diagnosis, according to the revised 2010 McDonald criteria [29]. All patients provided informed consent for participation and publication.

2.2. HLA-DPB1* and HLA-DRB1* Genotyping

HLA genotyping was performed at the Immunogenetics Laboratory of the 1st Department of Neurology, in Aeginition Hospital. High molecular weight DNA was extracted from peripheral blood samples (8 mL peripheral blood in sodium citrate, ACD Vacutainer® tube) using the DNA extraction, Maxi Kit (QIAGEN, Venlo, the Netherlands) as per manufacturer's guidelines in the commercial kit. HLA class II (*HLA-DRB1* and *HLA-DPB1*) frequencies were determined by molecular techniques for all the specificities included in the HLA Nomenclature of 2012 (we present only the first two or four digits of each allele, for low or high resolution respectively) [31]. *HLA-DRB1* genotyping had been previously performed, using a PCR-SSO (Polymerase-Chain-Reaction, PCR, Sequence-Specific Oligonucleotide, SSO) technique (Elpha Bio-Rad, High resolution), as described elsewhere [11]. *HLA-DPB1* genotyping was performed using a different PCR-SSO technique, based on a method that depends on reverse hybridization (Line Probe Assay, INNO-LiPA, Low Resolution, Innogenetics, Fujirebio, Europe) according to the manufacturer's protocol.

2.3. Statistical Analyses

The Hardy-Weinberg proportions (HWP) and linkage disequilibrium for *HLA-DPB1*, *HLA-DRB1* haplotypes were ascertained using the PyPoP software [32]. An Ewens-Watterson (EW) homozygosity test for neutrality was also performed. Calculation of the normalized deviate of the homozygosity (i.e., Fnd) was done, with positive and negative values implying directional and balancing selection, respectively. *HLA-DPB1** genotype frequency in patients with MS was compared with that reported in a previous study of Hellenic HC by using multiple binomial tests [33].

Separate analyses were performed in the EOMS and AOMS groups using the same expected genotype frequencies of the healthy controls [33]. A Fisher's exact test for categorical and Mann-Whitney U test for numerical variables were performed to allow us to make group comparisons. Mantel-Haenszel statistics were used to ascertain the role of MS groups in the association between *HLA-DPB1* genotypes and categorical clinical parameters. In *HLA-DPB1* genotype-related tests (except those for clinical parameters), p value correction was made according to the Benjamini–Yekutieli method (or B–Y) based on the following formula: p (B–Y) = a/(Σ1/i), where i denotes the number of comparisons and a = 0.05 [34,35]. Statistical analyses were performed using the SPSS v22.0 software (Armonk, NY, USA: IBM Corp).

3. Results

3.1. HWP and Linkage Disequilibrium of the Study's Sample

Twenty-three distinct *HLA-DPB1* alleles were identified (total alleles: 200). There were 28 homozygote and 72 heterozygote patients with MS. There were no deviations from the HWP (homozygotes: 29.71 expected, $F(1) = 0.1$, $p = 0.754$, heterozygotes: 70.29 expected, $F(1) = 0.04$, $p = 0.838$). The most common haplotypes were *HLA-DPB1*04/DPB1*04* (27%), followed by *HLA-DPB1*02/DPB1*04* (13%), *HLA-DPB1*03/DPB1*04* (11%) and *HLA-DPB1*10/DPB1*04* (6%). The EW homozygosity test of neutrality was found to be significantly positive (i.e., Fnd = 3.79, $p = 0.992$, i.e., over the limit 0.975), denoting a directional selection of the *HLA-DPB1*04* allele.

Twelve distinct *HLA-DRB1* alleles were identified (total alleles: 174) in 87 out of the 100 patients. There were 11 (12.6%) homozygote and 76 (87.3%) heterozygote patients. There were no deviations from the HWP (homozygotes: 10.84 expected, $F(1) = 0$, $p = 0.962$, heterozygotes: 76.16 expected, $F(1) = 0$, $p = 0.986$). The most common allele was *HLA-DRB1*11* (20.1%), followed by *HLA-DRB1*16* (15.5%), *HLA-DRB1*15* (13.2%), *HLA-DRB1*04* (12.1%) and *HLA-DRB1*13* (10.4%). The most common, but still of low frequency (4.6%), genotype was *HLA-DRB1*11/DRB1*16*. The EW homozygosity test of neutrality was found to be significantly negative (i.e., Fnd = −1.41, $p = 0.0033$, i.e., lower the limit 0.05), indicating a balancing selection.

The delta distance for the *HLA-DPB1* and *HLA-DRB1* haplotypes was 0.00938 ($p = 0.303$), denoting linkage equilibrium. This did not change when age of MS onset was taken into account (EOMS: delta 0.0128, $p = 0.954$, AOMS: delta 0.0133, $p = 0.351$). The most common (i.e., over 5%) *HLA-DPB1/HLA-DRB1* haplotypes were *HLA-DPB1*04/HLA-DRB1*11* (10.8%), *HLA-DPB1*04/HLA-DRB1*16* (7.7%), *HLA-DPB1*04/HLA-DRB1*04* (7%), *HLA-DPB1*02/HLA-DRB1*11* (6.6%) and *HLA-DPB1*04/HLA-DRB1*03* (5.5%).

3.2. Nongenetic Comparisons between Age of Onset Groups

Table 1 presents the main characteristics of the two MS groups. Patients with EOMS were significantly younger and had longer disease duration compared to AOMS, which primarily reflects the blood sampling timing, and has no specific clinical significance. Of most importance, patients with EOMS had significantly higher IgG indexes compared to AOMS. It should be noted that this difference reflects 60 out of the 100 patients with MS of this study, since, as mentioned in the methods section, no CSF testing was available for 40 patients.

Table 1. Nongenetic Comparisons Between Age of Multiple Sclerosis (MS) Onset Groups.

Characteristics	EOMS	AOMS	Sig [1]
Females	19/28 (67.9%)	43/72 (59.7%)	0.499
Age (years old)	29.9 ± 9.8	39.8 ± 10.8	0.001 *
Duration of MS (months)	148.1 ± 105.4	81.3 ± 69.1	0.011 *
EDSS	3.1 ± 1.7	3.3 ± 1.6	0.414
Primary Progressive	3/28 (10.7%)	7/72 (9.7%)	0.917
Relapses since onset	5.2 ± 6.1	3.7 ± 2.5	0.393
IgG index [2]	1.3 ± 0.7	0.8 ± 0.4	0.004 *
Presence of OCBs [2]	13/15 (86.7%)	31/45 (68.9%)	0.312
Subcortical lesions [2]	12/17 (70.6%)	27/43 (62.8%)	0.765
Periventricular lesions [2]	16/17 (94.1%)	42/43 (97.7%)	0.49
Infratentorial lesions [2]	12/17 (70.6%)	30/43 (69.8%)	1.000
Spinal cord lesions [2]	12/14 (85.7%)	29/31 (93.5%)	0.578

Numbers represent means ± standard deviation and absolute (%) frequencies; 1 Fisher exact test for categorical and Mann-Whitney U test for numerical characteristics. OCBs: Oligoclonal Bands, Sig.: significance. 2 Valid MRI and cerebrospinal fluid assessments were available for 60 (66%) and 60 (60%) patients, respectively. * $p \leq 0.05$.

3.3. HLA-DPB1 Allele Comparisons between the Age of Onset Groups

No significant *HLA-DPB1* allele differences were found between patients with EOMS and AOMS (Table 2). However, there were significantly fewer *HLA-DPB1*04*-positive patients in the EOMS group compared to HC (64.3% vs. 92.7%). The *HLA-DPB1*03* allele was found to be significantly increased in patients with AOMS compared to HC (23.6% vs. 13.4%). On the other hand, *HLA-DPB1*02* and *HLA-DPB1*04* were found to be significantly decreased ($p < 0.001$) in patients with AOMS compared to HC (22.2% vs. 36.6% and 79.2% vs.92.7%).

A total of 21 out of 87 patients (24.1%) were positive for the *HLA-DRB1*15* allele, which is significantly higher than the expected 11.4% allele frequency in HC ($p < 0.001$), confirming the well-established role of this allele in MS pathogenesis [33]. *HLA-DRB1*15* allele positivity was 20.8% (5/24) for EOMS and 25.4% (16/63) for AOMS ($p = 0.783$).

Table 3 presents the *HLA-DRB1*15* allele frequency among the different *HLA-DPB1** alleles. Only statistically significant associations are presented. The *HLA-DRB1*15* allele was statistically significantly absent among *HLA-DPB1*03* positive patients ($p = 0.001$) and among *HLA-DPB1*03* positive AOMS ($p = 0.003$), whereas it was significantly increased among *HLA-DPB1*04* ($p = 0.048$), *HLA-DPB1*14* ($p = 0.008$) -positive genotype patients. Finally, the *HLA-DRB1*15* allele was positive in the two *HLA-DPB1*14* positive patients with EOMS ($p = 0.036$).

In the 60 patients with available CSF examination, those with the *HLA-DPB1*02* allele had significantly higher IgG indexes than those who were negative for *HLA-DPB1*02* (mean 1.22 ± 0.70 vs. 0.75 ± 0.39, respectively, $p = 0.02$), irrespective of age of MS onset. There were no other significant associations between the *HLA-DPB1* or *HLA-DRB1* alleles (i.e., presence or not of each *HLA-DPB1** allele and *HLA-DRB1*15* allele) and gender, type of MS, MRI or CSF assessments (data not shown). Patients with AOMS who were positive for *HLA-DPB1*02* had significantly fewer relapses since onset than *HLA-DPB1*02* negative patients with AOMS (2.7 ± 2.5 vs. 3.9 ± 2.4, $p = 0.033$), corroborating the protective role of *HLA-DPB1*02* phenotype, as reported above (Figure 1). No other MS group effects on the HLA and clinical parameter associations were found.

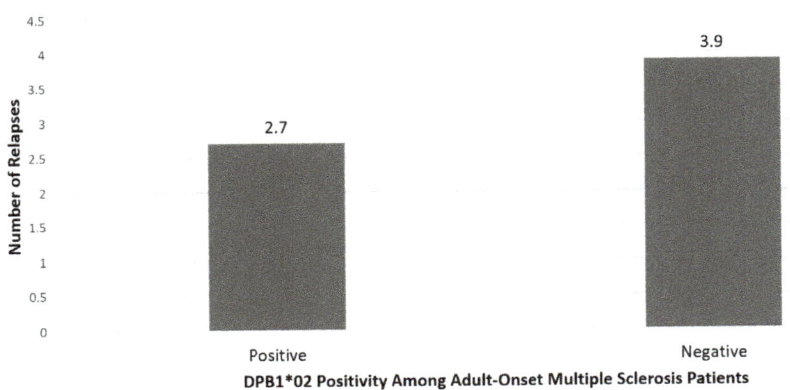

Figure 1. Number of relapses in adult-onset Multiple Sclerosis Patients with regards to HLA-*DPB1*02* genotype. Positive patients had significantly fewer relapses than negative ($p = 0.033$).

Table 2. *HLA-DPB1* Allele Frequencies Among Different Age of Multiple Sclerosis (MS) Onset Groups of Patients.

	Early MS	Adult MS	HCs	Early vs. Adult MS	Early MS vs. HCs	Adult MS vs. HCs
	(N = 28)	(N = 72)	(N = 246)	Sig [1]	Sig [2]	Sig [2]
HLA-DPB1*01	3.6	4.2	4.5	0.85 (0.09–8.55) 1.000	0.79 (0.1–6.37) 0.500	0.92 (0.25–3.42) 0.500
HLA-DPB1*02	39.3	22.2	36.6	2.27 (0.89–5.80) 0.131	1.12 (0.5–2.5) 0.461	0.5 (0.23–0.91) 0.008 **
HLA-DPB1*03	17.9	23.6	13.4	0.7 (0.23–2.13) 0.602	1.4 (0.50–3.95) 0.339	2.0 (1.04–3.84) 0.009 **
HLA-DPB1*04	64.3	79.2	92.7	0.47 (0.18–1.24) 0.132	0.14 (0.06–0.35) <0.001 **	0.3 (0.14–0.63) <0.001 **
HLA-DPB1*05	3.6	2.8	2.4	1.3 (0.11–14.89) 1.000	1.48 (0.17–12.77) 0.500	1.14 (0.23–5.79) 0.500
HLA-DPB1*06	0	1.4	0.8	- 1.000	- 0.500	1.72 (0.15–19.23) 0.500
HLA-DPB1*09	0	1.4	2.8	- 1.000	- 0.372	0.48 (0.06-3.97) 0.356
HLA-DPB1*10	10.7	9.7	4.9	1.11 (0.27–4.65) 1.000	2.34 (0.62–8.85) 0.162	2.1 (0.8–5.55) 0.052
HLA-DPB1*11	0	1.4	-	- 1.000	-	-
HLA-DPB1*13	3.6	2.8	6.1	1.3 (0.11–14.89) 1.000	0.57 (0.07–4.49) 0.435	0.44 (0.1–1.97) 0.176
HLA-DPB1*14	7.1	6.9	3.7	1 (0.18–5.48) 1.000	2.03 (0.42–9.88) 0.321	1.97 (0.64–6.06) 0.126
HLA-DPB1*15	0	4.2	2.0	- 0.557	- 0.468	2.1 (0.49–8.99) 0.186
HLA-DPB1*19	0	1.4	0.4	- 1.000	- 0.500	3.45 (0.21–55.87) 0.346
HLA-DPB1*22	3.6	0	-	- 0.280	-	-
HLA-DPB1*23	3.6	0	2.4	- 0.280	1.48 (0.17–12.77) 0.500	- 0.172

Table 2. *Cont.*

	Early MS	Adult MS	HCs	Early vs. Adult MS	Early MS vs. HCs	Adult MS vs. HCs
	(N = 28)	(N = 72)	(N = 246)	Sig [1]	Sig [2]	Sig [2]
HLA-DPB1*32	3.6	0	0.8	- 0.280	4.52 (0.4–51.49) 0.279	- 0.460
HLA-DPB1*33	3.6	2.8	0.4	1.3 (0.11–14.89) 1.000	9.07 (0.55–149.24) 0.123	7.0 (0.63–78.34) 0.130
HLA-DPB1*34	0	1.4	-	- 1.000	-	-
HLA-DPB1*35	0	4.2	0.8	- 0.557	- 0.500	10.65 (1.09–104.03) 0.005 **
HLA-DPB1*38	3.6	0	-	- 0.280	-	-
HLA-DPB1*46	3.6	0	-	- 0.280	-	-
HLA-DPB1*50	0	1.4	0.4	- 1.000	- 0.500	3.45 (0.21–55.87) 0.346
HLA-DPB1*56	0	1.4	-	- 1.000	-	-

Numbers represent frequencies (%). 1 Fishers's exact test (23 comparisons) 2 Binomial tests (17 comparisons). ** $p \leq 0.015$ (17 comparisons) or $p \leq 0.013$ (23 comparisons), according to the Benjamini–Yekutieli method for 17 comparisons, HCs: Healthy Control

Table 3. Significant DRB1*15 Positivity Differences among *HLA-DPB1* Alleles and Age of Onset Groups in Multiple Sclerosis (MS) Patients.

Total Sample of MS Patients		
*HLA-DPB1** Genotype	*HLA-DRB1*15* Positive	Sig [1]
*HLA-DPB1*03*	0/22 (0%)	0.001 *
*HLA-DPB1*04*	19/63 (30.2%)	0.048 *
*HLA-DPB1*14*	5/7 (71.4%)	0.008 *
Adult-Onset MS		
*HLA-DPB1** Genotype	*HLA-DRB1*15* Positive	Sig [1]
*HLA-DPB1*03*	0/17 (0%)	0.003 *
Early-Onset MS		
*HLA-DPB1** Genotype	*HLA-DRB1*15* Positive	Sig [1]
*HLA-DPB1*14*	2/2 (100%)	0.036 *

Values represent observed frequencies (%) of *HLA-DRB1*15* allele positivity among the different *HLA-DPB1* genotypes. 1 Fisher exact tests. * $p \leq 0.05$.

4. Discussion

HLA-immunogenetics is an old but still rapidly expanding field in MS pathogenesis. In order to keep abreast of rapid developments in this field, we investigated the role of the *HLA-DP* locus in MS pathophysiology. We genotyped 100 Hellenic patients with MS for *HLA-DR* and *HLA-DP* alleles, as described above, which is a rather small sample and the main limitation of this study. *HLA-DPB1* genotyping was performed for the first time on a Hellenic MS population and in patients with EOMS, which is the core novelty of our research, albeit on a small sample (28 patients); however, we highlight again that EOMS is a rare disease entity and represents only the 3–5% of all MS patients in Caucasian populations.

In our study, we replicated the well-established predominance of the *HLA-DRB1*15* genotype in Hellenic patients with MS compared to HC, independently of age at disease onset [11].

No statistically significant *HLA-DPB1* allele differences were found between patients with EOMS and AOMS. All statistically significant differences were investigated in the AOMS group, except for the *HLA-DPB1*04* allele, which is lower in EOMS and AOMS, compared to the HC group at a high statistical level ($p < 0.001$, Table 2), suggesting a possible protective role in the Hellenic population. This is in contrast with an early study in 1988 [12] where the frequencies of DPw4 were 93.3% in patients with MS and 72.3% in controls (relative risk, $R^2 = 5.4$, $p = 0.0014$). Nevertheless, we have to mention that in this early study, the HLA-DNA typing was carried out on a small sample of 45 patients with MS and 63 controls of different ethnic European groups (Swedish and Danish), using the Restriction Fragment Length Polymorphism (RFLP) technique for *HLA-DP* and *HLA-DR* genes. In this same study, the *HLA-DR2* antigen was present in 75.5% of patients and in 33.7% of the controls ($R^2 = 6.1$, p less than 10(-6)). *HLA-DPw4* was not associated (i.e., was not in linkage disequilibrium) with *HLA-DR2* in patients or controls. Thus, the researchers concluded that in MS, the associations with *HLA-DP* and *HLA-DR* are independent of each other, but the combined presence of *HLA-DPw4* (cellularly defined) and *HLA-DR2* represented a significantly higher risk than either antigen alone, indicating that synergism between *HLA-DP* and *HLA-DR* gene products may play a role in the genetic susceptibility to MS. On the other hand, a recent study on celiac disease showed that the *HLA-DPB1*04:01* allele protects genetically-susceptible children from celiac disease [36], a fact that is in line with our results, concerning children and adults with MS, while in another study in 2015, another *HLA-DPB1*04* allele, namely *HLA-DPB1*04:02*, conferred a strong protective effect against narcolepsy [37]. Finally, the worldwide risk *HLA-DRB1*15* allele in MS, in Caucasians, was found to be significantly increased among *HLA-DPB1*04* positive patients with MS ($p = 0.048$) in our sample.

In another early study in France in 1991, it was found that the distribution of *HLA-DPB1* alleles was not significantly different in patients with MS and controls [20]. Nowadays, it is perfectly clear that the *HLA-DP*03* allele is associated with MS and epitope spreading in MS [17,22], and in this study, we observed the risk susceptibility of this allele in our Hellenic MS sample, at a highly significant level ($p < 0.009$, Table 2).

Regarding AOMS, the *HLA-DPB1*03* allele could be a risk factor for the disease, as it was found to be significantly increased in patients with AOMS compared to HC. The percentage of *HLA-DPB1*03* positive patients with EOMS was higher than HC (17.9% vs. 13.4%), although at a nonstatistically significant level. The *HLA-DRB1*15* allele was absent among *HLA-DPB1*03* positive patients. This cannot be attributed to linkage disequilibrium, as this was tested. Linkage disequilibrium for the *HLA-DR* and *HLA-DP* genes was excluded in previous studies as well [12]. Since *HLA-DPB1*03* was found to be increased in AOMS, it may constitute a risk factor; this genotype may exert its risk factor effect only in the absence of *HLA-DRB1*15*, at least in AOMS. Moreover, the *HLA-DRB1*15* allele was found to be significantly increased among *HLA-DPB1*04*-positive patients, suggesting that *HLA-DPB1*04* exerts a protective effect only in the absence of *HLA-DRB1*15*. Despite the relatively small sample size in our study, these findings suggest that epistatic mechanisms between Class II *HLA-DR* and *HLA-DP* alleles may play a role in disease pathogenesis and risk of disease occurrence. This conclusion is in line with the results of Dekker et al. [19] who observed that in patients with MS who lacked *HLA-DQB1*06:02* allele, the *HLA-DPB1*03:01* allele frequency was significantly ($p = 0.006$) increased (50.0%) compared with *HLA-DQB1*06:02*-negative controls (9.1%). In parallel, in 2009, Lincoln et al. highlighted the role of epistasis between *HLA-DRB1*15* and *HLA-DQA1*01:02* alleles. More specifically, they proved that *HLA-DQA1*01:02*, which shows no primary MS association, increases disease risk when combined with *HLA-DRB1*15:01*, through transepistatic interactions [27]. Of note is the fact that the presented slight *HLA-DPB1* allele differences between AOMS and EOMS could also reflect the different clinical course of these two groups, given that patients with older age at onset are known to be more at risk of having secondary-progressive disease. For instance, predicting the onset of secondary-progressive multiple sclerosis is accomplished using genetic and nongenetic factors, with the *HLA-A*02:01* allele conferring a decreased risk for MS and also contributing to decreased hazards for SPMS [38].

Another finding in our study that is worthy of mention is the possible protective role of the *HLA-DPB1*02* allele in AOMS. *HLA-DPB1*02* was found to be significantly decreased in AOMS, while those who were *HLA-DPB1*02* positive had, in general, fewer relapses since onset compared to *HLA-DPB1*02*-negative patients with AOMS, corroborating the protective role of the *HLA-DPB1*02* allele reported above (Figure 1).

The *HLA-DPB1*35* allele was found to be significantly increased in patients with AOMS compared to HC, while increased prevalence of the *HLA- DRB1*15* allele in *HLA-DPB1*14*-positive patients with MS and patients with EOMS was also noted. Nevertheless, their possible genetic risk should be interpreted with caution, due to the very low frequency of these alleles.

At this point, we have to mention that the *HLA-DPB1*04* allele is the most frequent in the Hellenic population (92.7%), followed by *HLA-DP*02* (36.6%) and *HLA-DP*03* (13.4%) [30]. Additionally, according to our results, the most common *HLA-DPB1*-haplotypes in Hellenic patients with MS were *HLA-DPB1*04/DPB1*04* (27%), followed by *HLA-DPB1*02/DPB1*04* (13%), *HLA-DPB1*03/DPB1*04* (11%) and *HLA-DPB1*10/DPB1*04* (6%). Thus, the emerging protective role of the *HLA-DP*04* allele is in parallel with the *HLA-DR*11* allele, which is the most common in the Hellenic population, and the protective *HLA-DRB1* allele in Hellenic patients with MS [11].

The role of *HLA-DPB1* alleles has been studied in a range of other autoimmune diseases, especially NMOSD [23]. More specifically, *HLA-DPB1*05:01*, which is extremely rare in Caucasian populations, has the strongest association with opticospinal MS and anti-AQP4 seropositivity in Asian populations, while *HLA-DPB1*03* possibly offers genetic protection against the disease [23]. Moreover, *HLA-DPB1*02:01* has been associated with oligoarticular and rheumatoid factor-negative polyarticular juvenile idiopathic arthritis and childhood-onset diabetes type I in the Japanese population [24,25].

Therapeutic interventions in MS are sometimes difficult, because the patient's symptoms at the initial stages are not clearly suggestive of a definite demyelinating syndrome, especially in children. Furthermore, sometimes the neuroradiological (MRI) aspects and blood antibody tests are not helpful. In these situations, having a marker or a combination of markers that supports the differential diagnosis is of crucial importance, and has a direct impact on therapeutic decision making. The *HLA-DR* alleles, and especially the *HLA-DR*15* allele, are the most robust genetic markers for almost every clinical or paraclinical aspect of the disease [4] in Caucasians and for the therapeutic response to different Disease Modified Treatments (DMTs) [4]. Nowadays, the expansion and overlap of various demyelinating diseases, namely MS, NMOSD, ADEM (Acute Disseminating Encephalomyelitis), MOG-Demyelinating (Myelin Oligodendrocyte Glycoprotein-Demyelinating) disease, Optic Neuritis, etc., make the need for specific biomarkers more urgent than ever before, as noted in our previous critical review [39] and in this works of other researchers [40].

Apart from a genetic association with MS and other demyelinating diseases, *HLA-DP* molecules play a key role in MS pathogenesis and progression, as described many years ago [17,41].

Additionally, in our Hellenic cohort, the *HLA-DP* alleles seemed to play an independent role in patients with MS (risk/protective), apart from the *HLA-DR* alleles, a fact that has to be confirmed in larger cohorts in the future. This could pave the way for the usage of these alleles in patient stratification (carriers and noncarriers)—as already happens with various HLA-DRB1 alleles and especially with the *HLA-DRB1*15* allele [4,42]—for many MS characteristics and therapy responses in different DMTs in Caucasian populations [4,42].

Altogether, clarification of *HLA-DP* allele associations with both EOMS and AOMS is needed in every ethnic group to get a better idea of clinical features and MS phenotypes and disease progression, and as a form of future putative data for better therapeutics.

5. Conclusions-Limitations

In conclusion, our study supports the previously reported risk susceptibility role of the *HLA-DPB1*03* allele in AOMS in many Caucasian populations. Additionally, we report, for the first time in the international literature, the protective role of the *HLA-DPB1*04* allele for patients with both EOMS and AOMS, and the putative protective role of the *HLA-DPB1*02* allele in patients with AOMS in our sample. Another finding that is worthy of mention is the total absence of the well-established *HLA-DRB1*15* allele among patients having the most statistically frequent *HLA-DPB1*03* allele in our cohort.

A limitation of our study was the relatively small sample size (28 patients with EOMS and 72 patients with AOMS). Indeed, observed small effect sizes for DPB1 alleles from the different group comparisons were the following: EOMS vs. AOMS 7%, EOMS vs. HCs 21.9% and AOMS vs. HCs 25.8%. However, this is a first attempt towards clarifying the role of the *HLA-DPB1* alleles in MS in a Hellenic AOMS and EOMS cohort. Moreover, the small study sample did not allow us to conduct multivariable analyses, which would more readily reveal confounding effects in our analyses.

These novel data could also contribute to personalized MS-therapeutics in the near future, taking into account the rapid expansion of our knowledge of multiple sclerosis and other distinct demyelinating diseases in many ethnic groups.

Author Contributions: Conceptualization, methodology, patients' curating, manuscript's drafting, manuscript's proofreading, M.A., statistical analyses, manuscript's drafting, manuscript's proofreading, A.A., patients' curating, bibliography manuscript's drafting, M.G., project administration, bibliography, C.S., data curating, N.M., S.K., critical review of the manuscript, K.K., L.S., statistical suggestions, validation, critical review of the manuscript, I.D. All authors have read and agreed to the published version of the manuscript

Funding: No funding for this specific study and manuscript.

Acknowledgments: We want to gratefully thank the patients and their families for their valuable contribution in our research.

Conflicts of Interest: The authors declare no conflict of interest for this specific study and manuscript.

Availability of Data and Material: The datasets generated during and/or analyzed during the current study are available from the corresponding author on request.

References

1. Ramagopalan, S.V.; Dyment, D.A.; Ebers, G.C. Genetic epidemiology: The use of old and new tools for multiple sclerosis. *Trends Neurosci.* **2008**, *31*, 645–652. [CrossRef]
2. Gruen, J.R.; Weissman, S.M. Human MHC Class III and IV Genes and Disease Associations. *Front. Biosci.* **2001**, *6*, D960–D972. [CrossRef] [PubMed]
3. Katsavos, S.; Anagnostouli, M. Biomarkers in Multiple Sclerosis: An Up-to-Date Overview. *Mult. Scler. Int.* **2013**, *2013*, 340508. [CrossRef] [PubMed]
4. Stamatelos, P.; Anagnostouli, M. HLA-Genotype in Multiple Sclerosis: The Role in Disease onset, Clinical Course, Cognitive Status and Response to Treatment: A Clear Step Towards Personalized Therapeutics. *Immunogenet* **2017**, *2*, 116.
5. Goodin, D.; Khankhanian, P.; Gourraud, P.A.; Vince, N. Highly conserved extended haplotypes of the major histocompatibility complex and their relationship to multiple sclerosis susceptibility. *PLoS ONE* **2018**, *13*, e0190043. [CrossRef] [PubMed]
6. Bozikas, V.P.; Anagnostouli, M.C.; Petrikis, P.; Sitzoglou, C.; Phokas, C.; Tsakanikas, C.; Karavatos, A. Familial bipolar disorder and multiple sclerosis: A three-generation HLA family study. *Prog. Neuropsychopharmacol. Biol. Psychiatry* **2003**, *27*, 835–839. [CrossRef]
7. International Multiple Sclerosis Genetics Consortium; Wellcome Trust Case Control Consortium 2; Sawcer, S.; Hellenthal, G.; Pirinen, M.; Spencer, C.C.; Patsopoulos, N.A.; Moutsianas, L.; Dilthey, A.; Su, Z.; et al. Genetic risk and a primary role for cell-mediated immune mechanisms in multiple sclerosis. *Nature* **2011**, *476*, 214–219.
8. Kular, L.; Liu, Y.; Ruhrmann, S.; Zheleznyakova, G.; Marabita, F.; Gomez-Cabrero, D.; James, T.; Ewing, E.; Lindén, M.; Górnikiewicz, B.; et al. DNA Methylation as a Mediator of HLA-DRB1*15:01 and a Protective Variant in Multiple Sclerosis. *Nat. Commun.* **2018**, *9*, 2397. [CrossRef]
9. Gianfrancesco, M.; Stridh, P.; Shao, X.; Rhead, B.; Graves, J.S.; Chitnis, T.; Waldman, A.; Lotze, T.; Schreiner, T.; Belman, A.; et al. Genetic risk factors for pediatric–onset multiple sclerosis. *Mult. Scler.* **2018**, *24*, 1825–1834. [CrossRef]
10. Venkateswaran, S.; Banwell, B. Pediatric multiple sclerosis. *Neurologist* **2010**, *16*, 92–105. [CrossRef]
11. Anagnostouli, M.C.; Manouseli, A.; Artemiadis, A.; Katsavos, S.; Fillipopoulou, C.; Youroukos, S.; Efthimiopoulos, S.; Doxiadis, I. HLA-DRB1* Allele Frequencies in Pediatric, Adolescent and Adult-Onset Multiple Sclerosis Patients, in a Hellenic Sample. Evidence for New and Established Associations. *J. Mult. Scler.* **2014**, *1*, 104. [CrossRef]
12. Odum, N.; Hyldic-Nielsenn, J.J.; Morling, N.; Sandberg-Wollheim, M.; Platz, P.; Svejgaard, A. HLA-DP antigens are involved in the susceptibility to multiple sclerosis. *Tissue Antigens* **1988**, *31*, 235–237. [CrossRef] [PubMed]
13. Wu, X.M.; Wang, C.; Zhang, K.N.; Lin, A.Y.; Kira, J.; Hu, G.Z.; Qu, X.H.; Xiong, Y.Q.; Cao, W.F.; Gong, L.Y. Association of susceptibility to multiple sclerosis in Southern Han Chinese with HLA-DRB1, -DPB1 alleles and DRB1-DPB1 haplotypes: Distinct from other populations. *Mult. Scler.* **2009**, *15*, 1422–1430. [CrossRef]
14. Fukazawa, T.; Yamasaki, K.; Ito, H.; Kikuchi, S.; Minohara, M.; Horiuchi, I.; Tsukishima, E.; Sasaki, H.; Hamada, T.; Nishimura, Y.; et al. Both the HLA-DPB1 and -DRB1 alleles correlate with risk for multiple sclerosis in Japanese: Clinical phenotypes and gender as important factors. *Tissue Antigens* **2000**, *55*, 199–205. [CrossRef] [PubMed]
15. Fukazawa, T.; Kikuchi, S.; Miyagishi, R.; Miyazaki, Y.; Yabe, I.; Hamada, T.; Sasaki, H. HLA-DPB1*0501 is not uniquely associated with opticospinal multiple sclerosis in Japanese patients. Important role of DPB1*0301. *Mult. Scler.* **2006**, *12*, 19–23. [CrossRef] [PubMed]
16. Yoshimura, S.; Isobe, N.; Yonekawa, T.; Matsushita, T.; Masaki, K.; Sato, S.; Kawano, Y.; Yamamoto, K.; Kira, J.; South Japan Multiple Sclerosis Genetics Consortium. Genetic and infectious profiles of Japanese multiple sclerosis patients. *PLoS ONE* **2012**, *7*, e48592. [CrossRef]

17. Yu, M.; Kinkel, R.P.; Weinstock-Guttman, B.; Cook, D.J.; Tuohy, V.K. HLADP: A class II restriction molecule involved in epitope spreading during the development of multiple sclerosis. *Hum. Immunol.* **1998**, *59*, 15–24. [CrossRef]

18. Field, J.; Browning, S.R.; Johnson, L.J.; Danoy, P.; Varney, M.D.; Tait, B.D.; Gandhi, K.S.; Charlesworth, J.C.; Heard, R.N.; Australia and New Zealand Multiple Sclerosis Genetics Consortium; et al. A polymorphism in the HLA-DPB1 gene is associated with susceptibility to multiple sclerosis. *PLoS ONE* **2010**, *5*, e13454. [CrossRef]

19. Dekker, J.W.; Easteal, S.; Jakobsen, I.B.; Gao, X.; Stewart, G.J.; Buhler, M.M.; Hawkins, B.R.; Higgins, D.A.; Yu, Y.L.; Serjeantson, S.W. HLA-DPB1 alleles correlate with risk for multiple sclerosis in Caucasoid and Cantonese patients lacking the high-risk DQB1*0602 allele. *Tissue Antigens* **1993**, *41*, 31–36. [CrossRef]

20. Roth, M.P.; Coppin, H.; Descoins, P.; Ruidavets, J.B.; Cambon-Thomsen, A.; Clanet, M. HLA-DPB1 gene polymorphism and multiple sclerosis: A large case-control study in the southwest of France. *J. Neuroimmunol.* **1991**, *34*, 215–222. [CrossRef]

21. Marrosu, M.G.; Cocco, E.; Costa, G.; Murru, M.R.; Mancosu, C.; Murru, R.; Lai, M.; Sardu, C.; Contu, P. Interaction of loci within the HLA region influences multiple sclerosis course in the Sardinian population. *J. Neurol.* **2006**, *253*, 208–213. [CrossRef] [PubMed]

22. Patsopoulos, N.A.; Barcellos, L.F.; Hintzen, R.Q.; Schaefer, C.; van Duijn, C.M.; Noble, J.A.; Raj, T.; IMSGC; ANZgene; Gourraud, P.A.; et al. Fine-mapping the genetic association of the major histocompatibility complex in multiple sclerosis: HLA and non-HLA effects. *PLoS Genet.* **2013**, *9*, e1003926.

23. Gontika, M.P.; Anagnostouli, M.C. Human leukocyte antigens immunogenetics of neuromyelitis optica or Devic's disease and the impact on the immunopathogenesis, diagnosis and treatment: A critical review. *Neuroimmunol. Neuroinflamm.* **2014**, *1*, 44–50. [CrossRef]

24. Hersh, A.O.; Prahalad, S. Immunogenetics of juvenile idiopathic arthritis: A comprehensive review. *J. Autoimmun.* **2015**, *64*, 113–124. [CrossRef] [PubMed]

25. Nishimaki, K.; Kawamura, T.; Inada, H.; Yagawa, K.; Nose, Y.; Nabeya, N.; Isshiki, G.; Tatsumi, N.; Niihira, S. HLA DPB1*0201 gene confers disease susceptibility in Japanese with childhood onset type I diabetes, independent of HLA-DR and DQ genotypes. *Diabetes Res. Clin. Pract.* **2000**, *47*, 49–55. [CrossRef]

26. Sato, S.; Isobe, N.; Yoshimura, S.; Kanamori, Y.; Masaki, K.; Matsushita, T.; Kira, J. HLA-DPB1*0201 is associated with susceptibility to atopic myelitis in Japanese. *J. Neuroimmunol.* **2012**, *251*, 110–113. [CrossRef] [PubMed]

27. Lincoln, M.R.; Ramagopalan, S.V.; Chao, M.J.; Herrera, B.M.; Deluca, G.C.; Orton, S.M.; Dyment, D.A.; Sadovnick, A.D.; Ebers, G.C. Epistasis among HLA-DRB1, HLA-DQA1, and HLA-DQB1 loci determines multiple sclerosis susceptibility. *Proc. Natl. Acad. Sci. USA* **2009**, *106*, 7542–7547. [CrossRef] [PubMed]

28. Moutsianas, L.; Jostins, L.; Beecham, A.H.; Dilthey, A.T.; Xifara, D.K.; Ban, M.; Shah, T.S.; Patsopoulos, N.A.; Alfredsson, L.; Anderson, C.A.; et al. Class II HLA interactions modulate genetic risk for multiple sclerosis. *Nat. Genet.* **2015**, *47*, 1107–1113.

29. Polman, C.H.; Reingold, S.C.; Banwell, B.; Clanet, M.; Cohen, J.A.; Filippi, M.; Fujihara, K.; Havrdova, E.; Hutchinson, M.; Kappos, L.; et al. Diagnostic criteria for multiple sclerosis: 2010 revisions to the McDonald criteria. *Ann. Neurol.* **2011**, *69*, 292–302. [CrossRef]

30. Kurtzke, J.F. Rating neurologic impairment in multiple sclerosis: An expanded disability status scale (EDSS). *Neurology* **1983**, *33*, 1444–1452. [CrossRef]

31. Marsh, S.G. Nomenclature for factors of the HLA system, update June 2012. *Tissue Antigens* **2012**, *80*, 289–293. [CrossRef]

32. Lancaster, A.K.; Single, R.M.; Solberg, O.D.; Nelson, M.P.; Thomson, G. PyPop update—A software pipeline for large-scale multilocus population genomics. *Tissue Antigens* **2007**, *69* (Suppl. 1), 192–197. [CrossRef]

33. Papassavas, E.C.; Spyropoulou-Vlachou, M.; Papassavas, A.C.; Schipper, R.F.; Doxiadis, I.N.; Stavropoulos-Giokas, C. MHC class I and class II phenotype, gene, and haplotype frequencies in Greeks using molecular typing data. *Hum. Immunol.* **2000**, *61*, 615–623. [CrossRef]

34. Benjamini, Y.; Drai, D.; Elmer, G.; Kafkafi, N.; Golani, I. Controlling the false discovery rate in behavior genetics research. *Behav. Brain Res.* **2001**, *125*, 279–284. [CrossRef]

35. Narum, S.R. Beyond Bonferroni: Less conservative analyses for conservation genetics. *Conserv. Genet.* **2006**, *7*, 783–787. [CrossRef]

36. Hadley, D.; Hagopian, W.; Liu, E.; She, J.X.; Simell, O.; Akolkar, B.; Ziegler, A.G.; Rewers, M.; Krischer, J.P.; Chen, W.M.; et al. HLA-DPB1*04:01 Protects Genetically Susceptible Children from Celiac Disease Autoimmunity in the TEDDY Study. *Am. J. Gastroenterol.* **2015**, *110*, 915–920. [CrossRef]

37. Ollila, H.M.; Ravel, J.M.; Han, F.; Faraco, J.; Lin, L.; Zheng, X.; Plazzi, G.; Dauvilliers, Y.; Pizza, F.; Hong, S.C.; et al. HLA-DPB1 and HLA class I confer risk of and protection from narcolepsy. *Am. J. Hum. Genet.* **2015**, *96*, 852. [CrossRef]

38. Misicka, E.; Sept, C.; Briggs, F.B.S. Predicting onset of secondary-progressive multiple sclerosis using genetic and non-genetic factors. *J. Neurol.* **2020**. [CrossRef]

39. Gontika, M.P.; Anagnostouli, M.C. Anti-Myelin Oligodendrocyte Glycoprotein and Human Leukocyte Antigens as Markers in Pediatric and Adolescent Multiple Sclerosis: On Diagnosis, Clinical Phenotypes, and Therapeutic Responses. *Mult. Scler. Int.* **2018**, *2018*, 8487471. [CrossRef]

40. Uher, T.; McComb, M.; Galkin, S.; Srpova, B.; Oechtering, J.; Barro, C.; Tyblova, M.; Bergsland, N.; Krasensky, J.; Dwyer, M.; et al. Neurofilament levels are associated with blood-brain barrier integrity, lymphocyte extravasation, and risk factors following the first demyelinating event in multiple sclerosis. *Mult. Scler.* **2020**. [CrossRef]

41. Warabi, Y.; Matsumoto, Y.; Hayashi, H. Interferon beta1b exacerbates multiple sclerosis with severe optic nerve and spinal cord demyelination. *J. Neurol. Sci.* **2007**, *252*, 57–61. [CrossRef]

42. Werneck, L.C.; Lorenzoni, P.J.; Kay, C.S.K.; Scola, R.H. Multiple sclerosis: Disease modifying therapy and the human leukocyte antigen. *Arq. Neuropsiquiatr.* **2018**, *76*, 697–704. [CrossRef] [PubMed]

Article

A Multiple *N*-Glucosylated Peptide Epitope Efficiently Detecting Antibodies in Multiple Sclerosis

Francesca Nuti [1], Feliciana Real Fernandez [1], Giuseppina Sabatino [1,2], Elisa Peroni [1,3], Barbara Mulinacci [1,†], Ilaria Paolini [1,‡], Margherita Di Pisa [1,§], Caterina Tiberi [1,||], Francesco Lolli [4], Martina Petruzzo [5], Roberta Lanzillo [5], Vincenzo Brescia Morra [5], Paolo Rovero [2,6] and Anna Maria Papini [1,2,3,*]

1   Interdepartmental Research Unit of Peptide and Protein Chemistry and Biology, Department of Chemistry "Ugo Schiff", University of Florence, Via della Lastruccia 13, 50019 Sesto Fiorentino, Italy; francesca.nuti@unifi.it (F.N.); feliciana.realfernandez@unifi.it (F.R.F.); giuseppina.sabatino@cnr.it (G.S.); elisa.peroni@cyu.fr (E.P.); barbaramulinacci@gmail.com (B.M.); ilariapao@gmail.com (I.P.); marghedipi@gmail.com (M.D.P.); caterina.tiberi@gmail.com (C.T.)
2   CNR-IC Istituto di Cristallografia, Via Paolo Gaifami 18, 95126 Catania, Italy; paolo.rovero@unifi.it
3   UMR 8076 CNRS-BioCIS Team of Chemical Biology and PeptLab@UCP Platform of Peptide and Protein Chemistry and Biology, Neuville Campus, CY Cergy Paris Université, 5 mail Gay-Lussac, 95031 Cergy-Pontoise CEDEX, France
4   Department of Clinical and Experimental Biomedical Sciences "Mario Serio" and Careggi University Hospital, University of Florence, Largo Brambilla 3, 50134 Florence, Italy; lolli@unifi.it
5   Multiple Sclerosis Clinical Care and Research Centre, Department of Neurosciences, Reproductive Sciences and Odontostomatology, Federico II University, Via Sergio Pansini 5, 80131 Naples, Italy; martinapetruzzo@gmail.com (M.P.); roberta.lanzillo@unina.it (R.L.); vincenzo.bresciamorra2@unina.it (V.B.M.)
6   Interdepartmental Research Unit of Peptide and Protein Chemistry and Biology, Department of Neurosciences, Psychology, Drug Research and Child Health–Section of Pharmaceutical Sciences and Nutraceutics, University of Florence, Via Ugo Schiff 6, 50019 Sesto Fiorentino, Italy
*   Correspondence: annamaria.papini@unifi.it
†   Current address: Crelux GmbH, am Klopferspitz 19a, 82152 Martinsried, Germany.
‡   Current address: Isvea srl, Via Basilicata 1/3, Località Fosci, 53036 Poggibonsi, Siena, Italy.
§   Current address: Flamma SpA, Via Bedeschi 22, 24040 Chignolo d'Isola, Bergamo, Italy.
|| Current address: "Stabilimento Chimico Farmaceutico Militare", Via Reginaldo Giuliani 201, 50141 Firenze, Italy.

Received: 21 May 2020; Accepted: 10 July 2020; Published: 15 July 2020

Abstract: Diagnostics of Multiple Sclerosis (MS) are essentially based on the gold standard magnetic resonance imaging. Few alternative simple assays are available to follow up disease activity. Considering that the disease can remain elusive for years, identification of antibodies fluctuating in biological fluids as relevant biomarkers of immune response is a challenge. In previous studies, we reported that anti-*N*-glucosylated (*N*-Glc) peptide antibodies that can be easily detected in Solid-Phase Enzyme-Linked ImmunoSorbent Assays (SP-ELISA) on MS patients' sera preferentially recognize hyperglucosylated adhesin of non-typeable *Haemophilus Influenzae*. Since multivalency can be useful for diagnostic purposes to allow an efficient coating in ELISA, we report herein the development of a collection of Multiple *N*-glucosylated Peptide Epitopes (*N*-Glc MEPs) to detect anti-*N*-Glc antibodies in MS. To this aim, a series of *N*-Glc peptide antigens to be represented in the *N*-GlcMEPs were tested in competitive ELISA. We confirmed that the epitope recognized by antibodies shall contain at least 5-mer sequences including the fundamental *N*-Glc moiety. Using a 4-branched dendrimeric lysine scaffold, we selected the *N*-Glc MEP **24**, carrying the minimal epitope Asn(Glc) anchored to a polyethylene glycol-based spacer (PEG) containing a 19-atoms chain, as an efficient multivalent probe to reveal specific and high affinity anti-*N*-Glc antibodies in MS.

Keywords: Multiple Sclerosis; antibody detection; ELISA; multivalency; N-glucosylated peptide epitopes

---

1. Introduction

Multiple Sclerosis (MS) is the most frequent, chronic, inflammatory, demyelinating, disabling disease of the central nervous system, mainly caused by an autoimmune response to self-antigens in genetically susceptible individuals. MS diagnosis and prognosis are mainly supported by magnetic resonance imaging (MRI) that up to now is considered the gold standard diagnostic technique [1]. To the best of our knowledge, there is still no biological marker relevant not only for MS diagnosis but also for its prognosis [2–4]. In the last few years, the role of autoantibodies in MS and their identification have been re-evaluated [5–8]. In particular, a cell-based assay to detect antibodies to myelin oligodendrocyte glycoprotein (MOG), one of the candidate protein autoantigens in MS, has been proposed. However, the real antigen(s) responsible of anti-MOG antibody recognition in the assay remain elusive.

Covid-19, triggering the coronavirus pandemic era we have been living in 2020, marks the return of the old and familiar, but unfortunately misunderstood, enemy killing more human beings than natural disasters: viruses, bacteria, and parasites killers that our modern world tried to fight mainly by social distancing.

Historically, protein antigens isolated from biological material or reproduced by recombinant technologies were used to detect antibodies, but this approach has sometimes turned out as unrealistic. The main limiting factor depends on epitope recognition because of sequence mutations, incorrect folding, lack of post-translational modifications (aberrant versus native), and nonspecific binding.

Peptides mimicking the appropriate epitopes can be valuable tools. In fact, peptides that can be synthetically produced as unique molecules, can increase specificity of antibody recognition, eliminating or minimizing potential cross-reactivity with structurally similar fragments in non-relevant proteins [9–11].

Several Enzyme-Linked ImmunoSorbent Assays (ELISA) based on synthetic peptides have been proposed to detect antibodies in different diseases like Acquired ImmunoDeficiency Syndrome (AIDS), Infectious Bronchitis (IB), Severe Acute Respiratory Syndrome (SARS), and Bluetongue (BT) [12,13]. However, peptide-based ELISA can have some practical limitations, such as the antigen immobilization that can ultimately affects the sensitivity of the assay, particularly when short sequences (7–8 amino acids) are used [14,15]. Surface functionalization and biomolecular interactions can overcome these disadvantages, such as in the streptavidin-biotin system [16].

An interesting strategy to increase surface binding on the ELISA plate, improving sensitivity, is based on multimeric peptide dendrimers. In particular, Multiple Antigen Peptides (MAPs) are an optimal compromise between short peptide epitopes and recombinant or native antigens. It is widely recognized, that surface antibody binding is increased by multivalent presentation of the antigen. In fact, multivalent interactions can be collectively much stronger than the sum of the corresponding monovalent interactions [17,18]. Therefore, MAPs represent a useful chemically unambiguous system to explore antigen-antibody interaction, thanks to their shape and globular physical characteristics [19,20].

Moreover, aberrant Post-Translational Modifications (PTMs) of antigens, can play a fundamental role in triggering antibodies. Particularly, the *N*-glycosylation has been described as possible PTM involved in an antibody-mediated form of MS [21,22].

In previous studies, by a structure-based design, we developed a collection of synthetic glycopeptides, characterized by β-turn structures optimally exposing the sugars as minimal epitopes. We demonstrated that the β-D-glucopyranosyl moiety (Glc) linked to an Asn residue (*N*-Glc) on the tip of the turn is fundamental for antibody recognition in an MS patients' population. *N*-Glc is a prokaryote-specific modification that is found in selected Gram-negative bacteria, where it is most commonly found on cell-surface proteins such as (autotransporter) adhesins, biosynthesized as part of the three-protein HMW cluster including the *N*-glucosyl transferase HMW1C. We demonstrated that anti-*N*-Glc peptide antibodies, easily detected by a Solid-Phase (SP)-ELISA [22,23], preferentially recognize hyperglucosylated adhesin of non-typeable *Haemophilus influenzae* (NTHi) particularly the C-terminal portion HMW1(1205-1526) termed HMW1ct. This was the first example of an N-glucosylated native antigen that can be considered a relevant candidate for triggering pathogenic antibodies in MS [24]. The protein HMW1ct is expressed as a mixture of three *N*-Glc variants containing 7, 8, and 9 Glc moieties on Asn residues inside the consensus sequence NX(S/T) in a 1:1:1 ratio. Since the NTHi cell-surface adhesins are widely glucosylated, the *N*-Glc residues are likely to be exposed conceptually in vivo in a multivalent shape, thus potentially favoring the emergence of a rough immunological response.

In the present study, we aimed to reproduce multivalent exposure of *N*-Glc epitopes to increase coating efficiency on the ELISA microplate for the detection of anti-*N*-Glc antibodies in MS, reminiscent of an early infection. First of all, by ELISA experiments, both competitive and in solid-phase, we defined the *N*-Glc epitopes (assuring the specificity and selectivity of autoantibody recognition), decreasing the length of the originally developed *N*-glucosylated β-turn synthetic antigenic probes. Then, the best efficiency of coating to the polystyrene ELISA plate was guaranteed, considering the concept of the multivalency to increase the antibody binding affinity. In particular, the selected short epitopes were conjugated to 4-branched dendrimeric lysine scaffolds creating Multiple *N*-Glucosylated Peptide Epitopes (*N*-Glc MEPs). Therefore, the novel *N*-Glc MEPs were developed with the aim not only to enhance the diagnostic performance of the assay but also the coating efficiency of the minimal epitopes to the polystyrene ELISA plate.

2. Materials and Methods

2.1. Synthesis of the N-Glc Peptides

N-Glc peptide epitopes **2–22** (Tables 1 and 2) were synthesized following the Fmoc/tBu manual strategy. Experimental details and analytical data of the synthetic molecules are reported in the Supplementary Materials (Tables S1 and S2).

Table 1. Shortened *N*-glucosylated (*N*-Glc) sequences of CSF114(Glc).

Peptide Name Number	*N*-Glucosylated Peptide Sequence
CSF114(Glc) (**1**)	TPRVERN(Glc)GHSVFLAPYGWMVK
[Asn^7(Glc)] CSF114(1–18) (**2**)	Ac-TPRVERN(Glc)GHSVFLAPYGW-NH$_2$
[Asn^7(Glc)] CSF114(1–16) (**3**)	Ac-TPRVERN(Glc)GHSVFLAPY-NH$_2$
[Asn^7(Glc)] CSF114(1–14) (**4**)	Ac-TPRVERN(Glc)GHSVFLA-NH$_2$
[Asn^7(Glc)] CSF114(2–13) (**5**)	Ac-PRVERN(Glc)GHSVFL-NH$_2$
[Asn^7(Glc)] CSF114(4–11) (**6**)	Ac-VERN(Glc)GHSV-NH$_2$
[Asn^7(Glc)] CSF114(5–10) (**7**)	Ac-ERN(Glc)GHS-NH$_2$
[Asn^7(Glc)] CSF114(6–9) (**8**)	Ac-RN(Glc)GH-NH$_2$

Table 2. *N*-Glc tri- and pentapeptides as synthetic antigens.

Glucosylated Core Peptide	Glucosylated Tripeptides	Glucosylated Pentapeptides
	Ac-N(Glc)GS-NH$_2$ (**9**);	Ac-ERN(Glc)GS-NH$_2$ (**16**);
	Ac-N(Glc)GT-NH$_2$ (**10**);	Ac-ERN(Glc)GT-NH$_2$ (**17**);
Selected NXT/S sequences	Ac-N(Glc)KS-NH$_2$ (**11**);	Ac-ERN(Glc)KS-NH$_2$ (**18**);
	Ac-N(Glc)KT-NH$_2$ (**12**).	Ac-ERN(Glc)KT-NH$_2$ (**19**)
	Ac-N(Glc)GH-NH$_2$ (**13**);	Ac-ERN(Glc)GH-NH$_2$ (**20**);
CSF114(Glc)	Ac-N(Glc)KH-NH$_2$ (**14**)	Ac-ERN(Glc)KH-NH$_2$ (**21**)
[Asn^{31}(Glc)]hMOG(30–50)	Ac-N(Glc)AT-NH$_2$ (**15**)	Ac-KGN(Glc)AT-NH$_2$ (**22**)

2.2. Synthesis of N-Glc Multiple Epitope Peptides (N-Glc MEPs)

N-Glc MEPs **23–26** (see Section 3.4) were synthesized following the protocol described in the Supplementary Materials. Analytical data are reported in Table S3.

2.3. Immunological Assays

Multiple Sclerosis (MS) patients' sera samples were collected in the Multiple Sclerosis Clinical Care and Research Centre, Department of Neurosciences, Reproductive Sciences and Odontostomatology, Federico II University (Naples, Italy). Sera samples were obtained for diagnostic purposes, from patients and healthy blood donors who had given their informed consent, and stored at −20 °C until use. The present study was conducted in accordance with the Declaration of Helsinki. All experimental protocols performed were approved by the Ethics Committee 2006 and 2017 (protocol n. 120/06 and 160/17, respectively). The MS group consisted of relapsing-remitting MS (RR-MS) patients after a diagnostic lumbar puncture, cerebrospinal fluid analysis, and MRI examination and fulfilled established international diagnostic criteria [25,26]. Blood samplings in the patients' group were performed during the routine follow-up study, while the healthy control samples were carried out during routine health checks or blood donations.

2.3.1. Inhibition ELISA

Ninety-six-well activated polystyrene ELISA plates (NUNC Maxisorb, Sigma Aldrich, Milano, Italy) were coated with 1 μg per 100 μL of the type I′ beta turn glucosylated peptide CSF114(Glc) or MEPs per well, in pure carbonate buffer 0.05 M (pH 9.6) and incubated at 4 °C overnight. Washing steps were executed with an automatic Hydroflex microplate washer (Tecan Italia, Milano, Italy). After five washes with washing buffer containing 0.9% NaCl and 0.05% Tween 20, nonspecific binding sites were blocked with 100 μL per well fetal calf serum (FCS) buffer solution (10% in washing buffer) at room temperature for 60 min.

Antibody affinity was measured following the inhibition methods reported elsewhere [27,28]. Semi-saturating sera dilution was calculated in preliminary titration curves (absorbance 0.7). Six different concentrations of each synthetic antigenic peptide probe were used as inhibitors. Then, sera samples at the selected dilution were incubated in parallel with increasing concentrations of the synthetic shortened peptide sequences (range 1×10^{-10} to 1×10^{-4}) for 60 min at room temperature. All inhibition experiments were performed in duplicate or triplicate for each single MS patient's serum positive to CSF114(Glc) separately. All experiments were repeated at least twice on two different working days.

After three washes, uninhibited antibodies were identified by adding 100 μL/well of alkaline phosphatase-conjugated anti-human immunoglobulin G (IgG, Sigma-Aldrich, Milano, Italy) diluted 1:8000 in washing buffer containing 10% FCS. The microplates were then incubated 3 h at room temperature and, after three washes, 100 μL of substrate solution consisting of 1 mg/mL *p*-nitrophenyl phosphate (Sigma-Aldrich, Milano, Italy) in 10% diethanolamine buffer (pH 9.8) were added. After approximately 30 min, the reaction was stopped with 1M NaOH solution (50 μL/well), and the absorbance was read in a multichannel ELISA reader (Tecan Sunrise, Männedorf, Switzerland) at 405 nm. The selected ELISA microplates, coating conditions, reagent dilutions, buffers, and incubation times were previously tested [24,29]. The relationship between peptide concentrations and the absorbance values was represented graphically in absorbance inhibition percentage, and half-maximal response concentration values (IC50) were calculated.

2.3.2. Solid-Phase ELISA (SP-ELISA)

Immunoassays, to detect IgM or IgG antibodies in sera, were performed by SP-ELISA coating the synthetic peptides on 96-well plates (Nunc Maxisorp, Sigma−Aldrich, Milano, Italy).

Polystyrene 96-well ELISA plates were coated with a 10 μg/mL solution of diluted synthetic peptides, independently, in pure carbonate buffer 0.05 M (pH 9.6). After overnight incubation at

4 °C, plates were washed three times using washing buffer. Nonspecific binding sites were blocked with 100 µL/well of fetal calf serum buffer (10% FCS in washing buffer) at room temperature for 1 h. FCS buffer was removed and plates were incubated overnight with sera (diluted 1:100 in FCS buffer, 100 µL/well) at 4 °C. After three washes, plates were treated with 100 µL/well of anti-human IgG or IgM alkaline phosphatase-conjugated specific antibodies diluted in FBS buffer (1:8000 and 1:200, respectively). After 3 h of incubation at room temperature and three washes, 100 µL of substrate buffer was added to each well. After 15–30 min incubation at room temperature, the absorbance of each plate was read in a multichannel ELISA reader at 405 nm. The antibody levels are expressed as absorbance in arbitrary units at 405 nm.

2.4. Statistical Analysis

Data are expressed as measured absorbance values at 405 nm calculated as the mean ± SD. Statistical analysis was performed using the software GraphPad Prism version 6.01 (Graphpad Software Inc., La Jolla, CA, USA). Descriptive statistics was used to calculate mean and standard deviation for continuous variables and percentage for categorical variables. Mann–Whitney U tests were used to compare antibody response distributions. Differences were deemed statistically significant when p value < 0.05 (two-tailed test). Non-parametric Spearman's rho and related 95% confidence intervals were used to assess correlation between pair of tests. A p value < 0.05 (two-tailed test) was considered as significant. Receiver Operating Characteristic (ROC) curve analysis was employed to calculate cut-off values, establishing sensitivities and sensibilities.

3. Results

In previous studies, we reported the cross-reactivity between the *N*-glucosylated adhesin antigen HMW1ct-Glc and anti-CSF114(*N*-Glc) IgG antibodies in MS patients' sera by competitive ELISA [25]. The HMW1C from NTHi is one of the first examples of soluble bacterial protein glycosyltransferases capable of performing *N*-glycosylation with simple hexoses (i.e., glucose) on asparagine residues in conserved Asn-Xaa-Ser/Thr motifs [29,30]. On the other hand, CSF114(Glc) is a structure-based designed 21-mer peptide (TPRVERN(Glc)GHSVFLAPYGWMVK) that was optimized as a type I' beta-turn structure because of its ability to expose at the best, at position 7, the minimal, but fundamental moiety Asn(*N*-Glc), in the epitope for autoantibody recognition in the solid-phase conditions of the immunoenzymatic assay [31].

Starting from this assumption, preliminary experiments were performed to select the shortest peptide sequences corresponding to the epitope(s) to be presented in multiple copies in fully characterized multivalent Multiple Epitope Peptides (MEPs) to detect antibodies in MS patients' sera by SP-ELISA. The use of MEPs for specific antibody detection can pave the way for the development of a simple tool to identify immune responses to aberrant glycosylations, such as *N*-glucosylation in MS possibly linked to an early non-typeable *Haemophilus influenzae* bacterial infection.

3.1. Antibody Detection in Solid-Phase ELISA (SP-ELISA) Is Affected by the Length of the Peptide Antigen

First of all, we investigated the influence of the length of shortened peptide sequences of the synthetic antigenic probe CSF114(Glc) on the efficiency both in antibody recognition and on the coating in the SP-ELISA. At this purpose, the glucopeptide sequence was tightened down step-by-step and the shortened peptides **2–5** (Table 1) derived from CSF114(Glc) were synthesized (as described in the Supplementary Materials) and characterized using analytical Reverse-Phase High Performance Liquid Chromatography (RP-HPLC) and ElectroSpray Ionisation Mass Spectrometry (ESI-MS) (Table S1). The shortened peptide sequences were acetylated at the *N*-terminus and amides at the *C*-terminus, in order to dislodge free terminal charges, which are not present in the native protein sequence and might hamper the antibody recognition [32].

SP-ELISA, against pools of positive MS sera and healthy controls (Figure 1), clearly showed a progressive decrease in antibody titre in parallel with the decrease in peptide length. In fact, the 12-mer glucopeptide **5** completely lost its ability to identify antibodies in SP-ELISA, possibly because of an inefficient coating. Therefore 14 residues appeared to be the minimum length necessary for antibody identification in SP-ELISA, probably because shorter immobilized glucopeptide sequences (<14) are only partially prone to expose the epitope for antibody binding.

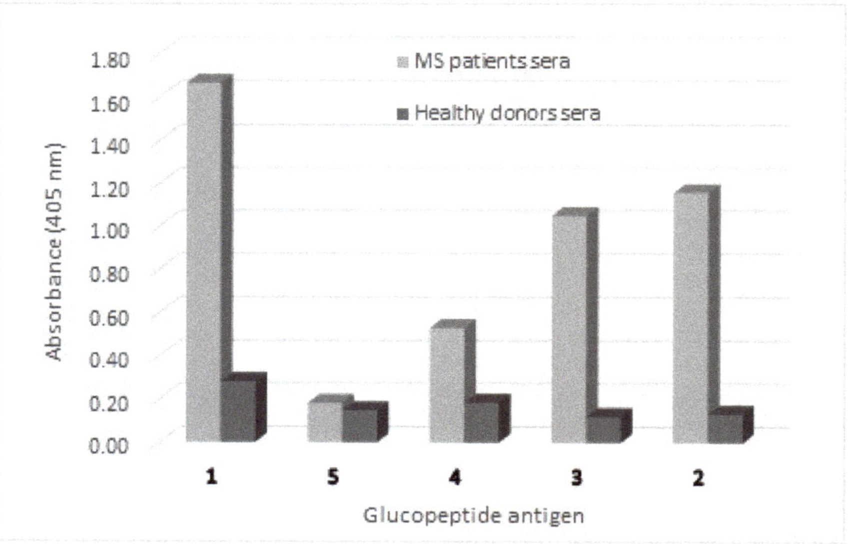

Figure 1. Antibody titres (expressed as Absorbance values) in Multiple Sclerosis (MS) patients' and healthy blood donors' sera against the glucopeptide antigens CSF114(Glc) (**1**), [Asn^7(Glc)]CSF114(1–18) (**2**), [Asn^7(Glc)]CSF114(1–16) (**3**), [Asn^7(Glc)]CSF114(1–14) (**4**), and [Asn^6(Glc)]CSF114(2–13) (**5**).

3.2. Antibody Affinity of Shortened CSF114(Glc) Glucopeptide Sequences in Competitive ELISA: Shortening CSF114(Glc) Does not Affect Antibody Epitope Recognition

Considering that SP-ELISA allows to evaluate fundamentally the relative antibody affinity that depends on the exposure of the peptide in the solid-phase conditions of the assay, we also investigated the absolute antibody affinity against the CSF114(Glc) shortened sequences by a competitive ELISA. For this purpose, the *N*-glucosylated β-turn peptide structure CSF114(Glc) was tightened down again step-by-step with the aim to identify the critical epitope displaying optimal antibody binding reactivity in the competitive in vitro assay.

Further, we investigated the role of the amino acids surrounding the previously identified minimal epitope Asn(*N*-Glc) [24]. In particular, we developed the synthetic shorter sequences **6–8** that were tested in competitive ELISA in parallel to **2–5** (Table 1).

All the synthetic glucopeptides were used as inhibitors in competitive ELISA using anti-CSF114(Glc) antibodies in one representative positive MS patient's serum. The IC50 s were calculated applying the non-linear regression least squares (ordinary) fit to the experimental data for each peptide, and values are summarized in the Supplementary Materials (Table S4). As shown in Figure 2, the full length CSF114(Glc) showed the highest degree of binding affinity (IC50 = 0.009 µM, Table S4 in the Supplementary Materials). Despite 14 residues are the minimum for antibody recognition in SP-ELISA, the inhibitory activity is still optimal when the epitope region is included in ca. 11 amino acid residues as **5** and **4** display (IC50 = 0.035 µM and 0.014 µM, respectively). On the other hand, the 4-mer glucopeptide **8** and the 6-mer glucopeptide **7** displayed the lowest inhibitory potency (IC50 = 2.2 µM and 3.5 µM, respectively), but, in any case, all are able to inhibit antibodies.

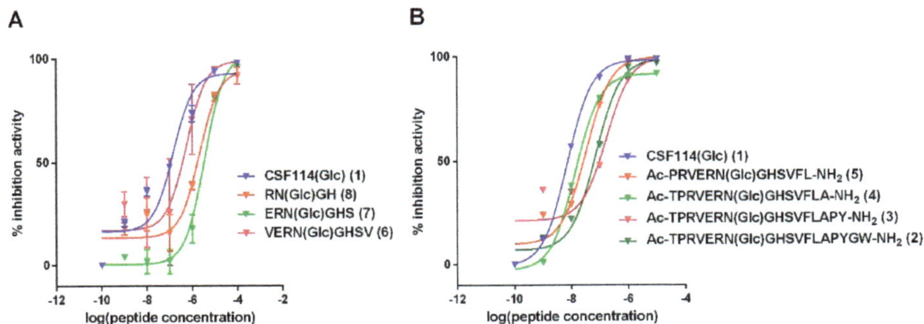

Figure 2. Inhibition curves: (**A**) Inhibition curve of anti-CSF114(Glc) IgG antibodies from a representative MS serum with *N*-glucosylated peptides, [Asn^7(Glc)]CSF114(4–11) (**6**), [Asn^7(Glc)]CSF114(5–10) (**7**), and [Asn^7(Glc)]CSF114(6–9) (**8**) in comparison with the *N*-glucosylated peptide CSF114(Glc) (**1**) in a competitive Enzyme-Linked ImmunoSorbent Assay (ELISA). (**B**) Inhibition curve of anti-CSF114(Glc) IgG antibodies from a representative MS serum with glucopeptides: [Asn^7(Glc)]CSF114 (1–18) (**2**), [Asn^7(Glc)]CSF114 (1–16) (**3**), [Asn^7(Glc)]CSF114 (1–14) (**4**), [Asn^7(Glc)]CSF114 (2–13) (**5**) in comparison with the glucopeptide CSF114(Glc) in a competitive ELISA. The results are expressed as percentage of inhibition activity of representative MS serum (ordinate axis) versus the peptide concentrations (M) in logarithmical scale.

3.3. Study of the Possible Role of a Consensus Sequence Surrounding N(Glc) in Antibody Recognition

Taking into consideration that shorter glucopeptides **6** and **8** displayed antibody affinity in competitive ELISA, we decided to investigate a series of *N*-Glc tri- and pentapeptides as synthetic antigens, undertaking a deductive approach to rule out *N*-glycosylation consensus sequences (sequons).

As a proof of concept, among the possible combinations of tripeptide sequons NX(T/S), we selected the ones containing X = Gly or Lys. In particular, we synthesized *N*(Glc)GS (**9**), *N*(Glc)GT (**10**), *N*(Glc)KS (**11**), and *N*(Glc)KT (**12**). Moreover, we synthesized *N*(Glc)GH (**13**) and *N*(Glc)KH (**14**), derived from the original glucosylated core of CSF114(Glc), that did not correspond to a consensus sequence. Then, we included the sequon *N*(Glc)AT (**15**), present both in the human myelin oligodendrocyte glycoprotein (MOG) sequence [Asn^{31}(Glc)]hMOG(30–50) (an antigenic immunodominant epitope recognizing also antibodies in MS) [33], but also highly represented (3 out of 12 positions) in the C-terminal portion HMW1(1205–1526), i.e., HMW1ct hyperglucosylated adhesin of non-typeable *Haemophilus influenzae* (NTHi) (N at positions 3, 7, and 9) [25] together with *N*(Glc)GS (**9**). Moreover, considering that a tripeptide sequence is probably not long enough for a good antigen-antibody interaction, the following step in the identification of the epitope was to test the pentapeptides **16–22**, including the sequons. At this purpose the two amino acids placed in positions –B2 and –B1 in the CSF114(Glc) sequence were added to allow the spatial hint in the shape of a β-turn, with Asn(Glc) on the tip. The same idea was applied to the sequon *N*(Glc)AT (**15**), inserting amino acids in positions 29 and 30 of hMOG(30–50) sequence, originally containing Asn(Glc) at position 31 [33].

Therefore, glucosylated tri- and pentapeptides **9–22** (Table 2) were synthesized, purified, and characterized as described in detail in the Supplementary Materials.

The antibody affinity of tripeptides **9–15** and pentapeptides **16–22** was evaluated by competitive ELISA compared to the original *N*-glucosylated β-turn peptide sequence (Figure 3). CSF114(Glc) was coated onto the ELISA microplate and patients' sera were incubated with different shortened peptide concentrations. IC50 values were reported in the Supplementary Materials (Table S5).

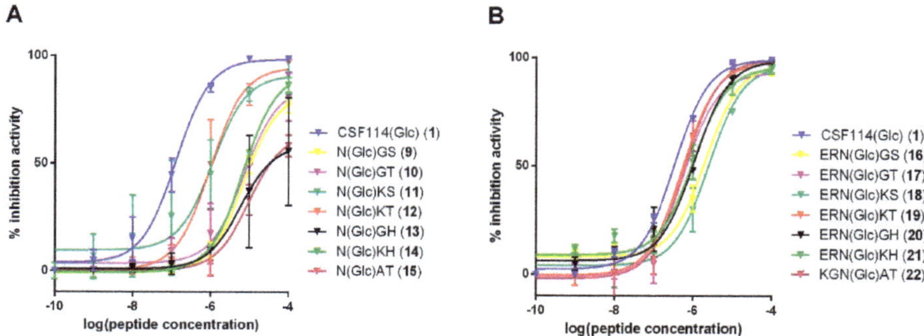

Figure 3. Inhibition curves: (**A**) Inhibition curve of anti-CSF114(Glc) IgG antibodies from a representative MS serum with the tripeptides **9–15**: Ac-N(Glc)GS-NH$_2$ (**9**), Ac-N(Glc)GT-NH$_2$ (**10**), Ac-N(Glc)KS-NH$_2$ (**11**), Ac-N(Glc)KT-NH$_2$ (**12**), Ac-N(Glc)GH-NH$_2$ (**13**) Ac-N(Glc)KH-NH$_2$ (**14**), Ac-N(Glc)AT-NH$_2$ (**15**), in comparison with the glucopeptide CSF114(Glc) (**1**) in a competitive ELISA. (**B**) Inhibition curve of anti-CSF114(Glc) IgG antibodies from a representative MS serum with the pentapeptides **16–22**: Ac-ERN(Glc)GS-NH$_2$ (**16**), Ac-ERN(Glc)GT-NH$_2$ (**17**), Ac-ERN(Glc)KS-NH$_2$ (**18**); Ac-ERN(Glc)KT-NH$_2$ (**19**), Ac-ERN(Glc)GH-NH$_2$ (**20**), Ac-ERN(Glc)KH-NH$_2$ (**21**); Ac-N(Glc)AT-NH$_2$ (**22**), in comparison with the glucopeptide CSF114(Glc) in a competitive ELISA. The results are expressed as the percentage of inhibition activity of a representative MS serum (ordinate axis) versus the peptide concentrations (M) in logarithmical scale.

Inhibition curves of the tripeptides **9–15** and pentapeptides **16–22** are shown in Figure 3. The original sequence of CSF114(Glc) presented the best inhibitory activity (IC50 = 0.34 µM, Table S5). The shortest tripeptide probes **9–15** recognized anti-CSF114(Glc) antibodies even if with an affinity lower than the synthetic glucosylated 21mer-peptide (Figure 1). Among all the tripeptides tested, the glucopeptides Ac-N(Glc)KS-NH$_2$ (**11**) and Ac-N(Glc)KT-NH$_2$ (**12**) showed the best inhibitory activity (IC50 = 1.1 and 0.97 µM, respectively, Table S5).

The longer linear sequences **16–22** exhibited high inhibition efficacy in competitive ELISA, better than their corresponding tripeptide analogs **9–15**. In particular, the pentapeptide Ac-ERN(Glc)KT-NH$_2$ (**19**) presented the best inhibitory activity (IC50 = 0.54 µM, Table 3) followed by the pentapeptides Ac-ERN(Glc)GT-NH$_2$ (**17**) (IC50 = 0.58 µM) and Ac-KGN(Glc)AT-NH$_2$ (**22**) (IC50 = 0.60 µM). This result indicates that the IC50 values for *N*-Glc pentapeptides are closely similar, and no substantial differences were observed among the pentapeptides tested, independently of the sequence.

Table 3. Receiver Operating Characteristic (ROC) analysis. Values obtained from ROC-analysis of *N*-Glc MEPs **23–26** and the glycopeptide CSF114(Glc) (**1**) for the area under the curve, *p* value, established cut-off and the corresponding sensitivity, specificity, and likelihood ratio.

Compound	Area Under Curve (AUC)				Criterion Value and Coordinates			
	AUC	Standard Error (SE)	95% Confidence Interval	*p*-Value	Cut-Off	Sensitivity (%)	Specificity (%)	Likelihood Ratio
N-Glc MEP (**23**)	0.5747	0.1073	0.3643 to 0.7850	0.4899	>1.057	17.65 (3.799 to 43.43)	92.31 (63.97 to 99.81)	2.294
N-Glc MEP (**24**)	0.5249	0.1100	0.3092 to 0.7406	0.8180	>0.6104	35.29 (14.21 to 61.67)	92.31 (63.97 to 99.81)	4.588
N-Glc MEP (**25**)	0.7511	0.09376	0.5673 to 0.9350	0.02023	>1.053	38.46 (13.86 to 68.42)	82.35 (56.57 to 96.20)	2.179
N-Glc MEP (**26**)	0.6380	0.1041	0.4339 to 0.8421	0.2018	>1.029	30.77 (9.092 to 61.43)	82.35 (56.57 to 96.20)	1.744
CSF114(Glc) (**1**)	0.6516	0.1049	0.4458 to 0.8573	0.1610	>0.7575	47.06 (22.98 to 72.19)	92.31 (63.97 to 99.81)	6.118

3.4. N-Glc Multiple Epitope Peptides (N-Glc MEPs) to Mimic Multivalency in SP-ELISA

In order to mimic a multivalent presentation of the minimal but fundamental epitope Asn(*N*-Glc), and increasing at the same time the efficiency in coating to the ELISA polystyrene plate of short linear peptides, we synthesized different *N*-Glc MEPs based on a dendrimeric lysine scaffold, having in common Asn(Glc) [34].

Starting from a lysine tetrameric core, we synthesized five *N*-Glc MEPs **23–26** bearing spacers of different length: β-Alanine and two different polyethylene glycol-based spacers (PEG), containing 9- and 19-atoms chain, respectively. The following antigens were selected: (i) the β-turn *N*-glucosylated pentapeptide ERN(Glc)GH (**12**) and (ii) the simple minimal *N*-glucosylated asparagine epitope Asn(Glc). The spacers were introduced to avoid steric hindrance possibly hampering antibody binding. The *N*-Glc MEP structures are reported in Figure 4.

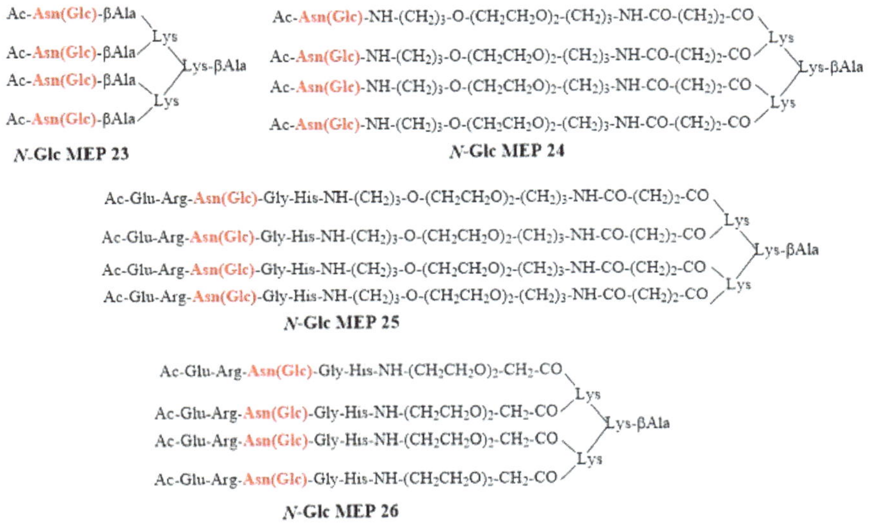

Ac-Asn(Glc)-βAla, Ac-Asn(Glc)-βAla → Lys; Ac-Asn(Glc)-βAla, Ac-Asn(Glc)-βAla → Lys; Lys-βAla

N-Glc MEP 23

Ac-Asn(Glc)-NH-(CH$_2$)$_3$-O-(CH$_2$CH$_2$O)$_2$-(CH$_2$)$_3$-NH-CO-(CH$_2$)$_2$-CO, Ac-Asn(Glc)-NH-(CH$_2$)$_3$-O-(CH$_2$CH$_2$O)$_2$-(CH$_2$)$_3$-NH-CO-(CH$_2$)$_2$-CO → Lys; Ac-Asn(Glc)-NH-(CH$_2$)$_3$-O-(CH$_2$CH$_2$O)$_2$-(CH$_2$)$_3$-NH-CO-(CH$_2$)$_2$-CO, Ac-Asn(Glc)-NH-(CH$_2$)$_3$-O-(CH$_2$CH$_2$O)$_2$-(CH$_2$)$_3$-NH-CO-(CH$_2$)$_2$-CO → Lys; Lys-βAla

N-Glc MEP 24

Ac-Glu-Arg-Asn(Glc)-Gly-His-NH-(CH$_2$)$_3$-O-(CH$_2$CH$_2$O)$_2$-(CH$_2$)$_3$-NH-CO-(CH$_2$)$_2$-CO, Ac-Glu-Arg-Asn(Glc)-Gly-His-NH-(CH$_2$)$_3$-O-(CH$_2$CH$_2$O)$_2$-(CH$_2$)$_3$-NH-CO-(CH$_2$)$_2$-CO → Lys; Ac-Glu-Arg-Asn(Glc)-Gly-His-NH-(CH$_2$)$_3$-O-(CH$_2$CH$_2$O)$_2$-(CH$_2$)$_3$-NH-CO-(CH$_2$)$_2$-CO, Ac-Glu-Arg-Asn(Glc)-Gly-His-NH-(CH$_2$)$_3$-O-(CH$_2$CH$_2$O)$_2$-(CH$_2$)$_3$-NH-CO-(CH$_2$)$_2$-CO → Lys; Lys-βAla

N-Glc MEP 25

Ac-Glu-Arg-Asn(Glc)-Gly-His-NH-(CH$_2$CH$_2$O)$_2$-CH$_2$-CO, Ac-Glu-Arg-Asn(Glc)-Gly-His-NH-(CH$_2$CH$_2$O)$_2$-CH$_2$-CO → Lys; Ac-Glu-Arg-Asn(Glc)-Gly-His-NH-(CH$_2$CH$_2$O)$_2$-CH$_2$-CO, Ac-Glu-Arg-Asn(Glc)-Gly-His-NH-(CH$_2$CH$_2$O)$_2$-CH$_2$-CO → Lys; Lys-βAla

N-Glc MEP 26

Figure 4. Collection of the synthetic *N*-Glc Multiple Epitope Peptides (MEPs).

The *N*-Glc MEPs were synthesized starting from Fmoc$_4$-Lys$_2$-Lys-β-Ala-Wang resin, according to the general procedure described in details in the Supplementary Materials. The Fmoc-Asn(βGlcAc$_4$)-OH building block was synthesized according to the method developed by Paolini et al. [35]. *N*-Glc MEPs were acetylated on the *N*-terminal function. All the *N*-Glc MEPs were purified by semi-preparative RP-HPLC, and characterized by RP-HPLC and ESI-MS (Table S3).

We assessed IgG isotype antibodies against *N*-Glc MEPs **23–26** in a cohort of 16 MS patients, previously selected on their reactivity to the original *N*-glucosylated type I' β-turn peptide structure CSF114(Glc), and compared to 14 healthy blood donors as controls (Figure 5).

The Receiver Operating Characteristic (ROC) analysis was employed to perform an accurate comparative investigation of the performances of the different *N*-Glc MEPs, evaluating their discrimination power at the different cut-off values. Sensitivity, specificity, and likelihood ratios were also calculated [36]. Selected cut-off values, sensitivities, and the rest of the statistical parameters are shown in Table 3 (ROC curves are reported in Figure S1 in the Supplementary Materials). The characteristics of the curves, their shape and steepness, and their underlying area provide evidence that the synthetic *N*-Glc MEPs are able to identify antibodies in MS patients' sera.

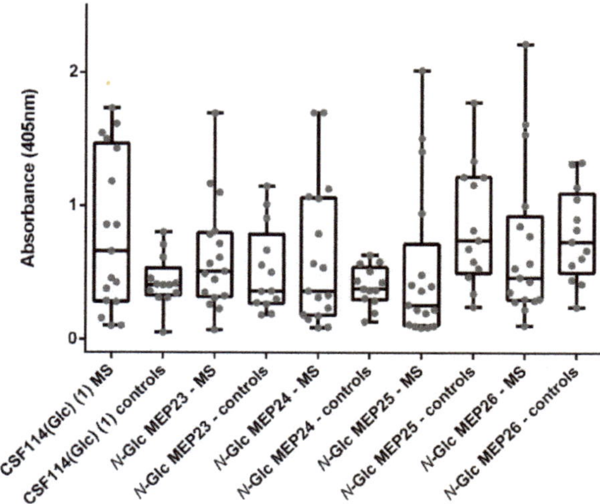

Figure 5. Antibody responses (expressed as Absorbance values) against the synthetic *N*-Glc MEPs. Data distribution of IgG antibodies against the glucopeptide CSF114(Glc) and the synthetic *N*-Glc MEPs **23–26** obtained by SP-ELISA (grey). Box and whiskers (min to max) are plotted for each data group (black).

Despite *N*-Glc MEP **25** and **26** presented the best Area Under the Curve (AUC) values (0.7511 and 0.6380, respectively), observing the data distribution we assume that their ability to identify specific MS antibodies is lower compared with *N*-Glc MEPs **23** and **24**. This is justified with the increased means of the antibody titers in controls compared to MS patients in both *N*-Glc MEP **25** and **26**, which apparently improved AUC and p values, but are due to non-specific interactions. In fact, the specificities calculated for *N*-Glc MEPs **25** and **26** are sensibly decreased compared with CSF114(Glc) (82.35% and 92.31%, respectively). Considering the sensitivity values, once the specificity is fixed in 92.31%, the *N*-Glc MEP **24** presented the best performance identifying as positive the 35.29% of the MS patients.

Consequently, we decided to deepen on *N*-Glc MEP **24** ability (containing only *N*-Glc moieties) to recognize specific MS antibodies in patients' sera. At this purpose, *N*-Glc MEP **24** was tested against an increased number of MS patients' sera (81) and healthy controls' (30) sera, in parallel, to the glucopeptide CSF114(Glc) as reference. Data distribution of IgG and IgM antibody titers identified by SP-ELISA are reported in Figure 6A,B.

The non-parametric Mann–Whitney test was applied to evaluate significant differences between MS patients' and controls' groups. The results showed significant differences (p value < 0.001, two-tailed test) for CSF114(Glc), both for IgM and IgG-type antibodies. Slightly different results were observed in the case of *N*-Glc MEP **24**. In fact, the differences among groups were statistically significant when IgM-isotype antibodies were detected (p value = 0.0001, two-tailed test). On the other hand, the IgGs against the multivalent *N*-Glc MEP **24** showed no level of significance (p value = 0.3611, Two-tailed test). In our opinion, the IgG antibody response as detected by *N*-Glc MEP **24** appears to be less specific because it detects a "noise" level in the control group, forcing a decreased specificity to maintain sensibility when selecting the corresponding cut-off.

Among the absorbance values of MS patients, the frequencies of the anti-CSF114(Glc) antibodies significantly correlated with the ones against the *N*-Glc MEP **24** (p value < 0.0001, two-tailored); the Spearman's correlation coefficients (rho values) were r = 0.7507 and 0.7424 for IgG and IgM, respectively (Figure 6C,D).

Then, we investigated the absolute antibody affinity of *N*-Glc MEP **24** in a competitive ELISA. In a set of three MS positive sera tested in parallel, the multivalent *N*-Glc MEP **24** inhibited the binding of antibodies to the glycopeptide CSF114(Glc), giving rise to contrasting inhibition curves among the different representative sera employed (Figure S2 in the Supplementary Materials). Data of serum MS1 (Figure S2A in the Supplementary Materials) showed that the affinity of *N*-Glc MEP **24** was lower than CSF114(Glc) (IC50 = 2.145×10^{-8} M and 5.200×10^{-7} M, respectively), whereas serum MS2 exhibited superimposable affinity (IC50 = 6.373×10^{-8} M and 6.088×10^{-8} M respectively). Moreover, in MS3 serum IC50 was lower for *N*-Glc MEP **24** compared to CSF114(Glc) (IC50 = 2.145×10^{-8} M and 5.116×10^{-9} M, respectively). This finding indicates that the *N*-Glc MEP **24** shares similar epitopes, all including the Asn(*N*-Glc) residue. In particular, its antibody affinity can be slightly different among the MS patients, probably because of the differential innate and adaptive immune responses typical of each subject.

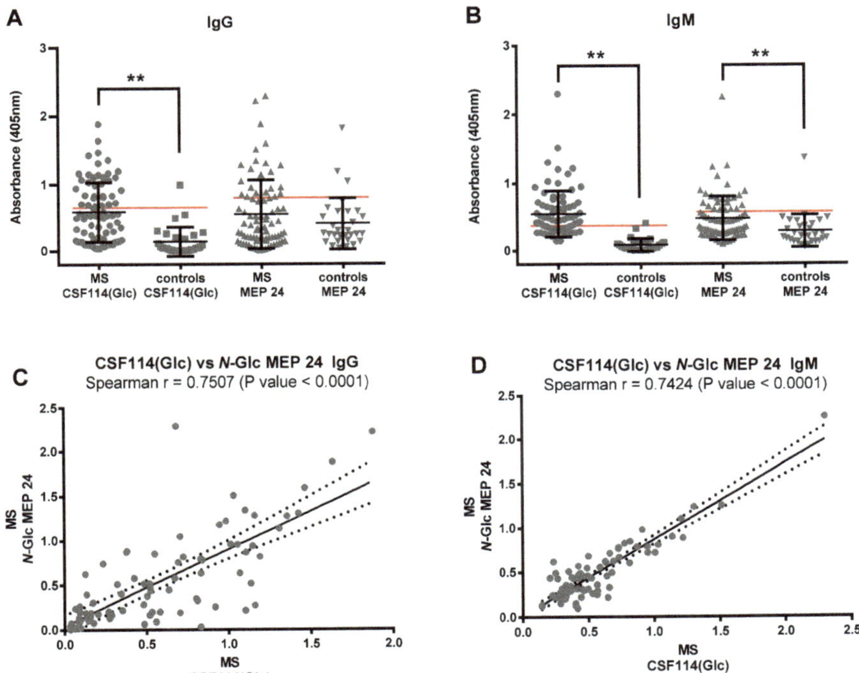

Figure 6. Data of antibody responses against the glucopeptide CSF114(Glc) and *N*-Glc MEP **24**. Data distribution of IgG (**A**) and IgM (**B**) antibody titers in 81 MS patients' sera and 30 controls' sera identified by SP-ELISA against the peptide sequences CSF114(Glc) and the synthetic multivalent *N*-Glc MEP **24**. Data are expressed as mean values ± standard deviation with a significance level ** $p < 0.001$ (two-tailed Mann–Whitney non-parametric test). Selected cut-off values for each compound are plotted in red. Correlation between CSF114(Glc) and *N*-Glc MEP **24** both for IgG (**C**) and for IgM (**D**) are shown. The Spearman's correlation coefficients (rho values) and the corresponding *p* values are reported in each plot. Regressions lines are plotted in black (dashed lines show the 95% confidence interval of the best-fit line).

4. Discussion

Multiple sclerosis diagnosis is still very challenging, relying on clinical and radiological criteria and in the absence of "better explanations" [37], the development of simple diagnostics detecting specific biomarkers is highly warranted. Moreover, native structures triggering specific antibodies in Multiple Sclerosis (MS) are still uncharacterized. Consequently, surrogate antigens used to identify antibodies in MS by ELISA are elusive. In spite of the fact that MS is considered mainly a T-cell mediated disease, the role of B-cells is increasingly appreciated. In this scenario, we demonstrated for the first time that an aberrant *N*-glucosylation is part of a relevant epitope that was identified by the structure-based designed β-turn 21-mer glucopeptide CSF114(Glc). This synthetic tool was instrumental for the discovery of antibodies in an MS patients' population preferentially recognizing the hyperglucosylated bacterial adhesin of non-typeable *Haemophilus influenzae*. With the idea in mind that multivalent presentation of glucosylated asparagine residues may occur in a variety of native antigens, as in the case of citrullination in rheumatoid arthritis [38], we focused on the development of a synthetic tool enhancing the role of multiple aberrant modifications versus amino acid sequences. We simplified the antigen to be synthetically produced in a multiple format, taking into consideration that peptide dendrimers are considered protein-like multivalent materials, whose architecture is a key parameter for activity [39,40]. Therefore, we selected the multivalent epitope peptide *N*-Glc MEP **24**, based on a lysine-dendritic scaffold (relatively simple to be produced), carrying four copies of the minimal glucosylated epitope Asn(Glc) anchored to a PEG-based spacer containing a 19-atoms chain [41]. In previous studies, we demonstrated that CSF114(Glc) shortened epitopes lost specificity and sensitivity for IgM antibodies in MS patients' sera by SP-ELISA [29], sensibly decreasing the diagnostic potential of the synthetic antigen. On the contrary, *N*-Glc MEP **24** displays an interesting ability to identify in SP-ELISA both IgG and IgM antibodies in a large number of MS patients' sera (81) compared to controls (30), with a sensitivity of 35% (95% CI: 14.21–61.67) and a specificity of 92.31 (95% CI: 63.97–99.81). This result is particularly relevant, since it offers the possibility to detect with the same tool both IgMs and IgGs in MS, thus increasing the diagnostic and prognostic value of the antigen thanks to multivalent presentation of the minimal epitope [24].

5. Conclusions

The present study reported the development of new *N*-glucosylated Multivalent Epitope Peptides for the detection of anti-*N*(Glc) antibodies in Multiple Sclerosis. After a preliminary screening of several short linear peptide epitopes and multiple epitope peptides, the promising results suggest that multivalent interaction can be useful in designing SP-ELISA to detect antibodies to aberrant *N*-glucosylation in Multiple Sclerosis for both diagnostic and prognostic purposes. Indeed, the multivalent *N*-Glc MEP **24** ligand, carrying the minimal glucosylated epitope Asn (Glc) anchored to the 19-atoms PEG-based spacer, can be an efficient probe to reveal both IgM and IgG autoantibodies as disease biomarkers.

Supplementary Materials: The following are available online at http://www.mdpi.com/2076-3425/10/7/453/s1. Details of the synthetic procedures, analytical and immunochemical characterization of the synthetic antigenic probes are reported online. Table S1: Shortened CSF114(Glc)-sequences of peptides **2–8.**, Table S2: Sequences and chemical data for the *N*-glucosylated (*N*-Glc) peptides **9–22**, Table S3: Analytical data of the Multiple *N*-glucosylated Peptide Epitopes (*N*-Glc MEPs) **23–26**, Table S4: Calculated half maximal inhibitory concentration (IC50). Calculated Log 1/IC50 with the Std errors and the IC50 with the Corresponding Confidence Interval (CI) for the shortened peptides versus CSF114(Glc), Table S5: Calculated Log 1/IC50 with the Std errors and the IC50 with the Corresponding Confidence Interval (CI) for the Peptides **9–22** and CSF114(Glc) (**1**), Figure S1: Received Operating Characteristic (ROC) analysis. ROC curve analysis of anti-CSF114(Glc) antibodies and anti-*N*-Glc MEPs **23–26** in Multiple Sclerosis versus controls determined by SP-ELISA, Figure S2: Competitive ELISA experiments with CSF114(Glc) and *N*-Glc MEP 24. Inhibition curves of anti-CSF114(Glc) antibodies with *N*-Glc MEP **24** compared with CSF114(Glc) in a competitive ELISA. The results are expressed as percentage of inhibition (ordinate axis) of three representative MS sera: MS1 (A), MS2 (B) and MS3 (C) versus the peptide concentrations (M) in logarithmical scale.

Author Contributions: F.R.F. and E.P. performed the immunoassays; I.P., B.M., M.D.P., C.T., F.N., F.R.F., G.S. performed the syntheses of the synthetic antigenic probes and glucosylated building blocks; M.P., R.L., and V.B.M. provided sera samples and oversaw the data collection; F.L., F.R.F. and E.P. performed data analyses of immunoassays; A.M.P. and P.R. conceived and designed the research project, supervised the research, reviewed and edited the final manuscript. All authors have read and agreed to the published version of the manuscript.

Funding: Authors gratefully acknowledge the Italian Multiple Sclerosis Foundation (FISM) (grant n. 2017/R/5). Moreover, Fondazione Ente Cassa di Risparmio di Firenze (Italy) (grant n. 2014/0306) is gratefully acknowledged for supporting the equipment to PeptLab of the University of Florence.

Acknowledgments: The authors would like to acknowledge Francis Ciolli and Giada Rossi for their technical assistance.

Conflicts of Interest: The authors declare no conflict of interest.

References

1. de Seze, J.; Vermersch, P. Sequential magnetic resonance imaging follow-up of multiple sclerosis before the clinical phase. *Mult. Scler.* **2005**, *11*, 395–397. [CrossRef]
2. Lanzillo, R.; Prinster, A.; Scarano, V.; Liuzzi, R.; Coppola, G.; Florio, C.; Salvatore, E.; Schiavone, V.; Brunetti, A.; Muto, M.; et al. Neuropsychological assessment, quantitative MRI and ApoE gene polymorphisms in a series of MS patients treated with IFN beta-1b. *J. Neurol. Sci.* **2006**, *245*, 141–145. [CrossRef] [PubMed]
3. Freedman, M.S.; Thompson, E.J.; Deisenhammer, F.; Giovannoni, G.; Grimsley, G.; Keir, G.; Ohman, S.; Racke, M.K.; Sharief, M.; Sindic, C.J.M.; et al. Recommended standard of cerebrospinal fluid analysis in the diagnosis of multiple sclerosis: A consensus statement. *Arch. Neurol.* **2005**, *62*, 865–870. [CrossRef] [PubMed]
4. Luque, F.A.; Jaffe, S.L. Cerebrospinal fluid analysis in multiple sclerosis. *Int. Rev. Neurobiol.* **2007**, *79*, 341–356. [PubMed]
5. Reindl, M.; Jarius, S.; Rostasy, K.; Berger, T. Myelin oligodendrocyte glycoprotein antibodies: How clinically useful are they? *Curr. Opin. Neurol.* **2017**, *30*, 295–301. [CrossRef] [PubMed]
6. Waters, P.J.; Komorowski, L.; Woodhall, M.; Lederer, S.; Majed, M.; Fryer, J.; Mills, J.; Flanagan, E.P.; Irani, S.R.; Kunchok, A.C. A multicenter comparison of MOG-IgG cell-based assays. *Neurology* **2019**, *92*, e1250–e1255. [CrossRef]
7. Papp, V.; Langkilde, A.R.; Blinkenberg, M.; Schreiber, K.; Jensen, P.E.H.; Sellebjerg, F. Clinical utility of anti-MOG antibody testing in a Danish cohort. *Mult. Scler. Relat. Disord.* **2018**, *26*, 61–67. [CrossRef]
8. de Seze, J. Myelin oligodendrocyte glycoprotein antibodies in neuromyelitis optica spectrum disorder. *Curr. Opin. Neurol.* **2019**, *32*, 111–114.
9. Meloen, R.H.; Puijk, W.C.; Langeveld, J.P.; Langedijk, J.P.; Timmerman, P. Design of synthetic peptides for diagnostics. *Curr. Protein Pept. Sci.* **2003**, *4*, 253–260. [CrossRef]
10. Papini, A.M. The use of post-translationally modified peptides for detection of biomarkers of immune-mediated diseases. *J. Pept. Sci.* **2009**, *15*, 621–628. [CrossRef]
11. Ballew, J.T.; Murray, J.A.; Collin, P.; Mäki, M.; Kagnoff, M.F.; Kaukinen, K.; Daugherty, P.S. Antibody biomarker discovery through in vitro directed evolution of consensus recognition epitopes. *Proc. Natl. Acad. Sci. USA* **2013**, *110*, 19330–19335. [CrossRef] [PubMed]
12. Gómara, M.J.; Fernández, L.; Pérez, T.; Ercilla, G.; Haro, I. Assessment of synthetic chimeric multiple antigenic peptides for diagnosis of GB virus C infection. *Anal. Biochem.* **2010**, *396*, 51–58. [CrossRef] [PubMed]
13. Tarradas, J.; Monsó, M.; Muñoz, M.; Rosell, R.; Fraile, L.; Frías, M.T.; Domingo, M.; Andreu, D.; Sobrino, F.; Ganges, L. Partial protection against classical swine fever virus elicited by dendrimeric vaccine-candidate peptides in domestic pigs. *Vaccine* **2011**, *29*, 4422–4429. [CrossRef]
14. Buus, S.; Rockberg, J.; Forsström, B.; Nilsson, P.; Uhlen, M.; Schafer-Nielsen, C. High-resolution mapping of linear antibody epitopes using ultrahigh-density peptide microarrays. *Mol. Cell. Proteom.* **2012**, *11*, 1790–1800. [CrossRef] [PubMed]
15. Hebbes, T.R.; Turner, C.H.; Thorne, A.W.; Crane-Robinson, C. A "minimal epitope" anti-protein antibody that recognises a single modified amino acid. *Mol. Immunol.* **1989**, *26*, 865–873. [CrossRef]
16. Waritani, T.; Chang, J.; McKinney, B.; Terato, K. An ELISA protocol to improve the accuracy and reliability of serological antibody assays. *MethodsX* **2017**, *4*, 153–165. [CrossRef]
17. Mammen, M.; Choi, S.K.; Whitesides, G.M. Polyvalent Interactions in Biological Systems: Implications for Design and Use of Multivalent Ligands and Inhibitors. *Angew. Chem. Int. Ed. Engl.* **1998**, *37*, 2754–2794. [CrossRef]
18. Liu, S.P.; Zhou, L.; Lakshminarayanan, R.; Beuerman, R.W. Multivalent Antimicrobial Peptides as Therapeutics: Design Principles and Structural Diversities. *Int. J. Pept. Res. Ther.* **2010**, *16*, 199–213. [CrossRef]
19. Joshi, V.G.; Dighe, V.D.; Thakuria, D.; Malik, Y.S.; Kumar, S. Multiple antigenic peptide (MAP): A synthetic peptide dendrimer for diagnostic, antiviral and vaccine strategies for emerging and re-emerging viral diseases. *Indian J. Virol.* **2013**, *24*, 312–320. [CrossRef] [PubMed]
20. Tam, J.P.; Zavala, F. Multiple antigen peptide. A novel approach to increase detection sensitivity of synthetic peptides in solid-phase immunoassays. *J. Immunol. Methods* **1989**, *124*, 53–61. [CrossRef]

21. Doyle, H.A.; Mamula, M.J. Post-translational protein modifications in antigen recognition and autoimmunity. *Trends Immunol.* **2001**, *22*, 443–449. [CrossRef]

22. Lolli, F.; Mazzanti, B.; Pazzagli, M.; Peroni, E.; Alcaro, M.C.; Sabatino, G.; Lanzillo, R.; Brescia Morra, V.; Santoro, L.; Gasperini, C.; et al. The glycopeptide CSF114(Glc) detects serum antibodies in multiple sclerosis. *J. Neuroimmunol.* **2005**, *167*, 131–137. [CrossRef]

23. Lolli, F.; Mulinacci, B.; Carotenuto, A.; Bonetti, B.; Sabatino, G.; Mazzanti, B.; D'Ursi, A.M.; Novellino, E.; Pazzagli, M.; Lovato, L.; et al. An N-glucosylated peptide detecting disease-specific autoantibodies, biomarkers of multiple sclerosis. *Proc. Natl. Acad. Sci. USA* **2005**, *102*, 10273–10278. [CrossRef]

24. Walvoort, M.T.; Testa, C.; Eilam, R.; Aharoni, R.; Nuti, F.; Rossi, G.; Lanzillo, R.; Papini, A.M.; Imperiali, B.; Rovero, P.; et al. Antibodies from multiple sclerosis patients preferentially recognize hyperglucosylated adhesin of non-typeable *Haemophilus influenzae*. *Sci. Rep.* **2017**, *7*, 44969. [CrossRef]

25. Polman, C.H.; Reingold, S.C.; Banwell, B.; Clanet, M.; Cohen, J.A.; Filippi, M.; Fujihara, K.; Havrdova, E.; Hutchinson, M.; Kappos, L.; et al. Diagnostic criteria for multiple sclerosis: 2010 revisions to the McDonald criteria. *Ann. Neurol.* **2011**, *69*, 292–302. [CrossRef] [PubMed]

26. Thompson, A.J.; Banwell, B.L.; Barkhof, F.; Carroll, W.M.; Coetzee, T.; Comi, G.; Correale, J.; Fazekas, F.; Filippi, M.; Freedman, M.S. Diagnosis of multiple sclerosis: 2017 revisions of the Mc Donald criteria. *Lancet Neurol.* **2018**, *17*, 162–173. [CrossRef]

27. Rath, S.; Stanley, C.M.; Steward, M.W. An inhibition enzyme immunoassay for estimating relative antibody affinity and affinity heterogeneity. *J. Immunol. Methods* **1988**, *106*, 245–249. [CrossRef]

28. Real Fernández, F.; Di Pisa, M.; Rossi, G.; Auberger, N.; Lequin, O.; Larregola, M.; Benchohra, A.; Mansuy, C.; Chessaing, G.; Lolli, F.; et al. Antibody recognition in multiple sclerosis and Rett syndrome using a collection of linear and cyclic N-glucosylated antigenic probes. *Biopolymers* **2015**, *104*, 560–576. [CrossRef]

29. Schwarz, F.; Fan, Y.Y.; Schubert, M.; Aebi, M. Cytoplasmic N-glycosyltransferase of Actinobacillus pleuropneumoniae is an inverting enzyme and recognizes the NX(S/T) consensus sequence. *J. Biol. Chem.* **2011**, *286*, 35267–35274. [CrossRef]

30. Choi, K.J.; Grass, S.; Paek, S.; St Geme, J.W.; Yeo, H.J. The Actinobacillus pleuropneumoniae HMW1C-like glycosyltransferase mediates N-linked glycosylation of the Haemophilus influenzae HMW1 adhesin. *PLoS ONE* **2010**, *5*, e15888. [CrossRef]

31. Carotenuto, A.; D'Ursi, A.M.; Mulinacci, B.; Paolini, I.; Lolli, F.; Papini, A.M.; Novellino, E.; Rovero, P. Conformation-activity relationship of designed glycopeptides as synthetic probes for the detection of autoantibodies, biomarkers of multiple sclerosis. *J. Med. Chem.* **2006**, *49*, 5072–5079. [CrossRef] [PubMed]

32. Van Regenmortel, M.H.V. Antigenicity and Immunogenicity of Synthetic Peptides. *Biologicals* **2001**, *29*, 209–213. [CrossRef] [PubMed]

33. Mazzucco, S.; Matà, S.; Vergelli, M.; Fioresi, R.; Nardi, E.; Mazzanti, B.; Chelli, M.; Lolli, F.; Giannenschi, M.; Pinto, F.; et al. Synthetic glycopeptide of human myelin oligodendrocyte glycoprotein to detect antibody responses in multiple sclerosis and other neurological diseases. *Bioorg. Med. Chem. Lett.* **1999**, *9*, 167–172. [CrossRef]

34. Araújo, P.R.; Ferreira, A.W. High diagnostic efficiency of IgM-ELISA with the use of multiple antigen peptides (MAP1) from T. gondii ESA (SAG-1, GRA-1 and GRA-7), in acute toxoplasmosis. *Rev. Inst. Med. Trop. Sao Paulo* **2010**, *52*, 63–68. [CrossRef] [PubMed]

35. Paolini, I.; Nuti, F.; de la Cruz Pozo-Carrero, M.; Barbetti, F.; Kolesinska, B.; Kaminski, Z.J.; Chelli, M.; Papini, A.M. A convenient microwave-assisted synthesis of N-glycosyl amino acids. *Tetrahedron Lett.* **2007**, *48*, 2901–2904. [CrossRef]

36. Habbema, J.D.; Eijkemans, R.; Krijnen, J.; Knottnerus, J.E. Analysis of data on the accuracy of diagnostic tests. In *The Evidence Base of Clinical Diagnosis*; Knottnerus, J.E., Ed.; BMJ Books: London, UK, 2002; pp. 117–143.

37. Calabrese, M.; Gasperini, C.; Tortorella, C.; Schiavi, G.; Frisullo, G.; Ragonese, P.; Fantozzi, R.; Prosperini, L.; Annovazzi, P.; Cordioli, C.; et al. "Better explanations" in multiple sclerosis diagnostic workup: A 3-year longitudinal study. *Neurology* **2019**, *92*, e2527–e2537. [CrossRef]

38. Migliorini, P.; Pratesi, F.; Tommasi, C.; Anzilotti, C. The immune response to citrullinated antigens in autoimmune diseases. *Autoimmun. Rev.* **2005**, *4*, 561–564. [CrossRef]

39. Gestwicki, J.E.; Cairo, C.W.; Strong, L.E.; Oetjen, K.A.; Kiessling, L.L. Influencing receptor-ligand binding mechanisms with multivalent ligand architecture. *J. Am. Chem. Soc.* **2002**, *124*, 14922–14933. [CrossRef]

40. Sadler, K.; Tam, J.P. Peptide dendrimers: Applications and synthesis. *J. Biotechnol.* **2002**, *90*, 195–229. [CrossRef]
41. Weimer, B.C.; Walsh, M.K.; Wang, X. Influence of a poly-ethylene glycol spacer on antigen capture by immobilized antibodies. *J. Biochem. Biophys. Methods* **2000**, *45*, 211–219. [CrossRef]

Article

The Use of Electrochemical Voltammetric Techniques and High-Pressure Liquid Chromatography to Evaluate Conjugation Efficiency of Multiple Sclerosis Peptide-Carrier Conjugates

Efstathios Deskoulidis [1], Sousana Petrouli [1], Vasso Apostolopoulos [2], John Matsoukas [2,3,4,*] and Emmanuel Topoglidis [1,*]

1   Materials Science Department, University of Patras, 26504 Patras, Greece; stathis.deskou@gmail.com (E.D.); sousanapetr@gmail.com (S.P.)
2   Institute for Health and Sport, Victoria University, Melbourne, VIC 3030, Australia; vasso.apostolopoulos@mail.com
3   Newdrug, Patras Science Park, 26500 Patras, Greece
4   Department of Physiology and Pharmacology, Cumming School of Medicine, University of Calgary, Alberta, AB T2N 4N1, Canada
*   Correspondence: imats1953@gmail.com (J.M.); etop@upatras.gr (E.T.)

Received: 30 June 2020; Accepted: 17 August 2020; Published: 21 August 2020

Abstract: Recent studies have shown the ability of electrochemical methods to sense and determine, even at very low concentrations, the presence and quantity of molecules or analytes including pharmaceutical samples. Furthermore, analytical methods, such as high-pressure liquid chromatography (HPLC), can also detect the presence and quantity of peptides at very low concentrations, in a simple, fast, and efficient way, which allows the monitoring of conjugation reactions and its completion. Graphite/SiO_2 film electrodes and HPLC methods were previously shown by our group to be efficient to detect drug molecules, such as losartan. We now use these methods to detect the conjugation efficiency of a peptide from the immunogenic region of myelin oligodendrocyte to a carrier, mannan. The HPLC method furthermore confirms the stability of the peptide with time in a simple one pot procedure. Our study provides a general method to monitor, sense and detect the presence of peptides by effectively confirming the conjugation efficiency. Such methods can be used when designing conjugates as potential immunotherapeutics in the treatment of diseases, including multiple sclerosis.

Keywords: mannan; peptide; conjugation; MOG_{35-55}; Graphite/SiO_2 electrode; voltammetry; HPLC; multiple sclerosis; immunotherapy; vaccine

1. Introduction

Voltammetric techniques, including differential pulse voltammetry (DPV) and cyclic voltammetry (CV), as well as high-performance liquid chromatography (HPLC), were applied to identify and detect a peptide to its conjugated carrier. This study describes for the first time an alternative, fast, low cost and reliable method for the adequate and reliable determination of an active pharmaceutical ingredient (API) in the biocompatible matrix. The performance of the voltammetric techniques is strongly dependent on the performance of the working electrode used. Film electrodes, such as the graphite/SiO_2 used in this study, are being used in electrochemistry, as it has a number of advantages over the standard metallic and glass carbon electrodes. These include ease of manufacture requiring lower temperatures, low cost, the high surface area that could be rapidly renovated, simple handling, and their increased conductivity in a wide range of potentials. In addition, these techniques exhibit a wide range of anodic and cathodic peaks and great electrocatalytic activity and stability. All these features are crucial for the correct choice of a working electrode, especially when direct electrochemistry is conducted [1]. We recently demonstrated that these film electrodes modified or not, could be used for electrochemical drug sensing, for validation in food chemistry, and for the immobilization of heme proteins for studying protein/electrode interaction [2–4]. The electrochemical analytical methods were recently applied effectively in the detection of anti-hypertensive drug losartan [3] and have applied this method to detect peptides in peptide-carrier conjugates. The peptide used was the multiple sclerosis (MS) immunogenic peptide from myelin oligodendrocyte (MOG_{35-55}).

Numerous methods have been established for the analytical determination of drugs at low concentrations, using state of the art systems, such as HPLC, high-performance thin-layer chromatography and capillary electrophoresis/capillary electrochromatography [3]. Although these methods provide very accurate and reliable data, they are costly, time consuming, and involve the use of expensive equipment and consumables. In addition, sample pre-treatment is usually necessary. In this sense, electrochemical methods have emerged as low cost, reliable alternatives for the characterization of peptides and drugs. Different electrochemical techniques, involving voltammetry or potentiometry, have been implemented for drug analysis, as they offer ease of preparation and operation, high sensitivity, fast response time, high quantification and detection limits, reasonable selectivity, wide linear range, and are cost effective [3]. In this regard, we applied voltammetry techniques to monitor the conjugation of a peptide to its carrier, for the first time as a proof of concept study.

MS is regarded an autoimmune disease where immune cells (such as, Th1, Th17, macrophages, B cells) and their constituents (pro-inflammatory cytokines) are involved in the pathophysiology of the disease, with destruction of myelin sheath and loss of neurological function [5–10]. In an attempt to develop immunotherapeutics against MS using immunogenic/agonist peptides is to either alter the peptide to make it an antagonist [11–16], make it cyclic [17–19], or conjugate it to an appropriate carrier, which would deliver the peptide in such a manner to either induce tolerance, or alter the profile of T cells from pro-inflammatory (Th1) to anti-inflammatory (Th2) [13–15]. One approach which our team has developed, is to use mannan, a poly-mannose carrier conjugated to MS peptides [20–24]. This approach was developed over 25 years ago by the group of Apostolopoulos et al., to be effective in targeting peptides and proteins to dendritic cells in a number of different cancer vaccine models, some of which were translated to human clinical trials [25–32]. As such, mannan was used as a carrier and conjugated to immunodominant MS peptides including MBP_{83-99}, $PLP_{139-141}$, and MOG_{35-55} or their analogues, and were shown in animal models to tolerize T cells or switch Th1 cells to Th2 cells, depending on the peptide analogue used and showed stimulation of Th2 cells in peripheral blood mononuclear cells from patients with MS [17,21,33–35]. The conjugation of mannan, in its oxidized form (OM), to MOG_{35-55} peptide (MOG_{35-55} was used as an example in this study) via a (Lys-Gly)$_5$ linker [$(KG)_5$] was used and evaluated (OM-$(KG)_5$-MOG_{35-55} conjugate) using voltammetric techniques [1–4,36–38]. The conjugation between OM and peptide (MOG_{35-55}) occurs via formation of Schiff bases between the free amines of the linker $(KG)_5$ and aldehydes of OM. The synthesis and efficacy of these conjugates have been described in numerous studies [23,24,39,40]. However, the extent of conjugation and the redox condition of the participating sugars, such as mannan, are most times assessed by high cost, complicated and lengthy analytical methods, such as capillary electrophoresis and polyacrylamide gel electrophoresis [23,24,39,40].

Among the approaches used in recent years for the immunomodulation of MS, the conjugation of mannan with myelin peptides has shown much promise, including that of OM-$(KG)_5$-MOG_{35-55}, which induces tolerance in mice, providing a promising conjugate for further studies. The electrochemical and HPLC analysis for identification of peptides or their mutants in mannan based conjugates requires specialized techniques, which differ significantly from those methods used for small molecules. In this study, novel analytical methods were developed and applied, that clearly, sense, detect, and confirm the conjugation of OM with MOG_{35-55}. Further, this study makes it possible to accurately evaluate the stability of the peptide component in the conjugate using HPLC [41,42].

2. Materials and Methods

2.1. Materials

Sodium metasilicate (Na_2SiO_3) (SiO_2, 50–53%), NaH_2PO_4, mannan isolated from yeast cells (*Saccharomyces cerevisiae*), potassium ferricyanide, ferrocyanide, and potassium chloride were obtained from Sigma Aldrich Chemie GmbH (Taufkirchen, Germany). MOG_{35-55} and MOG_{37-55} peptides were supplied by NewDrug S.A., Patras Science Park, Greece and purchased from China peptides Inc. The peptide analogue (Lys-Gly)$_5$-MOG_{35-55}, referred as (KG)$_5$-MOG_{35-55}, was synthesized using standard peptide chemistry techniques and previously published by our group. Briefly, Fmoc/tBu methodology was used which included 2-chlorotrityl chloride resin (CLTR-Cl) and N^a-Fmoc (9-fluorenylmethyloxycarboxyl) side chain protected amino acids [43,44]. The purity of the peptides were shown to be >97% by analytical HPLC. Graphite powder (synthetic, APS 7–11 µm, 99%) was obtained from Alfa Aesar. Soda lime glass slides (75 mm × 25 mm × 1.1 mm), with 15 Ohm/sqr Indium Tin Oxide (ITO) coating were obtained from PsiOTec, UK. All chemicals were of analytical grade and used without the need for further purification. All solutions were prepared in deionized water with resistance R = 18 MΩ cm.

2.2. Graphite/SiO$_2$ Film Electrodes Preparation

The graphite/SiO$_2$ film electrodes were prepared as described [2,3]. Briefly, silicate liquid polymer (50% Na$_2$SiO$_3$; pH 12–13) was gently mixed with 20% graphite powder at 23 °C, until the mixture became homogeneous and acquired a "sticky" texture. The mixture underwent ultrasonication for 2 min for the graphite powder to be fully soluble, and 100 µL of the silicate/graphite suspension were applied on the surface of a conductive ITO glass slide using the "Doctor Blade" technique. Prior to the deposition of the silicate/graphite suspension, the ITO glass slides were cleaned in a detergent solution using an ultrasonic bath for 15 min, and then rinsed with 18 MΩ distilled water and ethanol. Each glass slide was masked with 3M Magic Scotch tape (thickness 62.5 µm; type 810), in order to control the width and the thickness of the mixture spread area. For each graphite/SiO$_2$ film deposition, one layer of tape was used which provided a size 1 × 1 cm^2 and film thickness of ~66 µm. The films were allowed to dry for 30 min in a class 4000 room, prior to placing them in a preheated oven (330 °C) for 100 min. If required, the liquid suspension could be stored in an insulated flask at 25 °C for later usage. The resulting ITO substrates with the deposited graphite/SiO$_2$ films were cut in 10 mm × 25 mm pieces before use.

2.3. Characterization of Graphite/SiO$_2$ Film Electrodes

Field emission scanning electron microscopy (FE-SEM) using an FEI inspect microscope (25 kV) was used to determine morphology and thickness of the Graphite/SiO$_2$ film. The films were prepared by AU sputtering to increase the conductivity of the samples. Energy dispersive spectroscopy EDS was also used for the elemental analysis of the Graphite/SiO$_2$/ITO films.

2.4. Preparation of (KG)$_5$-MOG$_{35-55}$ Peptide

MOG_{35-55} agonist peptide was synthesized in our labs, >97% purity, with (KG)$_5$ extended at the N-terminus of the peptide. Peptide was prepared using our methods, either by coupling, catalyzed by microwave radiation in a CEM Liberty microwave system or by using the conventional step by step procedure by solid phase peptide methods (as described in [45]). (KG)$_5$-MOG_{35-55} peptide was also purchased by China Peptides Inc. In house synthesized peptides and purchased peptides were confirmed by HPLC and Mass Spectroscopy for purity and identity.

2.5. Preparation of Oxidized Mannan

Mannan (14 mg) was dissolved in 1 mL phosphate buffer (0.1 M sodium phosphate, pH 6.0), and was oxidized using 0.1 M sodium periodate and incubated at 4 °C for 1 h, after which 10 µL ethanediol was added for 30 min at 4 °C. Oxidized mannan (OM) was passed through a PD-10 column (Sigma Aldrich Chemie) pre-equilibrated in sodium bicarbonate buffer (sodium carbonate: Sodium bicarbonate, pH 9.0). Two ml of OM fraction (7 mg/mL) was collected and kept in the dark.

2.6. Conjugation of Oxidized Mannan to Peptide

To the OM fraction (2 mL; 7 mg/mL, sodium bicarbonate pH 9.0 buffer), 1 mg of $(KG)_5$-$MOG_{35\text{-}55}$ peptide was added and allowed to react overnight in the dark at 23 °C. A list of peptides and conjugates are summarized in Table 1.

Table 1. Peptides and conjugates used in this study.

Acronym	Specification
$MOG_{35\text{-}55}$	Myelin oligodendrocyte glycoprotein immunogenic epitope, region 35–55
$MOG_{37\text{-}55}$	Myelin oligodendrocyte glycoprotein immunogenic epitope, region 37–55
$(KG)_5$-$MOG_{35\text{-}55}$	Peptide analogue $MOG_{35\text{-}55}$ with $(KG)_5$ at the N-terminus
OM-$(KG)_5$-$MOG_{35\text{-}55}$	Oxidized mannan conjugated to $(KG)_5$-$MOG_{35\text{-}55}$

KG, lysine glycine; MOG, myelin oligodendrocyte glycoprotein; OM, oxidized mannan.

2.7. Monitoring of Conjugation by HPLC

We used a Waters 2695 HPLC (Alliance) system with a photodiode array detector equipped with a Lichrosorb RP-18 reversed phase analytical column (C18 35 µm, 4.6 × 50 mm PIN 186003034). Analysis was achieved with stepped linear gradient of solvent A (0.08% TFA in H_2O) and in solvent B (0.08% TFA in 100% acetonitrile) for 30 min with a flow rate 3 mL/min. The conjugation of OM with $(KG)_5$-$MOG_{35\text{-}55}$ peptide was evaluated by HPLC. The $(KG)_5$-$MOG_{35\text{-}55}$ HPLC peak disappeared within six hours indicating completion of conjugation to OM.

2.8. Electrochemical/Electrocatalytic Measurements

Electrochemical measurements were conducted using an Autolab PGStat-101 potentiostat (Metrohm, Utrecht, The Netherlands). The electrochemical cell comprised of a 10 mL, three-electrode stirring glass cell with a Teflon cap, a platinum mesh flag as the counter electrode, a Ag/AgCl/KCl_{sat} reference electrode and a Graphite/SiO_2 film on ITO conducting glass as the working electrode. The electrolyte contained a solution of NaH_2PO_4 (10 mM; pH 7.0), which was deoxygenated with argon prior to any measurements and an argon atmosphere was kept throughout the measurements. The DPV measurements took place in a potential range between −1 to +0.05 V. The optimized parameters of DPV correspond to a step potential at 5 mV, amplitude of 50 mV, modulation time of 25 ms with scan rate 100 mV s^{-1} and a frequency of 50 Hz. All potentials are reported against Ag/AgCl and all experiments were carried out at 23 °C.

3. Results and Discussion

3.1. FE-SEM Characterization

The general thickness and surface morphology of the graphite/SiO_2 films were demonstrated by FE-SEM. The top-view of the FE-SEM image (Figure 1a) shows that the surface of the graphite/SiO_2 film is rough and non-uniform with many wrinkles. It exhibits increased porosity and a high effective surface area. Figure 1b presents the cross section of a graphite/SiO_2 film electrode, with an estimated film thickness of ~65 µm as set by the adhesive tape used; the EDS for a graphite/SiO_2 film carried out during the FE-SEM analysis is shown in the Supplementary Materials (SM, Figure S1). The characteristic peaks of Na, O, and Si, due to the use of silicate glue (Na_2SiO_3), are presented in high intensity, thus, the peak of C is presented in lower intensity. Hence, the results validate the reduced concentration of carbon in the mixture used for the fabrication of the graphite/SiO_2 films.

(a)   (b)

Figure 1. SEM images of the graphite/SiO_2 working electrode from (**a**) top view and (**b**) a cross section.

3.2. UV Characterazation of (KG)$_5$-MOG$_{35-55}$ Peptide with Increasing Amounts of OM

It is known that most peptides exhibit strong absorbance at around 280 nm, due to aromatic amino acids (tyrosine and tryptophan) or disulfide bonds in the peptide sequences [46,47]. Figure 2 shows the UV-vis spectra of (KG)$_5$-MOG$_{35-55}$ with increasing amounts of OM. The increase of absorbance at 280 nm confirms the conjugation of MOG$_{35-55}$ peptide to OM. The intensity of the absorption peak at 280 increases until all of the free peptide in solution is conjugated to the OM. It should be noted that the conjugate of (KG)$_5$-MOG$_{35-55}$ with OM took place in solution and not on the surface of the graphite/SiO_2 film electrode as due to its non-transparency it is impossible to monitor the conjugation process on its surface. All the UV-visible absorption spectra of the peptide was recorded using a Shimadzu UV-1800 spectrophotometer.

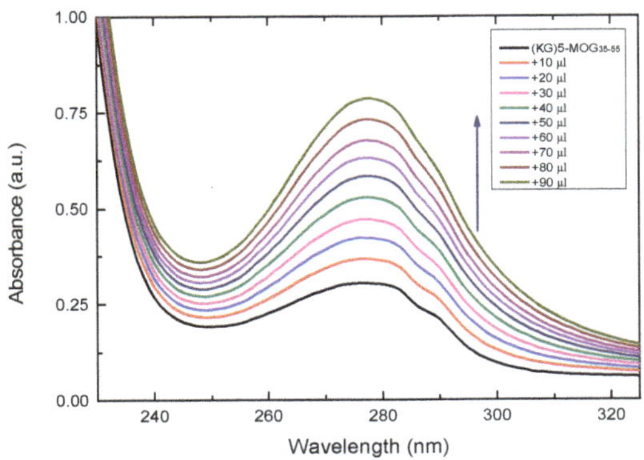

Figure 2. UV-Vis spectral changes of (KG)$_5$-MOG$_{35-55}$ in solution with increasing amounts of OM (10–90 µL).

3.3. Electrochemical Analysis Showing Conjugation of (KG)$_5$-MOG$_{35-55}$ to OM

Electrochemical characteristics of the graphite/SiO$_2$ film electrode were investigated by CV. Figure 3a shows the electrochemical behavior of a bare graphite/SiO$_2$ film electrode in a solution of 0.1 M KCl and 5 mM of [Fe(CN)$_6$]$^{3-/4-}$ through CV in the potential range of +1 to −1 V at different scan rates. Figure 3b shows the currents (anodic and cathodic) from the plots of I vs. square root of scan rate ($v^{1/2}$). Straight lines form for both the anodic and cathodic currents, confirming that a diffusional process has occurred in the reaction of ferrocyanide/ferricyanide. In addition, these results confirm that fast electron transfer occurs on the Graphite/SiO$_2$ film electrode due to its increased conductivity and surface area. In order to calculate the electroactive surface area of the film electrode, the Randles-Sevcik equation was used [36]:

$$i_p = \left(2.69 \times 10^5\right) \times A \times D^{1/2} \times n^{3/2} \times C \times v^{1/2} \tag{1}$$

where i_p corresponds to the maximum current (in Amperes), n is the number of electrons transferred ($n = 1$), D is the diffusion coefficient (cm^2 s^{-1}) of [Fe(CN)$_6$]$^{3-/4-}$ solution (7.6×10^{-6} cm^2 s^{-1}) [37], A is the electrode area (cm^2), C is the concentration (molcm^{-3}) and v is the scan rate (mV s^{-1}) and thus the electroactive surface area of the graphite/SiO$_2$ was estimated to be 0.0039 cm^2.

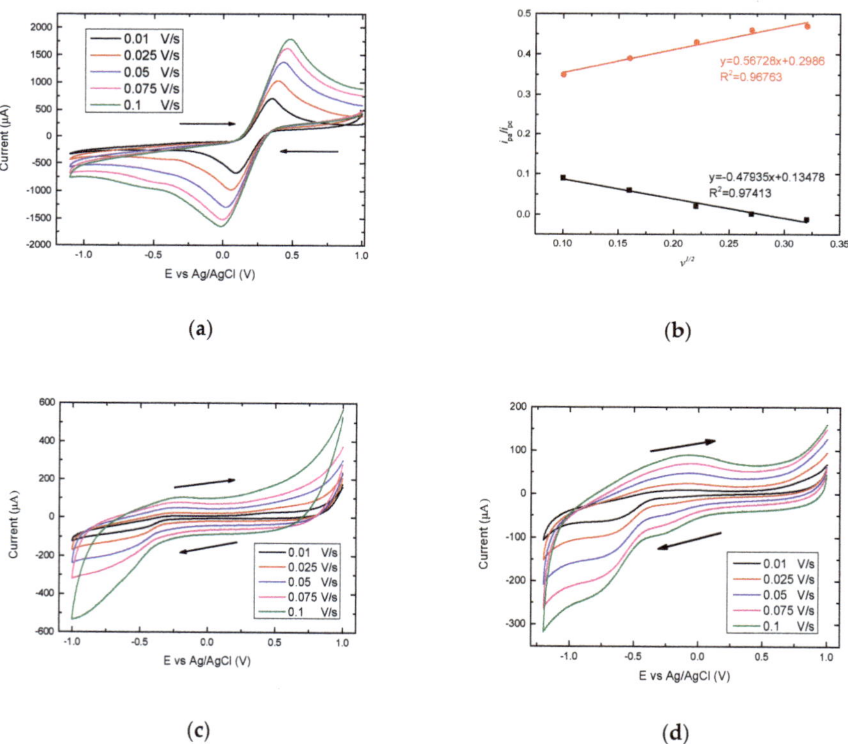

Figure 3. Cyclic voltammograms (CVs) of (**a**) a bare graphite/SiO$_2$ film electrode in 0.1 M KCl solution containing 5 mM of [Fe(CN)6]$^{3-/4-}$ at different scan rates. (**b**) Plot of anodic and cathodic peak current (Ipa/Ipc) vs. square root of scan rate ($v^{1/2}$). (**c**) A bare graphite/SiO$_2$ film electrode in 10 mM NaH$_2$PO$_4$, pH 7.0 at different scan rates and (**d**) the OM-(KG)$_5$-MOG$_{35-55}$ conjugate on graphite/SiO$_2$ in 10 mM NaH$_2$PO$_4$, pH 7.0 at different scan rates under an Argon atmosphere.

The electrochemical behavior of the graphite/SiO_2 film electrode was then investigated in the presence and absence of the MS myelin epitope peptide vaccine (OM-(KG)5-MOG$_{35-55}$). Figure 3c shows the effect of scan rate of a bare graphite/SiO_2 electrode, before the detection of the OM-(KG)5-MOG$_{35-55}$, at a scan rate range of 0.01 to 0.1 V s^{-1}. All electrochemical experiments were performed in a peptide free, anaerobic 10 mM NaH_2PO_4 (pH 7.0). The bare graphite/SiO_2 film electrode shows the characteristic charging/de-charging currents, and no cathodic or anodic peaks are observed even at the slowest scan rate (0.01 V s^{-1}). One of the advantages of using graphite paste electrodes is the increased conductivity, which allows a broader study of redox reactions occurring at very high or low biases (ranging from +1 V to −1 V). Further, the slower scan rate applied, the smaller the resulting current is obtained. Figure 3b, on the other hand, showing the CVs of OM-(KG)$_5$-MOG$_{35-55}$ on the graphite/SiO_2 film electrode, exhibits not only the characteristic charging/discharging currents assigned to electron injection into sub-band gap/conduction band states of the graphite/SiO_2 electrode, but also two reduction peaks around −0.22 V and −0.67 V and a broad re-oxidation peak at −0.1 V.

The redox peak currents were shown to be proportional to the scan rate, characteristic of quasi-reversible behavior. The rate of reaction between the graphite/SiO_2 electrode and the conjugate, OM-(KG)$_5$-MOG$_{35-55}$ was not fast enough to maintain equal concentrations of oxidized and reduced species at the surface of the electrode. In addition, the CV responses were shown to be stable, with the waveforms being unperturbed after being scanned several times, whilst no other consumption of the complex occurred nor other undesirable reactions in the phosphate buffer took place.

In Figure 3d, the two cathodic peaks at −0.27 V and −0.7 V and the wide anodic peak approximately at −0.1 V observed are due to the presence of the OM-(KG)$_5$-MOG$_{35-55}$. The two cathodic peaks correspond to the linker molecule (KG)$_5$ used to conjugate the MOG$_{35-55}$ peptide to OM, that contains 5 lysines and 5 glycines to its structure. Thus, the cathodic peaks attributed to the presence of lysines. On the other hand, the wide oxidation peak occurred probably due to superfluity of the free (KG)$_5$-MOG$_{35-55}$ peptide that was not able to conjugate to OM and created the final complex of the OM-(KG)$_5$-MOG$_{35-55}$ conjugate.

The CVs of the constituents of the OM-(KG)$_5$-MOG$_{35-55}$ conjugate are shown in Figure 4. According to Figure 4a, as mentioned earlier, the bare graphite/SiO_2 film electrode exhibited no reduction or oxidation peaks which is consistent with the currents being limited by the graphite conductivity at the voltage biases reported herein. On the other hand, the CV of the film electrode in the presence of mannan in 0.1 M buffer exhibited an oxidation peak at approximately 0.5 V, and the CV of the film electrode in the presence of 0.002 mg/mL OM displayed a slight cathodic peak at −0.56 V and the characteristic anodic peak at −0.1 V. At the same time, the electrochemical behavior of peptides MOG$_{35-55}$ and MOG$_{37-55}$ were examined. The main difference between these two peptides is that the MOG$_{35-55}$ peptide contained and additional linker with 5 lysines (KG)$_5$, whilst the MOG$_{37-55}$ peptide included a linker, which only contained 1 lysine. This was confirmed in Figure 4b, which displays the CVs of the Graphite/SiO_2 film electrode in the presence of each peptide. The two cathodic and anodic peaks observed are due to the presence of the lysine residues, however, the CV scan of the MOG$_{35-55}$ peptide exhibits a higher current and a wider electrochemical window compared to the CV scan of MOG$_{37-55}$ peptide, as the latter contained only 1 lysine residue.

DPV is a more sensitive approach compared to CV and hence, has been extensively used as a more sensitive method for the detection of molecules in low concentration [38]. In Figure 5, the DPVs are recorded for the bare film electrode, as well as for each part that constitutes the final structure of OM-(KG)$_5$-MOG$_{35-55}$ conjugate on the Graphite/SiO_2 working electrode. As can be seen in Figure 5a, the bare graphite/SiO_2 is free of any redox peaks. However, in Figure 5b, there are two peaks which correspond to (KG)$_5$-MOG$_{35-55}$ peptide, approximately at −0.65 V and −0.27 V, respectively. Figure 5c shows the DPV of mannan (in 0.1 M phosphate buffer) on the surface of the film electrode, displaying a clear sharper peak at around −0.26 V. The last step in order to evaluate the conjugation of peptide (KG)$_5$-MOG$_{35-55}$ with OM via DPV measurements is depicted in Figure 5d with a clear and distinct peak at −0.28 V and a shoulder peak at −0.62 V, which are actually due to the presence of OM-(KG)$_5$-MOG$_{35-55}$

on the graphite/SiO$_2$ film electrode (after the addition of 0.002 mg/mL of OM). This is a proof of concept study, and we intend to further study the quantification of this and other conjugates, focusing on the limit od detection (LOD) of these conjugates using voltammetric techniques.

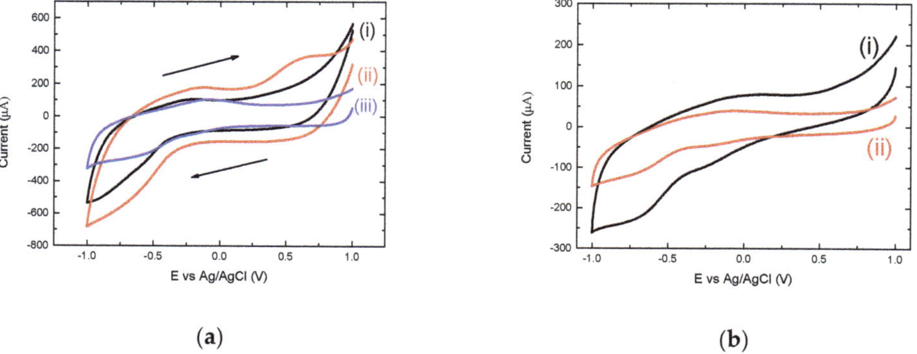

(**a**) (**b**)

Figure 4. (**a**) CV scans at a scan rate of 0.1 Vs^{-1} of (i) a bare graphite/SiO$_2$ film electrode, (ii) mannan and (iii) OM-(KG)$_5$-MOG$_{35\text{-}55}$ conjugate. (**b**) Depicts the comparison between the CV's of (i) MOG$_{35\text{-}55}$ and (ii) MOG$_{37\text{-}55}$, both on graphite/SiO$_2$ in 10 mM NaH$_2$PO4, pH 7.0 at scan rate of 0.075 Vs^{-1}.

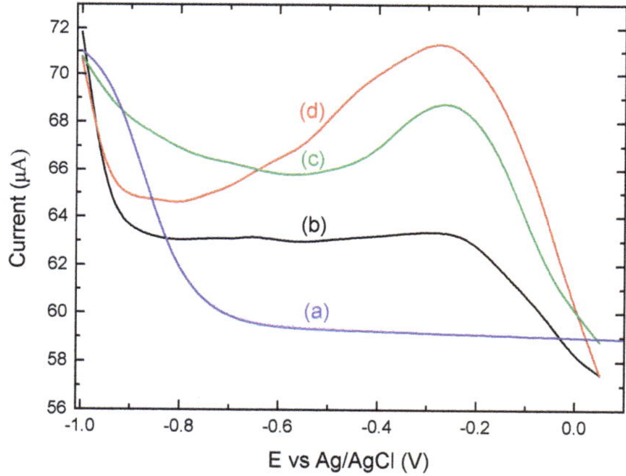

Figure 5. Differential pulse voltammetry (DPVs) comparison of (a) a bare graphite/SiO$_2$ film electrode, (b) (KG)$_5$-MOG$_{35\text{-}55}$, (c) mannan, and (d) OM-(KG)$_5$-MOG$_{35\text{-}55}$ conjugate on graphite/SiO$_2$ electrode in 10 mM NaH$_2$PO$_4$, pH 7.0.

3.4. Complete Conjugation between (KG)$_5$-MOG$_{35\text{-}55}$ Peptide to OM is Monitored by HPLC

Contrarily to the conjugation of MOG$_{35\text{-}55}$ peptide with mannan, which did not occur, the reaction of (KG)$_5$-MOG$_{35\text{-}55}$ with mannan (oxidized or not) resulted in gradual conjugation of (KG)$_5$-MOG$_{35\text{-}55}$ peptide within 6 h depicted in the gradual loss of the HPLC peak during this period (Figure 6). The amino groups of lysine residues within (KG)$_5$ forms a Schiff base reaction with the aldehyde groups of OM (resulting after the oxidation of mannan). The (KG)$_5$-MOG$_{35\text{-}55}$ peptide peak at 9.62 gradually disappears within this period, showing complete conjugation of (KG)$_5$-MOG$_{35\text{-}55}$ peptide to OM. Figure 6b shows the completion of conjugation within six hours.

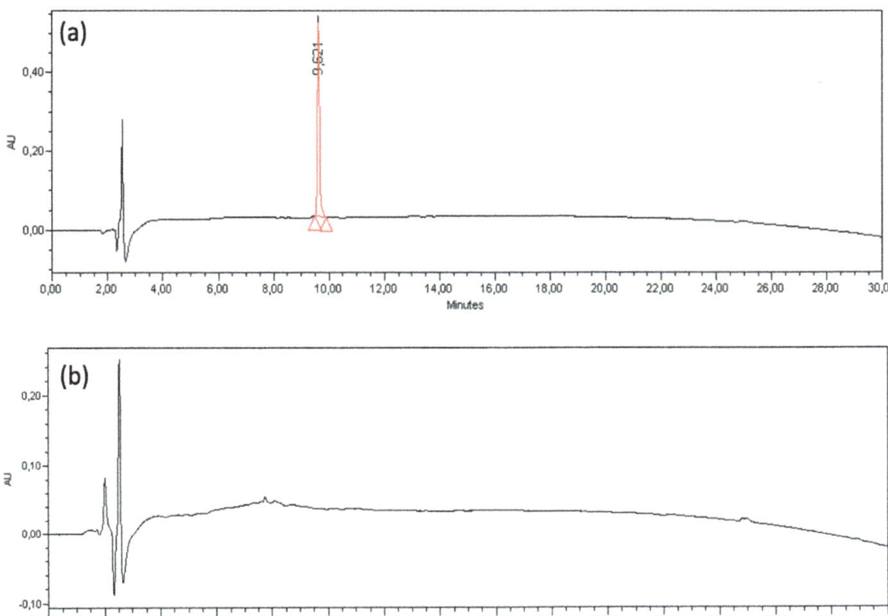

Figure 6. (**a**) High-performance liquid chromatography (HPLC) analysis of $(KG)_5$-MOG_{35-55}–214 nm at the beginning of the conjugation reaction and (**b**) HPLC analysis of OM-$(KG)_5$-MOG_{35-55} solution after 6 h.

3.5. The Importance of the Linker $(KG)_5$ for Conjugation of Peptides to OM

The conjugation of MOG_{35-55} peptide to OM was achieved through $(KG)_5$ linker, as previously described [23]. As demonstrated, this approach provides simple and efficient conjugation by the Schiff base reaction, where aldehyde groups of OM reacts with the amino groups of the lysine side chains of the $(KG)_5$-MOG_{35-55}, peptide. In previous similar studies using the linker KG of varying lengths, $(KG)_{n=1-5}$, we noted that the length of the linker plays a crucial role in the ability of peptides to be efficiently conjugated to the OM scaffold [48].

3.6. Mannan-Peptide Conjugate

In the OM-$(KG)_5$-MOG_{35-55} conjugate, unreacted aldehyde groups are necessary to immunoregulate the peptide to dendritic cells. This is a result of ethylene glycol addition to blockade further oxidation, and in line with previous studies on MUC1-mannan conjugates in cancer research, which required aldehyde groups in order to activate dendritic cells [39]. The matrix also contains intact mannose units, not oxidized, necessary to bind to the mannose receptor of the dendritic cells and their activation via toll-like receptor 4 [49–52]. In particular, the procedure we followed to produce the mannan-peptide conjugate allows: (i) the presence of antigen peptide MOG_{35-55} connected with aldehyde groups of the OM through immune bonds (Schiff base) with the amino groups of the lysine side chain in the $(KG)_5$-MOG_{35-55} peptide. The peptide-OM conjugate is delivered to dendritic cells via the mannan scaffold for regulation of the immune system; (ii) the presence of unreacted aldehyde groups are necessary to modulate dendritic cells; and (iii) the presence intact mannose units, not oxidized, necessary to bind to the mannose receptor of the dendritic cells.

3.7. Chemistry of the Mannose Cleavage

The cis-diols can form a cyclic complex upon oxidation with strong oxidizing agents as periodate. This allows the cleavage of the bond between the two carbons bearing the two hydroxyl groups, leading to the formation of aldehyde groups. Mannose is a carbohydrate, which holds two hydroxyl groups at positions 2,3 of the ring in a cis- position. This allows the oxidizing agent sodium periodate to form a cyclic complex, which finally leads to cleavage of the carbon-carbon bond bearing the cis-hydroxyl groups. This complex cannot be formed if the hydroxyl groups at the adjacent carbon atoms are in a trans position and subsequently this carbon-carbon bond cannot be cleaved. The formation of the cyclic mannose-periodate complex is leading finally to the cleavage of the ring and the formation of the two aldehyde groups. These groups react with the amino groups of the five lysines of the $(KG)_5$-MOG_{35-55} to form double bond imines (Schiff base reaction) thus, the MOG_{35-55} peptide attached to the mannan scaffold. Figure 7 shows the mechanism of cis diol cleavage.

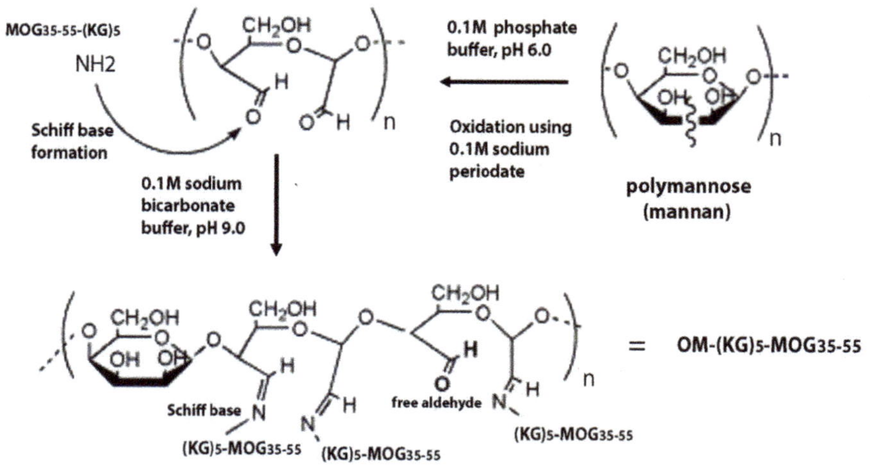

Figure 7. The mechanism of cis diol cleavage. Synthetic scheme of conjugation reaction of peptide with oxidized mannan [35].

4. Conclusions

We developed and confirm an analytical electrochemical method for monitoring the conjugation reaction of peptides to the carrier mannan; $(KG)_5$-MOG_{35-55} was used as the peptide example in this study. Peptide-OM conjugates can serve as potential vaccine candidates as has previously been shown by the group for cancer models and more recently in MS models. Electrochemical voltammetric techniques and HPLC experiments were used to confirm the conjugation of $(KG)_5$-MOG_{35-55} to the aldehyde groups of OM. It is shown that voltammetric technique and HPLC can be used to monitor the conjugation efficiency of peptide-carrier conjugates.

Supplementary Materials: The following are available online at http://www.mdpi.com/2076-3425/10/9/577/s1. Figure S1: EDS elemental microanalysis of a Graphite/SiO$_2$ film electrode.

Author Contributions: E.T. conceived and designed exclusively the electrochemical experiments. J.M. and V.A. conceived and designed the biochemical parts, the HPLC measurements and the chemistry of the mannose cleavage; E.D., S.P., E.T. and J.M. performed the experiments; E.T., J.M., E.D. and V.A. analyzed the data; E.T., J.M. and V.A. contributed to reagents/materials/analysis tools; E.D., E.T. and J.M. wrote their respective specialty parts of this paper; V.A. revised and edited the paper. All authors have read and agreed to the published version of the manuscript

Funding: This research was partially supported by Grant 80669 from the Research Committee of the University of Patras via C. CARATHEODORI program.

Acknowledgments: The authors would like to thank Elias Sakellis from NCSR Demokritos for the SEM images and EDS analysis. V.A. would like to thank the Institute for Health and Sport, Victoria University, for supporting her current efforts into MS research. J.M. would like to thank the General Secretariat for Research and Technology (GSRT) for supporting his MS research.

Conflicts of Interest: The authors declare no conflict of interest.

References

1. Nosrati, R.; Olad, A.; Maryami, F. The use of graphite/TiO_2 nanocomposite additive for preparation of polyacrylic based visible-light induced antibacterial and self-cleaning coating. *Res. Chem. Intermed.* **2018**, *44*, 6219–6237. [CrossRef]

2. Nikolaou, P.; Deskoulidis, E.; Topoglidis, E.; Kakoulidou, A.T.; Tsopelas, F. Application of chemometrics for detection and modeling of adulteration of fresh cow milk with reconstituted skim milk powder using voltammetric fingerpriting on a graphite/SiO_2 hybrid electrode. *Talanta* **2020**, *206*, 120223. [CrossRef] [PubMed]

3. Nikolaou, P.; Vareli, I.; Deskoulidis, E.; Matsoukas, J.; Vassilakopoulou, A.; Koutselas, I.; Topoglidis, E. Graphite/SiO_2 film electrode modified with hybrid organic-inorganic perovskites: Synthesis, optical, electrochemical properties and application in electrochemical sensing of losartan. *J. Solid State Chem.* **2019**, *273*, 17–24. [CrossRef]

4. Topoglidis, E.; Kolozoff, P.-A.; Tiflidis, C.; Papavasiliou, J.; Sakellis, E. Adsorption and electrochemical behavior of Cyt-c on carbon nanotubes/TiO_2 nanocomposite films fabricated at various annealing temperatures. *Colloid Polym. Sci.* **2018**, *296*, 1353–1364. [CrossRef]

5. Dargahi, N.; Katsara, M.; Tselios, T.; Androutsou, M.E.; De Courten, M.; Matsoukas, J.; Apostolopoulos, V. Multiple sclerosis: Immunopathology and treatment update. *Brain Sci.* **2017**, *7*. [CrossRef]

6. Katsara, M.; Apostolopoulos, V. Editorial: Multiple Sclerosis: Pathogenesis and Therapeutics. *Med. Chem.* **2018**, *14*, 104–105. [CrossRef]

7. Katsara, M.; Matsoukas, J.; Deraos, G.; Apostolopoulos, V. Towards immunotherapeutic drugs and vaccines against multiple sclerosis. *Acta Biochim. Biophys. Sin.* **2008**, *40*, 636–642. [CrossRef]

8. Katsara, M.; Tselios, T.; Deraos, S.; Deraos, G.; Matsoukas, M.T.; Lazoura, E.; Matsoukas, J.; Apostolopoulos, V. Round and round we go: Cyclic peptides in disease. *Curr. Med. Chem.* **2006**, *13*, 2221–2232. [CrossRef]

9. Steinman, L. Multiple sclerosis: A coordinated immunological attack against myelin in the central nervous system. *Cell* **1996**, *85*, 299–302. [CrossRef]

10. Steinman, L. Multiple sclerosis: A two-stage disease. *Nat. Immunol.* **2001**, *2*, 762–764. [CrossRef]

11. Candia, M.; Kratzer, B.; Pickl, W.F. On Peptides and Altered Peptide Ligands: From Origin, Mode of Action and Design to Clinical Application (Immunotherapy). *Int. Arch. Allergy Immunol.* **2016**, *170*, 211–233. [CrossRef] [PubMed]

12. Katsara, M.; Minigo, G.; Plebanski, M.; Apostolopoulos, V. The good, the bad and the ugly: How altered peptide ligands modulate immunity. *Expert Opin. Biol. Ther.* **2008**, *8*, 1873–1884. [CrossRef] [PubMed]

13. Katsara, M.; Yuriev, E.; Ramsland, P.A.; Deraos, G.; Tselios, T.; Matsoukas, J.; Apostolopoulos, V. A double mutation of MBP(83-99) peptide induces IL-4 responses and antagonizes IFN-gamma responses. *J. Neuroimmunol.* **2008**, *200*, 77–89. [CrossRef]

14. Katsara, M.; Yuriev, E.; Ramsland, P.A.; Deraos, G.; Tselios, T.; Matsoukas, J.; Apostolopoulos, V. Mannosylation of mutated MBP83-99 peptides diverts immune responses from Th1 to Th2. *Mol. Immunol.* **2008**, *45*, 3661–3670. [CrossRef]

15. Katsara, M.; Yuriev, E.; Ramsland, P.A.; Tselios, T.; Deraos, G.; Lourbopoulos, A.; Grigoriadis, N.; Matsoukas, J.; Apostolopoulos, V. Altered peptide ligands of myelin basic protein (MBP87-99) conjugated to reduced mannan modulate immune responses in mice. *Immunology* **2009**, *128*, 521–533. [CrossRef]

16. Trager, N.N.M.; Butler, J.T.; Harmon, J.; Mount, J.; Podbielska, M.; Haque, A.; Banik, N.L.; Beeson, C.C. A Novel Aza-MBP Altered Peptide Ligand for the Treatment of Experimental Autoimmune Encephalomyelitis. *Mol. Neurobiol.* **2018**, *55*, 267–275. [CrossRef]

17. Katsara, M.; Deraos, S.; Tselios, T.V.; Pietersz, G.; Matsoukas, J.; Apostolopoulos, V. Immune responses of linear and cyclic PLP139-151 mutant peptides in SJL/J mice: Peptides in their free state versus mannan conjugation. *Immunotherapy* **2014**, *6*, 709–724. [CrossRef]

18. Lourbopoulos, A.; Deraos, G.; Matsoukas, M.T.; Touloumi, O.; Giannakopoulou, A.; Kalbacher, H.; Grigoriadis, N.; Apostolopoulos, V.; Matsoukas, J. Cyclic MOG35-55 ameliorates clinical and neuropathological features of experimental autoimmune encephalomyelitis. *Bioorg. Med. Chem.* **2017**, *25*, 4163–4174. [CrossRef]

19. Lourbopoulos, A.; Matsoukas, M.T.; Katsara, M.; Deraos, G.; Giannakopoulou, A.; Lagoudaki, R.; Grigoriadis, N.; Matsoukas, J.; Apostolopoulos, V. Cyclization of PLP139-151 peptide reduces its encephalitogenic potential in experimental autoimmune encephalomyelitis. *Bioorg. Med. Chem.* **2018**, *26*, 2221–2228. [CrossRef]

20. Apostolopoulos, V.; Rostami, A.; Matsoukas, J. The Long Road of Immunotherapeutics against Multiple Sclerosis. *Brain Sci.* **2020**, *10*. [CrossRef]

21. Day, S.; Tselios, T.; Androutsou, M.E.; Tapeinou, A.; Frilligou, I.; Stojanovska, L.; Matsoukas, J.; Apostolopoulos, V. Mannosylated Linear and Cyclic Single Amino Acid Mutant Peptides Using a Small 10 Amino Acid Linker Constitute Promising Candidates Against Multiple Sclerosis. *Front. Immunol.* **2015**, *6*, 136. [CrossRef] [PubMed]

22. Deraos, G.; Rodi, M.; Kalbacher, H.; Chatzantoni, K.; Karagiannis, F.; Synodinos, L.; Plotas, P.; Papalois, A.; Dimisianos, N.; Papathanasopoulos, P.; et al. Properties of myelin altered peptide ligand cyclo(87-99)(Ala91,Ala96)MBP87-99 render it a promising drug lead for immunotherapy of multiple sclerosis. *Eur. J. Med. Chem.* **2015**, *101*, 13–23. [CrossRef] [PubMed]

23. Tapeinou, A.; Androutsou, M.E.; Kyrtata, K.; Vlamis-Gardikas, A.; Apostolopoulos, V.; Matsoukas, J.; Tselios, T. Conjugation of a peptide to mannan and its confirmation by tricine sodium dodecyl sulfate-polyacrylamide gel electrophoresis. *Anal. Biochem.* **2015**, *485*, 43–45. [CrossRef] [PubMed]

24. Tselios, T.V.; Lamari, F.N.; Karathanasopoulou, I.; Katsara, M.; Apostolopoulos, V.; Pietersz, G.A.; Matsoukas, J.M.; Karamanos, N.K. Synthesis and study of the electrophoretic behavior of mannan conjugates with cyclic peptide analogue of myelin basic protein using lysine-glycine linker. *Anal. Biochem.* **2005**, *347*, 121–128. [CrossRef]

25. Apostolopoulos, V.; Pietersz, G.A.; Tsibanis, A.; Tsikkinis, A.; Drakaki, H.; Loveland, B.E.; Piddlesden, S.J.; Plebanski, M.; Pouniotis, D.S.; Alexis, M.N.; et al. Pilot phase III immunotherapy study in early-stage breast cancer patients using oxidized mannan-MUC1 [ISRCTN71711835]. *Breast Cancer Res.* **2006**, *8*, R27. [CrossRef]

26. Apostolopoulos, V.; Pietersz, G.A.; Tsibanis, A.; Tsikkinis, A.; Stojanovska, L.; McKenzie, I.F.; Vassilaros, S. Dendritic cell immunotherapy: Clinical outcomes. *Clin. Transl. Immunol.* **2014**, *3*, e21. [CrossRef]

27. Karanikas, V.; Hwang, L.A.; Pearson, J.; Ong, C.S.; Apostolopoulos, V.; Vaughan, H.; Xing, P.X.; Jamieson, G.; Pietersz, G.; Tait, B.; et al. Antibody and T cell responses of patients with adenocarcinoma immunized with mannan-MUC1 fusion protein. *J. Clin. Investig.* **1997**, *100*, 2783–2792. [CrossRef]

28. Karanikas, V.; Lodding, J.; Maino, V.C.; McKenzie, I.F. Flow cytometric measurement of intracellular cytokines detects immune responses in MUC1 immunotherapy. *Clin. Cancer Res.* **2000**, *6*, 829–837.

29. Karanikas, V.; Thynne, G.; Mitchell, P.; Ong, C.S.; Gunawardana, D.; Blum, R.; Pearson, J.; Lodding, J.; Pietersz, G.; Broadbent, R.; et al. Mannan Mucin-1 Peptide Immunization: Influence of Cyclophosphamide and the Route of Injection. *J. Immunother* **2001**, *24*, 172–183. [CrossRef]

30. Loveland, B.E.; Zhao, A.; White, S.; Gan, H.; Hamilton, K.; Xing, P.X.; Pietersz, G.A.; Apostolopoulos, V.; Vaughan, H.; Karanikas, V.; et al. Mannan-MUC1-pulsed dendritic cell immunotherapy: A phase I trial in patients with adenocarcinoma. *Clin. Cancer Res.* **2006**, *12*, 869–877. [CrossRef]

31. Mitchell, P.L.; Quinn, M.A.; Grant, P.T.; Allen, D.G.; Jobling, T.W.; White, S.C.; Zhao, A.; Karanikas, V.; Vaughan, H.; Pietersz, G.; et al. A phase 2, single-arm study of an autologous dendritic cell treatment against mucin 1 in patients with advanced epithelial ovarian cancer. *J. Immunother Cancer* **2014**, *2*, 16. [CrossRef]

32. Vassilaros, S.; Tsibanis, A.; Tsikkinis, A.; Pietersz, G.A.; McKenzie, I.F.; Apostolopoulos, V. Up to 15-year clinical follow-up of a pilot Phase III immunotherapy study in stage II breast cancer patients using oxidized mannan-MUC1. *Immunotherapy* **2013**, *5*, 1177–1182. [CrossRef] [PubMed]

33. Deraos, G.; Chatzantoni, K.; Matsoukas, M.T.; Tselios, T.; Deraos, S.; Katsara, M.; Papathanasopoulos, P.; Vynios, D.; Apostolopoulos, V.; Mouzaki, A.; et al. Citrullination of linear and cyclic altered peptide ligands from myelin basic protein (MBP(87-99)) epitope elicits a Th1 polarized response by T cells isolated from multiple sclerosis patients: Implications in triggering disease. *J. Med. Chem.* **2008**, *51*, 7834–7842. [CrossRef] [PubMed]

34. Matsoukas, J.; Apostolopoulos, V.; Kalbacher, H.; Papini, A.M.; Tselios, T.; Chatzantoni, K.; Biagioli, T.; Lolli, F.; Deraos, S.; Papathanassopoulos, P.; et al. Design and synthesis of a novel potent myelin basic protein epitope 87-99 cyclic analogue: Enhanced stability and biological properties of mimics render them a potentially new class of immunomodulators. *J. Med. Chem.* **2005**, *48*, 1470–1480. [CrossRef] [PubMed]

35. Tselios, T.; Apostolopoulos, V.; Daliani, I.; Deraos, S.; Grdadolnik, S.; Mavromoustakos, T.; Melachrinou, M.; Thymianou, S.; Probert, L.; Mouzaki, A.; et al. Antagonistic effects of human cyclic MBP(87-99) altered peptide ligands in experimental allergic encephalomyelitis and human T-cell proliferation. *J. Med. Chem.* **2002**, *45*, 275–283. [CrossRef]

36. Song, M.-J.; Hwang, S.W.; Whang, D. Amperometric hydrogen peroxide biosensor based on a modified gold electrode with silver nanowires. *J. Appl. Electrochem.* **2010**, *40*, 2099–2105. [CrossRef]

37. Konopka, S.J.; McDuffie, B. Diffusion coefficients of ferri-and ferrocyanide ions in aqueous media, using twin-electrode thin-layer electrochemistry. *Anal. Chem.* **1970**, *42*, 1741–1746. [CrossRef]

38. Hussain, G.; Silvester, D.S. Comparison of Voltammetric Techniques for Ammonia Sensing in Ionic Liquids. *Electroanalysis* **2018**, *30*, 75–83. [CrossRef]

39. Apostolopoulos, V.; Pietersz, G.A.; Gordon, S.; Martinez-Pomares, L.; McKenzie, I.F. Aldehyde-mannan antigen complexes target the MHC class I antigen-presentation pathway. *Eur. J. Immunol.* **2000**, *30*, 1714–1723. [CrossRef]

40. Apostolopoulos, V.; Pietersz, G.A.; Loveland, B.E.; Sandrin, M.S.; McKenzie, I.F. Oxidative/reductive conjugation of mannan to antigen selects for T1 or T2 immune responses. *Proc. Natl. Acad. Sci. USA* **1995**, *92*, 10128–10132. [CrossRef]

41. Grunwald, J.; Rejtar, T.; Sawant, R.; Wang, Z.; Torchilin, V.P. TAT peptide and its conjugates: Proteolytic stability. *Bioconjug. Chem.* **2009**, *20*, 1531–1537. [CrossRef]

42. Lemus, R.; Karol, M.H. Conjugation of haptens. *Methods Mol. Med.* **2008**, *138*, 167–182. [CrossRef]

43. Berthet, M.; Martinez, J.; Parrot, I. MgI$_2$ -chemoselective cleavage for removal of amino acid protecting groups: A fresh vision for peptide synthesis. *Biopolymers* **2017**, *108*. [CrossRef]

44. Isidro-Llobet, A.; Alvarez, M.; Albericio, F. Amino acid-protecting groups. *Chem. Rev.* **2009**, *109*, 2455–2504. [CrossRef]

45. Apostolopoulos, V.; Deraos, G.; Matsoukas, M.T.; Day, S.; Stojanovska, L.; Tselios, T.; Androutsou, M.E.; Matsoukas, J. Cyclic citrullinated MBP87-99 peptide stimulates T cell responses: Implications in triggering disease. *Bioorg. Med. Chem.* **2017**, *25*, 528–538. [CrossRef] [PubMed]

46. Pagba, C.V.; McCaslin, T.G.; Veglia, G.; Porcelli, F.; Yohannan, J.; Guo, Z.; McDaniel, M.; Barry, B.A. A tyrosine-tryptophan dyad and radical-based charge transfer in a ribonucleotide reductase-inspired maquette. *Nat. Commun.* **2015**, *6*, 10010. [CrossRef] [PubMed]

47. Zhai, J.; Zhao, L.; Zheng, L.; Gao, F.; Gao, L.; Liu, R.; Wang, Y.; Gao, X. Peptide–Au Cluster Probe: Precisely Detecting Epidermal Growth Factor Receptor of Three Tumor Cell Lines at a Single-Cell Level. *ACS Omega* **2017**, *2*, 276–282. [CrossRef] [PubMed]

48. Tapeinou, A. Design, synthesis and evaluation of analogues of myelin protein immunodominant epitopes implemented in multiple sclerosis. *Eur. J. Med. Chem.* **2017**, *143*, 621–631. [CrossRef] [PubMed]

49. Apostolopoulos, V.; Barnes, N.; Pietersz, G.A.; McKenzie, I.F. Ex vivo targeting of the macrophage mannose receptor generates anti-tumor CTL responses. *Vaccine* **2000**, *18*, 3174–3184. [CrossRef]

50. Apostolopoulos, V.; Pietersz, G.A.; McKenzie, I.F. Cell-mediated immune responses to MUC1 fusion protein coupled to mannan. *Vaccine* **1996**, *14*, 930–938. [CrossRef]

51. Sheng, K.C.; Kalkanidis, M.; Pouniotis, D.S.; Wright, M.D.; Pietersz, G.A.; Apostolopoulos, V. The adjuvanticity of a mannosylated antigen reveals TLR4 functionality essential for subset specialization and functional maturation of mouse dendritic cells. *J. Immunol.* **2008**, *181*, 2455–2464. [CrossRef] [PubMed]

52. Sheng, K.C.; Pouniotis, D.S.; Wright, M.D.; Tang, C.K.; Lazoura, E.; Pietersz, G.A.; Apostolopoulos, V. Mannan derivatives induce phenotypic and functional maturation of mouse dendritic cells. *Immunology* **2006**, *118*, 372–383. [CrossRef] [PubMed]

MDPI

St. Alban-Anlage 66

4052 Basel

Switzerland

Tel. +41 61 683 77 34

Fax +41 61 302 89 18

www.mdpi.com

Brain Sciences Editorial Office

E-mail: brainsci@mdpi.com

www.mdpi.com/journal/brainsci